W0079822

Biology
The Dynamic Science

Peter J. Russell

Stephan L. Wolfe

Paul E. Hertz

Cecie Starr

Edited by

Paul Yancey
Whitman College

Contributors:

Scott Bowling
Auburn University

Frederick Essig
University of South Florida

Laurie Bradley
Hudson Valley Community College

Courtney Farrell

Carolyn Bunde
Idaho State University

Jacalyn Newman
University of Pittsburgh

Jose Egremy
Northwest Vista College

Laura Ritt
Burlington Country College

Johnny El-Rady
University of South Florida

Mark Sugalski
Southern Polytechnic University

Alexander Wait
Missouri State University

BROOKS/COLE
CENGAGE Learning

Australia • Brazil • Japan • Korea • Mexico • Singapore • Spain • United Kingdom • United States

BROOKS/COLE
CENGAGE Learning

ISBN-13: 978-0-534-40327-0
ISBN-10: 0-534-40327-1

Brooks/Cole
10 Davis Drive
Belmont, CA 94002-3098
USA

Cengage Learning is a leading provider of customized learning solutions with office locations around the globe, including Singapore, the United Kingdom, Australia, Mexico, Brazil, and Japan. Locate your local office at: **international.cengage.com/region**

Cengage Learning products are represented in Canada by Nelson Education, Ltd.

For your course and learning solutions, visit **academic.cengage.com**

Purchase any of our products at your local college store or at our preferred online store **www.ichapters.com**

Printed in the United States of America
1 2 3 4 5 6 7 11 10 09 08

TABLE OF CONTENTS

UNIT SIX: ANIMAL STRUCTURE AND FUNCTION

UNIT SEVEN: ECOLOGY AND BEHAVIORS

1

INTRODUCTION TO BIOLOGICAL CONCEPTS AND RESEARCH

Multiple-Choice

WHY IT MATTERS

1. The science of _____ explains the origin of life, the persistence of life, and studies the changes in living things.

a. nanotechnology
b. mathematics
c. biology
d. chemistry
e. pharmacology

Answer: c
Difficulty: Easy
Bloom's Taxonomy: Knowledge, Comprehension

2. It is through _____ that we further our knowledge of living things.

a. ideologies
b. research
c. philosophy
d. ethics
e. logic

Answer: b
Difficulty: Easy
Bloom's Taxonomy: Knowledge, Comprehension

1.1 WHAT IS LIFE? CHARACTERISTICS OF LIVING SYSTEMS

3. The difference between living and nonliving matter depends not only on the kinds of atoms and molecules present, it also depends on _____.

a. their electrons
b. their chemical interactions
c. their compounds
d. their organization and interactions
e. all of these

Answer: d
Difficulty: Moderate
Bloom's Taxonomy: Comprehension

4. Living organisms must gather energy and materials from their surroundings to _____.

a. build new biochemicals
b. grow
c. maintain and repair their parts
d. produce offspring
e. all of the above

Answer: e
Difficulty: Moderate
Bloom's Taxonomy: Knowledge, Comprehension

5. All cell consist of the following:

a. an organized chemical system.
b. specialized molecules.
c. a nucleus.
d. a membrane.
e. a, b, and d only

Answer: e
Difficulty: Moderate
Bloom's Taxonomy: Knowledge, Comprehension

6. The lowest level of biological organization that can survive and reproduce is the _____.

a. proton
b. DNA
c. nucleus
d. tissue
e. cell

Answer: e
Difficulty: Moderate
Bloom's Taxonomy: Knowledge, Comprehension

7. As long as a cell has access to a usable energy source, has the necessary raw materials, and is in appropriate environmental conditions, the cell will:

a. survive and reproduce.
b. stay in stasis.
c. not be in homeostasis.
d. remain dormant.

Answer: a
Difficulty: Moderate
Bloom's Taxonomy: Comprehension

8. Emergent properties are

a. characteristics of nonliving matter.
b. exclusive to atoms but not molecules.
c. neither exclusive to molecules nor compounds.
d. characteristics that depend on the level of organization of matter but do not exist at lower levels of organization.

Answer: d
Difficulty: Moderate
Bloom's Taxonomy: Knowledge, Comprehension

9. Bacteria and protozoans exist as _____.

a. multicellular organisms
b. unicellular organisms
c. both unicellular and multicellular organisms
d. the sole organisms of the oceans
e. precursors to cells

Answer: b
Difficulty: Easy
Bloom's Taxonomy: Knowledge, Comprehension

10.   A group of organisms of the same species that live together in the same place make up a(n) _____.

a.   cell
b.   ecosystem
c.   tissue
d.   population
e.   biosphere

Answer: d
Difficulty: Moderate
Bloom's Taxonomy: Knowledge, Comprehension

11.   All the populations of different organisms that live in the same place form a(n) _____.

a.   ecosystem
b.   community
c.   biosphere
d.   organ
e.   population

Answer: b
Difficulty: Moderate
Bloom's Taxonomy: Knowledge, Comprehension

12.   The highest level of the hierarchy of life is the _____.

a.   population
b.   biosphere
c.   ecosystem
d.   multicellular organism
e.   cell

Answer: b
Difficulty: Easy
Bloom's Taxonomy: Knowledge, Comprehension

13.   A(n) _____ includes the community and the nonliving environmental factors with which it interacts.

a.   community
b.   biosphere
c.   ecosystem
d.   multicellular organism
e.   earth

Answer: c
Difficulty: Easy
Bloom's Taxonomy: Knowledge, Comprehension

14.   The most fundamental and important molecule that distinguishes living systems from nonliving matter is _____.

a.   DNA
b.   deoxyribonucleic acid
c.   glucose
d.   fructose
e.   a and b only

Answer: e
Difficulty: Moderate
Bloom's Taxonomy: Knowledge, Comprehension

15.   _____ is a large, double stranded, helical molecule that contains instructions for assembling a living organism from simpler molecules.

a.   RNA
b.   DNA
c.   ATP
d.   NADPH
e.   Protein

Answer: b
Difficulty: Easy
Bloom's Taxonomy: Knowledge, Comprehension

16.   _____ contain DNA, but are not considered to be alive because they cannot reproduce independently of their host.

a.   Prions
b.   Eukaryotes
c.   Prokaryotes
d.   Viruses
e.   Cells

Answer: d
Difficulty: Moderate
Bloom's Taxonomy: Knowledge, Comprehension

17.   The information in DNA is copied into molecules of _____.

a.   protein
b.   carbohydrates
c.   hydrogen peroxide
d.   oxygen
e.   RNA

Answer: e
Difficulty: Easy
Bloom's Taxonomy: Knowledge, Comprehension

18.   What molecule carries out most of the activities of life, including the synthesis of all other biological molecules?

a.   carbohydrate
b.   lipid
c.   protein
d.   nucleid acid
e.   none of the above

Answer: c
Difficulty: Difficult
Bloom's Taxonomy: Knowledge, Comprehension

19.   Metabolism describes the ability of a cell or organism to:

a.   extract energy from its surroundings.
b.   maintain itself.
c.   grow.
d.   reproduce.
e.   all of the above

Answer: e
Difficulty: Moderate
Bloom's Taxonomy: Knowledge, Comprehension

20. Photosynthesis and cellular respiration are examples of _____.

a. anabolism
b. catabolism
c. synthesis
d. metabolism
e. cleavage

Answer: d
Difficulty: Difficult.
Bloom's Taxonomy: Knowledge, Comprehension

21. Which of the following is **not** matched correctly?

a. plant—primary producer
b. animal—consumer
c. fungus—decomposer
d. algae—primary producer
e. cyanobacteria—decomposer

Answer: e
Difficulty: Difficult
Bloom's Taxonomy: Knowledge, Comprehension

22. Which of the following statements is inaccurate?

a. Biological processes are not 100% efficient.
b. Some of the energy of biological processes is lost to the environment.
c. Biological processes are 100% efficient.
d. Some of the energy lost in biological processes is heat.
e. All statements are correct.

Answer: c
Difficulty: Difficult
Bloom's Taxonomy: Knowledge, Comprehension

23. Living systems have the capacity to detect environmental changes and compensate for them through controlled responses. This is possible because living systems have _____.

a. diverse and varied receptors
b. calcium ions
c. ion sinks
d. hydrogen containing molecules
e. potassium

Answer: a
Difficulty: Difficult
Bloom's Taxonomy: Knowledge, Comprehension

24. Maintaining your body's internal temperature within narrow tolerable range is one example of:

a. sebaceous glands working to lower your body temperature.
b. stasis.
c. homeostasis.
d. compensation.
e. hydrolysis.

Answer: c
Difficulty: Difficult
Bloom's Taxonomy: Knowledge, Comprehension

25. The process by which parents produce offspring is called:

a. reproduction.
b. homeostasis.
c. compensation.
d. feeding.
e. artificial selection.

Answer: a
Difficulty: Easy
Bloom's Taxonomy: Knowledge, Comprehension

26. Inheritance

a. is the process by which genetic information is transmitted to offspring.
b. occurs only in animals.
c. is a process by which proteins are transmitted to offspring.
d. is not a biological process.
e. a and b only

Answer: a
Difficulty: Easy
Bloom's Taxonomy: Knowledge

27. A series of programmed changes encoded in DNA, through which a fertilized egg divides into many cells that ultimately are transformed into an adult organism, is known as _____.

a. inheritance
b. compensation
c. homeostasis
d. transformation
e. development

Answer: e
Difficulty: Moderate
Bloom's Taxonomy: Knowledge

28. The sequential stages through which individuals develop, grow, maintain themselves, and reproduce are known as the _____.

a. life cycle
b. transformation
c. catabolic reactions
d. anabolic reactions
e. central dogma

Answer: a
Difficulty: Moderate
Bloom's Taxonomy: Knowledge

29. Populations of all organisms change from one generation to the next because their DNA changes over time. This is known as:

a. artificial selection.
b. biological evolution.
c. natural selection.
d. a and c only
e. none of the above

Answer: b
Difficulty: Difficult
Bloom's Taxonomy: Knowledge, Comprehension

1.2 BIOLOGICAL EVOLUTION

30. Our understanding of the evolutionary process reveals:

a. all populations change through time.
b. all organisms are related through a shared ancestry.
c. evolution has produced the spectacular diversity of life that we see around us.
d. all of the above
e. a and b only

Answer: d
Difficulty: Moderate
Bloom's Taxonomy: Knowledge

31. In the mid–nineteenth century Charles Darwin and Alfred Russel Wallace observed many organisms. Based on these observations they arrived to an explanation on how populations change through time. They termed it:

a. evolution.
b. natural selection.
c. creationism.
d. natural evolution.
e. genetics.

Answer: b
Difficulty: Moderate
Bloom's Taxonomy: Knowledge, Comprehension

32. Biological evolution according to Darwin and Wallace states that

a. organisms produce numerous offspring, but environmental factors limit the number of reproducing survivors.
b. heritable variations allow some individuals to out compete others.
c. the offspring of successful individuals inherit the favorable characteristics of their parents.
d. a, b, and c
e. none of the above

Answer: d
Difficulty: Moderate
Bloom's Taxonomy: Comprehension

33. Genes are

a. proteins and carbohydrates.
b. organized into functional units of DNA..
c. organized into functional units of RNA..
d. organized into functional units of proteins.
e. organized by nonpolar covalent bonds.

Answer: b
Difficulty: Moderate
Bloom's Taxonomy: Knowledge

34. Mutations

a. are the basis of variability among individuals.
b. are the basis of homogeneity in a population.
c. are always bad for populations.
d. are always good for populations.
e. none of the above

Answer: a
Difficulty: Moderate
Bloom's Taxonomy: Knowledge, Comprehension

35. Mutations

a. can be beneficial.
b. can be harmful.
c. can be neutral.
d. all of the above
e. none of the above

Answer: d
Difficulty: Moderate
Bloom's Taxonomy: Knowledge, Comprehension

36. Adaptations

a. are characteristics that arise during an organism's lifetime and that help an organism survive longer or reproduce more.
b. occur primarily in the form of useful molecules.
c. are characteristics that arise via natural selection and that help an organism survive longer or reproduce more.
d. are characteristics that mainly enhance feeding.
e. a and c only

Answer: c
Difficulty: Moderate
Bloom's Taxonomy: Knowledge, Comprehension

37. Cryptic coloration means

a. males and females look different.
b. blending with the background.
c. camouflage.
d. standing out against a background.
e. b and c only

Answer: e
Difficulty: Moderate
Bloom's Taxonomy: Knowledge, Comprehension

1.3 BIODIVERSITY

38. Why have scientists developed classification systems?

a. To arrange living and dead organisms into groups that reflect their relationships and evolutionary origins.
b. To make sense of the past and present diversity of life on Earth.
c. To help scientists distinguish two different groups of organisms with the same common name.
d. a nd b only
e. a, b and c

Answer: c
Difficulty: Moderate
Bloom's Taxonomy: Knowledge, Comprehension

39. A group of organisms in which the individuals are so closely related in structure, biochemistry, and behavior that they can successfully interbreed is a(n) _____.

a. kingdom
b. class
c. order
d. species
e. genus

Answer: d
Difficulty: Moderate
Bloom's Taxonomy: Knowledge, Comprehension

40. A group of similar species that share a recent common ancestry is a (n) _____.

a. kingdom
b. class
c. order
d. genus
e. species

Answer: d
Difficulty: Moderate
Bloom's Taxonomy: Knowledge, Comprehension

41. The scientific name of an organism is composed of two names. The first part identifies the _____ while the second part designates the _____.

a. genus; species
b. species; genus
c. genera; genus
d. phylum; species
e. family; genus

Answer: a
Difficulty: Easy
Bloom's Taxonomy: Knowledge, Comprehension

42. Scientific names

a. are always written in all capital letters and in italics.
b. are always written in italics in lower case.
c. are always written in lower case and underlined.
d. are always written in italics with both genus and species capitalized.
e. are always written in italics with only the genus capitalized.

Answer: e
Difficulty: Easy
Bloom's Taxonomy: Knowledge, Comprehension

43. Which of the following scientific names is written in the correct format?

a. *Canis Familiaris*
b. *Canis Lupus*
c. *Canis latrans*
d. *canis Familiaris*
e. *c. Latrans*

Answer: c
Difficulty: Moderate
Bloom's Taxonomy: Synthesis

Eukarya

Animalia

Chordata

Mammalia

Carnivora

Canidae

Canis

Canis familiaris

Use the figure above for questions 44 through 46.

44.   Which of the following represents a kingdom?

a.   Animalia
b.   Chordata
c.   Canis
d.   Canidae

Answer: a
Difficulty: Moderate
Bloom's Taxonomy: Synthesis, Application
Source: Figure 1.11

45.   In the illustration, which group is the most inclusive group?

a.   *Canis familiaris*
b.   Animalia
c.   Chordata
d.   Mammalia
e.   Eukarya

Answer: e
Difficulty: Moderate
Bloom's Taxonomy: Analysis
Source: Figure 1.11

46.   In the illustration, which group is the least inclusive?

a.   *Canis familiaris*
b.   Animalia
c.   Chordata
d.   Mammalia
e.   Chordata

Answer: a
Difficulty: Moderate
Bloom's Taxonomy: Analysis
Source: Figure 1.11

47. The group that is the most inclusive and has been recently been added to the classification scheme is the _____.

a. kingdom
b. protista
c. eukarya
d. domain
e. c and d only

Answer: e
Difficulty: Moderate
Bloom's Taxonomy: Application

48. Which of the following pairs would be classified as prokaryotes?

a. Animalia and Plantae
b. Bacteria and Archaea
c. Fungi and Plantae
d. all of the above
e. none of the above

Answer: b
Difficulty: Moderate
Bloom's Taxonomy: Application

49. A cell that is observed under the microscope is found to have no nucleus. The cell is a(n) _____.

a. eukaryote
b. proryote
c. animal cell
d. plant cell
e. prokaryote

Answer: e
Difficulty: easy
Bloom's Taxonomy: Application

50. A cell that is observed under the microscope is found to have its DNA enclosed in a nucleus, and has other specialized internal compartments. The cell is a(n) _____.

a. prokaryote
b. bacterium
c. eukaryote
d. *E. coli*
e. *S. aureus*

Answer: c
Difficulty: easy
Bloom's Taxonomy: Application

51. A researcher in a lab finds a microscopic organism that has no nucleus, but has distinctive structural molecules and mechanisms of photosynthesis. The researcher determines that the organism is a producer. The organisms are found everywhere on Earth. The researcher has identified an organism belonging to the domain _____.

a. Bacteria
b. Archaea
c. Eukarya
d. Animalia
e. Amoeba

Answer: a
Difficulty: Moderate
Bloom's Taxonomy: Analysis

52. A researcher in a lab finds a microscopic organism that is a producer. The organisms are found in extreme environments (i.e., hot springs). The researcher has identified an organism belonging to the domain _____.

a. Bacteria
b. Archaea
c. Eukarya
d. Animalia
e. Amoeba

Answer: b
Difficulty: Moderate
Bloom's Taxonomy: Analysis

53. A student encounters an organism whose cells contain a nucleus which resembles a plant. The organism is most likely classified as a(n) _____.

a. Bacteria
b. Archaea
c. Eukarya
d. Animalia
e. Amoeba

Answer: c
Difficulty: Moderate
Bloom's Taxonomy: Analysis

54. This kingdom includes the algae that are used to make sushi rolls.

a. Plantae
b. Fungi
c. Animalia
d. Protoctista
e. Bacteria

Answer: d
Difficulty: Difficult
Bloom's Taxonomy: Analysis, Application

55. The pages of your textbook consist mainly of material made by multicellular, photosynthetic organisms that function as producers in ecosystems. These organisms belong to the kingdom _____.

a. Plantae
b. Fungi
c. Animalia
d. Protoctista
e. Bacteria

Answer: a
Difficulty: Moderate
Bloom's Taxonomy: Analysis, Application

56. Shitake mushrooms are decomposers that break down biological molecules from dead organisms. These organisms belong to the _____ kingdom.

a. Plantae
b. Fungi
c. Animalia
d. Protoctista
e. Bacteria

Answer: b
Difficulty: Moderate
Bloom's Taxonomy: Analysis, Application

57. Cats, dogs, and fish are consumers that have the ability to move actively from one place to another. These organisms belong to the _____ kingdom.

a. Plantae
b. Fungi
c. Animalia
d. Protoctista
e. Bacteria

Answer: c
Difficulty: Easy
Bloom's Taxonomy: Analysi, Application

1.4 BIOLOGICAL RESEARCH

58. The observations you make and experimental data you collect in your biology laboratory class are examples of:

a. busy work.
b. homework.
c. biological dogma.
d. surface tension.
e. biological research.

Answer: e
Difficulty: easy
Bloom's Taxonomy: Analysis, Application

59. An approach in which scientists make observations about the natural world, develop tentative explanations about what they observe, and then test those explanations by collecting more information is referred to as _____.

a. science
b. education
c. the scientific method
d. the method
e. none of the above

Answer: c
Difficulty: Moderate
Bloom's Taxonomy: Comprehension

60. If a biologist searches for explanations about the natural phenomena to satisfy his or her curiosity and advance our collective knowledge of living systems, then, this researcher is a(n):

a. applied researcher.
b. basic researcher.
c. general researcher.
d. simple researcher.
e. scientist.

Answer: b
Difficulty: Easy
Bloom's Taxonomy: Analysis

61. Applied researchers conduct their work to:

a. solve specific practical problems.
b. solve any problem they face.
c. answer all questions.
d. advance our collective knowledge of living systems.
e. none of the above

Answer: a
Difficulty: Moderate
Bloom's Taxonomy: Analysis, Application

62. If a researcher collects basic information on biological structures or the details of biological processes, then the researcher's approach is considered to be:

a. science.
b. scientific.
c. descriptive science.
d. not scientific.
e. a and b only

Answer: c
Difficulty: Moderate
Bloom's Taxonomy: Analysis, Application

63. When conducting descriptive research, a scientist primarily uses:

a. control data.
b. data.
c. experiments.
d. observational data.
e. experimental data.

Answer: d
Difficulty: Moderate
Bloom's Taxonomy: Analysis, Application

64. When a student manipulates a system under study, he or she is collecting:

a. empirical data.
b. experimental data.
c. observational data.
d. data.
e. none of the above

Answer: b
Difficulty: Moderate
Bloom's Taxonomy: Analysis, Application

65. You are studying an ecosystem in your campus; after a solid base of carefully observed and described facts, your next step would be to:

a. make more observations.
b. share your data with others.
c. design an experiment.
d. wait for instructions.
e. make a hypothesis.

Answer: e
Difficulty: Moderate
Bloom's Taxonomy: Analysis, Application

66. While conducting an experiment in the lab, you collect data that help you demonstrate that your hypothesis is wrong. In other words, you have falsified your hypothesis. Your results are acclaimed by the:

a. ecclesiastic community.
b. the Pope.
c. scientific community.
d. student community.
e. nobody.

Answer: c
Difficulty: Moderate
Bloom's Taxonomy: Analysis, Application

67. Hypotheses that are falsifiable fall within the realm of:

a. science.
b. history.
c. not science.
d. philosophy.
e. English.

Answer: a
Difficulty: Easy
Bloom's Taxonomy: Knowledge

Short Answer

68. You are at a stage in your research in which you must design an experiment to test your hypothesis. What factors must you include to insure that your experimental data is good valid data?

Answer: Any experimental design must include a control group, an experimental group or variable, and must include replicates to validate data.
Difficulty: Moderate
Bloom's Taxonomy: Analysis, Application

69. Explain the need for a null hypothesis, especially in ecology and evolution. What does a null hypothesis accomplish?

Answer: A null hypothesis is a statement of what a researcher would see if the hypothesis being tested is wrong. Ecologists usually tackle systems that are too complex to control, so a null hypothesis anticipates, or provides alternative hypothesis to answer questions.
Difficulty: Moderate
Bloom's Taxonomy: Analysis, Application

70. Why do scientists use model organisms?

Answer: Model organisms have rapid development, short life cycles, and small adult sizes, making them ideal to work with in the laboratory setting.
Difficulty: Moderate
Bloom's Taxonomy: Analysis, Application

71. What are the contributions of Beadle, Tatum, Watson, and Crick to biology? What techniques did they use to revolutionize biological science?

Answer: Beadle and Tatum demonstrated that genes provide cells with instructions needed for the production of proteins, whereas Watson and Crick determined the structure of DNA. Molecular technologies have transformed biology.
Difficulty: Moderate
Bloom's Taxonomy: Analysis, Application

72. Scientific theories are of fundamental importance in science. Explain the difference between the term "theory" as employed in science versus "theory" as employed in everyday language.

Answer: Scientific theories have withstood the test of time and have been extensively confirmed by repeated experiments. The term as used in science has validity whereas in everyday context it takes the form of an opinion or a guess.
Difficulty: Moderate
Bloom's Taxonomy: Analysis, Application

2

LIFE, CHEMISTRY, AND WATER

Multiple-Choice

WHY IT MATTERS

1. According to studies by Norman Terry and coworkers, some plants can perform a version of bioremediation of selenium in wastewater by

a. converting selenium to a form that kills waterfowl.
b. using selenium to make a necessary supplement for humans.
c. converting selenium into a relatively nontoxic gas.
d. storing selenium in the soil.
e. increasing the selenium concentration in the water.

Answer: c
Difficulty: Moderate
Bloom's Taxonomy: Knowledge, Comprehension

2. The laws of chemistry and physics that govern living things are _____ the laws of chemistry and physics that govern nonliving things.

a. different from
b. the same as
c. roughly half the same as and half different from
d. mostly different from
e. mostly the same as

Answer: b
Difficulty: Easy
Bloom's Taxonomy: Knowledge

2.1 THE ORGANIZATION OF MATTER: ELEMENTS AND ATOMS

3. A substance that cannot be broken down into simpler substances by ordinary chemical or physical techniques is a(n) _____.

a. molecule
b. chemical
c. compound
d. element
e. all of these

Answer: d
Difficulty: Easy
Bloom's Taxonomy: Knowledge

4. Four elements make up more than 96% of the mass of most living organisms. Which of the following is NOT one of those four elements?

a. sodium
b. carbon
c. oxygen
d. nitrogen
e. hydrogen

Answer: a
Difficulty: Moderate
Bloom's Taxonomy: Knowledge

5. A trace element is found in organisms in _____ quantities and is _____ for normal biological functions.

a. moderate; unnecessary
b. moderate; vital
c. small; unnecessary
d. large; unnecessary
e. small; vital

Answer: e
Difficulty: Moderate
Bloom's Taxonomy: Knowledge

6. The smallest unit that retains the chemical and physical properties of an element is a(n) _____.

a. proton
b. compound
c. molecule
d. neutron
e. atom

Answer: e
Difficulty: Easy
Bloom's Taxonomy: Knowledge

7. The substance H_2O is considered to be

a. both a molecule and a compound.
b. a compound but not a molecule.
c. neither a molecule nor a compound.
d. a molecule but not a compound.

Answer: a
Difficulty: Easy
Bloom's Taxonomy: Comprehension

8. The substance O_2 is considered to be

a. both a molecule and a compound.
b. a compound but not a molecule.
c. neither a molecule nor a compound.
d. a molecule but not a compound.

Answer: d
Difficulty: Moderate
Bloom's Taxonomy: Comprehension

2.2 ATOMIC STRUCTURE

9.   An oxygen atom has _____ surrounding a nucleus composed of _____.

a.   neutrons; electrons and protons
b.   electrons; protons and neutrons
c.   protons and electrons; neutrons
d.   protons; neutrons and electrons
e.   electrons and neutrons; protons

Answer: b
Difficulty: Easy
Bloom's Taxonomy: Knowledge

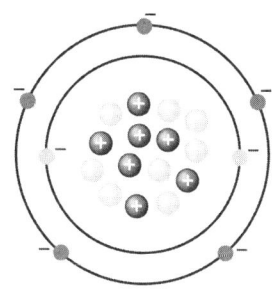

Use the figure above for questions 10–13.

10.   The mass number of the atom depicted in the figure is

a.   5.
b.   7.
c.   8.
d.   15.
e.   22.

Answer: d
Difficulty: Moderate
Bloom's Taxonomy: Analysis

11.   The atomic number of the atom depicted in the figure is

a.   5.
b.   7.
c.   8.
d.   15.
e.   22.

Answer: b
Difficulty: Moderate
Bloom's Taxonomy: Analysis

12.   The number of electrons for the atom depicted in the figure is

a.   5.
b.   7.
c.   8.
d.   15.
e.   22.

Answer: b
Difficulty: Easy
Bloom's Taxonomy: Analysis

13.   The number of neutrons for the atom depicted in the figure is

a.   5.
b.   7.
c.   8.
d.   15.
e.   22.

Answer: c
Difficulty: Easy
Bloom's Taxonomy: Analysis

14.   Which of the following are charged particles?

a.   electrons and protons
b.   neutrons only
c.   protons and neutrons
d.   electrons only
e.   protons, neutrons, and electrons

Answer: a
Difficulty: Moderate
Bloom's Taxonomy: Analysis

15.   Isotopes of the same element differ from each other in the number of

a.   electrons and protons.
b.   neutrons only.
c.   protons and neutrons.
d.   electrons only.
e.   protons, neutrons, and electrons.

Answer: b
Difficulty: Moderate
Bloom's Taxonomy: Knowledge, Comprehension

16.   A carbon atom with six protons, seven neutrons, and six electrons has a mass number of

a.   6.
b.   7.
c.   12.
d.   13.
e.   19.

Answer: d
Difficulty: Moderate
Bloom's Taxonomy: Application

17.   The isotope ^{14}C undergoes radioactive decay with a neutron splitting into an electron and a proton. This decay produces an atom of

a.   iron.
b.   carbon.
c.   hydrogen.
d.   oxygen.
e.   nitrogen.

Answer: e
Difficulty: Difficult
Bloom's Taxonomy: Knowledge

18.   An orbital describes the _____ of an electron.

a.   exact location
b.   exact path
c.   most frequent locations
d.   charge
e.   chemical bonds

Answer: c
Difficulty: Moderate
Bloom's Taxonomy: Comprehension

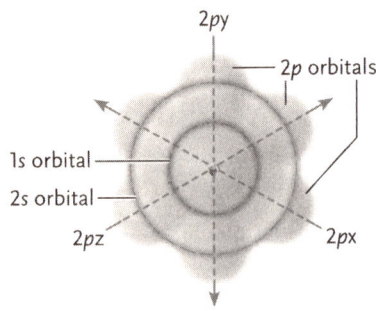

Use the figure above for questions 19–20.

19.   The electrons at the lowest energy level in the neon atom depicted in the figure above are found in which orbital?

a.   1*s*
b.   2*s*
c.   2*p*x
d.   2*p*y
e.   2*p*z

Answer: a
Difficulty: Moderate
Bloom's Taxonomy: Comprehension
Source: Figure 2.5

20.   All of the orbitals shown in the neon atom in the figure are completely filled with electrons. How many electrons does this neon atom have?

a.   5
b.   6
c.   8
d.   10
e.   16

Answer: d
Difficulty: Moderate
Bloom's Taxonomy: Application
Source: Figure 2.5

21.   Under the right conditions, an electron will

a.   move to a lower energy level.
b.   enter an orbital shared by two atoms.
c.   move to a higher energy level.
d.   move from one atom to another atom.
e.   all of these

Answer: e
Difficulty: Moderate
Bloom's Taxonomy: Synthesis

22.   Sodium has one valence electron in its third energy level. To reach a stable energy configuration, sodium will tend to

a.   take up an electron from another atom.
b.   move its valence electron to the second energy shell.
c.   give up an electron to another atom.
d.   share its valence electron with another atom.
e.   move an electron from the second energy level to the valence shell.

Answer: c
Difficulty: Moderate
Bloom's Taxonomy: Application

23.   Which of the following is most likely to share electrons with other atoms in joint orbitals?

a.   chlorine (7 valence electrons)
b.   calcium (2 valence electrons)
c.   argon (8 valence electrons)
d.   carbon (4 valence electrons)
e.   potassium (1 valence electron)

Answer: d
Difficulty: Difficult
Bloom's Taxonomy: Synthesis

24.   Which of the following is likely to be chemically unreactive?

a.   chlorine (7 valence electrons)
b.   calcium (2 valence electrons)
c.   argon (8 valence electrons)
d.   carbon (4 valence electrons)
e.   potassium (1 valence electron)

Answer: c
Difficulty: Difficult
Bloom's Taxonomy: Synthesis

25.   Which of the following is most likely to take up an electron from another atom?

a.   chlorine (7 valence electrons)
b.   calcium (2 valence electrons)
c.   neon (8 valence electrons)
d.   carbon (4 valence electrons)
e.   potassium (1 valence electron)

Answer: a
Difficulty: Moderate
Bloom's Taxonomy: Synthesis

26. Radioactive _____ is commonly used to treat patients with dangerously overactive thyroid glands.

a. carbon
b. radium
c. iodine
d. thallium
e. cobalt

Answer: c
Difficulty: Moderate
Bloom's Taxonomy: Knowledge

27. Melvin Calvin and his coworkers used a radioisotope of _____ to trace the reactions of photosynthesis.

a. carbon
b. radium
c. iodine
d. thallium
e. cobalt

Answer: a
Difficulty: Moderate
Bloom's Taxonomy: Knowledge

2.3 CHEMICAL BONDS

28. The chemical bonds that form when atoms that have lost electrons are electrically attracted to atoms that have gained electrons are called _____.

a. polar covalent bonds
b. van der Waals forces
c. ionic bonds
d. hydrogen bonds
e. nonpolar covalent bonds

Answer: c
Difficulty: Easy
Bloom's Taxonomy: Knowledge

29. The chemical bonds that are formed when atoms share electrons equally are called _____.

a. polar covalent bonds
b. van der Waals forces
c. ionic bonds
d. hydrogen bonds
e. nonpolar covalent bonds

Answer: e
Difficulty: Moderate
Bloom's Taxonomy: Knowledge, Comprehension

30. The chemical bonds that are formed when atoms share electrons unequally are called _____.

a. polar covalent bonds
b. van der Waals forces
c. ionic bonds
d. hydrogen bonds
e. nonpolar covalent bonds

Answer: a
Difficulty: Moderate
Bloom's Taxonomy: Knowledge, Comprehension

31. The chemical bonds that are formed when atoms with temporary zones of positive charge are attracted to other atoms with temporary zones of negative charge are called _____.

a. polar covalent bonds
b. van der Waals forces
c. ionic bonds
d. hydrogen bonds
e. nonpolar covalent bonds

Answer: b
Difficulty: Difficult
Bloom's Taxonomy: Knowledge, Comprehension

32. Chemical bonds that are formed when one atom with a partial positive charge (created from unequal sharing of electrons) is electrically attracted to another atom with a partial negative charge (also created from unequal sharing of electrons) are called _____.

a. polar covalent bonds
b. van der Waals forces
c. ionic bonds
d. hydrogen bonds
e. nonpolar covalent bonds

Answer: d
Difficulty: Moderate
Bloom's Taxonomy: Knowledge, Comprehension

33. Which of the following types of chemical linkages is the weakest?

a. polar covalent bonds
b. van der Waals forces
c. ionic bonds
d. hydrogen bonds
e. nonpolar covalent bonds

Answer: b
Difficulty: Difficult
Bloom's Taxonomy: Synthesis

34. The attraction between Na^+ cations and Cl^- anions forms _____ that hold together the compound NaCl.

a. polar covalent bonds
b. van der Waals forces
c. ionic bonds
d. hydrogen bonds
e. nonpolar covalent bonds

Answer: c
Difficulty: Easy
Bloom's Taxonomy: Application

35. Geckos are able to cling to vertical walls due to
_____.

a. polar covalent bonds
b. van der Waals forces
c. ionic bonds
d. hydrogen bonds
e. nonpolar covalent bonds

Answer: b
Difficulty: Difficult
Bloom's Taxonomy: Knowledge

36. Molecules such as H–H and O=O are held
together by _____.

a. polar covalent bonds
b. van der Waals forces
c. ionic bonds
d. hydrogen bonds
e. nonpolar covalent bonds

Answer: e
Difficulty: Moderate
Bloom's Taxonomy: Synthesis

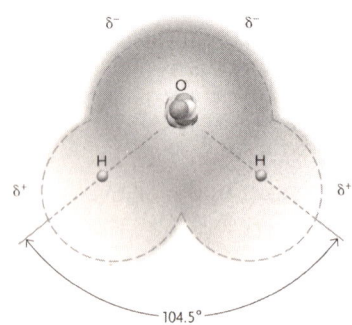

37. The molecule shown in the figure above is held
together by _____.

a. polar covalent bonds
b. van der Waals forces
c. ionic bonds
d. hydrogen bonds
e. nonpolar covalent bonds

Answer: a
Difficulty: Moderate
Bloom's Taxonomy: Application
Source: Figure 2.9

38. Metallic ions such as Ca^{2+}, Na^+, and Fe^{3+} readily
form _____.

a. polar covalent bonds
b. van der Waals forces
c. ionic bonds
d. hydrogen bonds
e. nonpolar covalent bonds

Answer: c
Difficulty: Moderate
Bloom's Taxonomy: Comprehension

39. The chemical linkages that exert an attractive
force over the greatest distance are _____.

a. polar covalent bonds
b. van der Waals forces
c. ionic bonds
d. hydrogen bonds
e. nonpolar covalent bonds

Answer: c
Difficulty: Difficult
Bloom's Taxonomy: Knowledge

40. In contrast to ionic bonds, covalent bonds
_____.

a. hold atoms together
b. have distinct, three-dimensional forms
c. transfer electrons from one atom to another
d. are relatively weak
e. are transient

Answer: b
Difficulty: Moderate
Bloom's Taxonomy: Synthesis

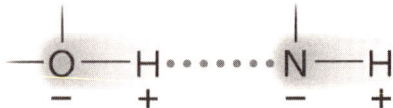

41. The dotted line in the figure above indicates
_____.

a. a polar covalent bond
b. van der Waals forces
c. an ionic bond
d. a hydrogen bond
e. a nonpolar covalent bond

Answer: d
Difficulty: Easy
Bloom's Taxonomy: Analysis
Source: Fig. 2.10

42. In a molecule of methane, CH_4, each hydrogen
atom shares an orbital with the carbon atom. The total
number of shared electrons in CH_4 is _____.

a. 4
b. 2
c. 1
d. 8
e. 5

Answer: d
Difficulty: Difficult
Bloom's Taxonomy: Analysis

43. A polar covalent bond would be most likely to
form between

a. atoms with different electronegativities.
b. cations and anions.
c. atoms with δ+ and δ- charges.
d. atoms with filled valence shells.
e. atoms of the same element.

Answer: a
Difficulty: Moderate
Bloom's Taxonomy: Evaluation

44. Which of these types of chemical bonds would you not expect to find in biological molecules?

a. covalent bonds
b. van der Waals forces
c. ionic bonds
d. hydrogen bonds
e. all of these types of bonds are found in biological molecules

Answer: e
Difficulty: Easy
Bloom's Taxonomy: Synthesis

45. In the presence of water, nonpolar associations form between molecules or regions of molecules that are _____.

a. partially charged
b. hydrophobic and hydrophilic
c. hydrophobic
d. fully charged
e. hydrophilic

Answer: c
Difficulty: Easy
Bloom's Taxonomy: Knowledge

46. A mixture of vegetable oil and water will separate into layers because oil is _____ and forms _____.

a. hydrophobic; nonpolar associations
b. hydrophilic; nonpolar associations
c. hydrophilic; polar associations
d. hydrophobic; polar associations

Answer: a
Difficulty: Moderate
Bloom's Taxonomy: Application

47. Analyze this chemical reaction:

$$6\ CO_2 + 6\ H_2O \rightarrow C_6H_{12}O_6 + 6\ O_2$$

Which of the following is FALSE?

a. Water is a reactant.
b. $C_6H_{12}O_6$ is a product.
c. Molecular oxygen is a product.
d. CO_2 is a reactant.
e. Molecular carbon is a reactant.

Answer: e
Difficulty: Moderate
Bloom's Taxonomy: Analysis

48. The formation and breaking of bonds between atoms requires

a. a chemical reaction.
b. van der Walls forces.
c. partial charges.
d. an empty valence shell.
e. an enzyme.

Answer: a
Difficulty: Moderate
Bloom's Taxonomy: Knowledge

2.4 HYDROGEN BONDS AND THE PROPERTIES OF WATER

49. A molecule of water in the middle of a chunk of ice will usually have _____ hydrogen bonds with other water molecules.

a. 3
b. 3.4
c. 6
d. 4
e. 2

Answer: d
Difficulty: Moderate
Bloom's Taxonomy: Application

50. Which of the following would have the most difficulty entering a water lattice?

a. table salt (NaCl)
b. a nonpolar molecule
c. a sodium ion
d. a proton (H^+)
e. an electron

Answer: b
Difficulty: Moderate
Bloom's Taxonomy: Application

51. Ice floats in liquid water because

a. ice forms hydrogen bonds with the surface of liquid water.
b. ice forms hydrogen bonds but liquid water does not.
c. the hydrogen bonds of liquid water are fixed in place.
d. liquid water forms hydrogen bonds but ice does not.
e. the distance between water molecules is maximized due to the hydrogen bonds which are fixed in place.

Answer: e
Difficulty: Difficult
Bloom's Taxonomy: Application

52. Biological membranes are held together mainly by

a. hydrogen bonds between lipid molecules.
b. hydration layers over lipid molecules.
c. exclusion of the nonpolar regions of lipids by water.
d. hydrogen bonds between water molecules.
e. surface tension at the interface between layers of water molecules.

Answer: c
Difficulty: Moderate
Bloom's Taxonomy: Knowledge

53. A _____ is formed when a _____ is dissolved in a _____.

a. solution; solute; solvent
b. solute; solvent; solution
c. solution; solvent; solute
d. solvent; solution; solute
e. solvent; solute; solution

Answer: a
Difficulty: Moderate
Bloom's Taxonomy: Knowledge

54. When sugar dissolves in water, water is acting as a _____ and the sugar molecules are acting as _____.

a. solution; solvents
b. solute; solutions
c. solvent; solutes
d. solute; solvents
e. solvent; solutions

Answer: c
Difficulty: Moderate
Bloom's Taxonomy: Application

Salt

55. When salt dissolves in water as illustrated in the figure above, the water molecules form _____ around the Na$^+$ and Cl$^-$ ions.

a. covalent bonds
b. hydration layers
c. nonpolar interactions
d. membranes
e. ionic bonds

Answer: b
Difficulty: Easy
Bloom's Taxonomy: Comprehension
Source: Fig. 2.14

56. Water has a molecular weight of 18 g per mole, and glucose has a molecular weight of 180 g per mole. Which of the following would have an approximately equal number of water and glucose molecules?

a. 1 g of water and 180 g of glucose
b. 90 g of water and 9 g of glucose
c. 180 g of water and 1 g of glucose
d. 9 g of water and 90 g of glucose
e. 90 g of water and 90 g of glucose

Answer: d
Difficulty: Difficult
Bloom's Taxonomy: Application

57. Water has a molecular weight of 18 Daltons or amu. Therefore, a mole of water would have a mass of _____.

a. 1 g
b. 6.02×10^{23} g
c. 36 g
d. 1.08×10^{25} g
e. 18 g

Answer: e
Difficulty: Moderate
Bloom's Taxonomy: Knowledge

58. Water has an unusually high boiling point for its molecular weight because water molecules

a. are very dense.
b. get much heavier as they are heated.
c. are held to each other by hydrogen bonds.
d. are held together by covalent bonds.
e. form hydration layers.

Answer: c
Difficulty: Moderate
Bloom's Taxonomy: Comprehension

59. The hydrogen-bond lattice causes water to have an unusually _____ specific heat and an unusually _____ heat of vaporization for its molecular weight.

a. high; high
b. low; high
c. high; low
d. low; low

Answer: a
Difficulty: Easy
Bloom's Taxonomy: Comprehension

60. Water is useful for cooling organisms mainly due to its

a. hydration layers.
b. specific heat.
c. low calories.
d. surface tension.
e. heat of vaporization.

Answer: e
Difficulty: Moderate
Bloom's Taxonomy: Knowledge

61. Water has an important stabilizing effect on temperature in living organisms and their environments because as water absorbs heat, much of the energy is used to _____ instead of raising the temperature.

a. create hydrogen bonds
b. create covalent bonds
c. break surface tension
d. break hydrogen bonds
e. create hydration layers

Answer: d
Difficulty: Moderate
Bloom's Taxonomy: Comprehension

62. The water strider shown in the figure above is able to stand on water because of the _____ of water.

a. covalent bonds
b. surface tension
c. van der Waals forces
d. density
e. hydration layer

Answer: b
Difficulty: Easy
Bloom's Taxonomy: Knowledge
Source: Fig. 2.15

2.5 WATER IONIZATION AND ACIDS, BASES, AND BUFFERS

63. When added to water, a base will act as a _____ and cause the pH of the solution to _____.

a. proton acceptor; rise
b. proton donor; rise
c. proton acceptor; fall
d. proton donor; fall
e. none of the above

Answer: a
Difficulty: Easy
Bloom's Taxonomy: Comprehension

64. When added to water at neutral pH (7.0), an acid will

a. act as a proton donor, raising the pH of the solution.
b. act as a proton acceptor, raising the pH of the solution.
c. act as a proton donor, lowering the pH of the solution.
d. act as a proton acceptor, lowering the pH of the solution.
e. do nothing since the aqueous solution is neutral.

Answer: c
Difficulty: Moderate
Bloom's Taxonomy: Application

65. A pH of 6 is _____ times more _____ than a pH of 2.

a. 3; acidic
b. 4; acidic
c. 3; basic
d. 10,000; basic
e. 40; basic

Answer: d
Difficulty: Moderate
Bloom's Taxonomy: Application

66. For pure water, which has a pH of 7.0, which of the following is true?

a. $[H^+] < [OH^-]$
b. $[H^+] = [OH^-]$
c. $[H^+] = 0$
d. $[OH^-] = 0$
e. $[H^+] > [OH^-]$

Answer: b
Difficulty: Moderate
Bloom's Taxonomy: Knowledge, Application

67. For acid rainwater, which has a pH as low as 3.0, which of the following is true?

a. $[H^+] < [OH^-]$
b. $[H^+] = [OH^-]$
c. $[H^+] = 0$
d. $[OH^-] = 0$
e. $[H^+] > [OH^-]$

Answer: e
Difficulty: Moderate
Bloom's Taxonomy: Knowledge, Application

68. Seawater typically is

a. highly basic.
b. neutral.
c. somewhat basic.
d. somewhat acidic.
e. highly basic.

Answer: c
Difficulty: Difficult
Bloom's Taxonomy: Knowledge, Analysis

69. Without _____, living organisms would often experience major changes in pH in their cells.

a. buffers
b. acids
c. surface tension
d. nonpolar bonds
e. bases

Answer: a
Difficulty: Easy
Bloom's Taxonomy: Knowledge

70. Most pH buffers are

a. strong acids.
b. weak acids or weak bases.
c. weak acids.
d. strong bases.
e. strong acids or strong bases.

Answer: b
Difficulty: Moderate
Bloom's Taxonomy: Knowledge

UNANSWERED QUESTIONS

71. A research group led by Joanne Santini has found a(n) _____ that potentially can be used in arsenic bioremediation.

a. amoeba
b. plant
c. alga
d. bacterium
e. fungus

Answer: d
Difficulty: Moderate
Bloom's Taxonomy: Knowledge

72. Which of the following statements about irradiation of food is true?

a. Irradiation does not affect viruses in food.
b. Irradiation kills many bacteria and parasites in food.
c. Studies have shown higher cancer rates in laboratory animals fed irradiated food.
d. Irradiation makes food radioactive.
e. Vitamins are completely destroyed when food is irradiated.

Answer: b
Difficulty: Moderate
Bloom's Taxonomy: Knowledge

Integrative Multiple-Choice

73. The most common isotope of carbon has an atomic number of 6 and a mass number of 12, while the most common isotope of oxygen has an atomic number of 8 and a mass number of 16. A molecule of CO_2 made up of these common isotopes has a molecular weight of _____.

a. 28
b. 44
c. 56
d. 14
e. 22

Answer: b
Difficulty: Difficult
Bloom's Taxonomy: Synthesis
Source: Sections 2.2, 2.3, and 2.4

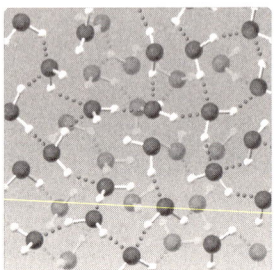

74. The water lattice illustrated in the figure above forms as a result of _____ between water molecules.

a. covalent bonds
b. hydrogen bonds
c. nonpolar interactions
d. ionic bonds
e. van der Walls forces

Answer: b
Difficulty: Easy
Bloom's Taxonomy: Application, Synthesis
Source: Fig. 2.12; Sections 2.3 and 2.4

75. The isotope of hydrogen most commonly found in water is protium, which has no neutrons. However, the form of hydrogen with one neutron, deuterium, can also be found in water. If you were to compare water that only has protium (protium water) with water that only has deuterium (deuterium water) you would find that

a. a mole of protium water weighs the same as a mole of deuterium water.
b. a mole of deuterium water weighs considerably less than half as much as a mole of protium water.
c. a mole of protium water weighs about twice as much as a mole of deuterium water.
d. a mole of deuterium water weighs about twice as much as a mole of protium water.
e. a mole of protium water weighs considerably less than half as much as a mole of deuterium water.

Answer: d
Difficulty: Moderate
Bloom's Taxonomy: Application, Synthesis
Source: Sections 2.2 and 2.4

Matching

Match each of the following terms with its correct definition.

76. _____ element

77. _____ compound

78. _____ matter

79. _____ orbital

80. _____ isotope

A. Anything that occupies space and has mass

B. A pure substance that cannot be broken down into simpler substances by ordinary chemical or physical techniques

C. An atom with the same number of protons as another atom but a different number of neutrons

D. The locations around an atomic nucleus where an electron occurs most frequently

E. A molecule whose component atoms are different from each other

Answers: **76. B   77. E   78. A   79. D   80. C**

Difficulty: Moderate
Bloom's Taxonomy: Knowledge
Source: Sections 2.1, 2.2

Choice

For each of the following situations, choose the correct type of chemical bond.

a. ionic bond(s)
b. nonpolar covalent bond(s)
c. polar covalent bond(s)
d. hydrogen bond(s)
e. van der Waals forces

81. _____ Occurs when electrons are shared equally between two atoms

82. _____ Used by geckos for clinging to and climbing up smooth vertical surfaces

83. _____ Formed by the attraction between partial positive and partial negative charges that were created due to unequal electron sharing

84. _____ Occurs in sodium chloride (NaCl)

85. _____ The weakest of the chemical linkages listed

86. _____ Occurs in a water molecule (H_2O)

87. _____ Characteristic of molecules that contain atoms of only one kind

88. _____ Forms when atoms gain or lose valence electrons completely

89. _____ Attraction that arises when the constant movement of electrons, by chance, produces temporary zones of partial positive and partial negative charges

90. _____ Occurs when electrons are shared unequally between two atoms

91. _____ Creates a region that is hydrophobic

92. _____ Occurs between water molecules

93. _____ Occurs in molecular oxygen (O_2)

Answers: **81. b   82. e   83. d   84. a   85. e**
86. c   87. b   88. a   89. e   90. c
91. b   92. d   93. b

Difficulty: Moderate
Bloom's Taxonomy: Knowledge, Application
Source: Section 2.3

Short Answer

94. Place a large amount of hydrogen gas and oxygen gas in the presence of a fire and you will get an explosion. In light of this, explain how it is possible that water, which is composed of hydrogen and oxygen, is often used to put out fires.

Answer: Water is a compound, and compounds typically have chemical and physical properties that are distinct from the atoms that make them up. So, water had different properties than the hydrogen and oxygen that it is made of and thus behaves differently from them in the presence of fire.
Difficulty: Moderate
Bloom's Taxonomy: Analysis
Source: Section 2.1

Essay

95. Oxygen generally forms two covalent bonds, while carbon generally forms four covalent bonds. In contrast, helium is inert (generally does not form any bonds). Explain the reason for the differences in chemical behavior between these three elements.

Answer: The number of valence electrons in the outermost energy level, or valence shell, determines chemical reactivity. Atoms of an element with a filled valence shell, such as helium, are nonreactive. In contrast, atoms with an unfilled valence shell are reactive; they will tend to gain, lose, or share electrons so that they wind up with a filled valence shell. Oxygen needs two electrons to fill its valence shell, so it tends to form two covalent bonds. Carbon needs four electrons to fill its valence shell so it tends to form four covalent bonds.
Difficulty: Moderate
Bloom's Taxonomy: Application, Synthesis
Source: Section 2.2

96.   Describe how the interaction of water with dual
       polarity lipid molecules establishes biological
       membranes.

*Answer: The hydrogen bonding between water
molecules forms a lattice that resists invasion by
nonpolar molecules. However, polar molecules can
interact with the hydrogen-bond lattice. Lipid
molecules with both polar and nonpolar regions can
align in a bilayer, with the lipid molecules oriented so
that their polar regions are on either side of the
bilayer and their nonpolar regions are buried in the
middle of the bilayer. In this arrangement only the
polar ends are exposed to the water. This creates a
membrane of lipid molecules that separates the watery
solution on one side of the bilayer from the watery
solution on the other side of the bilayer.*
Difficulty: Moderate
Bloom's Taxonomy: Synthesis
Source: Section 2.4

3

BIOLOGICAL MOLECULES: THE CARBON COMPOUNDS OF LIFE

Multiple-Choice

WHY IT MATTERS

1.   Most life on Earth depends, directly or indirectly, on a process that combines water and carbon dioxide to make carbon-based compounds. This process is called

a.   aerobic respiration.
b.   decomposition.
c.   photosynthesis.
d.   anaerobic respiration.
e.   fermentation.

Answer: c
Difficulty: Easy
Bloom's Taxonomy: Comprehension

2.   Which of the following does NOT add carbon dioxide to the atmosphere?

a.   forest fires
b.   decomposers
c.   burning of fossil fuels
d.   photosynthesis
e.   all of these add carbon dioxide to the atmosphere

Answer: d
Difficulty: Moderate
Bloom's Taxonomy: Application

3.1 CARBON BONDING

3.   Carbon has four electrons in its valence shell. It will typically form _____ covalent bonds to fill its valence shell.

a.   one
b.   two
c.   four
d.   six
e.   eight

Answer: c
Difficulty: Easy
Bloom's Taxonomy: Knowledge

4.   In general, molecules made mostly of carbon atoms covalently bound to each other and to other atoms are called _____.

a.   hydrocarbons
b.   carbohydrates
c.   polymers
d.   functional groups
e.   organic molecules

Answer: e
Difficulty: Easy
Bloom's Taxonomy: Comprehension

5.   In general, molecules consisting only of carbon and hydrogen atoms are called _____.

a.   hydrocarbons
b.   carbohydrates
c.   polymers
d.   functional groups
e.   organic molecules

Answer: a
Difficulty: Easy
Bloom's Taxonomy: Comprehension

6.   Which of the following is NOT found in hydrocarbons?

a.   branching
b.   double bonds
c.   ring formation
d.   triple bonds
e.   all of these can be found in hydrocarbons

Answer: e
Difficulty: Moderate
Bloom's Taxonomy: Comprehension

7.   The simplest hydrocarbon, CH_4, is called _____.

a.   methane
b.   ethyne
c.   cyclohexane
d.   ethane
e.   benzene

Answer: a
Difficulty: Moderate
Bloom's Taxonomy: Knowledge

8.   Which of the following is an example of a hydrocarbon?

a. C_2H_6O
b. C_8H_{18}
c. $C_6H_{12}O_6$
d. CO_2
e. C_2H_6S

Answer: b
Difficulty: Moderate
Bloom's Taxonomy: Application

3.2 FUNCTIONAL GROUPS IN BIOLOGICAL MOLECULES

9. Reactions that use the equivalent of a water molecule to break a molecule into smaller subunits are called _____ reactions.

a. equilibrium
b. hydration
c. hydrolysis
d. redox
e. dehydration synthesis

Answer: c
Difficulty: Moderate
Bloom's Taxonomy: Comprehension

10. Reactions that remove the equivalent of a water molecule when subunits are joined to make a larger molecule are called _____ reactions.

a. equilibrium
b. hydration
c. hydrolysis
d. redox
e. dehydration synthesis

Answer: e
Difficulty: Moderate
Bloom's Taxonomy: Comprehension

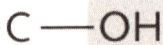

11. The functional group shown above is _____.

a. a carbonyl group
b. a hydroxyl group
c. a phosphate group
d. an amino group
e. a carboxyl group

Answer: b
Difficulty: Moderate
Bloom's Taxonomy: Application
Source: Table 3.1

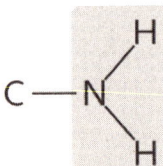

12. The functional group shown above is _____.

a. a carbonyl group
b. a hydroxyl group
c. a phosphate group
d. an amino group
e. a carboxyl group

Answer: d
Difficulty: Moderate
Bloom's Taxonomy: Application
Source: Table 3.1

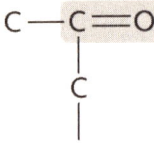

13. The functional group shown above is _____.

a. a carbonyl group
b. a hydroxyl group
c. a phosphate group
d. an amino group
e. a carboxyl group

Answer: a
Difficulty: Moderate
Bloom's Taxonomy: Application
Source: Table 3.1

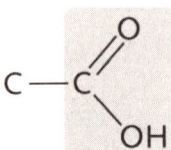

14. The functional group shown above is _____.

a. a carbonyl group
b. a hydroxyl group
c. a phosphate group
d. an amino group
e. a carboxyl group

Answer: e
Difficulty: Moderate
Bloom's Taxonomy: Application
Source: Table 3.1

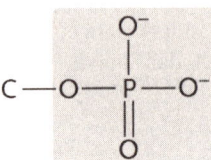

15. The functional group shown above is _____.

a. a carbonyl group
b. a hydroxyl group
c. a phosphate group
d. an amino group
e. a carboxyl group

Answer: c
Difficulty: Moderate
Bloom's Taxonomy: Application
Source: Table 3.1

C —SH

16. The functional group shown above often forms a sort of molecular fastener when two of them are joined by a covalent bond. This particular type of bond is found in many _____.

a. phospholipids
b. nucleic acids
c. steroids
d. proteins
e. carbohydrates

Answer: d
Difficulty: Difficult
Bloom's Taxonomy: Knowledge
Source: Table 3.1

17. Which functional group characteristic of organic acids such as acetic acid?

a. carbonyl group
b. hydroxyl group
c. phosphate group
d. amino group
e. carboxyl group

Answer: e
Difficulty: Moderate
Bloom's Taxonomy: Knowledge

18. Which functional group forms the highly reactive part of aldehydes and ketones?

a. carbonyl group
b. hydroxyl group
c. phosphate group
d. amino group
e. carboxyl group

Answer: a
Difficulty: Moderate
Bloom's Taxonomy: Knowledge

19. Which functional group is polar and a key component of alcohols?

a. carbonyl group
b. hydroxyl group
c. phosphate group
d. amino group
e. carboxyl group

Answer: b
Difficulty: Moderate
Bloom's Taxonomy: Knowledge

20. Which functional group acts as an organic base?

a. carbonyl group
b. hydroxyl group
c. phosphate group
d. amino group
e. carboxyl group

Answer: d
Difficulty: Moderate
Bloom's Taxonomy: Knowledge

3.3 CARBOHYDRATES

21. What organic molecules have a chemical formula that is (or is very nearly) a multiple of (CH_2O)?

a. proteins
b. lipids
c. nucleic acids
d. carbohydrates

Answer: d
Difficulty: Moderate
Bloom's Taxonomy: Knowledge

22. Monosaccharides and disaccharides are types of _____.

a. proteins
b. lipids
c. nucleic acids
d. carbohydrates

Answer: d
Difficulty: Easy
Bloom's Taxonomy: Knowledge

23. A monosaccharide with six carbons is called _____.

a. triose
b. hexose
c. pentose
d. heptose
e. tetrose

Answer: b
Difficulty: Moderate
Bloom's Taxonomy: Knowledge

24. Isomers are two or more molecules with _____ chemical formula and _____ molecular structures.

a. a different; different
b. the same; different
c. a different; the same
d. the same; the same

Answer: b
Difficulty: Easy
Bloom's Taxonomy: Comprehension

25. Structural isomers differ from each other

a. in the arrangement of their covalent bonds.
b. in their molecular formulas.
c. by being mirror images of each other that cannot be superimposed on each other.
d. by having double covalent bonds instead of single bonds.
e. by having different atomic isotopes in their molecules.

Answer: a
Difficulty: Moderate
Bloom's Taxonomy: Comprehension

26. When molecules are referred to as D- or L- (for example D-forms of sugars and L-forms of amino acids), the D- and L- designations refer to the specific _____.

a. functional group
b. structural isomer
c. covalent bond
d. secondary structure
e. enantiomer

Answer: e
Difficulty: Moderate
Bloom's Taxonomy: Application

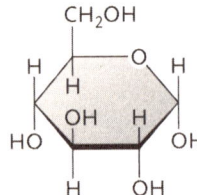

27. The molecule shown above is a(n) _____.

a. amino acid
b. monosaccharide
c. steroid
d. nucleotide
e. phospholipid

Answer: b
Difficulty: Moderate
Bloom's Taxonomy: Application
Source: Figure 3.5

28. The linkage commonly found between subunits in a chain of monosaccharides is called a _____ bond.

a. phosphodiester
b. disulfide
c. glycosidic
d. hydrogen
e. peptide

Answer: c
Difficulty: Easy
Bloom's Taxonomy: Knowledge

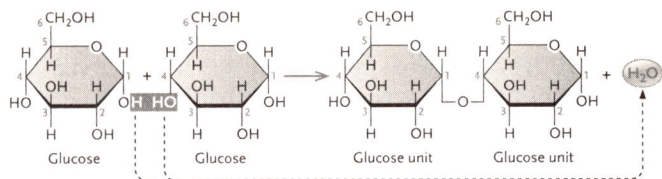

29. The disaccharide shown above is _____.

a. sucrose
b. maltose
c. lactose
d. cellulose
e. fructose

Answer: b
Difficulty: Moderate
Bloom's Taxonomy: Knowledge
Source: Figure 3.6

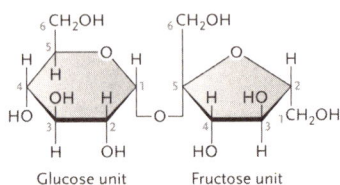

30. The disaccharide shown above is _____.

a. cellulose
b. fructose
c. maltose
d. sucrose
e. lactose

Answer: d
Difficulty: Moderate
Bloom's Taxonomy: Knowledge
Source: Figure 3.6

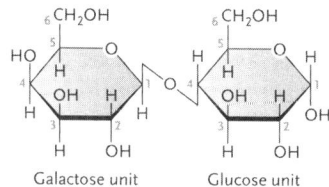

Galactose unit   Glucose unit

31. The disaccharide shown above is _____.

a. lactose
b. fructose
c. maltose
d. cellulose
e. sucrose

Answer: a
Difficulty: Moderate
Bloom's Taxonomy: Knowledge
Source: Figure 3.6

32. Amylose is a polymer made up of glucose monomers joined by _____.

a. β(1-4) linkages
b. β(1-6) linkages
c. α(1-4) linkages
d. α(1-6) linkages
e. both β(1-4) and α(1-6) linkages

Answer: c
Difficulty: Difficult
Bloom's Taxonomy: Knowledge

33. Probably the most abundant carbohydrate on Earth, this unbranched chain of β-glucose subunits is the primary structural fiber in plant cell walls.

a. chitin
b. amylopectin
c. cellulose
d. glycogen
e. amylose

Answer: c
Difficulty: Easy
Bloom's Taxonomy: Knowledge

34. This polysaccharide is a chain of glucose units that are modified by having nitrogen-containing groups. It is the main structural fiber in the external skeletons of arthropods and is also a structural material in the cell walls of fungi.

a. chitin
b. amylopectin
c. cellulose
d. glycogen
e. amylose

Answer: a
Difficulty: Easy
Bloom's Taxonomy: Knowledge

35. Animal cells commonly have polysaccharides attached to _____ in their surface membranes.

a. proteins
b. nucleic acids
c. lipids
d. both nucleic acids and lipids
e. both proteins and lipids

Answer: e
Difficulty: Difficult
Bloom's Taxonomy: Knowledge

36. The molecule shown above is glyceraldehyde, an example of a _____ sugar.

a. triose
b. hexose
c. pentose
d. heptose
e. tetrose

Answer: a
Difficulty: Easy
Bloom's Taxonomy: Application
Source: Figure 3.3

37. The molecule shown above is ribose, an example of a _____ sugar.

a. triose
b. hexose
c. pentose
d. heptose
e. tetrose

Answer: c
Difficulty: Moderate
Bloom's Taxonomy: Application
Source: Figure 3.3

H
 \
  C=O
 /
HO—C—H
HO—C—H
H—C—OH
H—C—OH
H—C—OH
      |
      H

38. The molecule shown above is mannose, an example of a _____ sugar.

a. triose
b. hexose
c. pentose
d. heptose
e. tetrose

Answer: b
Difficulty: Moderate
Bloom's Taxonomy: Application
Source: Figure 3.3

39. Suppose that an equal amount of each of the following polysaccharides was placed in a landfill. Which of them should last the longest before it is decomposed?

a. chitin
b. amylopectin
c. cellulose
d. glycogen
e. amylose

Answer: c
Difficulty: Moderate
Bloom's Taxonomy: Application

40. In many animals this polysaccharide is found in large quantities in liver and muscle cells. It is highly branched, with many $\alpha(1-4)$ and $\alpha(1-6)$ linkages.

a. chitin
b. amylopectin
c. cellulose
d. glycogen
e. amylose

Answer: d
Difficulty: Easy
Bloom's Taxonomy: Knowledge

3.4 LIPIDS

41. Although they have a diversity of structures, all members of this group of organic molecules are primarily nonpolar and thus water-insoluble.

a. proteins
b. lipids
c. nucleic acids
d. carbohydrates

Answer: b
Difficulty: Easy
Bloom's Taxonomy: Knowledge

42. The molecule shown above is _____, having _____.

a. unsaturated; a carbon-carbon double bond
b. saturated; a carbon-carbon double bond
c. unsaturated; no carbon-carbon double bonds
d. saturated; no carbon-carbon double bonds

Answer: d
Difficulty: Moderate
Bloom's Taxonomy: Application
Source: Figure 3.8

43. The molecule shown above is _____, having _____.

a. unsaturated; a carbon-carbon double bond
b. saturated; a carbon-carbon double bond
c. unsaturated; no carbon-carbon double bonds
d. saturated; no carbon-carbon double bonds

Answer: a
Difficulty: Moderate
Bloom's Taxonomy: Application
Source: Figure 3.8

44. Which of the following is the best explanation for why unsaturated fats tend to be more fluid at biological temperatures than saturated fats?

a. Double bonds in saturated fats cause them to be denser and thus more solid than unsaturated fats.
b. The presence of extra hydrogen atoms in saturated fats makes them lighter, so they float away from water and congeal together more easily than unsaturated fats.
c. The presence of extra hydrogen atoms in saturated fats makes them denser than unsaturated fats, and denser materials tend to be more solid.
d. Because unsaturated fats can still be hydrogenated but saturated fats cannot, the unsaturated fats are more reactive and thus more fluid.
e. Kinks in the chain of an unsaturated fat create more disorder or irregularity to the structure of the molecule, making unsaturated fats harder to pack together than saturated fats.

Answer: e
Difficulty: Difficult
Bloom's Taxonomy: Evaluation

45. A fatty acid has _____ group at the end of a hydrocarbon chain.

a. a carbonyl group
b. a hydroxyl group
c. a phosphate group
d. an amino group
e. a carboxyl group

Answer: e
Difficulty: Moderate
Bloom's Taxonomy: Knowledge

46. Among the following, which number of carbons would be the most likely number of carbons for a fatty acid from a living organism?

a. 8
b. 13
c. 16
d. 19
e. 26

Answer: c
Difficulty: Difficult
Bloom's Taxonomy: Application

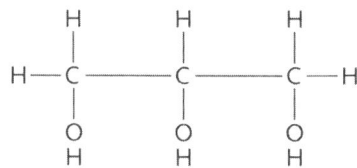

47. The molecule shown above is _____.

a. cholesterol
b. threonine
c. ethanol
d. glycerol
e. glucose

Answer: d
Difficulty: Moderate
Bloom's Taxonomy: Application
Source: Figure 3.9

48. Glycerol forms the backbone of

a. triglycerides.
b. polysaccharides and nucleic acids.
c. nucleic acids.
d. polypeptides.
e. triglycerides and phospholipids.

Answer: e
Difficulty: Moderate
Bloom's Taxonomy: Knowledge

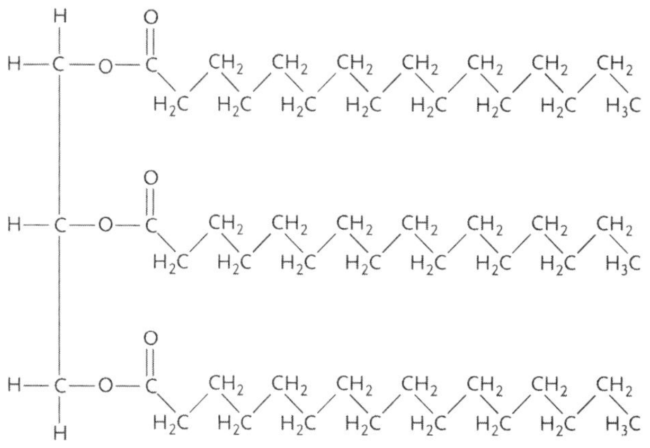

Use the figure above for questions 49 and 50.

49. The molecule shown above is _____.

a. a triglyceride
b. an amino acid
c. a steroid
d. a polysaccharide
e. a phospholipid

Answer: a
Difficulty: Moderate
Bloom's Taxonomy: Application
Source: Figure 3.9

50. Which of the following would the molecule shown above be called?

a. a protein
b. a lipid
c. a nucleic acid
d. a carbohydrate
e. none of these

Answer: b
Difficulty: Moderate
Bloom's Taxonomy: Application
Source: Figure 3.9

51. Which of these are the main structural components of biological membranes?

a. starches
b. triglycerides
c. proteins
d. phospholipids
e. steroids

Answer: d
Difficulty: Easy
Bloom's Taxonomy: Comprehension

$^{+}NH_3$—CH_2—CH_2—O—P—O—CH_2

(chemical structure of the molecule)

52. The molecule shown above is _____.

a. a triglyceride
b. an amino acid
c. a steroid
d. a polysaccharide
e. a phospholipid

Answer: e
Difficulty: Moderate
Bloom's Taxonomy: Application
Source: Figure 3.12

53. In a phospholipid, glycerol is bound to

a. three phosphate groups, each of which are bound to fatty acids.
b. two fatty acids chains and one phosphate group.
c. the hydrogen bonds of liquid water are fixed in place.
d. three phosphate groups, two of which are bound to fatty acids.
e. two phosphate groups on one side and three fatty acids on the other side.

Answer: b
Difficulty: Moderate
Bloom's Taxonomy: Comprehension

54. The molecule shown above is _____.

a. a triglyceride
b. an amino acid
c. a steroid
d. a polysaccharide
e. a phospholipid

Answer: c
Difficulty: Moderate
Bloom's Taxonomy: Application
Source: Figure 3.13

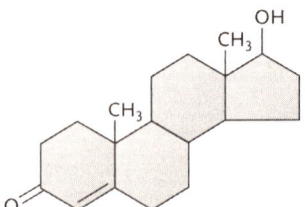

55. The molecule shown above is testosterone, a substance that has important regulatory functions in humans and many other animals. Molecules with regulatory functions like testosterone are called _____.

a. phytosterols
b. enzymes
c. lipoproteins
d. hormones
e. receptors

Answer: d
Difficulty: Easy
Bloom's Taxonomy: Knowledge

56. Waxy coatings, such as found on skin, hair, and feathers of some animals and the cuticle of some plants, are commonly used by living organisms for protection against water loss and for lubrication. Such waxes are considered to be a type of _____.

a. triglyceride
b. steroid
c. neutral lipid
d. phospholipid
e. fatty acid

Answer: c
Difficulty: Moderate
Bloom's Taxonomy: Comprehension

FOCUS ON RESEARCH: FATS, CHOLESTEROL, AND CORONARY ARTERY DISEASE

57. Two types of cholesterol are found in human blood, HDL and LDL. Clinical studies have shown that the risk of coronary heart disease is greatest for a person with a _____ level of HDL in their blood and a _____ level of LDL in their blood.

a. high; low
b. low; low
c. low; high
d. low; low

Answer: a
Difficulty: Moderate
Bloom's Taxonomy: Comprehension

3.5 PROTEINS

58. Which of the following is NOT a major function of proteins in living organisms?

a. speeding up biological reactions
b. transporting substances across membranes
c. providing structural support
d. regulating the activity of other cellular molecules
e. storing genetic information

Answer: e
Difficulty: Easy
Bloom's Taxonomy: Knowledge

59. Proteins are polymers of _____.

a. amino acids
b. monosaccharides
c. steroids
d. nucleotides
e. phospholipids

Answer: a
Difficulty: Easy
Bloom's Taxonomy: Knowledge

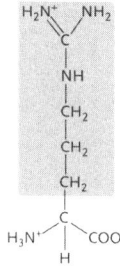 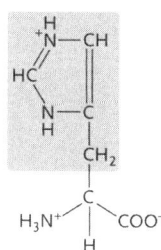

60. The two amino acids depicted above both have side groups (*R* groups) that are

a. acidic and polar.
b. uncharged and nonpolar.
c. basic and polar.
d. positively and negatively charged and nonpolar.
e. uncharged and polar.

Answer: c
Difficulty: Moderate
Bloom's Taxonomy: Application
Source: Figure 3.15

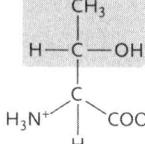

61. The two amino acids depicted above both have side groups (*R* groups) that are

a. acidic and polar.
b. uncharged and nonpolar.
c. basic and polar.
d. positively and negatively charged and nonpolar.
e. uncharged and polar.

Answer: e
Difficulty: Moderate
Bloom's Taxonomy: Application
Source: Figure 3.15

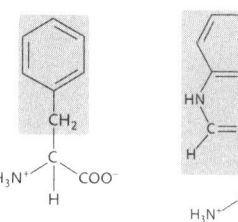

62. The two amino acids depicted above both have side groups (*R* groups) that are

a. acidic and polar.
b. uncharged and nonpolar.
c. basic and polar.
d. positively and negatively charged and nonpolar.
e. uncharged and polar.

Answer: b
Difficulty: Difficult
Bloom's Taxonomy: Application
Source: Figure 3.15

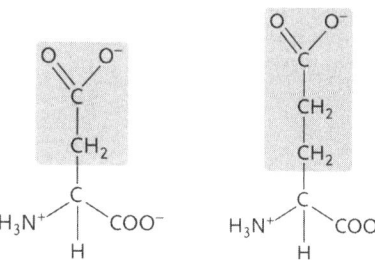

63. The two amino acids depicted above both have side groups (*R* groups) that are

a. acidic and polar.
b. uncharged and nonpolar.
c. basic and polar.
d. positively and negatively charged and nonpolar.
e. uncharged and polar.

Answer: a
Difficulty: Moderate
Bloom's Taxonomy: Application
Source: Figure 3.15

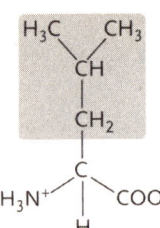 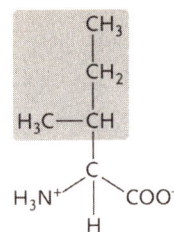

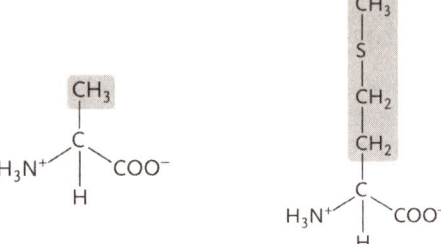

64. The two amino acids depicted above both have side groups (*R* groups) that are

a. acidic and polar.
b. uncharged and nonpolar.
c. basic and polar.
d. positively and negatively charged and nonpolar.
e. uncharged and polar.

Answer: b
Difficulty: Difficult
Bloom's Taxonomy: Application
Source: Figure 3.15

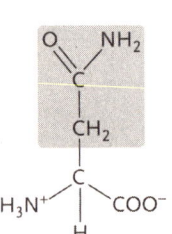 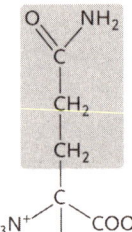

65. The two amino acids depicted above both have side groups (*R* groups) that are

a. acidic and polar.
b. uncharged and nonpolar.
c. basic and polar.
d. positively and negatively charged and nonpolar.
e. uncharged and polar.

Answer: e
Difficulty: Moderate
Bloom's Taxonomy: Application
Source: Figure 3.15

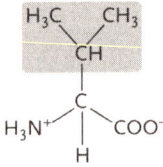 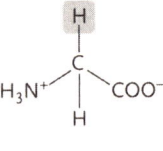

66. The two amino acids depicted above both have side groups (*R* groups) that are

a. acidic and polar.
b. uncharged and nonpolar.
c. basic and polar.
d. positively and negatively charged and nonpolar.
e. uncharged and polar.

Answer: b
Difficulty: Difficult
Bloom's Taxonomy: Application
Source: Figure 3.15

67. The two amino acids depicted above both have side groups (*R* groups) that are

a. acidic and polar.
b. uncharged and nonpolar.
c. basic and polar.
d. positively and negatively charged and nonpolar.
e. uncharged and polar.

Answer: b
Difficulty: Difficult
Bloom's Taxonomy: Application
Source: Figure 3.15

68. Which amino acid(s) can be involved in special covalent "disulfide bridges" that, when present, help to stabilize the tertiary and quaternary structure of proteins?

a. proline
b. alanine
c. leucine and isoleucine
d. cysteine
e. methionine

Answer: d
Difficulty: Moderate
Bloom's Taxonomy: Knowledge

69. The linkage commonly found between amino acids in a chain of amino acids is called a _____ bond.

a. phosphodiester
b. disulfide
c. glycosidic
d. hydrogen
e. peptide

Answer: e
Difficulty: Easy
Bloom's Taxonomy: Knowledge

70. The unique sequence of the monomer subunits in a protein is the _____ structure of the protein.

a. primary
b. secondary
c. tertiary
d. quaternary

Answer: a
Difficulty: Moderate
Bloom's Taxonomy: Knowledge

71.   A hydrogen bond involving only parts of the backbone in a single strand of a protein would be considered part of the _____ structure of the protein

a.   primary
b.   secondary
c.   tertiary
d.   quaternary

Answer: b
Difficulty: Moderate
Bloom's Taxonomy: Application

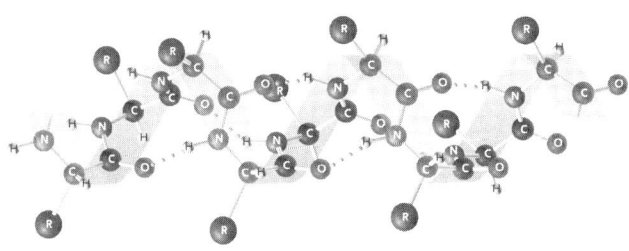

72.   The image shown above is an example of a(n) _____.

a.   random coil
b.   leucine zipper
c.   β sheet
d.   α helix
e.   β polysaccharide

Answer: d
Difficulty: Moderate
Bloom's Taxonomy: Application
Source: Figure 3.20

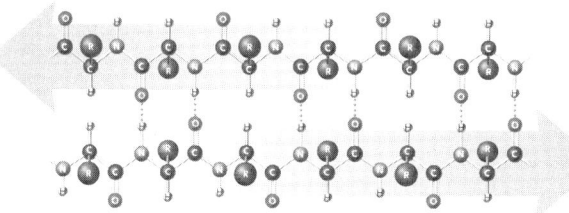

73   The image shown above is an example of a(n) _____.

a.   random coil
b.   leucine zipper
c.   β sheet
d.   α helix
e.   β polysaccharide

Answer: c
Difficulty: Moderate
Bloom's Taxonomy: Application
Source: Figure 3.21

74.   The amino acid shown above is proline, which is often found in a(n) _____.

a.   random coil
b.   leucine zipper
c.   β sheet
d.   α helix
e.   β polysaccharide

Answer: a
Difficulty: Difficult
Bloom's Taxonomy: Knowledge
Source: Figure 3.15

75.   A hydrogen bond between the side groups (*R* groups) of two different monomer subunits in a single strand of a protein would be considered part of the _____ structure of the protein.

a.   primary
b.   secondary
c.   tertiary
d.   quaternary

Answer: c
Difficulty: Difficult
Bloom's Taxonomy: Application

76.   The _____ structure of a protein refers to the conformation, or overall three-dimensional shape, of a polypeptide that has been folded into a functional protein..

a.   primary
b.   secondary
c.   tertiary
d.   quaternary
e.   none of these

Answer: c
Difficulty: Moderate
Bloom's Taxonomy: Knowledge

77.   A hydrogen bond between the amino acid side groups (*R* groups) from two different polypeptide chains in a multichain protein would be considered part of the _____ structure of the protein.

a.   primary
b.   secondary
c.   tertiary
d.   quaternary
e.   none of these

Answer: d
Difficulty: Difficult
Bloom's Taxonomy: Application

78. These structural segments in proteins provide flexibility that allows parts of the protein to bend, fold, or move.

a. random coil
b. leucine zipper
c. β sheet
d. α helix
e. β polysaccharide

Answer: a
Difficulty: Moderate
Bloom's Taxonomy: Knowledge

79. Excessive heat or extremes of pH often cause denaturation of proteins, which means that the proteins

a. have fallen apart into individual monomer subunits.
b. are no longer biological molecules.
c. have broken into many separate domains.
d. are no longer in a functional three dimensional structure.
e. are highly reactive.

Answer: d
Difficulty: Moderate
Bloom's Taxonomy: Comprehension

80. Chaperonins assist with

a. proteins synthesis.
b. assembly of DNA strands.
c. polysaccharide synthesis.
d. protein folding.
e. forming the DNA double helix.

Answer: d
Difficulty: Moderate
Bloom's Taxonomy: Comprehension

INSIGHTS FROM THE MOLECULAR REVOLUTION: GETTING GOOD VIBRATIONS FROM PROTEINS

81. Detection of conformational changes (or changes in shape) in proteins is important because

a. proteins that change their shape cannot function.
b. cancer is usually caused by protein shape changes.
c. protein shape changes are required to move proteins inside of cells.
d. protein shape changes prevent cell division.
e. many functions of proteins depend on shape changes.

Answer: e
Difficulty: Moderate
Bloom's Taxonomy: Comprehension

3.6 NUCLEOTIDES AND NUCLEIC ACIDS

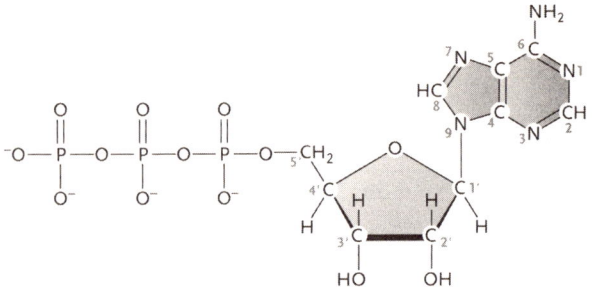

82. The molecule shown above is a(n) _____.

a. amino acid
b. monosaccharide
c. steroid
d. nucleotide
e. phospholipid

Answer: d
Difficulty: Moderate
Bloom's Taxonomy: Application
Source: Figure 3.26

83. The molecule shown above is the pyrimidine _____, which is typically found in DNA but not RNA.

a. thymine
b. adenine
c. uracil
d. guanine
e. cytosine

Answer: a
Difficulty: Moderate
Bloom's Taxonomy: Knowledge
Source: Figure 3.27

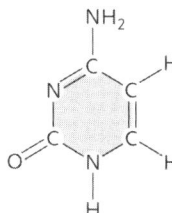

84. The molecule shown above is the pyrimidine _____, which is typically found in both DNA and RNA.

a. thymine
b. adenine
c. uracil
d. guanine
e. cytosine

Answer: e
Difficulty: Difficult
Bloom's Taxonomy: Application
Source: Figure 3.27

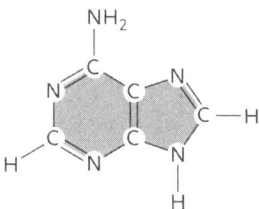

85. The molecule shown above is the pyrimidine _____, which is typically found in RNA but not DNA.

a. thymine
b. adenine
c. uracil
d. guanine
e. cytosine

Answer: c
Difficulty: Easy
Bloom's Taxonomy: Knowledge
Source: Figure 3.27

86. The molecule shown above is the purine _____.

a. thymine
b. adenine
c. uracil
d. guanine
e. cytosine

Answer: b
Difficulty: Difficult
Bloom's Taxonomy: Application
Source: Figure 3.27

87. Hereditary information in all eukaryotes, prokaryotes, and viruses is stored in _____.

a. proteins
b. lipids
c. nucleic acids
d. carbohydrates

Answer: c
Difficulty: Easy
Bloom's Taxonomy: Knowledge

88. Nucleotides are joined together to make a nucleic acid strand by _____ bonds.

a. phosphodiester
b. disulfide
c. glycosidic
d. hydrogen
e. peptide

Answer: a
Difficulty: Easy
Bloom's Taxonomy: Knowledge

89. The two strands of a DNA double helix are held to each other by _____ bonds between nitrogenous bases.

a. phosphodiester
b. disulfide
c. glycosidic
d. hydrogen
e. peptide

Answer: d
Difficulty: Moderate
Bloom's Taxonomy: Knowledge

90. Consider the DNA sequence GATTACA. If the strand with this sequence forms a double helix with another DNA strand, the sequence on the other strand should be _____.

a. TCGGCAC
b. GATTACA
c. ACATTAG
d. GTGCCGA
e. CTAATGA

Answer: e
Difficulty: Moderate
Bloom's Taxonomy: Application

91. RNA molecules are usually found as

a. a two-stranded double helix.
b. a single strand with some folds and twists where the strand bonds with itself to make some double-helical regions.
c. part of a pair with a single DNA strand.
d. a single strand that has no double-helical regions.
e. an extra strand joined to a DNA double helix.

Answer: b
Difficulty: Moderate
Bloom's Taxonomy: Comprehension

92.   Sterol regulatory element binding proteins (SREBPs) are involved in regulating the synthesis of _____.

a.   proteins
b.   lipids
c.   nucleic acids
d.   carbohydrates

Answer: b
Difficulty: Moderate
Bloom's Taxonomy: Knowledge

93.   Alzheimer disease, Parkinson disease, and non-insulin dependent (type 2) diabetes are all examples of diseases caused by _____.

a.   lipid biosynthesis
b.   improper blood sugar level
c.   protein misfolding
d.   high blood cholesterol level
e.   improper DNA base-pairing

Answer: c
Difficulty: Moderate
Bloom's Taxonomy: Knowledge

Integrative Multiple Choice

C —— SH

94.   The side group (*R* group) of the amino acid shown above contains which of the following functional groups?

a.   hydroxyl
b.   carbonyl
c.   carboxyl
d.   sulfhydryl
e.   amino

Answer: d
Difficulty: Moderate
Bloom's Taxonomy: Application
Source: Figure 3.15; Sections 3.2 and 3.5

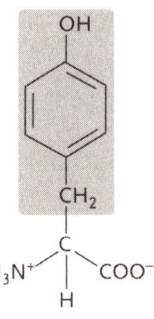

95.   The side group (*R* group) of the amino acid shown above contains which of the following functional groups?

a.   hydroxyl
b.   carbonyl
c.   carboxyl
d.   sulfhydryl
e.   amino

Answer: e
Difficulty: Moderate
Bloom's Taxonomy: Application
Source: Figure 3.15; Sections 3.2 and 3.5

96.   The side group (*R* group) of the amino acid shown above contains which of the following functional groups?

a.   hydroxyl
b.   carbonyl
c.   carboxyl
d.   sulfhydryl
e.   amino

Answer: a
Difficulty: Moderate
Bloom's Taxonomy: Application
Source: Figure 3.15; Sections 3.2 and 3.5

97. Which of the following is least likely to be found in a cell membrane?

a. phospholipid
b. nucleic acid
c. steroid
d. protein
e. carbohydrate

Answer: b
Difficulty: Moderate
Bloom's Taxonomy: Synthesis
Source: Sections 3.3–3.6

98. Which of the following types of molecules are most important for a cell to protect from damage if the cell is to stay alive and reproduce?

a. proteins
b. lipids
c. nucleic acids
d. carbohydrates

Answer: c
Difficulty: Difficult
Bloom's Taxonomy: Evaluation
Source: Sections 3.3-3.6

99. Which of the following types of molecules are the most diverse in terms of structure and types of roles in cells?

a. proteins
b. lipids
c. nucleic acids
d. carbohydrates

Answer: a
Difficulty: Difficult
Bloom's Taxonomy: Evaluation
Source: Sections 3.3-3.6

Matching

Match each of the functional groups below with the letter of the molecule that has the functional group highlighted.

100. _____ phosphate

101. _____ carboxyl

102. _____ sulfhydryl

103. _____ carbonyl

104. _____ hydroxyl

A.

B.

C.

D.

E.

Answers: 100. E   101. B   102. A   103. C   104. D

Difficulty: Moderate
Bloom's Taxonomy: Application
Source: Table 3.1; Section 3.2

Choice

For each of the following carbohydrates, select the choice that best represents one of its primary roles in living organisms.

a. energy source
b. structural fiber in plant cell walls
c. energy storage
d. structural fiber in the exoskeleton of arthropods

105. _____ starch

106. _____ glucose

107. _____ chitin

108. _____ glycogen

109. _____ cellulose

Answers: 105. c   106. a   107. d   108. c   109. b

Difficulty: Moderate
Bloom's Taxonomy: Synthesis
Source: Section 3.3

Choose the class of biological molecules that is most closely associated with each of the following items.

a. carbohydrate
b. lipid
c. protein
d. nucleic acid

110. _____ phosphodiester bond

111. _____ chemical formula at or close to a multiple of (CH_2O)

112. _____ peptide bond

113. _____ nitrogenous base

114. _____ triglyceride

115. _____ glycosidic bond

116. _____ sugar-phosphate backbone

117. _____ α helix

Answers: *110. d   111. a   112. c   113. d   114. b
115. a   116. d   117. c*

Difficulty: Moderate
Bloom's Taxonomy: Knowledge, Comprehension
Source: Sections 3.3–3.6

Essay

118. Describe the relationship between primary, secondary, tertiary, and quaternary structure in proteins.

Answer: The primary structure of a protein is the sequence of amino acids in a polypeptide chain; it is this sequence that determines what folds and bonds the chain can make for other levels of structure. The secondary structures are the twists and arrangements of the polypeptide chain into forms such as random coils, α helices, and β sheets based on what, if any, hydrogen bonds between parts of the polypeptide backbone can form. The secondary structure provides the framework for the tertiary structure, which is the overall three-dimensional shape of a single polypeptide chain. Interactions between the amino acid side groups combine with secondary structure to determine tertiary structure. If a protein is made up of more than one strand of amino acids, then the interactions between strands to provide the final three dimensional shape of the overall protein is the quaternary structure.
Difficulty: Moderate
Bloom's Taxonomy: Synthesis
Source: Section 3.5

119. Describe the general properties of each of the four major classes of organic molecules, including the major roles of each of them in cells.

Answer: The four major classes of organic molecules are carbohydrates, lipids, proteins, and nucleic acids. Carbohydrates contain carbon, hydrogen, and oxygen at or near the ratio of (CH_2O). Their major roles in cells are as energy sources and energy storage molecules, and as structural molecules. Lipids are primarily nonpolar, mainly hydrophobic organic molecules. Their main functions in cells are as structural components of cell membranes, as energy sources, as energy storage molecules, and as hormones. Proteins are polymers of amino acids that fold into specific structures that allow each type of protein to carry out its specific role or roles. Proteins play many major roles in cells. Some examples: enzymes that catalyze cellular reactions; motile molecules involved in movement of and within cells as well as for specialized groups of cells; transport of material across cell membranes; receptor molecules at cell surfaces; and regulation of the activity of other cellular molecules such as other proteins and DNA. Nucleic acids are polymers of nucleotides. There are two main types of nucleic acids, DNA and RNA. These differ from each other chemically in the exact type of sugar in their nucleotides (deoxyribose for DNA and ribose for RNA) and in whether thymine or uracil is used as a nitrogenous base (DNA uses thymine, RNA uses uracil). DNA is typically found as two strands connected by hydrogen bonds to form a double helix, while RNA is typically single-stranded. Hereditary information in cells is stored in DNA. There are many different types of RNA, but nearly all RNA molecules play either a direct or indirect role in the production of proteins using the information stored in DNA.
Difficulty: Moderate
Bloom's Taxonomy: Synthesis
Source: Sections 3.3–3.6

4

ENERGY, ENZYMES, AND BIOLOGICAL REACTIONS

Multiple-Choice

WHY IT MATTERS

1.   The term that best describes all of the chemical reactions of the cell, including acquisition and use of molecules and energy, is

a.   metabolism.
b.   anabolism.
c.   catabolism.
d.   energy budget.
e.   thermodynamics.

Answer: a
Difficulty: Easy
Bloom's Taxonomy: Knowledge

2.   Mushrooms such as the "old man of the woods mushrooms" acquire energy and nutrients from decaying organisms. By doing so they

a.   avoid competition with other organisms for energy resources.
b.   access nutrients unavailable to other fungi.
c.   cycle energy through the ecosystem when they also die and decay.
d.   manufacture hallucinogenic compounds as a defense from predation.
e.   absorb the dead cells and then digest them inside the mushroom's stalk.

Answer: c
Difficulty: Moderate
Bloom's Taxonomy: Comprehension

4.1 ENERGY, LIFE, AND THE LAWS OF THERMODYNAMICS

3.   Which of the following is *not* a form of energy?

a.   heat
b.   diffusion
c.   sound
d.   light
e.   gamma radiation

Answer: b
Difficulty: Easy
Bloom's Taxonomy: Knowledge

4.   A child swinging on a swing utilizes which type (or types) of energy?

a.   kinetic energy only because the child is in constant motion
b.   potential energy only, because the child has to invest energy to get the swing to go
c.   chemical energy only, because it is the child's metabolism that powers the muscles that make the swing move
d.   kinetic and potential energy only, but in constantly changing ratios: when changing direction it is pure potential energy; at the bottom of the arc, it is pure kinetic energy
e.   kinetic, potential, and chemical energy. The child powers the swing with chemical energy in the muscle cells and the swing moves like a pendulum with changing ratios of kinetic and potential energy.

Answer: e
Difficulty: Moderate
Bloom's Taxonomy: Knowledge, Comprehension

5.   Eating and digesting a candy bar for energy during a sports event is a good example of

a.   catabolism.
b.   anabolism.
c.   converting kinetic energy into potential energy.
d.   metabolism.
e.   converting potential chemical energy into kinetic chemical energy.

Answer: a
Difficulty: Easy
Bloom's Taxonomy: Comprehension

6.   Identify the closed system in the following list.

a.   a human
b.   a gas-powered automobile
c.   a single-celled organism
d.   a propane furnace
e.   none of the above

Answer: e
Difficulty: Moderate
Bloom's Taxonomy: Comprehension

7.   According to the first law of thermodynamics,

a.   matter can be created and destroyed.
b.   matter only changes forms.
c.   energy can be created and destroyed.
d.   energy only changes forms.
e.   matter and energy can be interconverted.

Answer: d
Difficulty: Easy
Bloom's Taxonomy: Knowledge

8. In general, the form of energy that is least useful to living organisms is

a. chemical.
b. electrical.
c. heat.
d. light.
e. sound.

Answer: c
Difficulty: Easy
Bloom's Taxonomy: Knowledge

9. We can calculate whether a reaction is spontaneous by calculating the change in free energy and accounting for entropy. Your paycheck always lists your gross pay, net (take home) pay, and tax withholdings. Which of the following best correlates the changes in free energy to your paycheck?

a. gross = net -tax; free energy = total energy - entropy
b. net salary = gross salary minus tax; free energy = total energy - entropy
c. tax = gross - net salary; entropy = free energy - total energy
d. gross salary = tax minus net salary; total energy = entropy - free energy
e. net salary = gross salary plus tax; total energy = free energy plus entropy

Answer: b
Difficulty: Moderate
Bloom's Taxonomy: Comprehension

10. During every energy transformation, it can be said that

a. the entropy of the universe increases.
b. the entropy of the universe decreases.
c. there is a change in the free energy of the universe.
d. there is a change in the total energy of the universe.
e. the system becomes more organized.

Answer: a
Difficulty: Easy
Bloom's Taxonomy: Knowledge

11. Identify the exergonic reaction in the list below.

a. burning wood for a campfire
b. folding laundry
c. building a tower out of blocks
d. synthesizing a protein
e. storing the third slice of pie in the form of fat

Answer: a
Difficulty: Easy
Bloom's Taxonomy: Comprehension

4.2 HOW LIVING ORGANISMS COUPLE REACTIONS TO MAKE SYNTHESIS SPONTANEOUS

12. When ATP is split into ADP and P_i,

a. the energy is released in the form of heat.
b. the energy is directly transferred to the target molecule by an unknown mechanism.
c. the binding of ADP or P_i to the target molecule allows the energy of ATP hydrolysis to be transferred to the target molecule.
d. the two remaining phosphates acquire the energy that had been present in the linkage of three phosphates.
e. the resulting delta G is positive.

Answer: c
Difficulty: Moderate
Bloom's Taxonomy: Comprehension

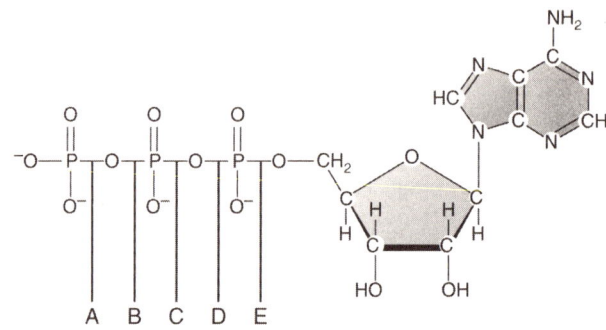

13. In the figure shown above, identify the bond that is cleaved when this molecule is coupled with an endergonic reaction to drive the endergonic reaction to completion. (The molecule of water also required for this reaction is not shown.)

a. A
b. B
c. C
d. D
e. E

Answer: a
Difficulty: Moderate
Bloom's Taxonomy: Knowledge

14. The coupling of endergonic to _____ reactions is used to generate _____ in a cell.

a. exergonic, ATP
b. endergonic, AMP
c. exergonic, AMP
d. endergonic, ATP
e. endergonic, ADP

Answer: a
Difficulty: Moderate
Bloom's Taxonomy: Knowledge

15. Hydrolysis of ATP is used to drive all of the following reactions in a cell except for

a. active transport of solutes.
b. catabolic reactions.
c. protein activation.
d. phosphorylation of target proteins.
e. all of the other answers reflect functions of ATP in a cell.

Answer: e
Difficulty: Easy
Bloom's Taxonomy: Knowledge

4.3 THERMODYNAMICS AND REVERSIBLE REACTIONS

16. When a reaction reaches equilibrium,

a. there is no longer entropy in the system.
b. the chemical reactions cease.
c. the rate of the forward and reverse reactions are equal.
d. the concentration of reactants equals the concentration of products.
e. ATP is no longer required to drive the reaction.

Answer: c
Difficulty: Moderate
Bloom's Taxonomy: Knowledge, Comprehension

17. Reversible reactions in a cell rarely reach equilibrium because

a. the products are generally reactants in other reactions and are thus immediately used.
b. a cell at equilibrium is dead.
c. most reactions in a cell are not reversible, allowing the cell to devote additional resources to regulating the few reversible reactions that do occur.
d. cells have no way of measuring the relative ratios of reactants and products.
e. conditions in the cell change too rapidly for any reaction to ever reach equilibrium.

Answer: a
Difficulty: Moderate
Bloom's Taxonomy: Knowledge

18. Metabolic pathways

a. are always catabolic.
b. are always anabolic.
c. can be either catabolic or anabolic.
d. are irreversible.
e. proceed until they reach equilibrium.

Answer: c
Difficulty: Easy
Bloom's Taxonomy: Knowledge

19. Which reaction is likely to have more products than reactants when the reaction reaches equilibrium?

a. ΔG = -25 kcal/mol
b. ΔG = -50 kcal/mol
c. ΔG = -75 kcal/mol
d. ΔG = -100 kcal/mol
e. They will all have the same ratio of products to reactants regardless of the ΔG value.

Answer: d
Difficulty: Moderate
Bloom's Taxonomy: Application

20. A reaction in progress has

a. more entropy than the same reaction at equilibrium.
b. less entropy than the same reaction at equilibrium.
c. the same entropy than the same reaction at equilibrium.
d. much less entropy than the same reaction at equilibrium.
e. an entropy level that is impossible to determine without additional information.

Answer: a
Difficulty: Moderate
Bloom's Taxonomy: Comprehension

21. When there are more reactants than products for a reaction,

a. the reaction is pulled towards generating more reactants.
b. the greater concentration of reactants pushes the reaction forward, toward generating more products.
c. the reaction is pulled in the forward direction by the high concentration of products.
d. the reaction is pushed toward the reactants by the low concentration of products.
e. the reaction will proceed to completion and no reactants will remain.

Answer: b
Difficulty: Moderate
Bloom's Taxonomy: Comprehension

4.4 ROLE OF ENZYMES IN BIOLOGICAL REACTIONS

22. Enzymes aid in metabolism by

a. changing the ΔG of the reaction.
b. adding additional reactants to the system.
c. slowing the rate of some reactions and increasing the rate of other reactions.
d. increasing the rate of a reaction.
e. removing unused reactants from the system.

Answer: d
Difficulty: Easy
Bloom's Taxonomy: Knowledge

23. Ribozymes
a. are RNA catalysts.
b. are proteins that catalyize RNA synthesis.
c. are RNA molecules that slow the rate of protein synthesis.
d. are the products of RNA degradation.
e. none of these

Answer: a
Difficulty: Easy
Bloom's Taxonomy: Knowledge

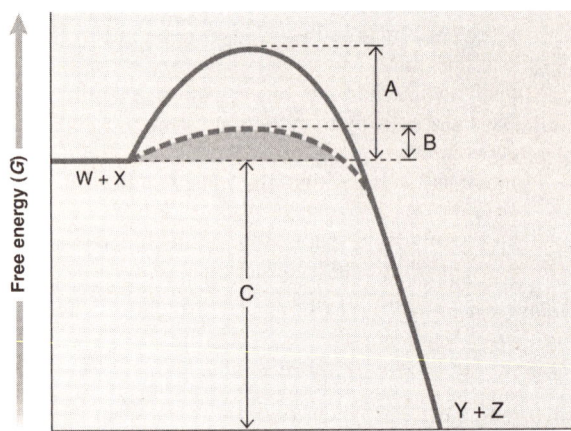

Refer to the above figure for questions 24–28

24. This portion of the graph shows the activation energy when there is no enzyme.

a. A
b. B
c. C
d. W + X
e. Y + Z

Answer: a
Difficulty: Easy
Bloom's Taxonomy: Knowledge
Source: Fig. 4.9

25. This portion of the graph shows the activation energy when there is an enzyme.

a. A
b. B
c. C
d. W + X
e. Y + Z

Answer: b
Difficulty: Easy
Bloom's Taxonomy: Knowledge
Source: Fig. 4.9

26. This portion of the graph shows the free energy of the reaction.

a. A
b. B
c. C
d. W + X
e. Y + Z

Answer: c
Difficulty: Easy
Bloom's Taxonomy: Knowledge
Source: Fig. 4.9

27. This portion of the graph shows the free energy of the reactants.

a. A
b. B
c. C
d. W + X
e. Y + Z

Answer: d
Difficulty: Easy
Bloom's Taxonomy: Knowledge
Source: Fig. 4.9

28. This portion of the graph shows the shows the free energy of the products.

a. A
b. B
c. C
d. W + X
e. Y + Z

Answer: e
Difficulty: Easy
Bloom's Taxonomy: Knowledge
Source: Fig. 4.9

29. The difference between cofactors and coenzymes is that

a. cofactors are not necessary and coenzymes are necessary.
b. cofactors can be inorganic or organic, coenzymes are always inorganic.
c. cofactors can be inorganic or organic, coenzymes are just another name for organic cofactors.
d. cofactors are always vitamins, coenzymes are always ions.
e. cofactors help with essential metabolic reactions, coenzymes assist with nonessential metabolic reactions.

Answer: c
Difficulty: Moderate
Bloom's Taxonomy: Knowledge

30. Enzymes

a. change the rate of a reaction.
b. change the direction of a reaction.
c. change the free energy of a reaction.
d. are used up in a reaction.
e. are present in concentrations approaching their substrates' concentrations.

Answer: a
Difficulty: Easy
Bloom's Taxonomy: Knowledge

31. Enzymes function primarily by

a. forcing the reactants into an altered environment which in turn creates a change in the free energy of the reactants relative to the products.
b. altering the equilibrium point of a particular reaction to favor the formation of products.
c. increasing the probability that the reactants will come into close proximity to each other in the proper orientation for forming the transition state molecule.
d. removing reactants from solution in a set ratio that enhances the chances of the remaining individual reactants interacting with each other.
e. changing the ratio of reactants to products so that the forward reaction is favored.

Answer: c
Difficulty: Easy
Bloom's Taxonomy: Knowledge

FOCUS ON RESEARCH: TESTING THE TRANSITION STATE

32. Who first *proposed* the idea that enzymes worked by pushing reactants towards the transition state?

a. Linus Pauling
b. W.P. Jencks
c. R.A. Lerner
d. Janet Smith
e. Josiah Gibbs

Answer: a
Difficulty: Easy
Bloom's Taxonomy: Knowledge

33. Designer enzymes are made by

a. customizing preexisting enzymes to force them to take on new functions.
b. injecting mock-transition state molecules into test subjects and selecting for antibodies that selectively bind these mock-transition state molecules.
c. altering an enzyme so it utilizes a new cofactor.
d. determining the primary sequence of an enzyme and replacing all of the alanine residues with threonine residues.
e. synthesizing new antibodies in bacterial cultures.

Answer: b
Difficulty: Easy
Bloom's Taxonomy: Knowledge

4.5 CONDITIONS AND FACTORS THAT AFFECT ENZYME ACTIVITY

34. Which of the following does not *always* alter the activity of the enzyme?

a. temperature
b. pH
c. reactant concentrations
d. inhibitors
e. lack of required cofactors

Answer: c
Difficulty: Difficult
Bloom's Taxonomy: Knowledge, Comprehension

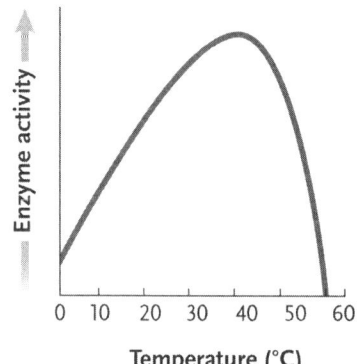

35. In the above figure, why does the curve sharply drop after approximately 45 °C instead of mirroring the slope of the line going from 0–40 °C?

a. At high temperatures, the reactions proceed so quickly that enzymes are no longer helpful or required.
b. This is true of all catalysts and is not due to any special features of enzymes.
c. The kinetic energy of the reactants is so great that it destabilizes the enzyme and diminishes the enzyme's activity.
d. The kinetic energy of the reactants is lower than that of the products, forcing a change in enzyme activity.
e. The enzyme's begins to denature above a certain temperature, eliminating all catalytic activity of the protein.

Answer: e
Difficulty: Moderate
Bloom's Taxonomy: Knowledge, Comprehension

36. If an enzyme's optimal temperature is 37 °C, you also know that

a. the enzyme has a cofactor.
b. the enzyme is likely also active at 30ºC and at the same rate.
c. the enzyme will probably be inactive at a pH below 4.5.
d. the enzyme will completely denature at 38º C.
e. the enzyme's activity will drop at temperatures above 37 °C and likely be eliminated by 45 °C.

Answer: e
Difficulty: Difficult
Bloom's Taxonomy: Application

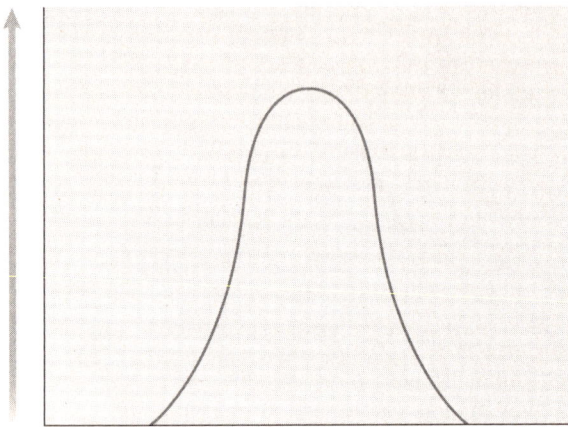

37. This curve *most likely* represents a graph of enzyme activity

a. as a function of temperature.
b. as a function of pH in a strongly acidic environment.
c. as a function of pH in a strongly basic environment.
d. as a function of pH in a fairly neutral environment.
e. as a function of temperature and pH combined.

Answer: d
Difficulty: Difficult
Bloom's Taxonomy: Comprehension

38. You decide to alter the rate of a reaction. Which of the following is *not* going to help you increase the rate of this reaction?

a. adding more reactants
b. adding more enzyme
c. adding heat
d. mechanically stirring the contents of a beaker of reactants
e. adding more product

Answer: e
Difficulty: Difficult
Bloom's Taxonomy: Application

39. If an enzyme is saturated, this means that

a. the enzymes cannot continue to catalyze the reaction.
b. the reaction is at equilibrium.
c. the rate of the reaction will slow and the reaction will stop.
d. the enzymes need more reactants.
e. the enzymes have sufficient reactants available for optimal activity.

Answer: e
Difficulty: Easy
Bloom's Taxonomy: Knowledge

40. You do an experiment in the laboratory and add increasing amounts of substrate to a solution containing an enzyme and a pH buffer. You incubate the container at the optimal temperature for your enzyme. Each time you add more substrate, you measure the rate of the reaction. If you graph the results where the x axis shows the substrate concentration and the y axis shows the resulting reaction rate, what will you find over time?

a. The rate of the reaction will proceed with a slope of 1 and continue in a linear fashion indefinitely or until you run out of reactants.
b. The rate of the reaction will increase rapidly, taper off, and plateau.
c. The rate of the reaction will increase slowly, plateau, and then drop sharply back to zero.
d. The resulting graph will be a perfect bell curve.
e. There is no way to predict what the graph will look like without more information.

Answer: b
Difficulty: Difficult
Bloom's Taxonomy: Application

41. In competitive inhibition

a. the products of the reaction block the active site of the enzyme.
b. the products of the reaction bind to a site other than the active site of the enzyme and block enzyme activity indirectly.
c. the substrate and cofactors compete for the active site.
d. the inhibitor binds to and directly blocks the active site of the enzyme.
e. none of these

Answer: d
Difficulty: Easy
Bloom's Taxonomy: Knowledge

42. In noncompetitive feedback inhibition

a. the products of the reaction block the active site of the enzyme.
b. the products of the reaction at the end of the pathway bind to a site other than the active site of an enzyme at or near the beginning of the pathway and block enzyme activity indirectly.
c. the substrate and cofactors compete for the active site.
d. the inhibitor binds to and directly blocks the active site of the enzyme.
e. none of these

Answer: b
Difficulty: Easy
Bloom's Taxonomy: Knowledge

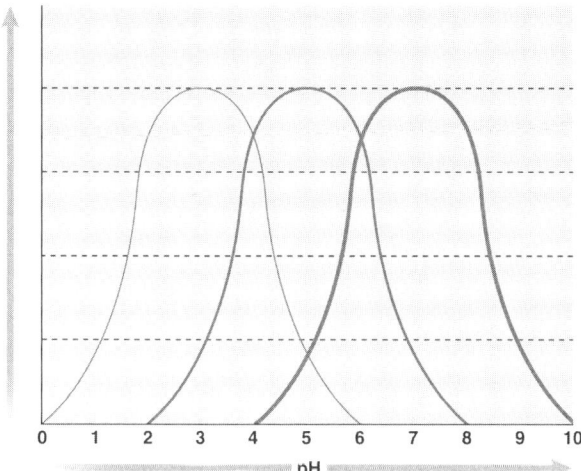

1 ———
2 ———
3 ———

The above graph should be used for questions 43–45.

43. The optimal pH for enzyme 1 is:

a. 2.
b. 3.
c. 4.
d. 5.
e. 6.

Answer: b
Difficulty: Moderate
Bloom's Taxonomy: Application

44. The optimal pH for enzyme 2 is:

a. 2.
b. 3.
c. 4.
d. 5.
e. 6.

Answer: d
Difficulty: Moderate
Bloom's Taxonomy: Application

45. If all three enzymes catalyize the same reaction and your conditions require a pH of 7, the best enzyme to use would be:

a. enzyme 1.
b. enzyme 2.
c. enzyme 3.
d. either enzyme 2 or 3.
e. both enzymes 2 and 3.

Answer: c
Difficulty: Moderate
Bloom's Taxonomy: Application

46. When an enzyme has an allosteric activator, it means that

a. a product of the enzyme, or other downstream product, will bind to the enzyme at the active site and inhibit enzyme activity.
b. a product of the enzyme, or other downstream product, will bind to the enzyme at the active site and stimulate enzyme activity.
c. a product of the enzyme, or other downstream product, will bind to the enzyme at a site other than the active site and inhibit enzyme activity.
d. a product of the enzyme, or other downstream product, will bind to the enzyme at a site other than the active site and stimulate enzyme activity.
e. a product of a different pathway (not a direct or indirect product of the enzyme) will stimulate the enzyme's activity.

Answer: d
Difficulty: Easy
Bloom's Taxonomy: Knowledge

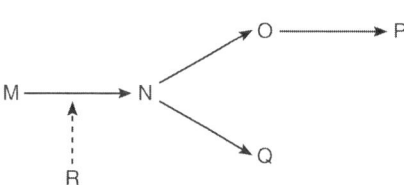

The above diagram of a metabolic pathway should be used for questions 47–50. Solid arrows represent enzyme catalyzed reactions. The dashed arrow represents molecule R interacting with an enzyme.

47. If the enzyme catalyzing the N to O reaction is inhibited, there will be a build-up of which compound?

a. M
b. N
c. O
d. P
e. Q

Answer: e
Difficulty: Difficult
Bloom's Taxonomy: Application

48. If R inhibits the enzyme catalyzing the M to N reaction, we call this

a. allosteric activation.
b. allosteric inhibition.
c. noncompetitive inhibition.
d. competitive inhibition.
e. either c or d, dependent on more information.

Answer: e
Difficulty: Moderate
Bloom's Taxonomy: Application

49. If the enzyme catalyzing the N to O reaction is allosterically stimulated, there will be a build-up of which compound(s)?

a. N
b. O
c. P
d. Q
e. P and Q

Answer: e
Difficulty: Difficult
Bloom's Taxonomy: Application

50. If the enzyme catalyzing the N to O reaction is inhibited by product P, what will be the final products and relative amounts if the reactions are permitted to go to completion? (You may assume that all of the enzymes have the same rate of activity.)

a. P and Q at first, then more Q
b. P and Q in equal amounts regardless of time
c. N and Q in equal amounts
d. N only
e. Q only

Answer: a
Difficulty: Difficult
Bloom's Taxonomy: Application

4.6 RNA-BASED BIOLOGICAL CATALYSTS: RIBOZYMES

51. What type or types of reactions can a ribozyme catalyze?

a. protein synthesis, but only with other proteins to assist in the process (like in a ribosome)
b. protein synthesis in the absence of other proteins
c. RNA splicing
d. a and c
e. b and c

Answer: e
Difficulty: Easy
Bloom's Taxonomy: Knowledge

52. Is a ribozyme a true enzyme?

a. Yes, because it is a catalyst.
b. No, because it is not a protein.
c. No, because it cannot catalyze a reaction.
d. Yes, because it can synthesize a protein.
e. No, because it doesn't require cofactors.

Answer: b
Difficulty: Moderate
Bloom's Taxonomy: Comprehension

INSIGHTS FROM THE MOLECULAR REVOLUTION: RIBOZYMES TAKE THE FIRST STEP IN PROTEIN SYNTHESIS

53. How do we know that ribozymes can catalyze the linkage of amino acids all by themselves?

a. Experiments have been done that completely removed all proteins from ribosomes. The residual proteins were removed by a protein-specific enzyme the degrades proteins to individual amino acids.
b. Scientists have artificially synthesized ribozymes by linking pure RNA nucleotides together in the proper sequence, eliminating the possibility of protein contamination. When tested, these artificial ribozymes were able to link amino acids together.
c. We don't know this for certain; more research remains to be done.
d. Harry Noller published a paper stating it as fact. Since the paper was reviewed by other scientists, the conclusions must be valid.
e. Ribozymes purified from a cell were shown to increase the rate of protein synthesis by 100,000 times.

Answer: b
Difficulty: Moderate
Bloom's Taxonomy: Comprehension

54. Why do we care about the catalytic activity of some RNA molecules?

a. If we want to understand everything about how cells work, we need to understand how all of the reactions in a cell are regulated and catalyzed.
b. It helps us understand gene expression since changes to a sequence of RNA can result in changes to the final protein product.
c. It helps explain how early organisms could have evolved.
d. a and b only.
e. a, b, and c.

Answer: e
Difficulty: Moderate
Bloom's Taxonomy: Comprehension

55. In Zhang and Chech's experiments, the ribozyme was linked to phenylalanine, and the other amino acid (methionine) was linked to biotin. What would happen if the methionine was not linked to biotin?

a. The ribozyme would not catalyze the formation of a peptide bond between phenylalanie and methionine.
b. The results of the experiment would not have changed.
c. The scientists would not have had a way to sort the unused reactants from the reaction's products.
d. Methionine would have linked to the ribozyme and not to the phenylalanine.
e. The biotin would not have been able to act as a cofactor in the reaction.

Answer: c
Difficulty: Difficult
Bloom's Taxonomy: Application

UNANSWERED QUESTIONS

56. Why is it useful to know the structure of a protein when trying to understand a protein's function?

a. Scientists still don't know exactly what features of a protein's structure determine how a protein works. By gathering information about many different protein structures and correlating them to their known functions, we can try to figure out this relationship.
b. Since all proteins have enzymatic activity, we hope to learn what features of a protein make an enzyme a catalyst.
c. Since x-rays are used on protein crystals to determine the structure, this research can help us understand the ways in which x-rays can be damaging to the proteins of living cells.
d. The enzymatic activity of a protein cannot be measured unless we also have the structure of that enzyme.
e. We need to know which amino acids are essential for a cell to function. If we know the structure of an enzyme, we can determine which amino acids are present and how they work in the cell.

Answer: a
Difficulty: Difficult
Bloom's Taxonomy: Comprehension

57. Why is it useful to know the structure of a ribozyme when trying to understand a particular ribozyme's function?

a. The way the RNA folds determines the activity of the ribozyme. We need to understand the structure in order to measure the rate of the reaction.
b. It's not very useful at all; it is just a matter of curiosity.
c. We need to know the structure in order to develop drugs that destroy that structure and prevent HIV infection.
d. We need to be able to determine if an RNA molecule is a ribozyme or not. The only way to do that is to determine the structure.
e. The way the RNA folds relates to the function of the ribozyme. If we can figure out the relationship, we might be able to develop new ribozymes that act as therapeutic agents.

Answer: e
Difficulty: Difficult
Bloom's Taxonomy: Comprehension

Integrative Multiple-Choice

58. This molecule has two functions in a cell: encoding genetic information and using two additional phosphates to transfer energy.

a. glucose
b. AMP
c. glutamic acid
d. aspartic acid
e. fat

Answer: b
Difficulty: Difficult
Bloom's Taxonomy: Knowledge
Section: 3.6, 4.2

59. When an enzyme-catalyzed reaction reaches equilibrium

a. the enzymes are now inhibited.
b. the chemical reactions cease.
c. the rate of the forward and reverse reactions are equal.
d. the concentration of reactants equals the concentration of products.
e. ATP is no longer required to drive the reaction.

Answer: c
Difficulty: Moderate
Bloom's Taxonomy: Knowledge, Comprehension
Source: Section 4.3, 4.4

60. Enzymes are

a. protein catalysts.
b. lipid catalysts.
c. nucleic acid catalysts.
d. inorganic catalysts.
e. ionic catalysts.

Answer: a
Difficulty: Moderate
Bloom's Taxonomy: Knowledge, Comprehension
Source: Section 4.3, 4.4

61. Enzymes

a. are catalytic lipids.
b. are catalytic proteins.
c. are RNA molecules that slow the rate of protein synthesis.
d. are catalytic carbohydrates.
e. are nonmacromolecule catalysts.

Answer: b
Difficulty: Moderate
Bloom's Taxonomy: Knowledge
Source: Section 3.5, 4.4

62. If a reaction in a cell requires an enzyme

a. we know that the reaction must be endergonic.
b. we know that the reaction must require ATP.
c. we know that the reaction is subject to allosteric regulation.
d. we know that the reaction is regulated by competitive inhibition.
e. we know that the reaction occurs more readily than it would if there were no enzyme.

Answer: e
Difficulty: Moderate
Bloom's Taxonomy: Knowledge
Source: Sections 4.2, 4.4, 4.5

63. The helix-turn-helix motif found in many regulatory proteins fits into the side of a DNA molecule. If a newly discovered enzyme has a helix-turn-helix motif, you would predict that

a. this enzyme has multiple peptide chain subunits.
b. this enzyme is probably able to bind to RNA, too.
c. this enzyme probably binds to DNA.
d. this enzyme has an allosteric regulator.
e. this enzyme is a polymerase that helps synthesize new DNA sequences.

Answer: c
Difficulty: Moderate
Bloom's Taxonomy: Knowledge, Comprehension
Source: Sections 3.5 4.4, 4.5

64. You modify the primary sequence of an enzyme in a region that will be the active site when the protein is properly folded. What is the predicted outcome of this change?

a. The enzyme will not properly bind to the substrate.
b. The enzyme will not be able to bind an allosteric inhibitor.
c. The enzyme will have an increased rate of activity.
d. There will be no change in the enzyme's function.
e. The enzyme will bind the substrate but not be able to release the products.

Answer: a
Difficulty: Moderate
Bloom's Taxonomy: Knowledge, Comprehension
Source: Sections 3.5, 4.5

Matching

Match each of the following terms with its correct definition.

65. _____ coupled reaction

66. _____ ATP

67. _____ phosphorylation

68. _____ equilibrium point

69. _____ metabolic pathway

70. _____ catalyst

71. _____ activation energy

72. _____ active site

73. _____ substrate

74. _____ transition state

75. _____ allosteric regulation

76. _____ ribozyme

A. Primary coupling agent in cellular reactions

B. The addition of a phosphate group to a target molecule

C. The product of the reaction interacts with an enzyme in a noncompetitive way to inhibit or enhance enzyme activity

D. The linking of an exergonic reaction with an endergonic reaction that allows a cell to drive a nonspontaneous reaction to completion

E. A series of chemical reactions where the products of one reaction are the reactants for a subsequent reaction

F. A substance that facilitates a chemical reaction without itself being consumed by the reaction

G. The energy needed to start a reaction, be it endergonic or exergonic

H. The portion of the enzyme that binds to a reactant or reactants

I. A state in which the rate of the forward reaction equals the rate of the reverse reaction

J. An RNA molecule that acts as a catalyst

K. An intermediate arrangement of unstable bonds between atoms that can proceed towards either the reactants or products of a reaction

L. The reactant molecule that binds to an enzyme

Answers: **65. D   66. A   67. B   68. I   69. E**
             70. F   71. G   72. H   73. L   74. K
             75. C   76. J

Difficulty: Moderate
Bloom's Taxonomy: Knowledge
Source: Sections 4.2, 4.3, 4.4. 4.5

Choice

For each of the following situations, choose the most appropriate term.

a. endergonic   b.   exergonic
c. equilibrium

77. _____   a beaker of water sitting on a bench

78. _____   folding laundry

79. _____   protein synthesis

80. _____   digestion of a candy bar

81. _____   a toddler dumping boxes of toys

82. _____   a dead cell

83. _____   a reaction where $\Delta G = 0$

84. _____   a reaction where ΔG is positive

85. _____   a reaction where $\Delta G=$ is negative

86. _____   the rate of synthesis equals the rate of degradation

Answers: **77. c   78. a   79. a   80. b   81. b**
             82. c   83. c   84. b   85. a
             86. c.

Difficulty: Easy
Bloom's Taxonomy: Knowledge, Comprehension
Source: Section 4.1, 4.2, 4.3

Short Answer

87. Define metabolism

Answer: Metabolism is the total of all of the chemical reactions in a cell. These reactions can be catabolic or anabolic and are used to synthesize new molecules or extract energy from molecules to power cellular work.
Difficulty: Moderate
Bloom's Taxonomy: Knowledge
Source: Section 4.1, 4.2

88. Explain how temperature can impact enzyme activity.

Answer: Temperature affects kinetic energy as well as the three-dimensional structure of an enzyme. Above the optimal temperature, a protein begins to denature and loses function. Below the optimal temperature, the kinetic energy keeps molecules from colliding with each other as rapidly, reducing the rate of the reaction.
Difficulty: Moderate
Bloom's Taxonomy: Knowledge, Comprehension
Source: Section 4.5

True/False

If the statement is true, write a "T" in the blank. If the statement is false, make it correct by changing the underlined word(s) and writing the correct word(s) in the answer blank.

89. ____ _____   AMP is the primary energy and phosphate source in coupled reactions.

90. _____   At equilibrium, the concentration of the reactants equals the concentration of the products.

91. _____   Reactions that reach an equilibrium point are reversible.

92. _____   All true enzymes are proteins.

93. _____   Activation energy is not required for nonspontaneous reactions.

94. _____   Enzymes don't change the ΔG of a reaction.

95. _____   Enzymes alter the equilibrium point of a reaction.

96. _____   Enzyme activity is increased by falling temperatures.

Answers: **89. F, ATP   90. F, rate of formation,**
             rate of formation   91. T   92. T
             93. F, activation energy is required
             94. T   95. F, activation energy
             96. F, decreased.

Difficulty: Difficult
Bloom's Taxonomy: Knowledge
Source: Section 4.1, 4.2, 4.3, 4.4

Select the Exception

97. Enzymes work with at least three mechanisms. Which of the following is *not* a mechanism by which enzymes function?

a. putting reactants in close proximity to each other
b. altering the free energy (ΔG) of the reaction
c. altering the immediate environment of the reactants to promote reactant interactions
d. orienting the reactants so they are positioned to favor the transition state

Answer: b
Difficulty: Easy
Bloom's Taxonomy: Knowledge
Source: Section 4.4

98. Which of the following does *not* impact an enzyme's function?

a. pH
b. temperature
c. absence of cofactors
d. allosteric interactions
e. excess product

Answer: e
Difficulty: Easy
Bloom's Taxonomy: Knowledge
Source: Section 4.5

Essay

99. Explain three ways in which other molecules regulate enzymes.

Answer: Enzymes can be inhibited by products of the reaction that bind to a site other than the active site. This is called feedback inhibition or allosteric inhibition. Enzymes can be inhibited by molecules that compete for access to the active site and when bound, block the substrate from entering the active site. This is called competitive inhibition. Enzymes can also be activated or stabilized by reaction products binding to a site other than the active site. This is called allosteric activation.
Difficulty: Moderate
Bloom's Taxonomy: Knowledge, Comprehension
Source: Section 4.5

100. Explain how a cell can use catabolic reactions to drive anabolic reactions despite energy loss in the form on entropy and heat.

Answer: Cells use coupled reactions to harness the free energy released from a catabolic reaction. The free energy is temporarily stored in the form on ATP. The energy in ATP is then transferred to reactants in the anabolic reaction. This energy powers the endergonic reaction using the energy that originally came from the exergonic catabolic reaction. Each time there is an energy transformation, some of the energy is lost in the form of heat. This means that not all of the free energy released from the catabolic reaction will be invested into the anabolic reaction. One way organisms can compensate is to consume more energy sources so they have sufficient energy to drive the necessary endergonic reactions despite entropy.
Difficulty: Moderate
Bloom's Taxonomy: Comprehension
Source: Section 4.1, 4.2

5
THE CELL: AN OVERVIEW

Multiple-Choice

WHY IT MATTERS

1.   The first observed cells were from _____.

a.   cork
b.   a maple leaf
c.   human skin
d.   pollen
e.   bacteria

Answer: a
Difficulty: Easy
Bloom's Taxonomy: Knowledge

2.   The individual credited with first observing the cell nucleus was _____.

a.   Anton van Leeuwenhoek
b.   Robert Brown
c.   Matthias Schleiden
d.   Theodor Schwann
e.   Rudolf Virchow

Answer: b
Difficulty: Easy
Bloom's Taxonomy: Knowledge

3.   Which early scientist proposed that cells arise only from preexisting cells?

a.   Anton van Leeuwenhoek
b.   Robert Brown
c.   Matthias Schleiden
d.   Theodor Schwann
e.   Rudolf Virchow

Answer: e
Difficulty: Easy
Bloom's Taxonomy: Knowledge

5.1 BASIC FEATURES OF CELL STRUCTURE AND FUNCTION

4.   The most commonly used unit for measuring cell size is a _____.

a.   decimeter (dm)
b.   centimeter (cm)
c.   millimeter (mm)
d.   micrometer (μm)
e.   nanometer (nm)

Answer: d
Difficulty: Moderate
Bloom's Taxonomy: Knowledge

5.   A human egg is approximately 100 μm in size. This would equal to _____ mm.

a.   0.10
b.   10.0
c.   0.010
d.   0.0010
e.   1.0

Answer: a
Difficulty: Moderate
Bloom's Taxonomy: Application

6.   _____ magnifies passing light directly through a specimen. Staining with a dye is typically used to enhance contrast and visualization of cellular structures.

a.   Confocal laser scanning microscopy
b.   Phase-contrast microscopy
c.   Bright field microscopy
d.   Scanning electron microscopy
e.   Fluorescence microscopy

Answer: c
Difficulty: Moderate
Bloom's Taxonomy: Knowledge, Comprehension

7.   The cell's hereditary information is stored in _____.

a.   genes
b.   protein
c.   RNA
d.   glucose
e.   amino acids

Answer: a
Difficulty: Easy
Bloom's Taxonomy: Knowledge

8.   The _____ regulates the movement of molecules in and out of the cell.

a.   nucleus
b.   cytoplasm
c.   ribosomes
d.   plasma membrane
e.   DNA

Answer: d
Difficulty: Moderate
Bloom's Taxonomy: Knowledge

9.   The majority of a cell's vital activities occur in _____.

a.   the mitochondria
b.   the nucleus
c.   the cytoplasm
d.   organelles
e.   the plasma membrane

Answer: c
Difficulty: Easy
Bloom's Taxonomy: Knowledge

5.2 PROKARYOTIC CELLS

10. Which one of the following is NOT found in prokaryotic cells?

a. nucleus
b. ribosomes
c. plasma membrane
d. cell wall
e. DNA

Answer: a
Difficulty: Moderate
Bloom's Taxonomy: Knowledge

11. Organisms in which one of the following groups are found in the greatest abundance on the Earth's surface?

a. Prokaryotes
b. Protists
c. Fungi
d. Plants
e. Animals

Answer: a
Difficulty: Moderate
Bloom's Taxonomy: Knowledge

5.3 EUKARYOTIC CELLS

12. Which one of the following groups does not belong to the domain of the eukaryotes?

a. Fungi
b. Protists
c. Bacteria
d. Animals
e. Plants

Answer: c
Difficulty: Moderate
Bloom's Taxonomy: Knowledge

13. Chromatin consists of _____.

a. only DNA
b. DNA and RNA
c. RNA only
d. DNA and associated proteins
e. proteins only in the nucleus

Answer: d
Difficulty: Moderate
Bloom's Taxonomy: Knowledge

14. All of the following are functions of proteins embedded in the plasma membrane EXCEPT

_____.

a. transport of substances in and out of the cell
b. generation of ATP
c. recognition of signal molecules
d. adherence to molecules on the surface of other cells
e. recognition of "like" cells

Answer: b
Difficulty: Difficult
Bloom's Taxonomy: Knowledge

15. A network of protein filaments called _____ line and reinforce the inner surface of the nuclear envelope in animal cells.

a. actins
b. chromatins
c. lamins
d. tubulins
e. lamellae

Answer: c
Difficulty: Moderate
Bloom's Taxonomy: Knowledge

16. The eukaryotic chromosome is composed of

_____.

a. DNA only
b. RNA only
c. DNA and carbohydrate
d. DNA and protein
e. RNA and protein

Answer: d
Difficulty: Moderate
Bloom's Taxonomy: Knowledge

17. The semiliquid substance within the nucleus is called _____.

a. nucleoplasm
b. nuclear gel
c. cytoplasm
d. chromatin
e. protoplasm

Answer: a
Difficulty: Moderate
Bloom's Taxonomy: Knowledge

18. The nucleoli are found within the nucleus and synthesize _____.

a. chromatin
b. mRNA
c. ribosomal subunits
d. genes
e. proteins

Answer: c
Difficulty: Moderate
Bloom's Taxonomy: Knowledge

19. The _____ is not a part of the endomembrane system.

a. endoplasmic reticulum
b. Golgi complex
c. lysosome
d. nucleolus
e. nuclear envelope

Answer: d
Difficulty: Difficult
Bloom's Taxonomy: Knowledge

20. The _____ is/are involved in the synthesis of lipids.

a. rough endoplasmic reticulum
b. Golgi complex
c. smooth endoplasmic reticulum
d. ribosomes
e. nucleoli

Answer: c
Difficulty: Moderate
Bloom's Taxonomy: Knowledge

21. The Golgi complex _____.

a. receives proteins made in the rough ER and chemically modifies them
b. synthesizes proteins for export from the cell
c. stores nucleic acids
d. synthesizes lipids
e. stores ribosomes

Answer: a
Difficulty: Moderate
Bloom's Taxonomy: Knowledge

22. Which one of the following cellular components is NOT directly involved in synthesis or secretion of molecules in the cell?

a. ribosomes
b. lysosome
c. rough endoplasmic reticulum
d. Golgi complex
e. smooth endoplasmic reticulum

Answer: b
Difficulty: Difficult
Bloom's Taxonomy: Knowledge

23. Cells active in secreting enzymes would likely exhibit a greater amount of _____ than other cells.

a. exocytosis
b. osmosis
c. endocytosis
d. receptor proteins
e. plasma membrane

Answer: a
Difficulty: Moderate
Bloom's Taxonomy: Comprehension

24. Molecules brought into the cell from the exterior are placed into vesicles for routing to other locations such as _____.

a. the nucleus
b. lysosomes
c. mitochondria
d. ribosomes
e. cholorplasts

Answer: b
Difficulty: Moderate
Bloom's Taxonomy: Comprehension

25. Lysosomes function best at a pH of _____.

a. 7.4
b. 6.5
c. 3.2
d. 5.0
e. 8.2

Answer: d
Difficulty: Moderate
Bloom's Taxonomy: Knowledge

26. The _____ contains hydrolytic enzymes for the digestion of proteins, lipids, nucleic acids, and polysaccharides.

a. Golgi complex
b. rough endoplasmic reticulum
c. nucleus
d. peroxisome
e. lysosome

Answer: e
Difficulty: Moderate
Bloom's Taxonomy: Knowledge

27. Cellular respiration occurs in the _____.

a. lysosomes
b. chloroplasts
c. mitochondria
d. peroxisomes
e. Golgi complex

Answer: c
Difficulty: Easy
Bloom's Taxonomy: Knowledge

28. Cellular respiration is the process by which _____ and _____ are converted to water and carbon dioxide during the formation of cellular energy.

a. CO_2; glucose
b. CO_2; fats
c. O_2; CO_2
d. O_2; glucose
e. ATP; O_2

Answer: d
Difficulty: Moderate
Bloom's Taxonomy: Knowledge

29. The interior surface area of mitochondria is greatly increased by _____.

a. cristae
b. the matrix
c. centrioles
d. nucleoli
e. microfilaments

Answer: a
Difficulty: Difficult
Bloom's Taxonomy: Comprehension

30. Scientists believe that mitochondria may have evolved from ancient bacteria because _____.

a. both are surrounded by a cell wall
b. both have their own DNA and ribosomes
c. the shapes and size of both are exactly the same
d. both have cristae to increase surface area
e. both have five chromosomes

Answer: b
Difficulty: Difficult
Bloom's Taxonomy: Knowledge

31. Select the correct path a protein synthesized on a ribosome attached to the rough ER would follow in the endomembrane system.

a. rough ER → smooth ER → Golgi complex → plasma membrane
b. rough ER → Golgi complex → vesicle → plasma membrane
c. rough ER → vesicle → lysosome → plasma membrane
d. rough ER → vesicle → smooth ER → plasma membrane
e. rough ER → smooth ER → lysosome → plasma membrane

Answer: b
Difficulty: Moderate
Bloom's Taxonomy: Comprehension

32. Cytoskeletal elements are assembled from _____.

a. proteins
b. phospholipids
c. glycogen
d. nucleotides
e. triglycerides

Answer: a
Difficulty: Moderate
Bloom's Taxonomy: Knowledge

33. Microfilaments are assembled from the protein _____.

a. keratin
b. myosin
c. laminin
d. actin
e. tubulin

Answer: d
Difficulty: Moderate
Bloom's Taxonomy: Knowledge

34. Microtubules are assembled from the protein _____.

a. keratin
b. myosin
c. laminin
d. actin
e. tubulin

Answer: e
Difficulty: Moderate
Bloom's Taxonomy: Knowledge

35. _____ radiate from the center of the cell and anchor the ER, Golgi complex, lysosomes, and secretory vesicles in place.

a. Microfilaments
b. Microtubules
c. Laminins
d. Actins
e. Cytokeratins

Answer: b
Difficulty: Moderate
Bloom's Taxonomy: Knowledge

36. A bundle of _____ extends from the base to the tip of a flagellum or cilium.

a. microtubules
b. microfilaments
c. actin
d. cytokeratin
e. intermediate filaments

Answer: a
Difficulty: Moderate
Bloom's Taxonomy: Knowledge

37. Cilia and flagella arise from which of the following cellular components?

a. nucleus
b. Golgi complex
c. centrioles
d. chromosomes
e. nucleolus

Answer: c
Difficulty: Moderate
Bloom's Taxonomy: Knowledge

38. The 9 + 2 complex refers to _____.

a. microtubules
b. Golgi complex
c. ribosomes
d. cilia
e. both microtubules and cilia

Answer: e
Difficulty: Moderate
Bloom's Taxonomy: Knowledge

INSIGHTS FROM THE MOLECULAR REVOLUTION: *AN OLD KINGDOM IN A NEW DOMAIN*

39. Which of the following information lead scientists to believe that archaeans belong to their own distinct domain?
a. They have a cell wall synthesized from a unique carbohydrate.
b. The genetic material of archaeans is RNA, not DNA.
c. Many archaeans live in extreme environments tolerated by no other organisms.
d. Archaeans do not utilize ATP to do cellular work.
e. Archaeans have unique cellular structures not found in either eukaryotic or prokaryotic cells.

Answer: c
Difficulty: Difficult
Bloom's Taxonomy: Knowledge

40. Genetically, Methanococcus is most closely related to _____.

a. bacteria
b. fungi
c. plants
d. neither bacterial nor eukaryotic cells
e. animals

Answer: d
Difficulty: Difficult
Bloom's Taxonomy: Comprehension

5.4 SPECIALIZED STRUCTURES OF PLANT CELLS

41. Cell walls are found in _____.
a. animal cells only
b. plant cells only
c. fungal cells only
d. plant and fungal cells
e. animal and plant cells

Answer: d
Difficulty: Moderate
Bloom's Taxonomy: Knowledge

42. Chloroplasts share many similarities with which one of the following organelles?
a. mitochondria
b. rough endoplasmic reticulum
c. nucleus
d. Golgi complex
e. lysosome

Answer: a
Difficulty: Moderate
Bloom's Taxonomy: Knowledge

43. Chloroplasts are the site of _____.
a. DNA synthesis
b. cellular digestion
c. protein synthesis
d. photosynthesis
e. lipid synthesis

Answer: d
Difficulty: Easy
Bloom's Taxonomy: Knowledge

44. Chloroplasts utilize light energy to make _____.
a. carbohydrates
b. fats
c. proteins
d. nucleic acids
e. steroids

Answer: a
Difficulty: Moderate
Bloom's Taxonomy: Knowledge

45. _____ store starch in plants.
a. Chloroplasts
b. Amyloplasts
c. Chromoplasts
d. Vacuoles
e. Leucoplasts

Answer: b
Difficulty: Moderate
Bloom's Taxonomy: Knowledge

46. Grana and thylakoids are structural components found in _____.
a. nucleoli
b. mitochondria
c. ribosomes
d. chromoplasts
e. chloroplasts

Answer: e
Difficulty: Moderate
Bloom's Taxonomy: Knowledge

47. In a mature plant cell, _____ may occupy more than 90% of a mature plant cell's volume.
a. the nucleus
b. chromoplasts
c. rough endoplasmic reticulum
d. the central vacuole
e. chloroplasts

Answer: d
Difficulty: Moderate
Bloom's Taxonomy: Knowledge

48.   Another name for the central vacuole is
_____.

a.   chloroplast
b.   tonoplast
c.   chromoplast
d.   amyloplast
e.   ionoplast

Answer: b
Difficulty: Moderate
Bloom's Taxonomy: Knowledge

49.   In plant cells, the _____ provides cellular
support and protects cells from pathogens.

a.   cytoplasm
b.   cell membrane
c.   cytoskeleton
d.   plasmodesmata
e.   cell wall

Answer: e
Difficulty: Moderate
Bloom's Taxonomy: Knowledge

50.   The cell wall is composed primarily of cellulose,
which is a network of highly branched _____.

a.   carbohydrates
b.   nucleic acids
c.   proteins
d.   steroids
e.   phospholipids

Answer: a
Difficulty: Moderate
Bloom's Taxonomy: Knowledge

51.   The walls of adjacent plant cells are held together
by the _____.

a.   secondary cell wall
b.   cell membrane
c.   primary cell wall
d.   middle lamella
e.   plasmodesmata

Answer: d
Difficulty: Moderate
Bloom's Taxonomy: Knowledge

52.   The correct sequence of plant cell wall layers,
beginning with the outermost layer and progressing
inward to the plasma membrane is _____.

a.   secondary cell wall, middle lamella, primary cell
wall
b.   middle lamella, primary cell wall, secondary cell
wall
c.   middle lamella, secondary cell wall, primary cell
wall
d.   secondary cell wall, primary cell wall, middle
lamella
e.   primary cell wall, middle lamella, primary cell
wall

Answer: b
Difficulty: Difficult
Bloom's Taxonomy: Comprehension

5.5 THE ANIMAL CELL SURFACE

53.   Over time, cancerous cells typically lose the cell
adhesion molecules embedded in their plasma
membrane. Loss of these molecules is best associated
with which of the following traits of cancer cells?

a.   increased rate of cell division
b.   production of new proteins
c.   angiogenesis
d.   migration to new locations in the body
e.   loss of chromosomes

Answer: d
Difficulty: Difficult
Bloom's Taxonomy: Knowledge

54.   In normal cells, cell adhesion molecules are
partially responsible for the ability of cells to
_____.

a.   migrate to new locations in the body
b.   recognize other cells as "self"
c.   exocytosis
d.   both a and b
e.   none of the above

Answer: d
Difficulty: Moderate
Bloom's Taxonomy: Knowledge

55.   Desmosomes are a type of _____.

a.   anchoring junction
b.   cell adhesion molecule
c.   tight junction
d.   intermediate filament
e.   gap junction

Answer: a
Difficulty: Moderate
Bloom's Taxonomy: Knowledge

56.   Tight junctions _____.

a.   seal the spaces between cells
b.   allow ions and small molecules to pass between
cells
c.   allow cells to communicate with each other
d.   give the cell its shape
e.   anchor the cell membrane to the cell wall

Answer: a
Difficulty: Moderate
Bloom's Taxonomy: Knowledge

57.   Gap junctions _____.

a.   seal the spaces between cells
b.   allow ions and small molecules to pass between
cells
c.   allow plant cells to communicate with each other
d.   give the cell its shape
e.   anchor the cell membrane to the cell wall

Answer: b
Difficulty: Moderate
Bloom's Taxonomy: Knowledge

58. _____ in heart muscle tissue allows for communication between the cells resulting in the coordinated beating of the heart.

a. Tight junctions
b. Desmosomes
c. Adherens junctions
d. Anchoring junctions
e. Gap junctions

Answer: e
Difficulty: Difficult
Bloom's Taxonomy: Comprehension

59. The main components of the extracellular matrix are _____.

a. glycolipids
b. phospholipids
c. glycoproteins
d. cellulose
e. glucose

Answer: c
Difficulty: Moderate
Bloom's Taxonomy: Knowledge

UNANSWERED QUESTIONS

60. Which organelle is Stephen High's group working with to study how proteins are inserted into the cell membrane?

a. endoplasmic reticulum
b. ribosomes
c. Golgi complex
d. nucleus
e. lysosome

Answer: a
Difficulty: Moderate
Bloom's Taxonomy: Knowledge

61. What happens to membrane proteins that do not function properly?

a. Nothing, they are just ignored.
b. They are secreted out of the membrane into the surrounded extracellular matrix.
c. They are removed from the membrane and degraded.
d. They are removed from the membrane and repaired.
e. None of these answers are correct.

Answer: c
Difficulty: Moderate
Bloom's Taxonomy: Knowledge
Difficulty: Easy

Integrative Multiple-Choice

62. Which one of the following characteristics is NOT true for all living cells?

a. All cells are derived from preexisting cells.
b. All cells contain a nucleus.
c. All cells utilize organic fuel molecules as energy sources for their activities.
d. All cells respond to outside stimulation.
e. All cells possess a plasma membrane.

Answer: b
Difficulty: Moderate
Bloom's Taxonomy: Comprehension
Source: Section Why It Matters, 5.1, and 5.2

63. What do mitochondria and chloroplasts have in common?

a. both are found in plants cells
b. both contain chlorophyll
c. DNA is present in both
d. ribosomes are present in both
e. more than one answer is correct

Answer: e
Difficulty: Moderate
Bloom's Taxonomy: Knowledge
Source: 5.3 and 5.4

64. Which of the following pairs is mismatched?

a. microfilaments : actin
b. cell membrane : phospholipid bilayer
c. plant cell wall : cellulose
d. chromosome : DNA
e. intermediate filaments : tubulin

Answer: e
Difficulty: Difficult
Bloom's Taxonomy: Application
Source: 5.2, 5.3, and 5.4

65. Comparison of prokaryotic and eukaryotic cells reveals that both possess _____.

a. DNA
b. ribosomes
c. a cell membrane
d. more than one answer is correct.
e. mitochondria

Answer: d
Difficulty: Difficult
Bloom's Taxonomy: Comprehension
Source: 5.2 and 5.3

Matching

Match each of the following types of microscopy with the description that best describes it.

66. _____ Scanning electron microscopy (SEM)

67. _____ Phase-contrast microscopy

68. _____ Bright field microscopy

69. _____ Confocal laser scanning microscopy

70. _____ Transmission electron microscopy (TEM)

A. Utilizes a thin beam of electrons to examine structures within a cell.

B. Utilizes lasers to scan a fluorescently stained specimen; a computer focuses the laser to show a single plane through a cell.

C. Utilizes differences in the way light is bent (refraction) in areas of various cellular density to visualize living cells.

D. Requires light passing through the specimen; typically involves staining with dye to enhance contrast. This treatment usually "fixes" and kills the cell.

E. A beam of electrons scanned over a whole cell allows visualization of surface structures; gives a 3D-appearing image.

Answers: **66. E   67. C   68. D   69. B   70. A**

Difficulty: Moderate
Bloom's Taxonomy: Knowledge
Source: Section Figure 5.4

Match each of the following cellular structures with the function that best describes it.

71. _____ ribosomes

72. _____ nucleus

73. _____ nucleoli

74. _____ lysosomes

75. _____ Golgi complex

76. _____ rough ER

77. _____ smooth ER

78. _____ mitochondria

79. _____ chloroplast

80. _____ vesicle

81. _____ central vacuole

82. _____ cell wall

A. Contain enzymes for intracellular digestion.

B. Location of genetic material.

C. Synthesize subunits that will be used to assemble ribosomes.

D. Site of protein synthesis.

E. Composed of cellulose; provides support and protection.

F. Synthesis of lipids.

G. Conversion of fuel molecules into energy.

H. Conversion of light energy into chemical energy.

I. Storage site in plant cells.

J. Synthesis of proteins for secretion.

K. Chemically modifies proteins.

L. Membrane-bound transport structure.

Answers: **71. D   72. B   73. C   74. A   75. K   76. J   77. F   78. G   79. H   80. L   81. I   82. E**

Difficulty: Difficult
Bloom's Taxonomy: Knowledge, Comprehension
Source: Sections 5.3 and 5.4

Classification

Match each of the following cellular structures to the cell type it would be found in. An answer may be used once, more than once, or not at all.

a. a feature of all living cells
b. found in prokaryotic cells only
c. found in eukaryotic cells only
d. found in plant cells only
e. found in animal cells only

83. _____ ribosome

84. _____ nucleus

85. _____ chloroplast

86. _____ cell membrane

87. _____ nucleoid

88. _____ mitochondria

Answers: **83. a   84. c   85. d   86. a   87. b   88. c**

Difficulty: Moderate
Bloom's Taxonomy: Comprehension
Source: Section 5.2, 5.3

For each of the following statements, choose the most appropriate structure of the cytoskeleton from the list below. An answer may be used once, more than once, or not at all.

a.   microfilaments   b.   microtubules
c.   intermediate filaments

89.   _____   Involved in the process of cytoplasmic streaming

90.   _____   Comprised of the hollow cylinders of tubulin monomers

91.   _____   Comprised of two helically coiled actin monomers

92.   _____   Involved in moving chromosomes during cell division

Answers: ***89. a   90. b   91. a   92. b***

Difficulty: Difficult
Bloom's Taxonomy: Knowledge, Comprehension
Source: Section 5.3

Short Answer

93.   Why are viruses not considered to be living organisms?.

Answer: Viruses consist only of a nucleic acid molecule surrounded by a protein coat. They are not capable of carrying out all the activities of life such as reproduction, response to external stimuli, growth, etc.

Difficulty: Moderate
Bloom's Taxonomy: Comprehension
Source: Section 5.1

94.   Explain how a cell isolated from the pancreas would be the same as a muscle cell. How would the two cell types be different?

Answer: Both cell types would contain the same organelles; however, due to the very different functions of the two cells, the proportion of certain organelles would be different. For example, the pancreatic cell which is involved in the production of digestive enzymes would have an extensive rough ER network while a muscle cell would have a large proportion of mitochondria to make the large amount of energy necessary for muscle contraction.

Difficulty: Difficult
Bloom's Taxonomy: Application
Source: Section 5.3

95.   If prokaryotic cells do not have mitochondria, where do they produce their cellular energy?

Answer: The cell membrane contains most of the molecular systems needed to metabolize food molecules to ATP.
Difficulty: Moderate
Bloom's Taxonomy: Knowledge
Source: Section 5.1

96.   In general, how are prokaryotic and eukaryotic cells different and how are they similar?

Answer: Both mitochondria and cloroplasts contain DNA, RNA, and ribosomes that resemble those found in bacteria In prokaryotic cells the genetic material is found in a central region called the nucleoid while in eukaryotic cells it is contained in the membrane-bound nucleus. Also, eukaryotic cells contain membrane systems that form organelles while prokaryotic cells do not. A plasma membrane surrounds both prokaryotic and eukaryotic cells.
Difficulty: Moderate
Bloom's Taxonomy: Knowledge, Comprehension
Source: Section 5.1

97.   Why are chloroplasts and mitochondria believed to have originated from ancient prokaryotes?

Answer: Both mitochondria and chloroplasts contain DNA, RNA, and ribosomes that resemble those found in bacteria.
Difficulty: Moderate
Bloom's Taxonomy: Knowledge, Comprehension
Source: Section 5.3and 5.4

98.   Compare animal and plant cells; how are they different? How are they the same?

Answer: Both animal cells and plant cells have a plasma membrane, nucleus, mitochondria, rndoplasmic reticulum, ribosomes, and Golgi complex. Animal cells, however, do not have a cell wall, central vacuole, or chloroplasts.
Difficulty: Moderate
Bloom's Taxonomy: Knowledge, Application
Source: Section 5.3and 5.4

Select the Exception

99.   The endomembrane system is involved in each of the following functions EXCEPT _____.
a.   DNA synthesis
b.   synthesis of proteins
c.   synthesis of lipids
d.   transport of proteins to outside the cell
e.   detoxification of some toxins

Answer: a
Difficulty: Moderate
Bloom's Taxonomy: Knowledge
Source: Section 5.3

100. The Golgi complex regulates transport of proteins to all of the following locations EXCEPT _____.

a. lysosomes
b. the cell membrane
c. the exterior of the cell
d. the nucleus
e. transport vesicles

Answer: d
Difficulty: Moderate
Bloom's Taxonomy: Knowledge
Source: Section 5.3

101. Functions of the cytoskeleton include all of the following EXCEPT _____.
a. sperm motility
b. amoeboid motion
c. skeletal muscle contraction
d. cell division
e. autophagy

Answer: e
Difficulty: Moderate
Bloom's Taxonomy: Knowledge
Source: Section 5.3

102. Activities that occur in the cytoplasm include all of the following EXCEPT _____.

a. conduction of stimulatory signals
b. division of genetic material
c. synthesis of DNA
d. protein synthesis
e. various metabolic reactions

Answer: c
Difficulty: Difficult
Bloom's Taxonomy: Comprehension
Source: 5.1

103. A supportive cell wall is found surrounding the plasma membrane of each of the following types of cells EXCEPT _____.

a. plant cells
b. fungal cells
c. many types of protists
d. animal cells
e. bacterial cells

Answer: d
Difficulty: Difficult
Bloom's Taxonomy: Comprehension
Source: 5.2 and 5.3

104. In the human body, cilia play an important role in the function of each of the following organs EXCEPT _____.

a. ventricles of the brain
b. oviducts of the female reproductive system
c. upper respiratory tract
d. ventricles of the heart

Answer: d
Difficulty: Difficult
Bloom's Taxonomy: Application
Source: Section 5.3

105. The functions of microbodies include all the following EXCEPT _____.
a. ATP synthesis
b. breakdown of amino acids
c. metabolism of alcohol
d. conversion of oils to sugars in plants
e. breakdown of hydrogen peroxide to water and oxygen

Answer: a
Difficulty: Moderate
Bloom's Taxonomy: Knowledge
Source: Section 5.3

106. The function of the central vacuole in plants may include all of the following EXCEPT _____.
a. protein synthesis
b. storage of salts, sugars, and proteins
c. enzymatic digestion
d. storage of chemical defenses against pathogens
e. storage of waste products

Answer: a
Difficulty: Moderate
Bloom's Taxonomy: Knowledge
Source: 5.4

107. When comparing an animal secretory cell and a photosynthetic leaf cell, both cell types are similar EXCEPT _____.

a. both have a Golgi complex
b. both have a cell membrane
c. both have mitochondria
d. both have a cell wall
e. both have ribosomes

Answer: d
Difficulty: Moderate
Bloom's Taxonomy: Application
Source: 5.4

108. All of the following characteristics describe glycoproteins EXCEPT _____.
a. Glycoproteins provide support and protection.
b. Glycoproteins are the main component of the extracellular matrix.
c. Glycoproteins include collagen and fibronectins.
d. Glycoproteins are produced only on free ribosomes in the cytoplasm.
e. Glycoproteins are found in tendons, cartilage, and bone.

Answer: d
Difficulty: Difficult
Bloom's Taxonomy: Comprehension
Source: 5.3 and 5.5

Labeling

Identify each of the cellular structures indicated in a eukaryotic animal cell.

109. _____ Nucleus
110. _____ Golgi complex
111. _____ Pair of centrioles
112. _____ Cytosol
113. _____ Lysosome
114. _____ Free ribosome
115. _____ Plasma membrane
116. _____ Rough ER
117. _____ Microtubules
118. _____ Mitochondrion
119. _____ Vesicle
120. _____ Attached ribosomes

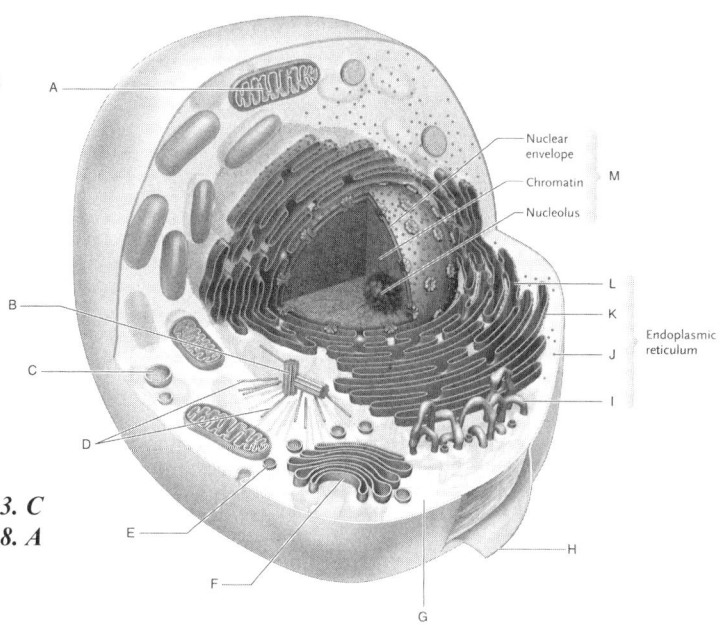

Answers: **109. M   110. F   111. B   112. G   113. C**
             114. J   115. H   116. L   117. D   118. A
             119. E   120. K

Difficulty: Moderate
Bloom's Taxonomy: Knowledge, Comprehension
Source: Section 5.3

Identify each of the cellular structures indicated in a eukaryotic plant cell.

121. _____ Chloroplast
122. _____ Plasma membrane
123. _____ Golgi complex
124. _____ Smooth ER
125. _____ Nucleus
126. _____ Free ribosomes
127. _____ Mitochondrion
128. _____ Cell wall
129. _____ Central vacuole
130. _____ Microtubules
131. _____ Vesicle

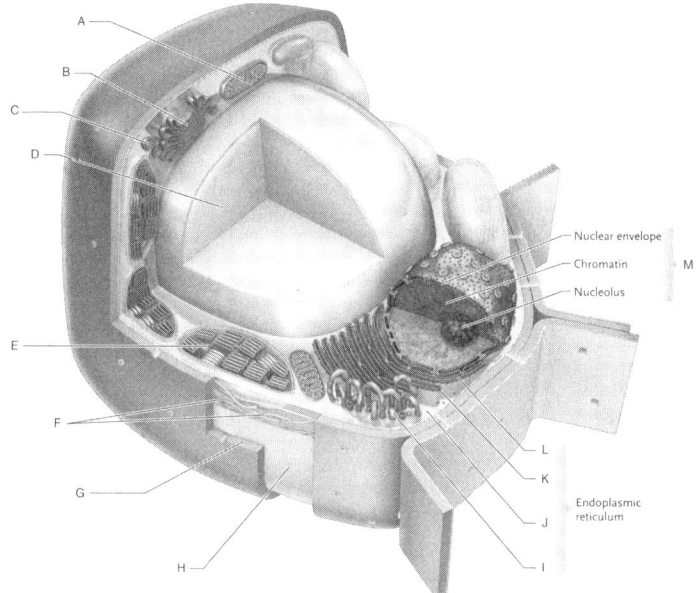

Answers: **121. E   122. H   123. B   124. I**
             125. M   126. J   127. A   128. G
             129. D   130. F   131. C

Difficulty: Moderate
Bloom's Taxonomy: Knowledge, Comprehension
Source: Section 5.3

Identify the structures of a mitochondrion.

132. _____ Matrix
133. _____ Intermembrane compartment
134. _____ Inner mitochrondrial
membrane
135. _____ Cristae
136. _____ Outer mitochondrial
membrane

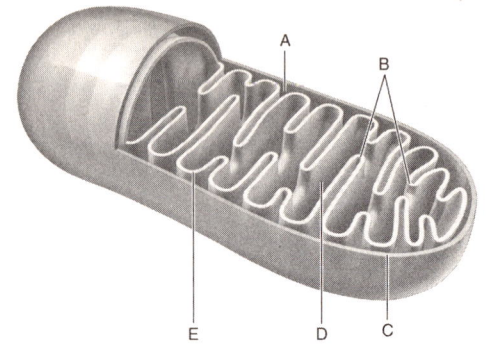

Answers: *132. D   133. A   134. E   135. B   136. C*

Difficulty: Moderate
Bloom's Taxonomy: Knowledge, Comprehension
Source: Section 5.3

Identify each of the cellular structures indicated in a
prokaryotic cell.

137. _____ Cytoplasm
138. _____ Cell wall
139. _____ Flagellum
140. _____ Pili
141. _____ Ribosomes
142. _____ Capsule
143. _____ Plasma membrane
144. _____ Nucleoid

Answers: *137. G   138. D   139. A   140. B*
              141. H   142. E   143. C   144. F

Difficulty: Moderate
Bloom's Taxonomy: Knowledge, Comprehension
Source: Section 5.2

6
MEMBRANES AND TRANSPORT

Multiple-Choice

WHY IT MATTERS

1.   Organisms must constantly bring in certain molecules and ions while keeping others out. This function is accomplished by _____.

a.   the nucleus
b.   lysosomes
c.   vesicles
d.   the plasma membrane
e.   the Golgi complex

Answer: d
Difficulty: Easy
Bloom's Taxonomy: Knowledge

2.   Certain organisms such as the striped bass (*Morone saxatilis*) are able to migrate between the ocean and freshwater streams. Consider when the striped bass migrates into freshwater streams; which one of the following situations would be necessary to maintain life?

a.   Ions must be kept within the cells while water is kept out.
b.   Water must be kept within the cells while ions are kept out.
c.   It is not possible for a living organism to survive in both the ocean and freshwater streams.
d.   The plasma membrane of the organism must change its phospholipid composition.
e.   The plasma membrane of the organism must change its protein composition.

Answer: a
Difficulty: Moderate
Bloom's Taxonomy: Applicaton

6.1 MEMBRANE STRUCTURE

3.   The major structural components of a cell membrane are _____.

a.   phospholipids and cellulose
b.   phospholipids, protein, and sterols
c.   protein and sterols
d.   glycolipids and proteins
e.   phospholipids and glycolipids

Answer: b
Difficulty: Moderate
Bloom's Taxonomy: Knowledge

4.   The polar end of a phospholipid _____.

a.   is in contact with an aqueous solution
b.   does not mix with water
c.   is composed of a phosphate group linked to an alcohol or amino acid
d.   is soluble in water
e.   more than one answer is correct

Answer: e
Difficulty: Moderate
Bloom's Taxonomy: Comprehension

5.   Membrane sterols such as cholesterol function in animal cell membranes to _____.

a.   increase the rate of diffusion
b.   store cellular energy
c.   facilitate ion transport
d.   maintain membrane fluidity
e.   receive chemical signals

Answer: d
Difficulty: Moderate
Bloom's Taxonomy: Comprehension

6.   According to the fluid mosaic model of cell membranes, the fluid part of the model refers to _____.

a.   the constant movement of the hydrophilic tails in the interior of the membrane
b.   a thin layer of water found sandwiched between the two layers of phospholipids
c.   the phospholipid molecules which vibrate, spin, and exchange places within the same layer, or *leaflet*, of the bilayer
d.   the free movement of cholesterol molecules within the membrane
e.   the frequent flip-flop of phospholipids from one side of the membrane to the other

Answer: c
Difficulty: Moderate
Bloom's Taxonomy: Comprehension

7.   Which type of lipid is most important in the structure of biological membranes?

a.   fat
b.   wax
c.   phospholipid
d.   cholesterol
e.   triglyceride

Answer: c
Difficulty: Moderate
Bloom's Taxonomy: Knowledge

8. In an aqueous environment, the phospholipids of a membrane _____.

a   are arranged in a single layer
b   are arranged in a bilayer with the polar heads of each layer located at the surface
c   are arranged in a bilayer with the fatty acid tails located at the surface
d   are arranged in a bilayer but the phospholipids have no specific orientation
e   dissolve

Answer: b
Difficulty: Moderate
Bloom's Taxonomy: Comprehension

9. Which of the following molecules demonstrate dual solubility characteristics?

a.   sterols only
b.   proteins only
c.   phospholipids only
d.   sterols and phospholids
e.   sterols, phospholipids, and proteins

Answer: e
Difficulty: Moderate
Bloom's Taxonomy: Application

10. When referring to membrane glycolipids and glycoproteins, the term "glyco-" indicates _____.

a.   nonpolar carbohydrate groups are attached to the molecules
b.   polar carbohydrate groups are attached to the molecules
c.   the molecules are found on both the interior and exterior of the membrane
d.   the molecules are attached to the membrane by ionic bonds
e.   the molecules are rarely found on the membrane exterior

Answer: b
Difficulty: Moderate
Bloom's Taxonomy: Knowledge

11. Our current view of membrane structure is based on the fluid mosaic model, proposed by _____ in 1972.

a.   H. Davson and J. Danielli
b.   I. Langmuir
c.   C. Overton
d.   S. Singer and G. Nicolson
e.   E. Gorter and F. Grendel

Answer: d
Difficulty: Easy
Bloom's Taxonomy: Knowledge

12. How are the membranes of winter rye able to remain fluid when the temperature becomes extremely cold?

a.   The percentage of unsaturated phospholipids in the membrane is increased.
b.   The percentage of saturated phospholipids in the membrane is increased.
c.   The percentage of cholesterol molecules in the membrane is increased.
d.   The percentage of both unsaturated phospholipids and cholesterol is increased.
e.   The percentage of both saturated phospholipids and cholesterol is increased.

Answer: d
Difficulty: Moderate
Bloom's Taxonomy: Application

13. The "mosaic" part of the fluid mosaic model refers to the membrane _____.

a.   proteins
b.   phospholipids
c.   cholesterol
d.   functions
e.   layers

Answer: a
Difficulty: Moderate
Bloom's Taxonomy: Knowledge

14. Which of the following is a reasonable explanation for why unsaturated fatty acids help keep a membrane more fluid at lower temperatures?

a.   The double bonds form a kink in the fatty acid tail, forcing adjacent lipids to be spaced further apart.
b.   Unsaturated fatty acids have a higher cholesterol content.
c.   Unsaturated fatty acids permit more water in the interior of the membrane.
d.   The double bonds block interaction among the hydrophilic head groups of the lipids.
e.   The double bonds result in a shorter fatty acid tail.

Answer: a
Difficulty: Moderate
Bloom's Taxonomy: Comprehension

15. When comparing the outside and inside halves of a cell membrane's phospholipid bilayer, the composition of lipids on the two surfaces is _____.

a.   asymmetrical
b.   identical
c.   not identical, but symmetrical
d.   highly random and varies throughout the cell
e.   a mirror image

Answer: a
Difficulty: Moderate
Bloom's Taxonomy: Knowledge

16. The _____ component of the cell membrane functions as a selective barrier, while the _____ component has specific functions such as transport, recognizing other cells, and binding to other cells.

a. carbohydrate, nucleic acid
b. protein, lipid
c. lipid, protein
d. lipid, carbohydrate
e. carbohydrate, protein

Answer: c
Difficulty: Moderate
Bloom's Taxonomy: Application

17. In what way do the various membranes of a cell differ?

a. Phospholipids are found only in certain membranes.
b. Certain proteins are unique to each membrane.
c. Only certain membranes are constructed from molecules with dual solubility.
d. Only certain membranes of a cell arc selectively permeable.
e. Some membranes have hydrophobic surfaces exposed to the cytoplasm, while others havehydrophilic surfaces facing the cytoplasm.

Answer: b
Difficulty: Moderate
Bloom's Taxonomy: Comprehension

18. The selective permeability of a membrane refers to _____.

a. the movement of a molecule from an area of greater concentration to an area of lesser concentration.
b. the ability of a substance to pass through a membrane
c. the ability of only certain molecules to pass across a membrane
d. the need for carrier proteins to transport some molecules
e. the ability of molecules to be transported across the membrane only certain times of the day

Answer: c
Difficulty: Moderate
Bloom's Taxonomy: Comprehension

19. Joe Scientist fused a mouse cell and a human cell, then treated the cell with specific antibodies covalently linked to fluorescent dyes (antibodies to mouse proteins—green; antibodies to human proteins—red). What does the cell look like immediately after fusion?

a. The cell is half red and half green.
b. The red and green fluorescent labels are uniformly distributed across the entire membrane.
c. The red and green labels are distributed in intermingled patches.
d. The red and green labels flash intermittently.
e. The red and green labels are distributed in a swirling pattern

Answer: a
Difficulty: Moderate
Bloom's Taxonomy: Application

20. Forty minutes later, Joe Scientist looked at the fused cell again. This time he observed _____.

a. the cell was still half red and half green.
b. the red and green fluorescent labels were uniformly distributed across the entire membrane
c. the red and green labels were distributed in intermingled patches
d. the red and green labels flashed intermittently
e. the red and green labels were distributed in a swirling pattern

Answer: b
Difficulty: Moderate
Bloom's Taxonomy: Application

FOCUS ON RESEARCH: **BASIC RESEARCH: KEEPING MEMBRANES FLUID AT COLD TEMPERATURES**

21. Reduction of environmental temperatures to freezing rapidly leads to cell death as a result of _____.

a. inhibition of cell division
b. inhibition of molecule transport across the cell membrane
c. increased cellular signaling
d. increase in metabolic reactions
e. decreased lipid synthesis

Answer: b
Difficulty: Moderate
Bloom's Taxonomy: Comprehension

22. Which of the following adaptations to the plasma membrane allows mammals to hibernate in subzero temperatures without their plasma membranes freezing?

a. Increase in cholesterol content only.
b. Increase in protein content only.
c. Increase in the number of double covalent bonds in phospholipids.
d. Increase in both cholesterol and protein content.
e. Increase in both cholesterol and double covalent bonds.

Answer: e
Difficulty: Moderate
Bloom's Taxonomy: Application

6.2 THE FUNCTIONS OF MEMBRANES IN TRANSPORT: PASSIVE TRANSPORT

23. The primary function of cellular membranes is _____.

a. the controlled transport of ions and molecules across the membrane
b. cell-cell binding
c. recognition of other cells as being "like"
d. reception of chemical signals form other cells
e. participation in metabolic reactions

Answer: a
Difficulty: Easy
Bloom's Taxonomy: Knowledge

24. Movement of a substance from an area of low concentration to an area of high concentration using energy obtained from ATP is called _____.

a. passive transport
b. diffusion
c. facilitated transport
d. osmosis
e. active transport

Answer: e
Difficulty: Easy
Bloom's Taxonomy: Knowledge

25. When a drop of food coloring is placed in a container of clear water, the colored dye molecules _____.

a. undergo osmosis to a different location
b. undergo active transport to a different location
c. diffuse to the top of the container
d. diffuse equally through out the container
e. stay at the bottom of the container

Answer: d
Difficulty: Easy
Bloom's Taxonomy: Application

26. The concentration gradient that drives diffusion is a form of _____.

a. heat
b. potential energy
c. kinetic energy
d. active transport
e. osmosis

Answer: b
Difficulty: Moderate
Bloom's Taxonomy: Knowledge

27. _____ molecules pass through a cell membrane most easily by diffusion.

a. Ionic
b. Large, polar
c. Large, hydrophilic
d. Small, hydrophobic
e. Large, hydrophobic

Answer: d
Difficulty: Moderate
Bloom's Taxonomy: Application

28. A channel that opens in response to changes in ionic charge across a membrane is called a _____.

a. voltage-gated channel
b. ligand-gated channel
c. charge-gated channel
d. electric-gated channel
e. positive-gated channel

Answer: a
Difficulty: Difficult
Bloom's Taxonomy: Knowledge

29. Transport of a molecule across a cell membrane by facilitated diffusion _____.

a. exhibits specificity for a particular type of molecule
b. requires the input of energy
c. depends on a concentration gradient
d. goes against the concentration gradient
e. more than one answer is correct

Answer: e
Difficulty: Difficult
Bloom's Taxonomy: Knowledge

30. Which one of the following is absolutely necessary for diffusion to occur?

a. a living cell
b. a phospholipid bilayer
c. a selectively permeable membrane
d. a concentration gradient
e. a solution

Answer: d
Difficulty: Moderate
Bloom's Taxonomy: Comprehension

31. Small polar and charged molecules typically cross the cell membrane by way of _____.

a. simple diffusion
b. osmosis
c. filtration
d. active transport
e. facilitated diffusion

Answer: e
Difficulty: Moderate
Bloom's Taxonomy: Knowledge

32. Facilitated diffusion is specific. This means _____.

a. a specific protein will transport certain polar or charged molecules but not others
b. that only one specific integral protein per membrane is involved in facilitated diffusion
c. that the energy molecule ATP is specifically required for transport
d. only specific hydrophobic molecules can be transported
e. that transport of molecules occurs only in specific cells

Answer: a
Difficulty: Moderate
Bloom's Taxonomy: Comprehension

33. Carrier proteins are often used by the cell to transport _____.

a. H_2O
b. proteins
c. glucose and amino acids
d. steroid hormones
e. CO_2

Answer: c
Difficulty: Moderate
Bloom's Taxonomy: Knowledge

INSIGHTS FROM THE MOLECULAR REVOLUTION: TRACKING GATING MOVEMENTS IN A CHANNEL PROTEIN

34. Gated channels are utilized by the cell _____.

a. to transport water
b. to stimulate muscle contraction
c. to generate nerve signals
d. for facilitated diffusion
e. more than one answer is correct

Answer: e
Difficulty: Moderate
Bloom's Taxonomy: Knowledge

6.3 PASSIVE WATER TRANSPORT AND OSMOSIS

35. The movement of water across a membrane from an area of high to low water concentration is an example of _____.

a. active transport
b. endocytosis
c. osmosis
d. both osmosis and active transport
e. both diffusion and osmosis

Answer: e
Difficulty: Moderate
Bloom's Taxonomy: Knowledge

36. The movement of water across a membrane from an area of lower solute concentration to a region of higher solute concentration is called _____.

a. osmosis
b. active transport
c. endocytosis
d. both diffusion and osmosis
e. both osmosis and active transport

Answer: d
Difficulty: Moderate
Bloom's Taxonomy: Application

37. A red blood cell was placed in a beaker of solution. The cell immediately began to swell and ultimately burst. This happened because the cytoplasm of the cell was _____ to the solution in the beaker which was _____.

a. hypertonic/hypotonic
b. hypotonic/hypertonic
c. hypotonic/isotonic
d. hypertonic/isotonic
e. isotonic/hypotonic

Answer: a
Difficulty: Difficult
Bloom's Taxonomy: Application

38. When a plant cell is placed in a hypotonic solution, the cell wall prevents _____.

a. plasmolysis
b. diffusion
c. active transport
d. the cell from bursting
e. the cell from shrinking

Answer: d
Difficulty: Moderate
Bloom's Taxonomy: Knowledge

39. Distilled water would be considered _____ to body cells.

a. isotonic
b. hypertonic
c. hypotonic
d. protonic
e. aquatonic

Answer: c
Difficulty: Moderate
Bloom's Taxonomy: Application

40. A(n) _____ environment is ideal for plant cells, while a(n) _____ environment is best for animal cells.

a. isotonic/hypotonic
b. hypotonic/isotonic
c. hypotonic/hypertonic
d. hypertonic/isotonic
e. isotonic/isotonic

Answer: b
Difficulty: Moderate
Bloom's Taxonomy: Comprehension

41. For osmosis to occur, _____.

a. a selectively permeable membrane must be present
b. a concentration gradient must exist
c. cellular energy must be expended
d. pure water must be on one side of the membrane
e. more than one answer is correct

Answer: e
Difficulty: Moderate
Bloom's Taxonomy: Comprehension

42. In plants, wilting of leaves and stems results from _____.

a. hemolysis
b. an increase in turgor pressure
c. plasmolysis
d. a lack of solutes in the cell
e. a higher than normal concentration of water in the cell

Answer: c
Difficulty: Difficult
Bloom's Taxonomy: Comprehension

43. An isotonic solution has a solute concentration _____ to the solute concentration inside the cell.

a. equal
b. greater than
c. less than
d. it would depend on the solute
e. it would depend on the type of cell

Answer: a
Difficulty: Moderate
Bloom's Taxonomy: Knowledge

6.4 ACTIVE TRANSPORT

44. The voltage across a membrane is called the _____.

a. electrochemical gradient
b. turgor pressure
c. membrane potential
d. chemical gradient
e. electron potential

Answer: c
Difficulty: Moderate
Bloom's Taxonomy: Knowledge

45. _____ is the net movement of uncharged molecules from a low concentration to a higher concentration.

a. Active transport
b. Facilitated diffusion
c. Exocytosis
d. Osmosis
e. Diffusion

Answer: a
Difficulty: Moderate
Bloom's Taxonomy: Knowledge

46. The _____ is responsible for maintaining the membrane potential across the cell membrane.

a. H^+ pump
b. Na^+/K^+ pump
c. diffusion gradient
d. Ca^{2+} pump
e. osmotic ratio

Answer: b
Difficulty: Moderate
Bloom's Taxonomy: Knowledge

47. The membrane potential across a cell creates a _____ charge inside the cell and a _____ charge outside the cell.

a. negative/negative
b. positive/negative
c. positive/positive
d. negative/positive
e. neutral/positive

Answer: b
Difficulty: Moderate
Bloom's Taxonomy: Knowledge

48. A(n) _____ is created as ions diffuse across membranes.

a. chemical gradient
b. electrochemical gradient
c. chemical gradient
d. electrical gradient
e. concentration gradient

Answer: b
Difficulty: Moderate
Bloom's Taxonomy: Comprehension

49. A transport system in which transport of an ion in one direction provides the energy for active transport in the opposite direction is known as _____.

a. antiport
b. exchange diffusion
c. symport
d. cotransport
e. more than one answer is correct

Answer: e
Difficulty: Moderate
Bloom's Taxonomy: Knowledge

6.5 EXOCYTOSIS AND ENDOCYTOSIS

50. Cells undergo exocytosis _____.

a. to pump protons down a concentration gradient
b. when replicating
c. to secrete protein and wastes from the cell
d. to ingest nutrients
e. as a means of cellular protection

Answer: c
Difficulty: Moderate
Bloom's Taxonomy: Knowledge

51. Eukaryotic cells import large molecules through the process of _____ and secrete larger molecules by _____.

a. endocytosis/exocytosis
b. diffusion/exocytosis
c. exocytosis/endocytosis
d. endocytosis/phagocytosis
e. phagocytosis/pinocytosis

Answer: a
Difficulty: Moderate
Bloom's Taxonomy: Knowledge

52. Pinocytosis and phagocytosis are accomplished in the cell by the _____.

a. nucleus
b. lysosome
c. endoplasmic reticulum
d. plasma membrane
e. mitocondria

Answer: d
Difficulty: Easy
Bloom's Taxonomy: Knowledge

UNANSWERED QUESTIONS

53. Aquaporins are _____.

a. pores made of water molecules
b. proteins utilized to transport water molecules
c. found only in bacterial cells
d. specific channels for water transport
e. useful for cell signaling

Answer: d
Difficulty: Moderate
Bloom's Taxonomy: Knowledge

54. Endocytosis of nanotubes has shown promise for _____.

a. the delivery of anticancer drugs to cancer cells
b. the delivery of cellular food to lysosomes
c. the delivery of cellular energy to weak cells
d. the removal of waste from rapidly metabolizing cells
e. the improved secretion of proteins from the cell

Answer: a
Difficulty: Moderate
Bloom's Taxonomy: Knowledge

Integrative Multiple-Choice

55. Carrier molecules are utilized for _____.

a. active transport only
b. passive transport only
c. both active and passive transport
d. osmosis
e. transport of all types of molecules

Answer: c
Difficulty: Moderate
Bloom's Taxonomy: Knowledge
Source:6.2 and 6.4

56. The selective permeability of a cell membrane is due to _____.

a. the hydrophobic core formed by the phospholipid tails
b. the hydrophilic end facing the cell exterior
c. the integral proteins of the membrane
d. the position of cholesterol in the membrane bilayer
e. the hydrophilic core formed by the phospholipid tails

Answer: a
Difficulty: Moderate
Bloom's Taxonomy: Application
Source: 6.1 and 6.2

Matching

Match each of the following mechanisms of cellular transport with its correct definition.

57. _____ osmosis

58. _____ facilitated diffusion

59. _____ phagocytosis

60. _____ pinocytosis

61. _____ receptor-mediated endocytosis

62. _____ diffusion

A. Movement of a molecule from an area of high concentration to an area of lower concentration.

B. Cells internalize molecules into a cell by the inward budding of vesicles possessing receptors specific to the molecule being transported.

C. Movement of water from a hypotonic solution into a hypertonic solution across a selectively permeable membrane.

D. Large particles are enveloped by the cell membrane and internalized.

E. A process in which liquid droplets are ingested by living cells.

F. Diffusion of molecules across the plasma membrane with the assistance of transport proteins.

Answers: *57. C*   *58. F*   *59. D*   *60. E*   *61. B*

   62 A.

Difficulty: Moderate
Bloom's Taxonomy: Knowledge
Source: Sections 6.2, 6.3, 6.4, and 6.5

Short Answer

63. Explain why the transport of molecules across the cell membrane is considered to be both specific and directional.

Answer: Transport is considered directional because only certain ions and molecules can move into the cell, while others can only move out of the cell. Transport is specific because only certain ions and molecules can actually move across the membrane.
Difficulty: Moderate
Bloom's Taxonomy: Comprehension
Source: Section 6.2

64. Water is a strongly polar molecule, so how does it cross the plasma membrane?

Answer: Water molecules are small enough to slip through spaces transiently created between the hydrocarbon tails of phospholipid molecules as they flex and rotate in the fluid bilayer. This type of water movement is relatively slow however.
Difficulty: Moderate
Bloom's Taxonomy: Knowledge, Comprehension
Source: Section6.2, 6.3

65. Carrier proteins share several characteristics with enzymes used to catalyze metabolic reactions. In what ways are carrier proteins and enzymes similar?

Answer: Both carrier proteins and enzymes are specific; carrier proteins only bind and transport molecules that specifically fit the binding site while enzymes only react with molecules that 'fit' the enzyme active site. Both carrier proteins and enzymes can become saturated when there are more molecules than protein to interact with.
Difficulty: Difficult
Bloom's Taxonomy: Knowledge, Application
Source: Section macromolecule chapter, 6.2

Sequence

Arrange the following steps involved in secondary active transport in the correct order with question 70. being step one and question 74 step 5.

66. _____   A. Transporter undergoes a folding change that exposes the binding site to the opposite side of the membrane.

67. _____   B. Transport protein hydrolyzes ATP to ADP + phosphate.

68. _____   C. Ion is released to the side of higher concentration

69. _____   D. Attachment of free phosphate to the transport protein allows binding of ion.

70. _____   E. Protein reverts to its original shape.

Answers: *66. B*   *67. D*   *68. A*   *69. C*   *70. E*

Difficulty: Difficult
Bloom's Taxonomy: Knowledge
Source: Section 6.4

True/False

If the statement is true, write a "T" in the blank. If the statement is false, make it correct by changing the underlined word(s) and writing the correct word(s) in the answer blank.

71. _____   A water concentration gradient is influenced by the number of solute molecules present on both sides of the membrane.

72. _____   An animal cell placed in a hypertonic solution will swell and perhaps burst.

73. _____ Physiological saline is 0.9 percent NaCl; red blood cells placed in such a solution will not gain or lose water; therefore, one could state that the fluid in red blood cells is <u>hypertonic</u>.

74. _____ A solution of 65 percent water, 35 percent solute is <u>more</u> concentrated with respect to solute than a solution of 70 percent water, 30 percent solute.

75. _____ A <u>hypertonic</u> environment would be ideal for a healthy plant.

Answers: **71. T**   **72. F, hypotonic**   **73. F, isotonic**
           74. T   **75. F, hypotonic**

Difficulty: Difficult
Bloom's Taxonomy: Knowledge, Application
Source: Section 6.3

Select the Exception

76. All of the following are functions of integral membrane proteins EXCEPT _____.

a. transport of polar molecules
b. reception of chemical signals form other cells
c. cell division
d. recognition of other cells as being "like"
e. cell-cell binding

Answer: c
Difficulty: Moderate
Bloom's Taxonomy: Knowledge
Source: Section 6.1

77. Each of the following characteristics is true for both facilitated diffusion and active transport EXCEPT

_____.

a. the use of transport proteins is required
b. transport proteins are specific for the molecules being transported
c. transport proteins can become saturated
d. a concentration gradient is present
e. the direction of transport is always with the concentration gradient

Answer: e
Difficulty: Moderate
Bloom's Taxonomy: Comprehension
Source: Section 6.2, Table 6.1

78. Each of the following small molecules easily diffuses through the cell membrane EXCEPT _____.

a. steroid hormones
b. O_2
c. H_2O
d. glucose
e. CO_2

Answer: d
Difficulty: Moderate
Bloom's Taxonomy: Application
Source: Section 6.2

79. All of the following describe facilitated diffusion EXCEPT _____.

a. it is specific
b. it is dependent on a concentration gradient
c. it requires ATP
d. it can involve either carrier proteins or channel proteins
e. it involves integral proteins

Answer: c
Difficulty: Moderate
Bloom's Taxonomy: Application
Source: Section 6.2

80. Examples of molecules secreted from a eukaryotic cell include all of the following EXCEPT _____.

a. insulin
b. ions
c. milk proteins
d. mucus
e. digestive enzymes

Answer: b
Difficulty: Moderate
Bloom's Taxonomy: Application
Source: Section 6.5

Labeling

Identify each structure found in a typical plasma membrane in the figure below.

81. _____ Microfilament
82. _____ Carbohydrate groups
83. _____ Integral proteins
84. _____ Peripheral proteins
85. _____ Cholesterol

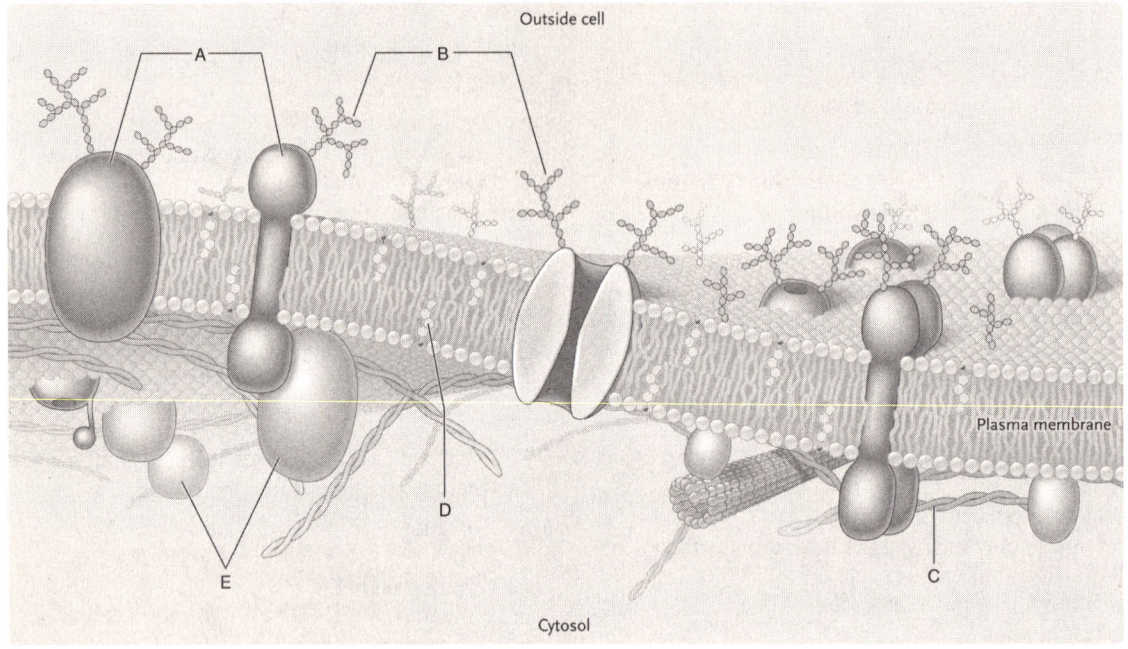

Answers: **81. C   82. B   83. A   84. E   85. D**

Difficulty: Moderate
Bloom's Taxonomy: Knowledge, Comprehension
Source: Section 6.1

The micrographs of animal cells depict the effects of various aqueous environments on red blood cells placed in a hypotonic, hypertonic and isotonic environment. Identify the correct term below.

86. _____ Isotonic
87. _____ Hypertonic
88. _____ Hypotonic

Answers: **86 C   87. A   88 B**

Difficulty: Moderate
Bloom's Taxonomy: Knowledge, Comprehension
Source: Section 6.3

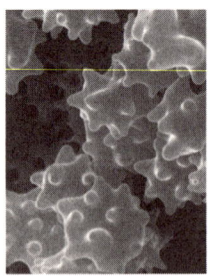
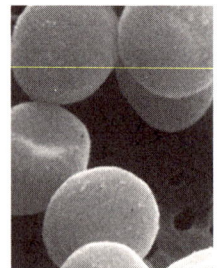
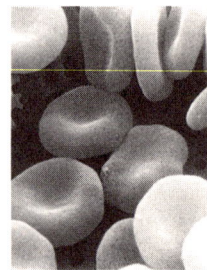

A   B   C

7

CELL COMMUNICATION

Multiple-Choice

WHY IT MATTERS

1. In order for a cell to receive a signal, it must possess an appropriate _____.
a. glycolipid
b. glycoprotein
c. receptor
d. transfer protein
e. hormone

Answer: c
Difficulty: Easy
Bloom's Taxonomy: Knowledge

2. When a cell binds a signal molecule, it may affect _____.
a. gene activity
b. protein synthesis
c. cell division
d. secretion
e. more than one answer is correct

Answer: e
Difficulty: Moderate
Bloom's Taxonomy: Knowledge

7.1 CELL COMMUNICATION: AN OVERVIEW

3. Adjacent animal cells utilize _____ to rapidly communicate with each other.
a. gap junctions
b. plasmodesmata
c. hormones
d. desmosomes
e. neurotransmitters

Answer: a
Difficulty: Moderate
Bloom's Taxonomy: Knowledge

4. Adjacent plant cells utilize _____ to rapidly communicate with each other.
a. gap junctions
b. plasmodesmata
c. transport proteins
d. desmosomes
e. neurotransmitters

Answer: b
Difficulty: Moderate
Bloom's Taxonomy: Knowledge

5. In order for a cell to respond to the signaling molecule epinephrine, it must _____.
a. it must have ion channels
b. it must have a lipid bilayer that allows the response to occur
c. have receptors exposed on the inner plasma membrane surface
d. have receptors exposed on the plasma membrane surface
e. have nuclear membrane receptors

Answer: d
Difficulty: Difficult
Bloom's Taxonomy: Knowledge, Comprehension

6. Receptors for polar molecules are found _____ while receptors for nonpolar molecules are located _____.
a. on the cell surface/within the cell
b. within the cell/on the cell surface
c. on the cell surface/on the nuclear membrane
d. on the nuclear membrane/on the cell surface
e. on the cell surface/within the membrane interior

Answer: a
Difficulty: Difficult
Bloom's Taxonomy: Knowledge, Comprehension

7. Nonpolar signal molecules enter the cell by _____.
a. facilitated diffusion
b. simple diffusion
c. osmosis
d. active transport
e. receptor mediated endocytosis

Answer: b
Difficulty: Moderate
Bloom's Taxonomy: Comprehension

8. Examples of nonpolar signal molecules include _____.
a. epinephrine
b. insulin
c. estrogen and testosterone
d. growth factors
e. neurotransmitters

Answer: c
Difficulty: Moderate
Bloom's Taxonomy: Knowledge

9. What is the response of a cell surface receptor to the binding of a signal molecule?

a. The cell surface receptor denatures.
b. The signal is transduced through the plasma membrane into the cell.
c. The receptor relays a signal to another location of the cell surface.
d. The cell surface receptor flips through the membrane to the inside of the cell.
e. More than one answer is correct.

Answer: b
Difficulty: Moderate
Bloom's Taxonomy: Knowledge

10. The overall process in which information carried by a signal molecule is translated into changes that occur inside the cell is called _____.

a. signal digestion
b. signal digression.
c. signal induction
d. signal interaction
e. signal transduction

Answer: e
Difficulty: Moderate
Bloom's Taxonomy: Knowledge

11. You have recently identified a molecule you believe to be a signal molecule associated with signal transduction. All you know about this signal molecule is that chemically it is hydrophilic. As a result you expect it to interact with _____.

a. a receptor on the cell surface
b. a receptor within the cytoplasm of the cell
c. a receptor in the nucleus of the cell
d. a receptor on the cytoplasmic surface of the plasma membrane
e. a receptor associated with the endoplasmic reticulum

Answer: a
Difficulty: Difficult
Bloom's Taxonomy: Application

12. How do cells in the body of a multicellular organism communicate with each other?

a. by way of signaling molecules that interact with specific receptors
b. through long projections that directly connect cells to each other
c. through electrical signals passed between cells
d. by the transport of ions between cells in different parts of the organism
e. a and c both

Answer: e
Difficulty: Moderate
Bloom's Taxonomy: Comprehension

13. In the 1950s, Earl Sunderland and colleagues discovered epinephrine _____.

a. leads to release of a second messenger that ultimately results in hydrolysis of glycogen to glucose
b. leads to release of a second messenger that lowers blood glucose by causing it to bind to liver cells
c. interacts directly with the enzyme glycogen phosphorylase
d. interacts directly with the cell membrane to help transport glucose into the cell
e. is a signaling molecule that does not require a cell surface receptor

Answer: a
Difficulty: Moderate
Bloom's Taxonomy: Knowledge

INSIGHTS FROM THE MOLECULAR REVOLUTION: SURVIVING SOMETHING BAD BY TAKING A RISK

14. _____ is a normal cellular process in which a sequence of events eventually results in the death of a cell.

a. Thanatopis
b. Apopotsis
c. Autophagy
d. Autolysis
e. Amphipathis

Answer: b
Difficulty: Moderate
Bloom's Taxonomy: Knowledge

7.2 CHARACTERISTICS OF CELL COMMUNICATION SYSTEMS WITH SURFACE RECEPTORS

15. The two major categories of extracellular signal molecules that bind to cell surface receptors are _____.

a. peptide hormones and steroid hormones
b. steroid hormones and neurotransmitters
c. neurotransmitters and vitamins
d. growth hormones and vitamins
e. peptide hormones and neurotransmitters

Answer: e
Difficulty: Moderate
Bloom's Taxonomy: Knowledge

16. Cells that release peptide hormones are know as _____.

a. gonads
b. germ cells
c. gland cells
d. connective cells
e. glial cells

Answer: c
Difficulty: Moderate
Bloom's Taxonomy: Knowledge

17. The surface receptors that recognize and bind signal molecules are _____.

a. glycoproteins
b. glycolipids
c. phospholipids
d. integrins
e. cadherins

Answer: a
Difficulty: Moderate
Bloom's Taxonomy: Knowledge

18. Recognition of a chemical signal by a receptor protein in a membrane is most similar to _____.

a. mRNA specifying the sequence of amino acids in a specific metabolic pathway operating in an organelle
b. binding of a specific substrate to the active site of an enzyme
c. turning on transcription of a gene
d. allosteric regulation of proteins
e. an enzyme requiring a specific optimum pH and temperature for activity

Answer: b
Difficulty: Difficult
Bloom's Taxonomy: Comprehension

19. In most cases, protein kinases _____.

a. add phosphate groups to proteins
b. bind cGMP
c. stimulate adenylyl cyclase
d. polymerize amino acids
e. hydrolyze proteins

Answer: a
Difficulty: Difficult
Bloom's Taxonomy: Comprehension

20. The effects of protein kinases are reversed by another group of enzymes called _____.

a. hydrolases
b. catalases
c. isomerases
d. phosphatases
e. proteases

Answer: d
Difficulty: Moderate
Bloom's Taxonomy: Knowledge

21. Once transduction of a signal is complete, the receptor and its bound signal molecule are removed from the cell surface by _____.

a. hydrolysis
b. diffusion
c. endocytosis
d. pinocytosis
e. exocytosis

Answer: c
Difficulty: Moderate
Bloom's Taxonomy: Knowledge

7.3 SURFACE RECEPTORS WITH BUILT-IN PROTEIN KINASE ACTIVITY: RECEPTOR TYROSINE KINASES

22. The protein kinase activity of the receptor protein-tyrosine kinase is located _____.

a. on the extracellular surface of the membrane
b. on the cytoplasmic side of the cell membrane
c. within the hydrophobic portion of the lipid bilayer
d. in the nuclear membrane
e. None of the answers are correct.

Answer: b
Difficulty: Moderate
Bloom's Taxonomy: Knowledge

23. Receptor protein-tyrosine kinases add phosphate groups to which amino acid on the receptor?

a. serine
b. glycine
c. threonine
d. tryptophan
e. tyrosine

Answer: e
Difficulty: Moderate
Bloom's Taxonomy: Knowledge

24. What is the immediate result once a signal molecule binds to a receptor protein-tyrosine kinase?

a. receptor dimerization
b. receptor trimerization
c. receptor denaturation
d. receptor polymerization
e. receptor hydrolysis

Answer: a
Difficulty: Moderate
Bloom's Taxonomy: Knowledge

25. The receptors for a group of peptide signaling molecules known as growth factors are typically _____.

a. receptor tyrosine kinases
b. cyclic AMP
c. neurotransmitters
d. G-protein-linked receptors
e. ligand-gated ion channels

Answer: a
Difficulty: Moderate
Bloom's Taxonomy: Knowledge

26. Which one of the following would be the most likely result of a peptide growth factor binding to its receptor?

a. phosphorylase activity
b. adenylyl cyclase activity
c. protein kinase activity
d. GTPase activity
e. protein phosphatase activity

Answer: c
Difficulty: Moderate
Bloom's Taxonomy: Comprehension

7.4 G PROTEIN-COUPLED RECEPTORS

27.   The toxin of *Vibrio cholerae* causes severe diarrhea because it _____.

a.   modifies DAG and activates a cascade of protein kinases
b.   activates phospholipase to produce inositol triphosphate
c.   decreases the cytosolic concentration of Ca^{++}
d.   modifies a G protein involved in regulating salt and water secretion
e.   binds with adenylyl cyclase and triggers formation of cAMP

Answer: d
Difficulty: Moderate
Bloom's Taxonomy: Knowledge

28.   Which one of the following is a common second messenger?

a.   cGTP
b.   cMHC
c.   cATP
d.   cAMP
e.   more than one answer is correct

Answer: d
Difficulty: Moderate
Bloom's Taxonomy: Knowledge

29.   Many signal transduction pathways utilize second messengers to _____.

a.   transport a signal through the lipid bilayer of the plasma membrane
b.   relay a signal from the outside of the cell to the inside
c.   relay a signal from the inside of a cell membrane to the outside of the cell
d.   decrease the message once the signal molecules have left the receptor
e.   relay the message from the inner surface of the plasma membrane throughout the cytoplasm

Answer: e
Difficulty: Moderate
Bloom's Taxonomy: Comprehension

30.   In the cAMP pathway, the G protein stimulates _____.

a.   adenylyl cyclase
b.   diacylglycerol (DAG)
c.   phospholipase C
d.   inositol triphosphate
e.   phosphodiesterase

Answer: a
Difficulty: Moderate
Bloom's Taxonomy: Knowledge

31.   Once activated, cAMP is quickly degraded to AMP by _____ which results in switching off the signal pathway.

a.   adenylyl cyclase
b.   diacylglycerol (DAG)
c.   phospholipase C
d.   acetylcholinesteras
e.   phosphodiesterase

Answer: e
Difficulty: Moderate
Bloom's Taxonomy: Knowledge

FOCUS ON RESEARCH: DETECTING CALCIUM RELEASE IN CELLS

32.   Several hundred types of G-protein coupled receptors are associated with the sense of _____.

a.   vision
b.   taste
c.   smell
d.   hearing
e.   touch

Answer: c
Difficulty: Moderate
Bloom's Taxonomy: Knowledge

7.5 PATHWAYS TRIGGERED BY INTERNAL RECEPTORS: STEROID HORMONE RECEPTORS

33.   The reason that some individual hormones have so many different effects is that

a.   they influence gene transcription.
b.   they trigger a second messenger system that produces a cascade effect.
c.   there are a great many different cells in different tissues that have specific receptors for the hormone.
d.   the hormone is carried throughout the body and only a small amount is needed to produce its effect.
e.   all of these

Answer: c
Difficulty: Moderate
Bloom's Taxonomy: Knowledge

34.   Steroid hormones and thyroid hormones do not require a membrane receptor because they

a.   are small enough to pass directly through the membrane.
b.   are soluble in the lipid bilayer.
c.   pass through special membrane channels.
d.   are water-soluble.
e.   dissolve in the cholesterol of the membranes.

Answer: b
Difficulty: Moderate
Bloom's Taxonomy: Knowledge

35. Steroid hormones

a. are proteins.
b. include testosterone, estrogens, and cortisol.
c. never activate second messengers.
d. never alter membrane transport of ions.
e. all of these

Answer: b
Difficulty: Moderate
Bloom's Taxonomy: Knowledge

36. The release of cyclic AMP as a second messenger is a response to _____ hormones.

a. peptide
b. steroid
c. protein
d. peptide and protein
e. peptide, steroid, and protein

Answer: a
Difficulty: Moderate
Bloom's Taxonomy: Knowledge

7.6 INTEGRATION OF CELL COMMUNICATION PATHWAYS

37. A common second messenger molecule is

a. a steroid compound.
b. cyclic AMP.
c. ADP.
d. prostaglandin.
e. intermedin.

Answer: b
Difficulty: Moderate
Bloom's Taxonomy: Knowledge

Integrative Multiple-Choice

38. In general, a cell receiving a message undergoes three stages of cell signaling. The stages are _____.

a. the paracrine, autocrine, and local stages.
b. signal reception, signal transduction, and cellular response
c. signal reception, nucleus disintegration, and apoptosis
d. signal reception, cellular response, and cell division
e. the alpha, beta, and gamma stages

Answer: b
Difficulty: Moderate
Bloom's Taxonomy: Knowledge
Source: Section 7.1

39. The insulin receptor is an example of _____.

a. a G-protein coupled receptor
b. a hydrophobic receptor
c. a hormone receptor
d. a receptor tyrosine kinase
e. an ion channel receptor

Answer: d
Difficulty: Moderate
Bloom's Taxonomy: Knowledge
Source: Section 7.3, 7.4, and 7.5

Sequence

Arrange the following steps in the pathway activated by G-protein-coupled receptors. Label 1 as the first step through 5 as the last step.

40. _____   Activation of effector
41. _____   Activation of protein kinases
42. _____   Receptor binds first messenger
43. _____   Production of second messenger
44. _____   Activation of G protein

Answers: 40. 3   41. 5   42. 1   43. 4   44. 2

Difficulty: Difficult
Bloom's Taxonomy: Knowledge
Source: Section 7.4

Short Answer

45. Why is it so necessary for cells to communicate with each other in a regulated way?

Answer: Regulated communication is responsible for the controlled growth and development of an animal as well as the integrated activities of tissues and organs.
Difficulty: Moderate
Bloom's Taxonomy: Knowledge
Source: Section 7.1

46. What is the function of signal transduction?

Answer: Signal transduction refers to the process by which a cell converts a chemical signal into an intracellular signal typically through a series of biochemical reactions involving enzymes. This ultimately results a specific cellular response occurring.
Difficulty: Moderate
Bloom's Taxonomy: Knowledge, Comprehension
Source: Section 7.1

47. Are some of the functions of apoptosis in vertebrate biology?

Answer: Apoptosis is a natural part of many developmental pathways. These include loss of webbing between fingers and toes during embryonic development as well as proper brain development.
Difficulty: Moderate
Bloom's Taxonomy: Knowledge, Comprehension
Source: Section: Insight from the Molecular Revolution

48. How does a signal molecule bring about specific changes in the cell to which it binds?

Answer: When a signal molecule binds to a surface receptor, the molecular structure of the receptor is changed so that it transmits the signal across the cell membrane, consequently activating the cytoplasmic end of the receptor. The activated receptor then initiates the first step in a cascade of molecular events that triggers a cellular response.
Difficulty: Moderate
Bloom's Taxonomy: Knowledge, Comprehension
Source: Section: 7.2

49. Amplification is an important characteristic of signal transduction pathways involving surface receptors. What is amplification and how is it accomplished in the cell?

Answer: Amplification is an increase in the magnitude of each step in a signal transduction pathway. Amplification occurs because many of the proteins that carry out individual steps in the pathway are enzymes. Once activated, each enzyme can activate hundreds or thousands of proteins in the signal transduction cascade.
Difficulty: Moderate
Bloom's Taxonomy: Knowledge
Source: Section: 7.2

Select the Exception

50. Cells communicate with each other by various mechanisms. All of the following are examples of ways cells communicate EXCEPT _____.

a. gap junctions
b. transfer chain proteins
c. recognition of specific membrane surface molecules
d. cell adhesion molecules
e. chemical messengers

Answer: b
Difficulty: Moderate
Bloom's Taxonomy: Knowledge

51. Each of the following is a type of cell surface receptor EXCEPT _____.

a. receptor tyrosine kinases
b. enzyme-linked receptors
c. G-protein coupled receptors
d. steroid hormone receptors
e. chemically-gated ion channels

Answer: d
Difficulty: Difficult
Bloom's Taxonomy: Knowledge, Comprehension
Source: Section

52. Binding of the peptide hormone insulin to its receptor tyrosine kinase results in each of the following cellular activities EXCEPT _____.

a. cell division
b. glucose uptake
c. regulation of metabolic reactions
d. cell growth
e. photosynthesis

Answer: e
Difficulty: Moderate
Bloom's Taxonomy: Knowledge
Source: Section 7.3

53. All of the following are true of cell communication systems EXCEPT _____.

a. cell signaling occurred early in the evolution of life
b. communicating cells may be found far a part or adjacent to each other
c. most signal receptors are found bound to the outer nuclear membrane
d. protein phosphorylation is a major mechanism of signal transduction
e. in response to a signal, a cell can impact transcription of RNA

Answer: c
Difficulty: Moderate
Bloom's Taxonomy: Comprehension
Source: Section 7.3

54. Each of the following are steps in the receptor kinase signaling pathway EXCEPT _____.

a. a signal molecule binds to a membrane receptor protein
b. protein kinase is activated
c. GTP donates a phosphate group to an inactive protein kinase
d. a specific protein is activated via phosphorylation
e. a cellular response is initiated

Answer: c
Difficulty: Difficult
Bloom's Taxonomy: Knowledge
Source: Section 7.3

55. All of the following statements apply to G proteins EXCEPT _____.

a. G proteins transmit a signal from the cell surface to the interior of the cell
b. all G proteins have a similar structure
c. G proteins do not use a second messenger but transmit the signal directly to the nucleus
d. G proteins act to amplify the signal creating a cascade response in the cell
e. G proteins are the target of toxins released by some pathogenic bacteria

Answer: c
Difficulty: Difficult
Bloom's Taxonomy: Knowledge

Labeling

Label the following components of a surface receptor.

56. _____ extracellular signal molecule
57. _____ site triggering cellular response, in inactive state
58. _____ transmembrane segment
59. _____ cytoplasmic segment
60. _____ extracellular segment of receptor
61. _____ signal binding site

Answer: 56. A 57. C   58. E   59. D   60. F   61. B
Difficulty: Moderate
Bloom's Taxonomy: Knowledge
Source: Figure 7.4a, Section 7.2

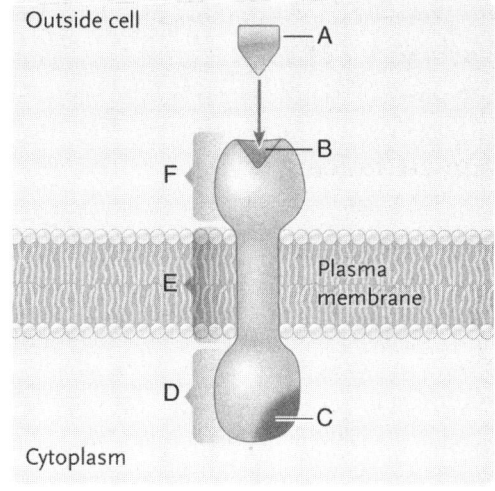

8

HARVESTING CHEMICAL ENERGY: CELLULAR RESPIRATION

Multiple-Choice

WHY IT MATTERS

1.   It has been thought that many diseases now associated with aging are related to malfunctioning mitochondria. Why are the mitochondria so important to all cells?

a.   They produce energy in the form of ATP.
b.   They generate heat to keep the body warm.
c.   They are the source of all human disease.
d.   They are extremely large.
e.   They are located only in vital organs.

Answer: a
Difficulty: Easy
Bloom's Taxonomy: Comprehension

2.   When individuals have mitochondrial disorders, why are the skeletal and heart muscles and the brain most often affected?

a.   They are the most important organs.
b.   They have the highest energy needs.
c.   They are generally very fragile.
d.   They have fewer mitochondria in the cell.
e.   They are the most complex organs.

Answer: b
Difficulty: Easy
Bloom's Taxonomy: Comprehension

3.   Which of the following types of organisms do NOT depend on mitochondria to survive?

a.   humans
b.   plants
c.   fungus
d.   protists
e.   bacteria

Answer: e
Difficulty: Easy
Bloom's Taxonomy: Analysis

4.   What is the ultimate source of chemical energy for all living organisms?

a.   photosynthesis
b.   cellular respiration
c.   ATP
d.   GTP
e.   glucose

Answer: c
Difficulty: Easy
Bloom's Taxonomy: Knowledge

5.   Some organisms are not able to live in an environment where there is oxygen; these types of organisms are called obligate anaerobes. Which of the following proposals is the most likely for how they survive without oxygen?

a.   They are able to survive using less energy than aerobes.
b.   All of their ATP is imported into the cell from an external source.
c.   Sulfur is used instead of oxygen because it is chemically similar.
d.   These organisms use photosynthesis to produce energy.
e.   Their mitochondria are damaged, and consequently they are short-lived.

Answer: c
Difficulty: Difficult
Bloom's Taxonomy: Synthesis

8.1 OVERVIEW OF CELLULAR ENERGY METABOLISM

6.   For a molecule to be reduced, it can gain electrons from the environment. How else is a molecule reduced?

a.   gains hydrogens
b.   loses hydrogens
c.   gains oxygen
d.   loses oxygen
e.   both loses hydrogens and gains oxygen

Answer: a
Difficulty: Easy
Bloom's Taxonomy: Knowledge

7.   Where would you expect to find proteins responsible for transporting glucose into the cell?

a.   plasma membrane of the cell
b.   inner mitochondrial membrane
c.   outer mitochondrial membrane
d.   nuclear membrane
e.   lysosomal membrane

Answer: a
Difficulty: Moderate
Bloom's Taxonomy: Analysis

8.   What is the ultimate fate of oxygen gas (O_2) in cellular respiration?

a.   It is respired as CO_2.
b.   It is converted to water.
c.   It is attached to glucose.
d.   It is attached to pyruvate.
e.   It accepts electrons in glycolysis.

Answer: b
Difficulty: Moderate
Bloom's Taxonomy: Comprehension

9.   In the process of aerobic metabolism, carbon containing molecules are broken down and the energy from the electrons is used for what purpose?

a.   to directly supply the energy needs of an organism
b.   to generate a proton gradient
c.   to alter enzyme structure
d.   to heat the organism in a cold environment

Answer: b
Difficulty: Moderate
Bloom's Taxonomy: Comprehension

10.   Oxygen is able to allow electrons of a very low energy level to combine with it at the end of cellular respiration and ultimately make water. How are the specific properties of oxygen beneficial to the organism that uses it as a final electron acceptor?

a.   Oxygen is highly reactive and readily accepts electrons.
b.   Oxygen is strongly electronegative and helps pull the electrons though the electron transport chain.
c.   Oxygen allows a maximum output of energy from ATP synthesis.
d.   Oxygen is the only molecule that can act as a final electron acceptor.

Answer: b
Difficulty: Moderate
Bloom's Taxonomy: Comprehension

11.   Oxidative phosphorylation is the process by which _____.

a.   high energy NADH is made to supply the cell with its needed energy
b.   a final electron acceptor is used indirectly to facilitate the production of ATP
c.   ATP is made using high energy intermediates of cellular respiration
d.   specific enzymes are regulated to control cellular respiration
e. NAD^+ is regenerated to allow glycolysis to continue

Answer: b
Difficulty: Easy
Bloom's Taxonomy: Knowledge

12.   During which of the following processes is ATP NOT made?

a.   glycolysis
b.   pyruvate oxidation
c.   citric acid cycle
d.   electron transfer system
e.   both b and d

Answer: e
Difficulty: Moderate
Bloom's Taxonomy: Knowledge

13.   Which answer best describes energy flow in biological systems as described in the text?

a.   glucose $\rightarrow$ G3P $\rightarrow$ NADH $\rightarrow$ ATP
b.   bacteria $\rightarrow$ archaea $\rightarrow$ plants $\rightarrow$ animals
c. NAD^+ $\rightarrow$ NADH $\rightarrow$ ADP $\rightarrow$ ATP
d.   G3P $\rightarrow$ glucose $\rightarrow$ ATP $\rightarrow$ NAD^+
e.   pyruvate oxidation $\rightarrow$ glycolysis $\rightarrow$ fermentation $\rightarrow$ citric acid cycle

Answer: a
Difficulty: Moderate
Bloom's Taxonomy: Analysis

14.   In the absence of ATP synthase, animal cells would not be able to _____.

a.   create a proton gradient
b.   hydrolyze glucose to G3P
c.   carry out oxidative phosphorylation
d.   produce ATP
e.   carry out pyruvate oxidation

Answer: c
Difficulty: Difficult
Bloom's Taxonomy: Evaluation

15.   How are NADH and $FADH_2$ related?

a.   They both directly produce ATP.
b.   They are both used in glycolysis.
c.   They both contain high energy phosphates.
d.   They both contain high energy electrons.
e.   They are both in the oxidized form.

Answer: d
Difficulty: Moderate
Bloom's Taxonomy: Comprehension

16.   How does cellular respiration differ in prokaryotes and eukaryotes?

a.   Eukaryotes use substrate-level phosphorylation; prokaryotes use oxidative phosphorylation.
b.   Eukaryotes use mitochondria; prokaryotes use their plasma membrane.
c.   Eukaryotes use NAD^+/NADH; prokaryotes use FAD^+/$FADH_2$.
d.   Eukaryotes do not all use oxygen; prokaryotes only use oxygen.
e.   Eukaryotes use only glucose; prokaryotes use only galactose.

Answer: b
Difficulty: Easy
Bloom's Taxonomy: Comprehension

17. What type of chemical reaction must occur for electrons to flow from one molecule to the next and supplies the energy for metabolism?

a. acid/base
b. reduction/oxidation
c. exothermic
d. trimolecular
e. phosphorylation

Answer: b
Difficulty: Easy
Bloom's Taxonomy: Comprehension

8.2 GLYCOLYSIS

18. During glycolysis, glucose molecules are broken down by breaking the carbon-hydrogen bonds that are present and forming carbon-oxygen bonds. What is this chemical phenomena called?

a. oxidation
b. reduction
c. protonation
d. electrolysis

Answer: a
Difficulty: Easy
Bloom's Taxonomy: Application

19. The enzymes responsible for hydrolyzing glucose into G3P are found in what part of the cell?

a. cytosol
b. mitochondria
c. rough ER
d. nucleus
e. cell membrane

Answer: a
Difficulty: Moderate
Bloom's Taxonomy: Application

20. If the inner membrane of the mitochondria were compromised in some way, what effect would this have on cellular respiration?

a. The transport of electrons across the inner mitochondrial membrane would not occur.
b. The proton gradient across the inner mitochondrial membrane would dissipate.
c. The ATP synthase enzyme would be denatured.
d. The cell would generate more ATP than before.
e. ATP would no longer be made anywhere in the cell by any mechanism.

Answer: b
Difficulty: Moderate
Bloom's Taxonomy: Synthesis

21. The final product of glycolysis is _____.

a. glucose
b. fructose
c. glyceraldehyde-3-phosphate
d. pyruvate
e. carbon dioxide

Answer: d
Difficulty: Easy
Bloom's Taxonomy: Knowledge

22. The various steps of glycolysis require a number of enzymes and small molecules. Which of the following molecules are NOT required for glycolysis to take placc?

a. FAD
b. NAD^+
c. ATP
d. ADP
e. phosphofructokinase

Answer: a
Difficulty: Difficult
Bloom's Taxonomy: Knowledge

23. A toxic substance (compound x) had been found to inhibit glucose transport into mammalian cells. If this substance was administered to a hamster, the likely cause of death would be from a lack of _____.

a. NAD^+ production
b. ATP production
c. GTP production
d. CO_2 production
e. FAD production

Answer: b
Difficulty: Easy
Bloom's Taxonomy: Synthesis

24. If all of the NADH inside a cell was depleted, which of the following parts of cellular respiration could you NOT go through?

a. glycolysis
b. pyruvate oxidation
c. electron transfer system
d. citric acid cycle
e. substrate-level phosphorylation

Answer: c
Difficulty: Moderate
Bloom's Taxonomy: Analysis

25. The small molecule intermediate phosphoenol-pyruvate (PEP) is unusual due to which chemical characteristic?

a. supplies a high energy phosphate to phosphorylate ADP
b. provides high energy electrons to make ATP
c. its ability to carry out redox reactions
d. regulates the activity of phosphofructokinase
e. donates an electron to NAD^+ and FAD

Answer: a
Difficulty: Difficult
Bloom's Taxonomy: Knowledge

26. Which enzyme in the glycolysis pathway acts as a switch that can be regulated by ATP, ADP, and NADH?

a. pyruvate kinase
b. triosephosphate isomerase
c. aldolase
d. ATP synthase
e. phosphofructokinase

Answer: e
Difficulty: Moderate
Bloom's Taxonomy: Knowledge

27. What directly supplies the energy for oxidative phosphorylation?

a. ATP
b. NADH
c. glucose
d. proton gradient
e. ATP synthetase

Answer: d
Difficulty: Moderate
Bloom's Taxonomy: Analysis, Knowledge

28. What supplies the electrons for oxidative phosphorylation?

a. ATP
b. NADH and $FADH_2$
c. glucose
d. proton gradient
e. ATP synthetase

Answer: b
Difficulty: Moderate
Bloom's Taxonomy: Analysis, Knowledge

29. An enzyme that is able to transfer a phosphate group to another molecule is called what?

a. kinase
b. enolase
c. dehydrogenase
d. polymerase
e. phosphotase

Answer: a
Difficulty: Easy
Bloom's Taxonomy: Knowledge

30. Which of these molecules has the most potential energy?

a. glucose
b. pyruvate
c. ATP
d. NADH
e. $FADH_2$

Answer: a
Difficulty: Moderate
Bloom's Taxonomy: Comprehension

8.3 PYRUVATE OXIDATION AND THE CITRIC ACID CYCLE

31. During which stages of cellular respiration is CO_2 released?

a. glycolysis
b. pyruvate oxidation
c. citric acid cycle
d. electron transport system
e. both pyruvate oxidation and citric acid cycle

Answer: e
Difficulty: Easy
Bloom's Taxonomy: Knowledge

32. Ultimately, the carbon molecules in pyruvate end up as what molecule?

a. CO_2
b. acetate
c. ATP
d. CoA
e. NADH

Answer: a
Difficulty: Moderate
Bloom's Taxonomy: Comprehension

33. What is the function of NADH and $FADH_2$?

a. Both release energy for glycolysis to proceed forward.
b. Both provide electrons to the electron transfer system.
c. Both produce ATP by substrate-level phosphorylation.
d. NADH delivers electrons, while $FADH_2$ supplies H^+.
e. NADH is found only in the cytosol and $FADH_2$ only in the matrix.

Answer: b
Difficulty: Moderate
Bloom's Taxonomy: Analysis

34. The electrons that are present in NADH and $FADH_2$ are very high energy and theoretically sufficient to produce ATP alone. Why then have biological systems evolved in a way to use the various steps of cellular respiration to produce ATP?

a. It would be less efficient to produce ATP directly from NADH and $FADH_2$.
b. There is insufficient amounts of ADP to phosphorylate and make ATP.
c. All living organisms must use oxygen and therefore must carry out cellular respiration.
d. The steps in respiration allow for a controlled release of energy in small increments.
e. Not all cells have NADH and $FADH_2$ present in their cytoplasm.

Answer: d
Difficulty: Moderate
Bloom's Taxonomy: Evaluation

35. What molecule is responsible for carrying the acetyl group from pyruvate into the citric acid cycle?

a. NADH
b. $FADH_2$
c. ATP
d. CoA
e. oxaloacetate

Answer: d
Difficulty: Easy
Bloom's Taxonomy: Knowledge

36. When starting with a single molecule of glucose, how many times does oxidation occur during the conversion of pyruvate to acetyl CoA in the process of cellular respiration?

a. 1
b. 2
c. 3
d. 4
e. 5

Answer: b
Difficulty: Easy
Bloom's Taxonomy: Comprehension

37. There are eight total steps in the citric acid cycle, and pyruvate is completely oxidized in steps three and four. What is the purpose of the last four steps that occur in the cycle?

a. to replenish the supplies of NAD^+ and FAD
b. to break down glucose into a three-carbon molecule
c. to regenerate oxaloacetate to attach another acetate molecule
d. to produce ATP by substrate-level phosphorylation
e. to produce ATP by oxidative phosphorylation

Answer: c
Difficulty: Moderate
Bloom's Taxonomy: Analysis

38. What is the fate of CoA after it delivers an acetyl group into the citric acid cycle?

a. It is degraded and used for energy.
b. It is recharged with another acetate.
c. It is used in protein synthesis.
d. It remains in an inactive form until the cell dies.
e. It is reused to start glycolysis.

Answer: b
Difficulty: Moderate
Bloom's Taxonomy: Comprehension

39. Citrate synthase is the first enzyme in the citric acid cycle and is also able to be regulated to control the amount of ATP that is produced as a result of the citric acid cycle. Why is early regulation the most beneficial method for the cell?

a. There is little wasted energy by regulating early steps.
b. More ATP can be produced before the system is shut off.
c. NADH is rarely found in the cell and needs to be conserved.
d. The first enzyme is more amenable to regulation.
e. There is no possible way to control the last enzyme in a pathway.

Answer: a
Difficulty: Moderate
Bloom's Taxonomy: Analysis

40. For every glucose molecule that goes through cellular respiration, how many times is a carbon molecule fully oxidized to CO_2 in the citric acid cycle?

a. 1
b. 2
c. 3
d. 4
e. 5

Answer: d
Difficulty: Moderate
Bloom's Taxonomy: Comprehension

41. Which molecule(s) is/are responsible for delivering the high energy electrons from the citric acid cycle to the electron transfer system?

a. NADH only
b. $FADH_2$ only
c. Both NADH and $FADH_2$
d. Cyt C and Q
e. ATP and ADP

Answer: c
Difficulty: Easy
Bloom's Taxonomy: Knowledge

42. We study cellular respiration because it is one of the most important pathways in biology. In fact, nearly all carbohydrates at some point in their catabolism are directed through cellular respiration. Why is it NOT necessary to have multiple independent pathways to break down different molecules?

a. Using cellular respiration is theoretically the most efficient way to break down sugars and other molecules.
b. Oxygen must be used in the breakdown of all molecules in order to yield ATP.
c. Greater complexity would lead to an eventual failure of the biological system.
d. Most biological cells only catabolize one or two different types of sugars and only need one main pathway.
e. The molecules that are degraded are all structurally similar and relatively easily interconverted.

Answer: e
Difficulty: Moderate
Bloom's Taxonomy: Analysis

43. Why is there more energy stored in fats as opposed to sugars?

a. Fats are more prevalent in the body.
b. Fats are more chemically related to ATP.
c. Fats are processed differently after the citric acid cycle.
d. Fats are in a more reduced state.
e. Fats are in a more oxidized state.

Answer: d
Difficulty: Moderate
Bloom's Taxonomy: Analysis

44. What are the two main steps that must happen to a protein in order for it to be catabolized and broken down by the citric acid cycle?

a. The protein must be phophorylated and broken into amino acids.
b. The protein must be glycosylated and transported to the liver.
c. The protein must be hydrolyzed and the amino group removed.
d. The amino group must be removed followed by the carboxyl group.
e. The protein is broken into monomers and converted to acetate.

Answer: c
Difficulty: Easy
Bloom's Taxonomy: Knowledge

8.4 THE ELECTRON TRANSFER SYSTEM AND OXIDATIVE PHOSPHORYLATION

45. Oxygen acts as a final electron acceptor in respiration and is ultimately converted into which molecule?

a. water
b. ATP
c. CO_2
d. glucose
e. phosphate

Answer: a
Difficulty: Moderate
Bloom's Taxonomy: Comprehension

46. Glycolysis, pyruvate oxidation, and the citric acid cycle all have this molecule in common as one of their products.

a. CO_2
b. H_2O
c. ATP
d. $FADH_2$
e. NADH

Answer: e
Difficulty: Easy
Bloom's Taxonomy: Analysis

47. Compared to the cytoplasm, the mitochondrial matrix could be described as having _____.

a. a low pH and high pyruvate concentration
b. a high pH and high pyruvate concentration
c. a low pH and low pyruvate concentration
d. a high pH and low pyruvate concentration
e. the same pH and pyruvate concentration

Answer: d
Difficulty: Difficult
Bloom's Taxonomy: Evaluation

48. What must be true when electrons are passed between protein complexes?

a. Electrons are directly passed from one complex to the next by physical contact.
b. The oxidized electron carrier passes on the electron and becomes reduced.
c. A proton is translocated to the inner mitochondrial space.
d. The matrix decreases in pH.
e. The reduced electron carrier passes on the electron and becomes oxidized.

Answer: e
Difficulty: Difficult
Bloom's Taxonomy: Comprehension

49. What are the functions of cytochrome *c* and ubiquinone?

a. They translocate protons from the matrix to the inner mitochondrial space.
b. They shuttle electrons between the protein complexes.
c. They synthesize water from molecular oxygen.
d. They produce ATP by substrate-level phosphorylation.
e. They produce ATP by oxidative phosphorylation.

Answer: b
Difficulty: Easy
Bloom's Taxonomy: Knowledge

50. Coenzyme Q (also known as ubiquinone or CoQ) is often found in lotions and moisturizers. Given what you know about its role in cellular respiration, infer what it might be doing in these products.

a. It helps glycolysis proceed faster in the cytosol of skin cells.
b. It absorbs dangerous free radicals (loose electrons) that can cause aging.
c. It decreases the rate of metabolism and slows cellular growth.
d. It is absorbed and used by cells with a high metabolism.
e. It causes older cells to spontaneously die and regenerates new cells.

Answer: b
Difficulty: Difficult
Bloom's Taxonomy: Synthesis

51. The enzyme succinate dehydrogenase is located in the inner mitochondrial membrane and is directly involved in what two steps of cellular respiration?

a. glycolysis and pyruvate oxidation
b. pyruvate oxidation and citric acid cycle
c. citric acid cycle and electron transfer system
d. electron transfer system and glycolysis
e. electron transfer system and fermentation

Answer: c
Difficulty: Moderate
Bloom's Taxonomy: Comprehension

52. What is directly responsible for pumping protons out of the matrix in the mitochondria?

a. protein complexes I, II, III, and IV
b. cytochrome *c* and ubiquinone
c. protein complexes I and III
d. protein complexes I, III, and IV
e. NADH and FADH$_2$

Answer: d
Difficulty: Moderate
Bloom's Taxonomy: Knowledge

53. What is the proton-motive force?

a. the force needed to move protons into the inner mitochondrial space
b. the amount of energy required to protonate a glucose molecule
c. the free energy associated with the removal of hydrogen from NADH
d. the combination of a proton and voltage gradient across the membrane
e. the synthesis of ATP from a proton gradient

Answer: d
Difficulty: Moderate
Bloom's Taxonomy: Knowledge

54. What is chemiosmosis?

a. the synthesis of ATP by ATP synthase, using a proton gradient
b. the production of NADH and FADH$_2$ in cellular respiration
c. the build up of acetate in the cytoplasm in the absence of oxygen
d. the breakdown of oxygen as it enters the mitochondria
e. the release of CO_2 as glucose is oxidized in cellular respiration

Answer: a
Difficulty: Easy
Bloom's Taxonomy: Knowledge

55. What powers ATP synthase directly?

a. electron transfer
b. NADH and FADH$_2$
c. carbohydrate metabolism
d. proton gradient
e. protein complexes

Answer: d
Difficulty: Easy
Bloom's Taxonomy: Knowledge

56. Where is ATP synthase located in non-photosynthetic eukaryotes?

a. outer membrane of the cell
b. nuclear envelope
c. rough endoplasmic reticulum
d. matrix of the mitochondria
e. inner mitochondrial membrane

Answer: e
Difficulty: Easy
Bloom's Taxonomy: Knowledge

57. What directly supplies the electrons for the electron transfer system?

a. ATP and ADP
b. FADH$_2$ and NADH
c. pyruvate and acetate
d. various enzymes
e. oxygen and water

Answer: b
Difficulty: Easy
Bloom's Taxonomy: Knowledge

58. What piece of the ATP synthase contains the channel for H^+ to flow through?

a. the basal unit
b. the headpiece
c. the stalk
d. the lollipop
e. the three catalytic sites

Answer: a
Difficulty: Moderate
Bloom's Taxonomy: Knowledge

59. Assuming the pH of the matrix was significantly lower than the inner mitochondrial space, how would the ATP synthase function differently?

a. It would require an ion to stabilize it.
b. ATP synthase would no longer function properly.
c. It would hydrolyze ATP to form ADP.
d. It would function exactly the same.
e. The electron transport chain would be uncoupled.

Answer: c
Difficulty: Moderate
Bloom's Taxonomy: Evaluation

60. What part of the ATP synthase is responsible for catalyzing ATP formation?

a. the basal unit
b. the headpiece
c. the stalk
d. the lollipop
e. the electrons

Answer: b
Difficulty: Moderate
Bloom's Taxonomy: Knowledge

61. If all the protons that powered the ATP synthase were not dealt with in some fashion, the proton gradient would be destroyed and ATP could no longer be produced by oxidative phosphorylation. What happens to these protons as they reenter the matrix?

a. They are attached to NAD^+ and FAD.
b. They combine with oxygen to form water.
c. They synthesize ATP by substrate-level phosphorylation.
d. They help in the production of CO_2.
e. They regenerate Coenzyme A.

Answer: b
Difficulty: Easy
Bloom's Taxonomy: Comprehension

62. How many catalytic sites does ATP synthase have?

a. 1
b. 2
c. 3
d. 4
e. 5

Answer: c
Difficulty: Easy
Bloom's Taxonomy: Knowledge

63. Why does NADH produce more energy than $FADH_2$?

a. $FADH_2$ donates electrons to protein complex III as opposed to complex II.
b. $FADH_2$ requires more ATP to produce it and gives more energy back.
c. NADH and $FADH_2$ are synthesized in different steps of cellular respiration.
d. NADH donates electrons to a higher energy acceptor in the electron transfer chain.
e. NADH supplies fewer electrons that are of a higher energy state than $FADH_2$.

Answer: d
Difficulty: Moderate
Bloom's Taxonomy: Analysis

64. When you consider the contribution of NADH and $FADH_2$, how many of the 32 total ATP molecules produced in cellular respiration come from the citric acid cycle?

a. 32
b. 28
c. 24
d. 20
e. 16

Answer: d
Difficulty: Difficult
Bloom's Taxonomy: Knowledge

65. How efficient is cellular respiration in extracting the energy stored in the bonds of glucose?

a. 20 percent
b. 30 percent
c. 40 percent
d. 50 percent
e. 80 percent

Answer: b
Difficulty: Easy
Bloom's Taxonomy: Knowledge

66. There are different shuttle systems to get high energy electrons from the NADH in the cytoplasm to the NADH in the mitochondria. Why is this beneficial?

a. This allows for 100% efficiency in cellular respiration.
b. Each step of cellular respiration can happen faster.
c. It allows for regulation of ATP production by individual cells.
d. It diverts more electrons to ATP synthase to produce more ATP.
e. This is actually a detrimental event that the cell must overcome.

Answer: c
Difficulty: Moderate
Bloom's Taxonomy: Comprehension

INSIGHTS FROM THE MOLECULAR REVOLUTION: KEEPING THE POTATOES HOT

67. If it is true that potato plants use uncoupling proteins (UCPs) in a similar way as mammals, one would expect the following in these plants:

a. increased amounts of ATP production.
b. decreased sugar metabolism.
c. increased internal tissue temperature.
d. decreased mitochondrial catabolism.
e. increased cytosolic pH.

Answer: c
Difficulty: Moderate
Bloom's Taxonomy: Synthesis

8.5 FERMENTATION

68. Why do cells go through fermentation?

a. to replenish ADP
b. to replenish NADH
c. to replenish FAD
d. to replenish O_2
e. to replenish NAD^+

Answer: e
Difficulty: Easy
Bloom's Taxonomy: Knowledge

UNANSWERED QUESTIONS

69. If a new human metabolic disease is found that only allows electrons to be used from NADH and not $FADH_2$, what is a probable explanation for the cause of the disease?

a. a defect in assembly protein genes for complex II of the electron transfer system
b. enzyme defects in glycolysis and the citric acid cycle
c. a deficient amount of cytochrome *c* and coenzyme Q
d. improper regulation of phosphofructokinase
e. inability of oxygen to act as a final electron acceptor

Answer: a
Difficulty: Difficult
Bloom's Taxonomy: Synthesis

Integrative Multiple-Choice

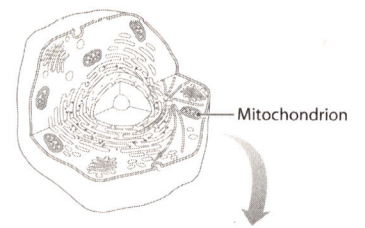

Mitochondrion

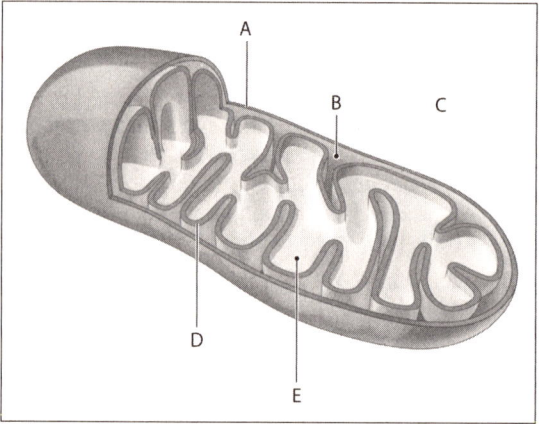

Refer to the illustration above for questions 70–75. In what part of the mitochondrion or cell are the following steps of cellular respiration taking place?

70. _____ glycolysis

71. _____ pyruvate oxidation

72. _____ citric acid cycle

73. _____ electron transfer system

74. _____ ATP synthesis

75. _____ oxidative phosphorylation

A. outer mitochondrial membrane
B. intermembrane compartment
C. cytosol
D. inner mitochondrial membrane
E. matrix

Answer: **70. C   71. E   72. E   73. D**
74. D   75. D

Difficulty: Easy
Bloom's Taxonomy: Knowledge
Source: Section 8.2, 8.3, and 8.4, and Figure 8.4

Sequence

Arrange the following steps of cellular respiration in their proper order. Write the letter of the first step next to 76 and the letter of the last step next to 80.

76. _____   A.   pyruvate oxidation
77. _____   B.   citric acid cycle
78. _____   C.   electron transfer system
79. _____   D.   transport of glucose into cell
80. _____   E.   glycolysis

Answer: ***76. D   77. E   78. A   79. B   80. C***
Difficulty: Easy
Bloom's Taxonomy: Knowledge
Source: Section 8.1

True/False

For questions 81–90, if the statement is true, write a "T" in the blank. If the statement is false, make it correct by changing the underlined word(s) and writing the correct word(s) in the answer blank.

81. _____   Metabolism is best described as <u>how fast mitochondria work</u>.

82. _____   Molecular oxygen is breathed in and expired as <u>carbon dioxide</u>.

83. _____   If during a chemical reaction, a molecule loses hydrogens, it could be described as being <u>oxidized</u>.

84. _____   Both <u>plants and animals</u> possess mitochondria, which are used for cellular respiration to make ATP.

85. _____   Glycolysis takes place in the <u>mitochondria</u>.

86. _____   During glycolysis, ATP is produced by <u>oxidative phosphorylation</u>.

87. _____   During the process of pyruvate oxidation, pyruvate is broken down and the remaining two carbons are attached to <u>Coenzyme A</u>.

88. _____   During pyruvate oxidation, one molecule of <u>ATP</u> is produced.

89. _____   Two mobile electron carriers that are important for the electron transfer system are <u>cytochrome *c* and coenzyme Q</u>.

90. _____   The main purpose of fermentation is to regenerate <u>NADH and FADH$_2$</u>.

*Answer: **81. F, the sum of anabolism and catabolism
82. F, H$_2$O   83. T   84. T   85. F, cytosol
86. F, substrate-level phosphorylation
87. T   88. F, CO$_2$ 89. T   90. F,
NAD$^+$ only***
Difficulty: Moderate
Bloom's Taxonomy: Knowledge, Analysis
Source: Sections Why It Matters, 8.1, 8.2, 8.3, 8.4, 8.5

9
PHOTOSYNTHESIS

Multiple-Choice

WHY IT MATTERS

1.   In Engelmann's classic experiment, why were the oxygen-requiring bacteria clustered around the regions of *Spirogyra* algae that were bathed in red, blue, and violet light?

a.   Chlorophyll is green and thus reflects green light.
b.   The bacteria were immobile and that is where Engelmann happened to place them.
c.   Photosynthesis is most active in those wavelengths of light and thus more oxygen is consumed by the algae in those regions.
d.   Photosynthesis is most active in those wavelengths of light and thus more oxygen is produced by the algae in those regions.
e.   The most energy-rich wavelengths of light are found at the edges of visible and nonvisible light (ultra violet and infrared).

Answer: d
Difficulty: Easy
Bloom's Taxonomy: Knowledge

2.   If you want to buy a colored light bulb for your indoor plants, the *least* effective color would be

a.   red.
b.   blue.
c.   green.
d.   yellow.
e.   orange.

Answer: c
Difficulty: easy
Bloom's Taxonomy: Knowledge

3.   Suppose you explored a new planet and found a photosynthetic organism unlike any on Earth. You repeat Englemann's classic experiment using this new organism in place of *Spirogyra* and find that oxygen-dependent bacteria cluster near the green and yellow portions of the spectrum. What does this tell you?

a.   The sun of this new planet emits different wavelengths of light than our own.
b.   This new organism is using yellow and green light to drive photosynthesis.
c.   This organism is green just like plants on Earth.
d.   This organism utilizes the most energy-rich photons of the spectrum.
e.   All of these statements are correct.

Answer: b
Difficulty: Moderate
Bloom's Taxonomy: Application

9.1 PHOTOSYNTHESIS: AN OVERVIEW

4.   If the hydrogen atoms of water are radioactively labeled, which products of photosynthesis, if any, will also have this radio-labeled hydrogen?

a.   water and oxygen gas both
b.   water only
c.   sugar only
d.   water and sugar both
e.   water, sugar, and oxygen gas

Answer: d
Difficulty: Moderate
Bloom's Taxonomy: Application

5.   Which is the most correct statement about the role of photosynthesis in an ecosystem?

a.   Primary producers convert chemical energy to light energy.
b.   Primary consumers rely directly on light energy.
c.   Primary producers convert light energy to chemical energy as a service to the ecosystem, not for their own gain.
d.   Primary consumers rely on primary producers to make light energy available in a usable form.
e.   Primary producers are heterotrophs and thus make no contribution to the flow of energy into the ecosystem.

Answer: d
Difficulty: Easy
Bloom's Taxonomy: Knowledge, Comprehension

6.   Where does photosynthesis occur in eukaryotes?

a.   plasma membrane
b.   Golgi bodies
c.   cytosol
d.   mitochondria
e.   chloroplasts

Answer: e
Difficulty: Easy
Bloom's Taxonomy: Knowledge

7. Choose the most correct statement about the two phases of photosynthesis.

a. The light-dependent reactions occur in the cytosol; the light-independent reactions occur in the stroma.

b. The products of the light-dependent reactions are ATP, NADPH, and O_2; the products of the light-independent reactions are ADP, $NADP^+$, and sugar.

c. The light-dependent reactions occur during the daylight hours; the light-independent reactions occur when it is dark.

d. The light-dependent reactions produce water as a by-product; the light-independent reactions produce carbon dioxide as a waste product.

e. The products of the light-dependent reactions are ADP, $NADP^+$, and O_2; the products of the light-independent reactions are ATP, NADPH, and sugar.

Answer: b
Difficulty: Easy
Bloom's Taxonomy: Knowledge

9.2 THE LIGHT-DEPENDENT REACTIONS OF PHOTOSYNTHESIS

8. What absorbs the photons of light in photosynthesis?

a. carotenoids
b. chlorophyll
c. chlorophyll and carotenoids
d. the thylakoid membrane
e. inner membrane of the chloroplast

Answer: c
Difficulty: Easy
Bloom's Taxonomy: Knowledge

9. Fluorescence occurs when

a. a high-energy electron returns to its ground state by releasing energy in a photon.

b. a low-energy electron moves to a high-energy state by absorbing heat.

c. a high-energy electron leaves its nucleus and moves to a different molecule.

d. a pigment molecule accepts a high energy electron and releases a photon of energy

e. the energy from an electron is transferred to a different molecule while the electron returns to the ground state.

Answer: a
Difficulty: Easy
Bloom's Taxonomy: Knowledge

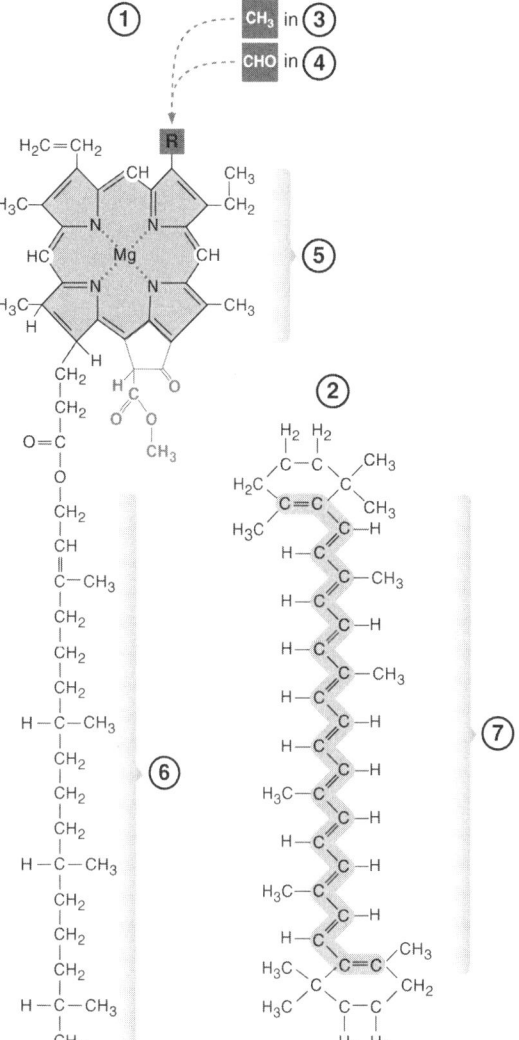

Use the figure above for questions 10–16.

10. In the figure above, diagram 1 represents

a. a molecule of chlorophyll.
b. a phospholipid in the thylakoid membrane.
c. a primary electron acceptor.
d. an accessory pigment.
e. a carotenoid.

Answer: a
Difficulty: Easy
Bloom's Taxonomy: Knowledge

11. In the figure above, diagram 2 represents

a. a molecule of chlorophyll.
b. a phospholipid in the thylakoid membrane.
c. a primary electron acceptor.
d. a nonlight-reactive accessory pigment.
e. a carotenoid

Answer: e
Difficulty: Easy
Bloom's Taxonomy: Knowledge

12. In the figure above, identify the photon absorbing region of chlorophyll.

a. 3
b. 4
c. 5
d. 6
e. 7

Answer: c
Difficulty: Easy
Bloom's Taxonomy: Knowledge

13. In the figure above, region 7 aids photosynthesis by

a. splitting water.
b. absorbing photons.
c. transferring hydrogen ions to chlorophyll.
d. absorbing energy from chlorophyll.
e. none of the above

Answer: b
Difficulty: Easy
Bloom's Taxonomy: Knowledge

14. In the figure above, what absorbs photons of both blue and red light?

a. 5 with 3
b. 5 with 4
c. 5, regardless of the R group
d. 6
e. 7

Answer: c
Difficulty: Easy
Bloom's Taxonomy: Knowledge

15. In the figure above, what absorbs photons of light *only* in the blue/violet range and *not* light in the red portion of the spectrum?

a. 5 with 3
b. 5 with 4
c. 5, regardless of the R group
d. 6
e. 7

Answer: e
Difficulty: Easy
Bloom's Taxonomy: Knowledge

16. Identify the hydrophobic side chain of chlorophyll in the figure above.

a. 5 with 3
b. 5 with 4
c. 5, regardless of the R group
d. 6
e. 7

Answer: d
Difficulty: Easy
Bloom's Taxonomy: Knowledge

17. The molecule of chlorophyll a in photosystem II's reaction center is known as P680 because

a. it absorbs 680 photons per minute.
b. it absorbs photons with a wavelength of 680 nm.
c. there are exactly 680 accessory pigments in the photosystem.
d. it will generate 680 molecules of ATP per photon absorbed.
e. there are 680 electrons that can be energized by light.

Answer: b
Difficulty: Easy
Bloom's Taxonomy: Knowledge

18. The molecule of chlorophyll a photosystem I's reaction center is known as P700 because

a. it absorbs 700 photons per minute.
b. it absorbs photons with a wavelength of 700 nm.
c. there are exactly 700 accessory pigments in the photosystem.
d. there are 700 molecules of NADPH generated per photon of light absorbed.
e. there are 700 electrons that can be energized by light.

Answer: b
Difficulty: Easy
Bloom's Taxonomy: Knowledge

19. P700 is located in photosystem _____ and is comprised of chlorophyll _____ absorbing light with a wavelength of 700 nm.

a. I; b
b. I; a
c. II; b
d. II; a
e. I and II; a

Answer: b
Difficulty: Moderate
Bloom's Taxonomy: Knowledge

20. All of the following processes are associated with photosystem I EXCEPT

a. ATP synthesis.
b. NADPH synthesis.
c. cyclic electron flow.
d. accepting electrons flowing from the electron transport chain.
e. splitting of water.

Answer: e
Difficulty: Moderate
Bloom's Taxonomy: Comprehension

21. All of the following are associated with photosystem II EXCEPT for

a. the excitement of electrons to generate a proton gradient that will drive ATP synthesis.
b. the transfer of electrons to NADP+ to generate NADPH.
c. the splitting of water.
d. the transfer of electrons to a primary electron acceptor.
e. noncyclic electron flow.

Answer: b
Difficulty: Moderate
Bloom's Taxonomy: Knowledge

22. The primary purpose of the light-dependent reactions is to

a. provide electrons and energy for the light-independent reactions.
b. generate O_2 gas.
c. create a proton gradient across the thylakoid membrane.
d. transfer electrons to the primary electron acceptors.
e. produce sugars such as glucose.

Answer: a
Difficulty: Moderate
Bloom's Taxonomy: Comprehension

23. The reason the light reactions have both cyclic and noncyclic electron pathways is

a. because the light-independent reactions require ATP and NADPH in different amounts than are generated by noncyclic electron flow.
b. to ensure that ATP and NAPDH are generated in a 1:1 molar ratio.
c. because only the combination of pathways can generate sufficient NADPH for the light-independent reactions.
d. to provide more electrons from water than would be released by the noncyclic pathway.
e. because a single photon of light can't energize a ground state electron in photosystem II to the level necessary for tranfer to photosystem I's primary electron acceptor.

Answer: a
Difficulty: Moderate
Bloom's Taxonomy: Comprehension

24. Trace the path of a single electron through the noncyclic electron pathway by putting the following in order:

1. The electron is excited in photosystem I.
2. The electron is passed to the plastoquinone pool.
3. The electron is excited in photosystem II.
4. The electron is released from water and added to the photosystem.
5. A proton gradient is generated by the cytochrome complex as the electron arrives and then departs.
6. The electron is transferred to NADP+.

a. 1, 4, 3, 6, 2, 5
b. 3, 4, 2, 5, 1, 6
c. 4, 3, 2, 5, 1, 6
d. 1, 4, 5, 2, 1, 6
e. 1, 4, 6, 2, 1, 5

Answer: c
Difficulty: Moderate
Bloom's Taxonomy: Comprehension

25. Suppose you recreate Jagendorf and Uribe's chloroplast experiment with one small change: after placing the chloroplasts (in the dark) in an acidic medium to fill the thylakoid membrane with protons, you move the chloroplasts to a *stronger* acid rather than the basic solution used in the original experiment. What would be the outcome of your experiment?

a. ATP synthesis would still occur because a proton gradient was created.
b. ATP synthesis would occur but at a slower rate because of the strength of the acid.
c. ATP synthesis would not occur because the protons are on the "wrong" side of the membrane and unable to drive the activity of ATP synthase.
d. ATP synthesis would not occur because the acid is too strong and would rupture the thylakoid membrane.
e. ATP synthesis would occur, but only until the protons reached equilibrium.

Answer: c
Difficulty: Difficult
Bloom's Taxonomy: Application

9.3 THE LIGHT-INDEPENDENT REACTIONS OF PHOTOSYNTHESIS

26. The following are all products of the light-independent reactions EXCEPT for

a. O_2 gas.
b. ADP.
c. $NADP^+$.
d. G3P.
e. RuBP.

Answer: a
Difficulty: Easy
Bloom's Taxonomy: Knowledge

27. The purpose of the Calvin cycle is to

a. produce sugars using CO_2 as a carbon source.
b. recover electrons lost when water was split.
c. capture photons of light.
d. counteract increasing atmospheric CO_2 concentrations (global warming).
e. generate O_2 gas for cellular respiration.

Answer: a
Difficulty: Moderate
Bloom's Taxonomy: Knowledge

28. Where do the electrons from NADPH go in the Calvin cycle?

a. They are put onto oxygen, just like in cellular respiration.
b. They are used to regenerate RuBP from G3P
c. They are added to 3PGA.
d. They are transferred to rubisco.
e. They remain on NADPH to help drive the light reactions

Answer: c
Difficulty: Moderate
Bloom's Taxonomy: Knowledge, Comprehension

29. Of the G3P generated in the Calvin cycle, 5/6 will be used to regenerate RuBP. A company manufacturing widgets would go out of business if 5/6 of the widgets were recycled at the manufacturing site rather than being sold. And yet plants survive this extreme inefficiency. Why?

a. Actually, 5/6 of the G3P is used to build glucose and only 1/6 of the G3P is needed to regenerate RuBP.
b. Plants are extremely efficient in other ways and that compensates for this anomaly.
c. There is unlimited CO_2 for the plant to use, so it's okay to "waste" some by regenerating RuBP.
d. The plant expends no energy to acquire the light energy that drives the Calvin cycle, so there is no pressure on the system to become more efficient.
e. The cyclic pathway of the light reactions compensates for the energy lost in RuBP regeneration.

Answer: d
Difficulty: Difficult
Bloom's Taxonomy: Application, Evaluation

30. Where does the Calvin cycle occur?

a. thylakoid membrane
b. thylakoid lumen
c. stromal lamellae
d. cytosol
e. stroma

Answer: e
Difficulty: Easy
Bloom's Taxonomy: Knowledge

31. The Calvin cycle requires all of the following molecules EXCEPT for

a. NADPH.
b. ATP.
c. CO_2.
d. rubisco.
e. glucose.

Answer: e
Difficulty: Easy
Bloom's Taxonomy: Knowledge

32. If photosynthetic eukaryotic cells are provided with CO_2 synthesized with heavy oxygen (^{18}O), the ^{18}O label will be found in all but one compound. That compound is

a. 3-phosphoglycerate.
b. cellulose.
c. glucose.
d. O_2 gas.
e. RuBP (Ribulose bisphosphate).

Answer: d
Difficulty: Difficult
Bloom's Taxonomy: Application

33. You extract all of the proteins of a leaf and measure the percentage that is rubisco. You find that rubisco comprises what percentage of the leaf proteins?

a. 10%
b. 25%
c. 50%
d. 75%
e. 90%

Answer: c
Difficulty: Easy
Bloom's Taxonomy: Knowledge

FOCUS ON RESEARCH: BASIC RESEARCH: TWO-DIMENSIONAL PAPER CHROMATOGRAPHY AND THE CALVIN CYCLE

34. When Melvin Calvin and Andrew Benson reduced the CO_2 available to cells in the photosynthesis experiments, they discovered that RuBP accumulated. Why did this molecule accumulate?

a. There wasn't sufficient light to drive photosynthesis.
b. The cells had too little ATP.
c. There were not enough reactants to generate 3PGA.
d. The RuBP was defective in structure and unable to form a bond with CO_2.
e. Rubisco was inhibited by the RuBP excess.

Answer: c
Difficulty: Moderate
Bloom's Taxonomy: Knowledge, Comprehension

35. 2D chromatography is called that because

a. there are two steps: labeling with a radioactive marker and separation via chromatography.
b. the molecules are separated into a linear arrangment, meaning two dimensions.
c. the process separates the molecules in two directions, first along the x axis and then along the y axis.
d. two different solutions are used in the paper chromatography process.
e. two different types of radioactive markers must be used.

Answer: c
Difficulty: Moderate
Bloom's Taxonomy: Knowledge, Comprehension

INSIGHTS FROM THE MOLECULAR REVOLUTION: SMALL BUT PUSHY

36. Rubisco has 16 subunits joined together to make a functional unit. Eight of the subunits are large, the other 8 are small. Where on this enzyme are the active sites located?

a. on each of the 8 large subunits
b. on each of the 8 small subunits
c. on one of the 8 large subunits
d. on one of the 8 small subunits
e. on each of the 16 subunits

Answer: a
Difficulty: Moderate
Bloom's Taxonomy: Knowledge

37. Rubisco has 16 subunits joined together to make a functional unit. Eight of the subunits are large, the other 8 are small. What is the function of the small subunit?

a. The small subunits contain the reactant binding sites.
b. It stabilizes the overall enzyme's structure but has no function by itself.
c. The function of the small subunit is still unknown.
d. It regulates the enzyme's rate of catalysis.
e. The small subunits contain allosteric regulator binding sites.

Answer: d
Difficulty: Moderate
Bloom's Taxonomy: Knowledge

9.4 PHOTORESPIRATION AND THE C_4 CYCLE

38. Rubisco can add O_2 to RuBP, which generates

a. two molecules of 3PGA.
b. two molecules of phosphoglycolate.
c. three molecules of 3PGA.
d. one molecule of G3P.
e. a molecule of 3PGA and a molecule of phosphoglycolate.

Answer: e
Difficulty: Easy
Bloom's Taxonomy: Knowledge

39. Which molecule associated with photorespiration is toxic to plant cells?

a. 3PGA
b. G3P
c. phosphoglycolate
d. glycolate
e. none of these

Answer: d
Difficulty: Easy
Bloom's Taxonomy: Knowledge

40. Why is photorespiration more likely in warm weather?

a. Plants require warm weather (at least 23° C) to drive photosynthesis.
b. Plants are more likely to close their stomata in the daytime heat than at night when it is cool.
c. Plants are more likely to dehydrate in warm weather, forcing them to close the stomata to conserve water; this prevents CO_2 from entering the leaf.
d. The rubisco enzyme is very temperature sensitive and becomes less selective in warmer temperatures, allowing it to fix O_2 instead of CO_2.
e. Plants use up the water in their central vacuoles in warm weather which inhibits the water-requiring light reactions; this in turn alters the activity of rubisco.

Answer: c
Difficulty: Moderate
Bloom's Taxonomy: Knowledge, Comprehension

41. The C_4 cycle

a. replaces the carbon fixation stage of the Calvin cycle.
b. supplements the activity of rubisco by providing a second source of 3PGA for the reduction stage of the Calvin cycle.
c. is more efficient than the Calvin cycle because less ATP is consumed in the process.
d. ensures that CO_2 is provided to rubisco and thus prevents photorespiration.
e. is most commonly associated with plants living in humid climates.

Answer: d
Difficulty: Moderate
Bloom's Taxonomy: Comprehension

42. The temperature at which the C_4/Calvin cycle combination becomes more advantageous than the unmodified pathway is

a. 10° C
b. 20° C
c. 23° C
d. 30° C
e. 33° C

Answer: d
Difficulty: Easy
Bloom's Taxonomy: Knowledge

43. During which of the following time periods does a CAM plant use rubisco?

a. daylight only
b. in darkness only
c. noon to midnight
d. midnight to noon
e. all 24 hours of a standard calendar day

Answer: a
Difficulty: Moderate
Bloom's Taxonomy: Knowledge, Comprehension

44. During which of the following time periods does a C_4 plant use rubisco?

a. daylight only
b. in darkness only
c. noon to midnight
d. midnight to noon
e. all 24 hours of a standard calendar day

Answer: a
Difficulty: Moderate
Bloom's Taxonomy: Knowledge, Comprehension

45. During which of the following time periods does a CAM plant take in atmospheric CO_2?

a. daylight only
b. in darkness only
c. noon to midnight
d. midnight to noon
e. all 24 hours of a standard calendar day

Answer: b
Difficulty: Moderate
Bloom's Taxonomy: Knowledge, Comprehension

46. I'm a botanist looking for new species of C_4 plants. Where should I focus my search?

a. Maine
b. Quebec, Canada
c. equatorial rainforest
d. The Everglades (in Florida)
e. Arizona

Answer: e
Difficulty: Difficult
Bloom's Taxonomy: Application

47. The products of the C_4 cycle

a. are immediately used in CAM plants.
b. are stored by CAM plants in the central vacuole.
c. are stored by CAM plants in the stroma.
d. diffuse to bundle sheath cells for immediate use in C_4 plants.
e. are stored in mesophyll cells of C_4 plants.

Answer: b
Difficulty: Moderate
Bloom's Taxonomy: Knowledge, Comprehension

UNANSWERED QUESTIONS

48. The efficiency of photosynthesis is reduced

a. by the damaging effects of excess heat.
b. by the damaging effects of excess intermediates in the light-dependent reactions.
c. by the low efficiency of ATP synthase.
d. by inhibition of rubisco by excess ATP.
e. all of the above

Answer: b
Difficulty: Moderate
Bloom's Taxonomy: Knowledge

49. The genes encoding for the proteins embedded in the thylakoid membrane

a. are all in the nucleus.
b. are all located in the chloroplast.
c. are found in the mitochondria, chloroplast, and nucleus.
d. are split between the nucleus and chloroplast.
e. are duplicated in the nucleus and chloroplast.

Answer: d
Difficulty: Easy
Bloom's Taxonomy: Knowledge

Integrative Multiple-Choice

50. Where can you find ATP synthase in a plant cell?

a. in the thylakoid membrane
b. in the mitochondrial inner membrane
c. embedded in the plasma membrane
d. only in the nucleus
e. thylakoid and inner mitochondrial membranes

Answer: e
Difficulty: Moderate
Bloom's Taxonomy: Knowledge
Source: Sections 8.4, 9.2

51. Which products of the light reactions are used in the Calvin cycle?

a. ADP and $NADP^+$
b. Water, O_2, ATP.
c. CO_2 and RuBP
d. electrons and photons
e. ATP and NADPH

Answer: e
Difficulty: Moderate
Bloom's Taxonomy: Knowledge, Comprehension
Source: Section 9.2, 9.3

52. Rubisco has 16 subunits joined together to make a functional unit. Eight of the subunits are large, the other 8 are small. What does this tell you about this protein's structure?

a. It is an example of a protein having quaternary structure.
b. The protein has tertiary structure and domains but no quaternary structure.
c. The primary sequence of all 16 subunits is identical, which is a unique property of enzymes participating in photosynthesis.
d. The subunits each have 8 alpha helices in order to bind to CO_2 and RuBP.
e. It's an enzyme that is extremely resistant to denaturation.

Answer: a
Difficulty: Moderate
Bloom's Taxonomy: Knowledge, Comprehension
Source: Section 3.5, 9.3

53. If you put chloroplasts in the dark, they can continue to make sugar if provided with

a. NADPH, ATP, and CO_2.
b. NADPH and ATP.
c. CO_2 and ATP.
d. NADPH, CO_2, and ATP.
e. water and ATP.

Answer: a
Difficulty: Moderate
Bloom's Taxonomy: Application
Source: Sections 9.2, and 9.3

54. Which Calvin Cycle products are used in the light reactions?

a. G3P, ATP, and NADPH
b. O_2, water, and ATP.
c. ADP, P_i, and NADP+
d. electrons from CO_2
e. protons and P_i.

Answer: c
Difficulty: Moderate
Bloom's Taxonomy: Knowledge
Source: Section 9.2 and 9.3

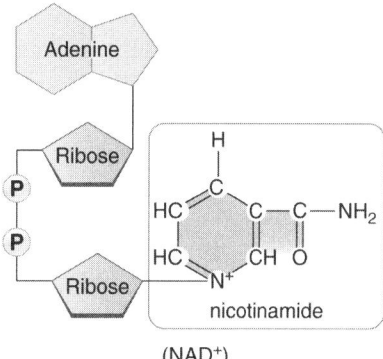

(NAD⁺)

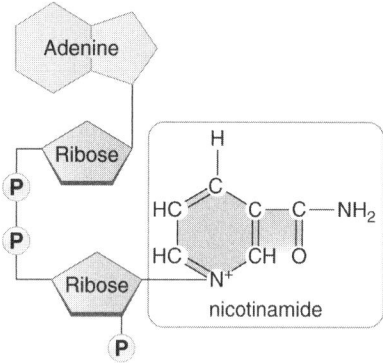

(NADP⁺)

55. What is the primary difference between NAD^+ and $NADP^+$ (shown in the figure above)?

a. NAD^+ functions as an electron transporter, whereas NADP does not.
b. NAD^+ functions as an electron transporter in chloroplasts while $NADP^+$ functions as an electron transporter in mitochondria.
c. Both function as electron carriers, but $NADP^+$ has a third phosphate group and NAD^+ does not.
d. Both transport electrons to the electron transport chain (ETC) found on the inner mitochondrial membrane, but $NADP^+$ transfers its electrons to the ETC at a higher energy level.
e. NAD^+ functions as a free energy source for cells while $NADP^+$ does not.

Answer: c
Difficulty: Easy
Bloom's Taxonomy: Knowledge, Comprehension
Source: Section 8.2, 9.2

56. Standard photosynthesis (C3) plants are more susceptible to photorespiration than either C_4 or CAM plants, so why do most plants lack the C_4 pathway?

a. Rubisco is more efficient in C3 plants when compared to rubisco C_4 and CAM plants.
b. Photorespiration is not a serious problem for most plants.
c. Most plant species do not have vacuoles in their cells to store the malate generated from CO_2 taken in during the night.
d. There is an energetic cost to C_4 and CAM pathways which only makes them advantageous in very hot and/or arid environments.
e. Switching between C3, C_4, and CAM photosynthesis is very difficult for most plants.

Answer: d
Difficulty: Moderate
Bloom's Taxonomy: Comprehension
Source: Section 9.2 and 9.4

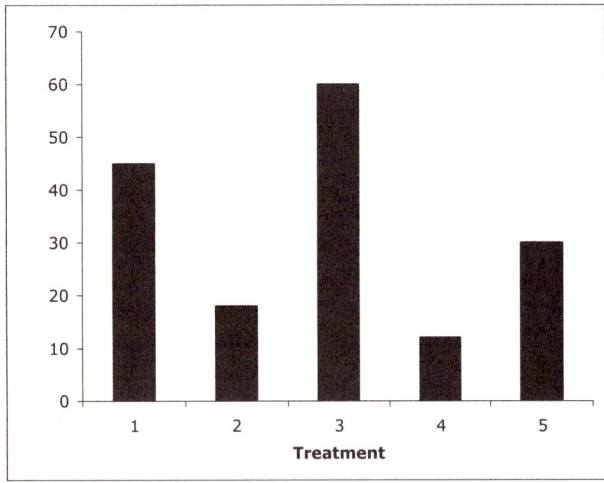

Use the above figure and the following information for questions 57–61.

You grow 25 plants under 5 different experimental treatment conditions to determine the effects of light and CO_2 on plants. Each plant started out with the same biomass and was measured again in 10 days; the average results for each plant group are in the graph. The five conditions are as follows:

I. Normal light, 0.5x CO_2 concentration
II. Normal light, 1x (normal) CO_2 concentration
III. Normal light, 2x CO_2 concentration
IV. No light, 1x CO_2 concentration
V. No light, 2x CO_2 concentration

57. In the figure above, this group of plants was given treatment I.

a. 1
b. 2
c. 3
d. 4
e. 5

Answer: e
Difficulty: Difficult
Bloom's Taxonomy: Application
Source: Sections 9.2 and 9.3

58. In the figure above, this group of plants was given treatment II.

a. 1
b. 2
c. 3
d. 4
e. 5

Answer: a
Difficulty: Difficult
Bloom's Taxonomy: Application
Source: Sections 9.2 and 9.3

59. In the figure above, this group of plants was given treatment III.

a. 1
b. 2
c. 3
d. 4
e. 5

Answer: c
Difficulty: Difficult
Bloom's Taxonomy: Application
Source: Sections 9.2 and 9.3

60. In the figure above, this group of plants was given treatment IV.

a. 1
b. 2
c. 3
d. 4
e. 5

Answer: d
Difficulty: Difficult
Bloom's Taxonomy: Application
Source: Sections 9.2 and 9.3

61. In the figure above, this group of plants was given treatment V.

a. 1
b. 2
c. 3
d. 4
e. 5

Answer: b
Difficulty: Difficult
Bloom's Taxonomy: Application
Source: Section 9.2 and 9.3

Matching

Match each of the following terms with its correct definition.

62. _____ photoautotrophs

63. _____ photosystems

64. _____ photophosphorylation

65. _____ stroma

66. _____ photorespiration

67. _____ inductive resonance

68. _____ stromal lamellae

69. _____ malate

A. A process by which an electron transfers its energy to a different molecule and returns to a ground state but doesn't itself move to a new molecule

B. The generation of ATP via a proton gradient created by light-energized electrons moving down an electron transport chain

C. The site of the light-independent reactions in the chloroplast

D. Complexes of light-absorbing molecules clustered in the thylakoid and stromal membranes of chloroplasts

E. Membranous connections between granum that allow individual thylakoid lumens to form a single, contiguous compartment

F. The addition of O_2 to RuBP by rubisco, resulting in a net loss of carbon from the Calvin cycle

G. The molecule used to store carbon overnight in CAM plants

H. Organisms that use CO_2 as their carbon source and light as their energy source.

Answers: **62. H** **63 D** **64. B** **65. C** **66. F** **67. A** **68. E** **69. G**

Difficulty: Moderate
Bloom's Taxonomy: Knowledge
Source: Sections 9.1-9.4

Classification

Use the five processes listed below for questions 70–74.

    a. noncyclic electron flow
    b. cyclic electron flow
    c. carbon fixation stage of the Calvin cycle
    d. carbon reduction stage of Calvin cycle
    e. RuBP regeneration stage of Calvin cycle

70. _____ This converts G3P into a molecule that can bind to CO_2.

71. _____ This form of the light dependent reactions produces ATP and NADPH in equal molar ratios.

72. _____ This reaction uses rubisco, the most abundant enzyme on Earth.

73. _____ This is when the G3P product is actually synthesized in the chloroplast.

74. _____ This allows for the production of additional ATP to drive the RuBP regeneration process.

Answers: **70. e** **71. a** **72. c** **73. d** **74. b**

Difficulty: Easy
Bloom's Taxonomy: Knowledge
Source: Sections 9.2 and 9.3

Choice

For each of the following statements, choose the most appropriate process.

a. light-dependent reactions only
b. Calvin cycle only
c. light-dependent reactions and the Calvin cycle both
d. neither the light-dependent reactions nor the Calvin cycle

75. _____ produces O_2

76. _____ consumes NADPH

77. _____ creates a proton gradient

78. _____ requires ATP

79. _____ produces ADP

80. _____ requires G3P to continue

81. _____ produces ATP

82. _____ consumes O_2

83. _____ requires water

84. _____ produces NADP+

85. _____ requires glucose

86. _____ requires rubisco

87. _____ occurs only during the daytime

Answers: **75. a** **76. b** **77. a** **78. b** **79. b**
          80. b **81. a** **82. d** **83. a.** **84. b**
          85. d **86. b** **87. c**

Difficulty: Moderate
Bloom's Taxonomy: Knowledge, Comprehension
Source: Sections 9.2 and 9.3

Short Answer

88. Explain why we say nearly all life on Earth ultimately depends on the sun for energy.

Answer: Photosynthetic organisms convert light energy into chemical energy. While the photosynthetic organisms use this chemical energy to drive their own cellular processes, other organisms consume plants (or consume the herbivores that consumed the plants) to acquire chemical energy. Through this process, nearly all organisms on earth can trace their energy back to the original photosynthetic process, which was dependent on sunlight.

Difficulty: Moderate
Bloom's Taxonomy: Knowledge, Comprehension
Source: Section 9.1

True/False

If the statement is true, write a "T" in the blank. If the statement is false, make it correct by changing the underlined word(s) and writing the correct word(s) in the answer blank.

89. _____ The process of photosynthesis, including all of the enzymes and chemical intermediates, is <u>completely</u> understood.

90. _____ Atmospheric O_2 is actually a <u>waste product</u> of photosynthesis.

91. _____ Glucose is the <u>only</u> sugar produced from photosynthesis.

92. _____ The genes encoding for proteins involved in photosynthesis are located in the <u>chloroplast</u>.

93. _____ The <u>arrangement</u> of photosystems I and II in the thylakoid membrane is exactly like drawn in the Z-pathway of most textbooks

94. _____ Not all plants can use the <u>C_4 pathway</u> to avoid photorespiration.

95. _____ Glycolate is released into the <u>atmosphere</u> to prevent it from killing the plant.

Answers: 89. F, incompletely   90. T   91. F, primary 92. F, chloroplast and nucleus 93. F, sequence of use   94. T   95. F, microbodies

Difficulty: Difficult
Bloom's Taxonomy: Knowledge
Source: Sections 9.1-9.4

Essay

96. Explain why it is important, from a global standpoint, to try to improve the efficiency of photosynthesis in common crops. Specify which aspect(s) of photosynthesis we should target.

Answer: The land available for agriculture is limited and the population of the planet continues to increase. By understanding how plants can become more efficient in photosynthesis, the harvest of a set acre of land could be increased. This would allow us to better utilize the land available.

Photorespiration is the logical target of our research since photorespiration is known to decrease the efficiency of photosynthesis. If we can enable C3 plants to better avoid photorespiration, we can increase the productivity of crops.

Difficulty: Difficult
Bloom's Taxonomy: Analysis, Synthesis
Source: Section 9.4

97. What are some of the problems that must be addressed when planting C3 plants in hot, arid climates for ornamental purposes?

Answer: C3 plants are prone to photorespiration, especially at high temperatures or in arid climates. In order to allow these plants to survive, they must be provided with a reliable, adequate water supply. Protection from intense sunlight should be considered and can be accomplished by position plants in areas where they will get morning or evening light, but be in partial shade during the afternoon.

Difficulty: Difficult
Bloom's Taxonomy: Synthesis, Evaluation
Source: Section 9.2 and 9.4

98. Compare and contrast the role of O_2 and H_2O in photosynthesis and cellular respiration.

Answer: The role of oxygen gas and water are almost the reverse in the two processes. In cellular respiration, electrons from the electron transport chain are put onto oxygen gas. This attracts protons and creates water. In photosynthesis, water is used as a source of electrons, releasing oxygen gas into the atmosphere while the protons are kept and used in production of a sugar.

Difficulty: Difficult
Bloom's Taxonomy: Synthesis
Source: Sections 8.4 and 9.3

Labeling

Identify each part of the following illustration.

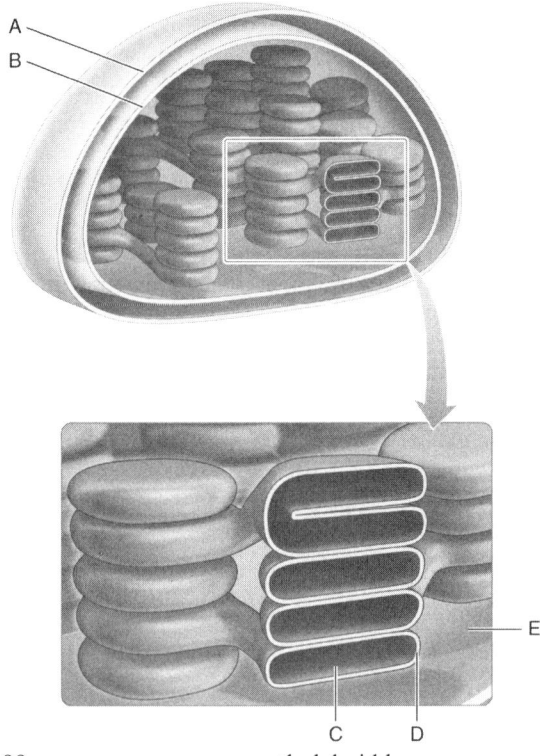

99. _____ thylakoid lumen
100. _____ outer membrane
101. _____ stroma
102. _____ thylakoid membrane
103. _____ inner membrane
104. _____ site of light-dependent reactions
105. _____ site of light-independent reactions
106. _____ where you would search for rubisco
107. _____ membrane most important in generating a proton gradient for ATP synthesis
108. _____ membrane most likely to surround an acidic solution
109. _____ where you would find chlorophyll
110. _____ where photorespiration occurs
111. _____ this membrane is in contact with the cytosol
112. _____ prevents the stroma from coming into contact with the outer membrane
113. _____ where you would find carotenoids

Answers: 99. C   100. A   101. E   102. D   103. B   104. D   105. E
106. E   107. D   108. D   109. D   110. E   111. A   112. B
113. D

Difficulty: Moderate
Bloom's Taxonomy: Knowledge, Comprehension
Source: Figure 9.3, Sections 9.2, 9.3, and 9.4

10
CELL DIVISION AND MITOSIS

Multiple-Choice

WHY IT MATTERS

1.   Mitosis results in _____.

a.   growth of an organism
b.   wound healing
c.   gamete formation
d.   replacement of old tissue
e.   all but one of these answers

Answer: e
Difficulty: Easy
Bloom's Taxonomy: Knowledge

2.   The duplication of the complete set of chromosomes in an organism's cell, followed by the separation of the duplicated chromosomes into two new cells is known as _____.

a.   mitotic cell division
b.   zygote formation
c.   binary fission
d.   meiotic cell division
e.   fertilization

Answer: a
Difficulty: Easy
Bloom's Taxonomy: Knowledge

10.1 THE CYCLE OF CELL GROWTH AND DIVISION: AN OVERVIEW

3.   A somatic cell divides to form two genetically identical daughter cells during mitosis. Prior to mitosis occurring, which of the following must occur?

a.   The cell must replicate its DNA.
b.   The cell must first be fertilized.
c.   The nucleus must divide.
d.   Chromatids must be separated.
e.   The nuclear envelope must disintegrate.

Answer: a
Difficulty: Moderate
Bloom's Taxonomy: Application

4. At the conclusion of mitosis, each daughter cell has _____.

a.   twice the amount of DNA and half the cytoplasm of the parent cell
b.   DNA identical to the parent cell
c.   half the DNA and half the cytoplasm found in the parent cell
d.   twice the cytoplasm and the same amount of DNA as the parent cell
e.   DNA genetically different from the parent cell

Answer: b
Difficulty: Easy
Bloom's Taxonomy: Knowledge, Comprehension

FOCUS ON RESEARCH: MODEL RESEARCH ORGANISMS: THE YEAST *SACCHAROMYCES CEREVISIAE*

5.   Which of the following is not a characteristic of *Saccharomyces* cultures?

a.   The cells are haploid.
b.   The cells have a relatively short generation time.
c.   Cells reproduce asexually by budding.
d.   The cells can be induced to reproduce sexually.
e.   All answers are true of *Saccharomyces* cultures.

Answer: e
Difficulty: Moderate
Bloom's Taxonomy: Knowledge

6.   *Saccharomyces* is a _____ cell.

a.   prokaryotic
b.   eukaryotic
c.   bacterial
d.   plant
e.   protistan

Answer: b
Difficulty: Moderate
Bloom's Taxonomy: Knowledge

7.   The first eukaryotic genome to be elucidated was from _____.

a.   *Escherichia coli*
b.   *Caenorhabditis elegans*
c.   *Drosophila melanogaster*
d.   *Saccharomyces cerevisiae*
e.   *Schizosaccharomyces pombe*

Answer: d
Difficulty: Moderate
Bloom's Taxonomy: Knowledge

10.2 THE MITOTIC CELL CYCLE

8.   At the beginning of interphase in G1, a cell has 36 chromosomes. How many chromosomes would be found in that same cell in G2 of interphase?

a.   36
b.   18
c.   72
d.   64
e.   44

Answer: c
Difficulty: Moderate
Bloom's Taxonomy: Application

9.	Replication of DNA occurs during _____.

a.	S phase
b.	G1 phase
c.	G0 phase
d.	prophase
e.	cytokinesis

Answer: a
Difficulty: Moderate
Bloom's Taxonomy: Knowledge

10.	Early in mitosis chromatin condenses into the compact, rodlike structures known as chromosomes. Which of the following processes takes place more easily as a result?

a.	disappearance of the nuclear envelope
b.	replication of DNA
c.	orderly distribution of DNA into the two new nuclei
d.	formation of the mitotic spindle
e.	disintegration of nucleoli

Answer: c
Difficulty: Moderate
Bloom's Taxonomy: Application

11.	Chromosomes decondense into chromatin at which point in the cell cycle?

a.	at the beginning of interphase
b.	at the end of interphase
c.	at the beginning of prophase
d.	at the beginning of metaphase
e.	at the end of telophase

Answer: e
Difficulty: Moderate
Bloom's Taxonomy: Knowledge

12.	A cell is committed to undergoing mitosis once it transitions from _____.

a.	G2 to prophase
b.	G1 to G0
c.	G1 to S
d.	S to G2
e.	telophase to cytokinesis

Answer: c
Difficulty: Moderate
Bloom's Taxonomy: Knowledge

13.	DNA is found in its condensed form known as a chromosome in which of the following stages of the cell cycle?

a.	prophase only
b.	prophase and metaphase
c.	interphase
d.	throughout the cell cycle
e.	throughout mitosis until late telophase

Answer: e
Difficulty: Moderate
Bloom's Taxonomy: Synthesis

14. Once human nerve cells become mature, they normally exit the cell cycle and remain in _____.

a.	G0
b.	G1
c.	G2
d.	S phase
e.	prophase

Answer: a
Difficulty: Moderate
Bloom's Taxonomy: Knowledge

15.	Which sequence of the cell cycle is correct?

a.	prophase, metaphase, interphase, telophase, anaphase
b.	prometaphase, anaphase, prophase, telophase, interphase, metaphase
c.	anaphase, interphase, telophase, prometaphase, prophase, metaphase
d.	interphase, prophase, prometaphase, metaphase, anaphase, telophase
e.	interphase, metaphase, prometaphase, prophase, telophase, anaphase

Answer: d
Difficulty: Easy
Bloom's Taxonomy: Knowledge

16. Comparison of mitosis and cytokinesis in animal versus plant cells reveals _____.

a.	sister chromatids are identical in animal cells, but they differ from one another in plants
b.	a cleavage furrow is initiated in an animal cell, while a cell plate begins to form in plant cells during telophase
c.	in animal cells, chromosomes do not become attached to the spindle until anaphase, whereas chromosomes become attached to the spindle at prophase in plant cells
d.	the nuclear envelope disintegrates in prophase in animal cells, but remains intact in plant cells

Answer: b
Difficulty: Moderate
Bloom's Taxonomy: Synthesis

17. Embryonic development begins with a single egg fertilized by a single sperm forming a zygote. The zygote then undergoes mitosis; how many cells would be present at the conclusion of 4 mitotic divisions?

a.	2
b.	4
c.	8
d.	16
e.	32

Answer: d
Difficulty: Moderate
Bloom's Taxonomy: Application

18. During which phases of mitosis are chromosomes composed of two sister chromatids?

a. prophase only
b. prophase and metaphase
c. prophase, metaphase, and anaphase
d. anaphase and telophase
e. chromosomes are composed of two sister chromatids during all phases of mitosis

Answer: b
Difficulty: Moderate
Bloom's Taxonomy: Synthesis

19. Each of two daughter cells that result from the normal mitotic division of the original parent cell contains _____.

a. the same number of chromosomes but different genes than the parent cell
b. the same number of chromosomes and genes identical to those of the parent cell
c. one half the number of chromosomes but different genes than those of the parent cell
d. one-half of the number of chromosomes and genes identical to those of the parent cell

Answer: b
Difficulty: Moderate
Bloom's Taxonomy: Application/Synthesis

20. Which of the following does NOT occur during prophase of mitosis?

a. the nuclear envelope disappears.
b. the mitotic spindle forms.
c. chromatin condenses into chromosomes.
d. centrioles migrate to opposite poles of the cell.
e. chromosomes decondense into chromatin.

Answer: e
Difficulty: Easy
Bloom's Taxonomy: Knowledge

21. Which one of the following stages of mitosis is utilized for karyotype analysis?

a. prophase
b. prometaphase
c. metaphase
d. anaphase
e. telophase

Answer: c
Difficulty: Moderate
Bloom's Taxonomy: Knowledge

22. Cytokinesis typically begins during which stage of mitosis?

a. prophase
b. prometaphase
c. metaphase
d. anaphase
e. telophase

Answer: e
Difficulty: Easy
Bloom's Taxonomy: Knowledge

23. Where in the cell would the centromere be found?

a. position where metaphase chromosomes align
b. location where the mitotic spindle forms
c. central region of a chromosome where the spindle microtubules attach
d. location where chromosomes cluster during telophase
e. center of the cell where the nucleus is found during prophase

Answer: c
Difficulty: Moderate
Bloom's Taxonomy: Knowledge

24. A cell containing 84 chromatids at metaphase of mitosis would produce two nuclei containing how many chromosomes each in late telophase?

a. 168
b. 84
c. 42
d. 21
e. cannot be determined from the information given

Answer: c
Difficulty: Moderate
Bloom's Taxonomy: Application

25. Which statement best describes the difference between cell division in plant and animal cells?

a. In animal cells but not plant cells, cytokinesis is accomplished by formation of a cleavage furrow.
b. In plant cells but not animal cells, cytokinesis is accomplished by formation of a cleavage furrow.
c. In plant cells, centrosomes have an important role in spindle formation, while in animal cells centrosomes do not function during cell division.
d. In animal cells, replication of chromosomes occurs during interphase, while in plant cells replication occurs when the nuclear envelope disintegrates.
e. none of the above

Answer: a
Difficulty: Moderate
Bloom's Taxonomy: Knowledge

26. Certain cell types in humans, such as skeletal muscle cells, have several nuclei per cell. Based on your understanding of mitosis, how could this happen?

a. The cell undergoes repeated cytokinesis but not mitosis.
b. The cell undergoes repeated mitosis with concomitant cytokinesis.
c. The cell undergoes repeated mitosis but not cytokinesis.
d. The cell undergoes anaphase twice before entering telophase.
e. The cell goes through multiple S phases before entering mitosis.

Answer: c
Difficulty: Moderate
Bloom's Taxonomy: Application/Synthesis

27. The cells produced by mitotic divisions are considered to be _____.

a. genetically identical
b. genetically different
c. clones
d. similar but not exactly the same
e. a and c both

Answer: e
Difficulty: Easy
Bloom's Taxonomy: Comprehension

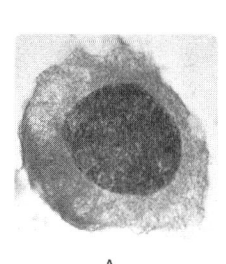 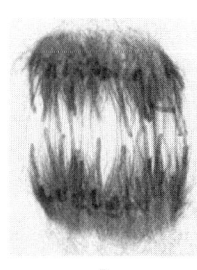 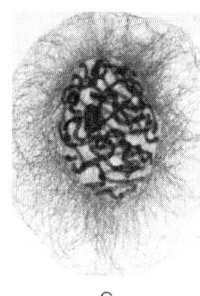 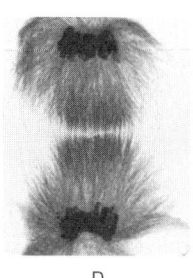 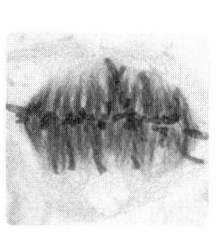

A   B   C   D   E

Refer to the figure above for questions 28–32, which shows the various stages of mitosis in the blood lily *Haemanthus*.

28. Identify the stage of mitosis in photo A.

a. interphase
b. prophase
c. metaphase
d. anaphase
e. telophase

Answer: a
Difficulty: Easy
Bloom's Taxonomy: Comprehension
Source: Figure 10.5

29. Identify the stage of mitosis in photo B.

a. interphase
b. prophase
c. metaphase
d. anaphase
e. telophase

Answer: d
Difficulty: Easy
Bloom's Taxonomy: Knowledge
Source: Figure 10.5

30. Identify the stage of mitosis in photo C.

a. interphase
b. prophase
c. metaphase
d. anaphase
e. telophase

Answer: b
Difficulty: Easy
Bloom's Taxonomy: Knowledge
Source: Figure 10.5

31. Identify the stage of mitosis in photo D.

a. interphase
b. prophase
c. metaphase
d. anaphase
e. telophase

Answer: e
Difficulty: Easy
Bloom's Taxonomy: Knowledge
Source: Figure 10.5

32. Identify the stage of mitosis in photo E.

a. interphase
b. prophase
c. metaphase
d. anaphase
e. telophase

Answer: c
Difficulty: Easy
Bloom's Taxonomy: Knowledge
Source: Figure 10.5

33. Organisms that reproduce asexually usually do so through the process of _____.

a. mitosis
b. meiosis
c. gametogenesis
d. spore formation
e. fertilization

Answer: a
Difficulty: Moderate
Bloom's Taxonomy: Knowledge

34.   When plants are produced by cloning, which process is most directly involved?

a.   mitotic cell division
b.   binary fission
c.   meiotic cell division
d.   gamete production
e.   budding

Answer: a
Difficulty: Moderate
Bloom's Taxonomy: Knowledge

35.   Scientists often use cancer cells for many types of research. Why are cancer cells preferred over normal cells?

a.   Cancer cells do not need as many ingredients to grow as normal cells.
b.   Cancer cells are better models for demonstrating cellular activity than normal cells.
c.   Cancer cells are immortal; that is, they continue to divide indefinitely.
d.   It is unethical to use normal cells for research.

Answer: c
Difficulty: Moderate
Bloom's Taxonomy: Knowledge

10.3 FORMATION AND ACTION OF THE MITOTIC SPINDLE

36.   When cells undergoing mitosis are exposed to colchicine, a chemical extracted from the autumn crocus (*Colchicum autumnale*), cell division is suppressed. Specifically, at which stage will mitosis be arrested?

a.   anaphase
b.   interphase
c.   prophase
d.   telophase
e.   metaphase

Answer: e
Difficulty: Moderate
Bloom's Taxonomy: Knowledge

37.   The mitotic spindle is primarily composed of _____.

a.   actin
b.   intermediate filaments
c.   microtubules
d.   keratin
e.   motor proteins

Answer: c
Difficulty: Moderate
Bloom's Taxonomy: Knowledge

38.   The microtubule organizing center (MTOC) of an animal cell is an identifiable structure known as the _____.

a.   cell plate
b.   centrosome
c.   kinetochore
d.   centromere
e.   chromosome

Answer: b
Difficulty: Easy
Bloom's Taxonomy: Knowledge

39.   Centriole replication occurs during _____.

a.   G1
b.   S
c.   G2
d.   prophase
e.   telophase

Answer: b
Difficulty: Easy
Bloom's Taxonomy: Knowledge

40.   The primary function of centrioles in animal cells is _____.

a.   generation of flagella and cilia
b.   spindle formation
c.   to connect the two sister chromatids
d.   to provide a point of attachment for spindle fibers on chromosomes
e.   cell plate formation

Answer: a
Difficulty: Moderate
Bloom's Taxonomy: Knowledge

41.   Duplicated centrioles move to opposite poles of a(n) _____ cell during _____ of the cell cycle.

a.   animal; prophase
b.   plant; prophase
c.   animal; metaphase
d.   plant; metaphase
e.   both animal and plant cells; prophase

Answer: a
Difficulty: Moderate
Bloom's Taxonomy: Application

42.   The microtubules making up the mitotic spindle attach to specialized structures called _____ that are found in the centromere region of the chromosome.

a.   nucleosome
b.   centrosome
c.   kinetochore
d.   chromatids
e.   centrioles

Answer: c
Difficulty: Moderate
Bloom's Taxonomy: Knowledge

43. Where do the microtubles of the mitotic spindle originate in both plant and animal cells?

a. microtubule organizing centers
b. centromere
c. chromatid
d. centriole
e. centrosome

Answer: a
Difficulty: Moderate
Bloom's Taxonomy: Knowledge

44. Research conducted by G.J Gorbsky and colleagues suggests that during anaphase, chromosomes move as a consequence of _____.

a. a combined action of actin and microtubules
b. the disassembly of kinetochore microtubules
c. the movement of kinetochore microtubules
d. constriction of the contractile protein actin
e. sliding over or along kinetochore microtubules

Answer: e
Difficulty: Moderate
Bloom's Taxonomy: Comprehension

10.4 CELL CYCLE REGULATION

45. Addition of mitosis-promoting factor (MPF) to germinal cells in the skin arrested in G2 would lead to _____.

a. cessation of cell division
b. the cells entering prophase
c. the cells undergoing cytokinesis
d. cell death
e. nothing happening

Answer: b
Difficulty: Moderate
Bloom's Taxonomy: Application

46. CDKs are protein kinases. They function to _____.

a. add phosphate groups to target proteins
b. interact with cyclin proteins
c. remove phosphate groups from target proteins
d. regulate cell division
e. more than one of these answers is correct

Answer: e
Difficulty: Moderate
Bloom's Taxonomy: Knowledge

47. At the G1-to-S checkpoint, _____ has reached a concentration great enough to complex with _____ and initiate DNA synthesis.

a. cyclin E; CDK2
b. cyclin E; CDK1
c. cyclin B; CDK2
d. cyclin B; CDK1
e. cyclin B; SPF

Answer: a
Difficulty: Moderate
Bloom's Taxonomy: Synthesis

48. At the G2-to-M checkpoint, _____ has reached a concentration great enough to complex with _____ and initiate DNA synthesis.

a. cyclin E; CDK2
b. cyclin E; CDK1
c. cyclin B; CDK2
d. cyclin B; CDK1
e. cyclin B; SPF

Answer: d
Difficulty: Moderate
Bloom's Taxonomy: Synthesis

49. What effect would a mutation in the CDK2 gene that resulted in a nonfunctional gene product have on a cell in the root tip of a plant?

a. The cell would be unable to undergo DNA replication.
b. The cell would be unable to enter G2.
c. The cell would be unable to enter G1.
d. The cell would be unable to enter mitosis.
e. more than one of these answers is correct

Answer: e
Difficulty: Difficult
Bloom's Taxonomy: Synthesis

50. During _____ cyclin B is degraded resulting in the transition to G1.

a. interphase
b. prophase
c. metaphase
d. anaphase
e. telophase

Answer: d
Difficulty: Moderate
Bloom's Taxonomy: Knowledge

51. If a cell's DNA becomes damaged by radiation or chemicals, it is unlikely to _____.

a. pass the G2 checkpoint
b. synthesize cyclin-dependent kinases
c. enter G1 from mitosis
d. enter S from G1
e. activate DNA repair mechanisms

Answer: a
Difficulty: Moderate
Bloom's Taxonomy: Comprehension

52. Progression through the phases of the cell cycle is regulated by fluctuating concentrations of _____.

a. microtubules
b. actin
c. cyclins
d. cyclin-dependent kinases
e. histones

Answer: c
Difficulty: Moderate
Bloom's Taxonomy: Knowledge

53. Which of the following are characteristics of cyclin-dependent kinases (CDKs)?

a. They are active only when combined with a cyclin molecule.
b. The levels of CDKs remain constant throughout the cell cycle.
c. The levels of CDKs fluctuate throughout the cell cycle.
d. CDKs add phosphate groups to target proteins.
e. more than one of these answers is correct

Answer: e
Difficulty: Moderate
Bloom's Taxonomy: Comprehension

54. Signaling molecules such as peptide hormones and growth factors bind to receptors on the cell surface, which subsequently triggers reactions in the cell. These reactions include _____.

a. initiation of cell division
b. stopping cell division
c. slowing the rate of cell division
d. causing cells to enter the G0 state
e. a and d both

Answer: a
Difficulty: Moderate
Bloom's Taxonomy: Knowledge

55. Contact inhibition is an important mechanism for maintaining cell growth in developed organs and tissues. As long as the cells remain in contact with each other they remain in _____ and are prevented from dividing.

a. prophase
b. G1
c. G2
d. G0
e. S

Answer: d
Difficulty: Moderate
Bloom's Taxonomy: Comprehension

56. Contact inhibition is best explained by which of the following statements?

a. Contact between neighboring cells triggers reactions leading to inhibition of mitosis.
b. As neighboring cells become more tightly packed together, their size is restricted and cytokinesis can no longer occur.
c. As cell number increases, the protein kinases they produce compete with neighboring cells, inhibiting mitosis.
d. As cell number increases, the level of waste products increase, consequently slowing metabolism leading to mitosis.
e. As cell number increases, size restrictions inhibit protein synthesis.

Answer: a
Difficulty: Moderate
Bloom's Taxonomy: Analysis

57. Which one of the following characterizes cancer cells?

a. Mitosis is strictly regulated.
b. Cancer cells form tumors that strongly adhere to surrounding tissues.
c. Cancer cells display uncontrolled cell division.
d. Tumors formed by cancer cells remain encapsulated and only grow in one location.
e. Cancer cells behave like normal cells.

Answer: c
Difficulty: Moderate
Bloom's Taxonomy: Evaluation

58. Vinblastine, a chemotherapeutic drug used to treat breast and testicular cancers, interferes with the assembly of microtubules. Speculate how this drug works to inhibit cancer cell growth.

a. Vinblastine inhibits transition from G1 to S.
b. Vinblastine inhibits transition from S to mitosis.
c. Vinblastine inhibits cytokinesis.
d. Vinblastine disrupts mitotic spindle formation and consequently mitosis.
e. Vinblastine inhibits cyclin production.

Answer: d
Difficulty: Moderate
Bloom's Taxonomy: Evaluate

59. Comparison of a cancer cell and a normal cell reveals that _____.

a. cancer cells undergo contact inhibition, normal cell do not
b. cancer cells do not undergo contact inhibition, normal cells do
c. cancer cells cannot metastasize, normal cells can
d. regulation of mitosis is strictly regulated in cancer cells, mitosis is unregulated in normal cells.
e. cancer cells remain adherent to other cells, normal cells do not

Answer: b
Difficulty: Moderate
Bloom's Taxonomy: Analysis

INSIGHTS FROM THE MOLECULAR REVOLUTION: HERPES VIRUSES AND UNCONTROLLED CELL DIVISION

60. Research with the herpes virus *herpesvirus 8* focused on the transition from _____.

a. G1 to S
b. S to G2
c. G1 to G0
d. G2 to prophase
e. telophase to cytokinesis.

Answer: a
Difficulty: Moderate
Bloom's Taxonomy: Knowledge

61. *Herpesvirus 8* encodes a protein that acts as a(n) _____.

a. tumor suppressor
b. transcription factor
c. cyclin
d. polymerase
e. enzyme

Answer: c
Difficulty: Moderate
Bloom's Taxonomy: Knowledge

10.5 CELL DIVISION IN PROKARYOTES

62. Hereditary information is encoded in a single, circular DNA molecule in most prokaryotes. Which of the following is FALSE regarding the bacterial chromosome?

a. The DNA of the bacterial chromosome is replicated prior to segregation.
b. Bacterial cells can replicate their DNA very rapidly.
c. Bacterial cells utilize a mitotic spindle to segregate their replicated DNA.
d. The bacterial chromosome has genes that control binary fission.
e. DNA replication begins at the origin of replication.

Answer: c
Difficulty: Moderate
Bloom's Taxonomy: Knowledge

63. Some bacteria produce the enzyme Beta-lactamase, which results in resistance to certain antibiotics such as penicillin. Since these same organisms reproduce asexually, they produce offspring that _____.

a. can be killed by penicillin
b. have an abnormally high rate of mutation
c. are resistant to penicillin
d. have variable numbers of chromosomes

Answer: c
Difficulty: Moderate
Bloom's Taxonomy: Synthesis

64. During prokaryotic cell division, two chromosomes separate and are distributed to the two ends of the cell by _____.

a. the action of the mitotic spindle
b. an unknown mechanism
c. attachment to actin
d. attachment to separating membrane regions
e. formation of a newly made cell wall

Answer: b
Difficulty: Moderate
Bloom's Taxonomy: Knowledge

65. Replication of a bacterial chromosome begins at a specific region called the _____.

a. *ter*
b. replication fork
c. *beg*
d. *ori*
e. *rep*

Answer: d
Difficulty: Easy
Bloom's Taxonomy: Knowledge

66. The mechanism of prokaryotic growth, DNA replication, and cell division resulting in two identical daughter cells is called _____.

a. mitosis
b. meiosis
c. binary fission
d. budding
e. zygote formation

Answer: c
Difficulty: Easy
Bloom's Taxonomy: Knowledge

UNANSWERED QUESTIONS

67. Loss of function of which of the following genes is involved in 50% of all human cancers?

a. p53
b. p27
c. EGFR
d. IL2
e. MYC

Answer: a
Difficulty: Moderate
Bloom's Taxonomy: Knowledge

Integrative Multiple-Choice

68. In general, microtubules disassemble and consequently pull the chromatids to the ends of the spindle during _____.

a. prophase
b. interphase
c. anaphase
d. metaphase
e. telophase

Answer: c
Difficulty: Moderate
Bloom's Taxonomy: Synthesis
Source: Sections 10.2, 10.3

69. Which of the following processes is not necessary for highly accurate cell division to occur in eukaryotic cells?

a.   cell cycle regulation
b.   formation of the mitotic spindle
c.   DNA replication
d.   condensation of genetic material into chromosomes
e.   all are necessary for accurate cell division

Answer: e
Difficulty: Moderate
Bloom's Taxonomy: Synthesis
Source: Section 10.2, 10.3, and 10.4

70. Bacterial cells typically have _____ while eukaryotic cells have _____.

a.   a single circular chromosome; many linear chromosomes.
b.   several circular chromosomes; many linear chromosomes.
c.   one linear chromosome; many circular chromosomes.
d.   two circular chromosomes; numerous circular chromosomes depending on the species.
e.   numerous circular chromosomes depending on the species; many linear chromosomes.

Answer: a
Difficulty: Moderate
Bloom's Taxonomy: Synthesis
Source: Sections 10.2, 10.5

Choice

For questions 71–78 identify the stage of mitosis where each of the following activities occurs.

a.   prophase
b.   prometaphase
c.   metaphase
d.   anaphase
e.   telophase

71. _____   Nuclear envelope disappears

72. _____   Duplicated chromosomes condense into chromosomes

73. _____   Mitotic spindle disassembles

74. _____   RNA synthesis shuts down

75. _____   Spindle fiber begins to form

76. _____   Chromosomes align at the center of the cell

77. _____   Sister chromatids are pulled to opposite spindle poles

78. _____   Daughter chromosomes decondense and the nuclear envelope reforms

Answers: 71. b   72. a   73. e   74. a   75. a   76. c   77. d   78. e

Difficulty: Easy
Bloom's Taxonomy: Knowledge
Source: Section 10.2

For each of the following statements in questions 79–82, choose the most appropriate component of cell cycle regulation from the choices below.

a.   CDK
b.   cyclins
c.   CDK/cyclin complex
d.   SPF

79. _____   Levels remain constant throughout the cell cycle

80. _____   Turns CDK on or off

81. _____   Levels fluctuate throughout the cell cycle

82. _____   Regulates cell division

Answers: 79. a   80. b   81. b   82. c

Difficulty: Moderate
Bloom's Taxonomy: Knowledge
Source: Section 10.4

Select the Exception

83. The stages of nuclear division include all of the following activities EXCEPT _____.

a.   movement of chromosomes into alignment on the metaphase plate
b.   separation of sister chromatids
c.   condensation of DNA into chromosomes
d.   cytokenesis
e.   disappearance of the nucleolus

Answer: d
Difficulty: Difficult
Bloom's Taxonomy: Synthesis/Evaluation

84. All of the following events occur during mitosis EXCEPT _____.

a.   alignment of chromosomes at the spindle midpoint
b.   disappearance of the nuclear envelope
c.   formation of the spindle
d.   DNA synthesis
e.   separation of sister chromatids

Answer: d
Difficulty: Easy
Bloom's Taxonomy: Knowledge

85. When growing animal cells in culture, all of the following must be added to culture medium for growth to occur EXCEPT _____.

a. amino acids
b. glucose
c. DNA
d. growth factors
e. all are necessary for growth to occur

Answer: c
Difficulty: Moderate
Bloom's Taxonomy: Comprehension

Short Answer

86. What is the difference between a chromosome and a chromatid?

Answer: A chromosome is a linear DNA molecule. Once replicated, each chromosome is composed of two exact copies called sister chromatids.
Difficulty: Moderate
Bloom's Taxonomy: Analysis
Source: Section 10.1, 10.2

87. How is cell division different in plants versus animal cells?

Answer: 1. Unlike animal cells, no centrosome or centrioles are present in most types of plants. Instead, the spindle forms from microtubules originating from multiple MTOCs surrounding the nucleus. 2. Cytokinesis occurs by different pathways in plants and animals; animals (including protists and many fungi) form a cleavage furrow composed of microfilaments of actin, which surround the cell and gradually deepen until it cuts the cytoplasm in two. In plants, a new cell wall called the cell plate forms between the daughter nuclei and grows until it divides the cytoplasm.
Difficulty: Analysis
Bloom's Taxonomy: Knowledge
Source: Section 10.1

88. What is the purpose of mitosis in living organisms?

Answer: Multicellular organisms mainly use cell division for growth and for cell replacement and repair.
Difficulty: Moderate
Bloom's Taxonomy: Knowledge
Source: Section 10.1, 10.2

89. What external factors influence a cell to divide?

Answer: A variety of external chemical and physical factors can influence cell division. Particularly important for mammalian cells are growth factors, proteins released by one group of cells that stimulate other cells to divide. Each cell type responds specifically to a certain growth factor or combination of factors.
Difficulty: Moderate
Bloom's Taxonomy: Knowledge
Source: Section 10.4

90. Why are cancer cell lines frequently used to study biological processes rather than normal cells?

Answer: Scientific researchers frequently prefer to use cancer cell lines over normal cells for a number of reasons. Most importantly, cancer cells are immortal, that is they can continue to undergo cell division indefinitely provided the proper nutrients and environmental conditions are met. Normal or "primary" cell lines undergo a finite number of divisions that limit the duration of their use. Other reasons include: cancer cells have less strict nutrient requirements for growth and often have a faster division time.
Difficulty: Moderate
Bloom's Taxonomy: Analysis
Source: Section 10.4

11
MEIOSIS: THE CELLULAR BASIS OF SEXUAL REPRODUCTION

Multiple-Choice

WHY IT MATTERS

1. The advantage of sexual reproduction over asexual reproduction is

a. sexual reproduction requires an interaction between two individuals.
b. there is a mixing of genetic information into a new combination in the next generation.
c. it ensures a greater number of progeny.
d. it keeps the number of chromosomes constant more effectively.
e. it allows for the production and use of gametes.

Answer: b
Difficulty: Easy
Bloom's Taxonomy: Comprehension

2. _____ reproduction generates more _____ than does _____ reproduction.

a. Asexual, genetic diversity, sexual
b. Sexual, genetic stability, asexual
c. Sexual, genetic diversity, asexual
d. Asexual, haploid cells, sexual
e. Sexual, identical offspring, asexual

Answer: c
Difficulty: Easy
Bloom's Taxonomy: Knowledge

11.1 THE MECHANISMS OF MEIOSIS

3. A homologous chromosome pair is best described as two chromosomes having

a. the same genes in the same order but having different alleles.
b. the same alleles of the same genes in the same order.
c. different genes in the same order and possibly having different alleles of some genes.
d. different alleles of the same genes arranged in a different order.
e. an identical DNA sequence.

Answer: a
Difficulty: Moderate
Bloom's Taxonomy: Knowledge

4. Homologous chromosomes undergo recombination during

a. prophase II.
b. metaphase I.
c. metaphase II.
d. both prophase I and II.
e. prophase I.

Answer: e
Difficulty: Easy
Bloom's Taxonomy: Knowledge

5. Sister chromatids are best described as two chromosomes having

a. the same genes in the same order but having different alleles.
b. the same alleles of the same genes in a different order.
c. different genes in the same order and possibly having different alleles of some genes.
d. different alleles of the same genes arranged in a different order.
e. an identical DNA sequence.

Answer: e
Difficulty: Easy
Bloom's Taxonomy: Knowledge

6. This stage of meiosis is characterized by DNA condensation into compact chromosomes.

a. prophase II
b. anaphase I
c. prophase I
d. prometaphase I
e. metaphase I

Answer: c
Difficulty: Easy
Bloom's Taxonomy: Knowledge

7. Sex chromosomes are

a. completely different between the two sexes.
b. partially homologous but also have unique regions.
c. found only in males.
d. unable to line up properly at the metaphase plate.
e. completely homologous but always have different alleles.

Answer: b
Difficulty: Easy
Bloom's Taxonomy: Knowledge

8. Recombination occurs

a. during prophase I and II and involves swapping of chromosome fragments between all 4 chromatids.
b. only during prophase I and involves swapping between homologous chromatids.
c. only during prophase I and involves swapping of chromosome fragments between sister chromatids.
d. during prophase I and II and involves swapping of chromosome fragments between sister chromatids.
e. during prophase I, prophase II, and prometaphase; it involves exchange of alleles between homologous chromosomes.

Answer: b
Difficulty: Moderate
Bloom's Taxonomy: Knowledge

9. The *results* of nondisjunction include

a. a failure of spindle fibers to separate a homologous pair of chromosomes.
b. a change in the status of a daughter cell from diploid to haploid.
c. a change in the status of a daughter cell from haploid to diploid.
d. one pole receiving neither member of a homologous pair of chromosomes.
e. a gamete that cannot fuse with another gamete.

Answer: d
Difficulty: Moderate
Bloom's Taxonomy: Knowledge

10. Meiosis results in the generation of

a. four haploid cells.
b. two diploid cells.
c. two diploid cells and two haploid cells.
d. four diploid cells.
e. one haploid and three diploid cells.

Answer: a
Difficulty: Easy
Bloom's Taxonomy: Knowledge

INSIGHTS FROM THE MOLECULAR REVOLUTION: FERTILE FIELDS IN THE HUMAN Y CHROMOSOME

11. The part of the Y chromosome that is non-homologous to the X chromosome

a. contains so-called housekeeping genes.
b. contains 15 genes with no known function.
c. does not code for any genes but does provide a template for synthesizing ribosomal RNA.
d. contains genes only for ribosomal proteins.
e. is an evolutionary artifact with no current function.

Answer: a
Difficulty: Moderate
Bloom's Taxonomy: Knowledge

12. Lahn and Page's research with the non-homologous regions of the Y chromosome

a. refutes the modern theory that the Y chromosome doesn't contain any functional genes.
b. demonstrated that the Y chromosome is still not very well understood.
c. led to the discovery of the SRY gene.
d. proved that male fertility results from mutations to the SRY gene.

Answer: b
Difficulty: Moderate
Bloom's Taxonomy: Knowledge

11.2 MECHANISMS THAT GENERATE GENETIC VARIABILITY

13. The purpose of the synaptonemal complex is to

a. hold the homologous chromosomes tightly together.
b. support homologous chromosomes as they undergo recombination.
c. prevent chromosome fragments from floating free in the cytosol after DNA breakage for recombination.
d. ensure that the process of crossing over is random.
e. a, b, and c only

Answer: e
Difficulty: Moderate
Bloom's Taxonomy: Knowledge, Comprehension

14. Recombination at the synaptonemal complex always results in

a. two changed and two unchanged chromatids.
b. four unchanged chromatids.
c. four changed chromatids.
d. a random number of changed vs. unchanged chromatids.
e. one unchanged and three changed chromatids from two subsequent recombination events.

Answer: a
Difficulty: Easy
Bloom's Taxonomy: Knowledge

15. Random segregation (by itself) means that a particular daughter cell

a. has some paternal chromosomes and some maternal chromosomes.
b. has only paternal chromosomes.
c. only maternal chromosomes.
d. has chromosomes containing genes from the paternal and maternal chromosomes.
e. a and d both

Answer: a
Difficulty: Moderate
Bloom's Taxonomy: Knowledge

16. If a species has 42 pairs of chromosomes, which value represents the number of combinations of maternal and paternal chromosome combinations that will be sorted to the poles?

a. 42^2
b. 2^{42}
c. 2^{21}
d. 21^2
e. 84^2

Answer: b
Difficulty: Moderate
Bloom's Taxonomy: Knowledge

17. Identical twins can result from

a. the fusion of two sets of identical gametes.
b. a division of a zygote into two separate cells that develop into two separate embryos.
c. the fusion of two paternal gametes with a single maternal gamete that then divides.
d. a lack of chromosomal separation during meiosis resulting in gametes that are diploid and don't need to be fertilized.
e. a mechanism that is not understood.

Answer: b
Difficulty: Easy
Bloom's Taxonomy: Knowledge

18. The process of fertilization

a. is random in the selection of which gametes fuse.
b. is not random but selects for gametes with a greater number of paternal chromosomes.
c. is not random but selects for gametes with a greater number of maternal chromosomes.
d. is not a source of genetic variability in progeny.
e. appears to be random in the selection of gametes to fuse, but only because we don't yet understand the forces at work.

Answer: a
Difficulty: Easy
Bloom's Taxonomy: Knowledge

19. Two genes on each of four homologous chromosomes (HC) are described here, but only one of each pair of sister chromatids is listed (HC1 and HC2). Capital and lower case indicate different alleles of the same gene. What four gametes will result from a single chiasma if the chromosomes have the following genes: HC1: Ab; HC2: aB?

a. Ab, AB, ab, aB
b. Ab, Ab, aB, aB
c. Aa, Bb, ab, Ab.
d. Ab, Ab, AB, AB
e. AB, AB, ab, ab

Answer: a
Difficulty: Difficult
Bloom's Taxonomy: Application

20. Two genes on each of four homologous chromosomes (HC) are described here, but only one of each pair of sister chromatids is listed (HC1 and HC2). Capital and lower case indicate different alleles of the same gene. What four gametes will result from a single chiasma if the chromosomes have the following genes: HC1: ab; HC2: AB?

a. Ab, AB, ab, aB
b. Ab, Ab, aB, aB
c. Aa, Bb, ab, Ab.
d. Ab, Ab, AB, AB
e. AB, AB, ab, ab

Answer: a
Difficulty: Difficult
Bloom's Taxonomy: Application

11.3 THE TIME AND PLACE OF MEIOSIS IN ORGANISMAL LIFE CYCLES

21. Which pattern of diploid and haploid phases reflect the life cycle in animals?

a. alternation between haploid and diploid generations
b. two haploid generations followed by a diploid generation
c. one haploid generation followed by two diploid generations
d. a single generation that limits the haploid state to gametes while the rest of the organism is diploid
e. a single generation that limits the diploid state to gametes while the rest of the organism is haploid

Answer: d
Difficulty: Moderate
Bloom's Taxonomy: Knowledge

22. Which pattern of diploid and haploid phases reflect the life cycle in plants?

a. alternation between haploid and diploid generations
b. two haploid generations followed by a diploid generation
c. one haploid generation followed by two diploid generations
d. a single generation that limits the haploid state to gametes while the rest of the organism is diploid
e. a single generation that limits the diploid state to gametes while the rest of the organism is haploid

Answer: a
Difficulty: Moderate
Bloom's Taxonomy: Knowledge

23. Which pattern of diploid and haploid phases reflect the life cycle in some fungi and algae, but not plants or animals?

a. alternation between haploid and diploid generations
b. two haploid generations followed by a diploid generation
c. one haploid generation followed by two diploid generations
d. a life cycle that limits the diploid state to a single cell produced by fertilization
e. a life cycle that limits the haploid state to a single cell which is immediately fertilized

Answer: d
Difficulty: Difficult
Bloom's Taxonomy: Knowledge

24. An unknown organism from an alien planet is observed and found to have a diploid state only following fertilization. Meiosis follows and the majority of this organism's life is spent in a haploid state. Which multicellular organism on earth most closely resembles this type of life cycle strategy?

a. humpback whale
b. maple tree
c. fern
d. some algae
e. all plants and some fungi

Answer: d
Difficulty: Difficult
Bloom's Taxonomy: Comprehension, Synthesis

25. An unknown organism from an alien planet is observed and found to have a diploid state in every other generation and results from fertilization. The haploid state is able to grow into a gametophyte-like structure. This alien organism most closely mimics the life cycle strategy of which of the following earth organisms?

a. humpback whale
b. maple tree
c. fungi and algae
d. algae
e. all plants and some fungi

Answer: e
Difficulty: Difficult
Bloom's Taxonomy: Comprehension, Synthesis

26. In plants, the gametes produced by a particular gametophyte are

a. always identical because they result from meiosis.
b. always identical because they result from mitosis.
c. sometimes identical because they result from meiosis.
d. sometimes identical because they result from mitosis.
e. never identical because they result from meiosis.

Answer: b
Difficulty: Moderate
Bloom's Taxonomy: Knowledge

UNANSWERED QUESTIONS

27. The nematode *Caenorhabditis elegans* has been used to research how chromosomes pair up for synapsis. Recent findings include the discovery of

a. synaptic centers on each chromosome.
b. pairing centers located near one end of each chromosome.
c. pairing centers located at both ends of individual chromosomes.
d. synaptic centers that are located on paternal chromosomes only.
e. pairing centers located in the center of each chromosome.

Answer: b
Difficulty: Moderate
Bloom's Taxonomy: Knowledge

28. Two species that have been used to study the genetic and molecular controls of meiosis are

a. *Escherichia coli* (bacteria) and *Caenorhabditis elegans* (nematode).
b. *Escherichia coli* (bacteria) and *Saccaromyces cerevisiae* (yeast).
c. *Drosophila melanogaster* (fruit fly) and *Saccaromyces cerevisiae* (yeast).
d. *Crepidula fornicata* (slipper limpet) and *Caenorhabditis elegans* (nematode).
e. *Crepidula fornicata* (slipper limpet) and *Saccaromyces cerevisiae* (yeast).

Answer: c
Difficulty: Easy
Bloom's Taxonomy: Knowledge

Integrative Multiple-Choice

29. The primary difference between mitosis and meiosis in animals is

a. the generation of haploid cells from diploid cells.
b. the number of daughter cells that result (2 or 4).
c. the number of times chromosomes align at the metaphase plate.
d. the purpose of generating reproductive cells as opposed to generating somatic cells.
e. all of the above

Answer: e
Difficulty: Moderate
Bloom's Taxonomy: Knowledge, Comprehension
Source: Section 10.2, 11.1

30. If an organism has the same life strategy as a maple tree, the synaptonemal complex will be observed

a. only in alternating generations.
b. in each generation.
c. only in a unicellular haploid phase.
d. only in a multicellular haploid phase.
e. only in gametophyte.

Answer: a
Difficulty: Moderate
Bloom's Taxonomy: Knowledge, Comprehension
Source: Sections 11.1 and 11.3

Matching

Match each of the following terms with its correct definition.

31. _____ gametes

32. _____ somatic cells

33. _____ sexual reproduction

34. _____ zygote

35. _____ fertilization

36. _____ meiosis

37. _____ haploid

38. _____ diploid

39. _____ allele

40. _____ synapsis

41. _____ chiasmata

A. Cell division that is modified to generate haploid cells

B. The haploid products of meiosis

C. The matching of homologous pairs of chromosomes during prophase I

D. A term used to describe cells having two sets of chromosomes

E. Diploid body cells that do not undergo meiosis

F. A term used to describe cells having a single set of chromosomes

G. A life strategy that involves fusion of haploid gametes from two source organisms to form a diploid cell

H. Another name for crossover sites

I. A variant of a particular gene having a slightly different DNA sequence than another variant of the same gene

J. The diploid product of gamete fusion

K. The process of gamete fusion

Answers: *31. B*   *32. E*   *33. G*   *34. J*   *35. K*
                 36. A   *37. F*   *38. D*   *39. I*   *40. C*
                 41. H

Difficulty: Moderate
Bloom's Taxonomy: Knowledge
Source: Sections 11.1, 11.2, 11.3

Classification

Use the processes listed below for questions 42–48.

I.	Prophase I	II.	Prometaphase I
III.	Metaphase I	IV.	Anaphase I
V.	Telophase I	VI.	Prophase II
VII.	Prometaphase II	VIII.	Metaphase II
IX.	Anaphase II	X.	Telophase II

42. This stage is always characterized by the breakdown of the nuclear envelope.

43. Homologous chromosomes undergo recombination.

44. Tetrads align at the metaphase plate.

45. Spindles are disassembled and replaced by new spindles.

46. In some species, the nuclear envelope may reform during this stage.

47. Nuclear envelope reforms in ALL species.

48. Sister chromatids are separated.

Answers: *42. I*   *43. I*   *44. III*   *45. V*   *46. V*
                 47. X   *48. IX*

Difficulty: Moderate
Bloom's Taxonomy: Knowledge
Source: Section 4.1

Short Answer

49. Explain the purpose of meiosis.

Answer: Meiosis is a process that reduces the chromosomal content of a cell to a haploid state so that a daughter can fuse with another haploid cell from a different member of the same species. This increases genetic variation in the species.
Difficulty: Moderate
Bloom's Taxonomy: Knowledge
Source: Section 11.1

50. List and briefly explain the three ways in which genetic variability is increased during meiosis.

Answer: Meiosis increases genetic variability in three ways: recombination, random segregation, and random fertilization. Recombination is where homologous chromosomes swap segments, creating a new combination of alleles on a single chromosome. Random segregation refers to the sorting of homologous chromosomes to the daughter cells. This results in gametes that have some paternal and maternal chromosomes in a haploid cell. Fertilization is random in the selection of individual alleles that fuse to form a zygote.
Difficulty: Moderate
Bloom's Taxonomy: Knowledge, Comprehension
Source: Section 11.2

Sequence

Arrange the following in the order in which they occur. Write the letter of the first step next to 51. The letter of the last step should be written next to 62.

51. _____   A. Anaphase I
52. _____   B. Anaphase II
53. _____   C. Interkinesis
54. _____   D. Metaphase I
55. _____   E. Metaphase II
56. _____   F. Premeiotic Interphase
57. _____   G. Prometaphase I
58. _____   H. Prometaphase II
59. _____   I. Prophase I
60. _____   J. Prophase II
61. _____   K. Telophase I
62. _____   L. Telophase II

Answers: *51. F*   *52. I*   *53. G*   *54. D*   *55. A*
          56. K   *57 C*   *58. J*   *59. H*   *60. E*
          61. B   *62. L*

Difficulty: Moderate
Bloom's Taxonomy: Knowledge
Source: Section 11.1

True/False

If the statement is true, write a "T" in the blank. If the statement is false and has an underlined term or phrase, make it correct by changing the underlined word(s) and writing the correct word(s) in the answer blank.

63. _____ A slipper limpet can change its sex from <u>female to male</u>.

64. _____ <u>Prokaryotic</u> cells can undergo meiosis.

65. _____ The portion of the Y chromosome <u>not homologous</u> to the X chromosome encodes several housekeeping genes.

66. _____ Crossing over is limited to a single occurrence for a chromosome.

67. _____ Spore is another name for gamete.

68. _____ Haploid organisms growing via mitosis from spores are called <u>gametophytes</u>.

69. _____ Even if chromosomes are not a perfect match, <u>pairing centers</u> sometimes allow synapsis.

70. _____ <u>Haploid</u> cells have homologous chromosomes.

71. _____ The mechanisms that control the number of crossover events in a cell has been well characterized.

Answers: *63. F, male to female*   *64. F, eukaryotic*
          65. T   *66. F*   *67. F*   *68. T*   *69. T*
          70. F, diploid   *71. F.*

Difficulty: Difficult
Bloom's Taxonomy: Knowledge
Source: Sections 11.1, 11.2, 11.3

Select the Exception

72. In the following list, pick the one that is *not* a source of genetic variability from meiosis and fertilization.

a. recombination of homologous chromosomes
b. segregation of chromosomes
c. genes contained in the gametes that fuse to form a zygote
d. recombination between sister chromatids
e. physical arrangement of chromosomes along the metaphase plate in preparation for anaphase

Answer: d
Difficulty: Moderate
Bloom's Taxonomy: Knowledge, Comprehension
Source: Section 11.2

73. Which pattern of diploid and haploid phases is *not* a life cycle discussed in the text?

a. alternation between haploid and diploid generations
b. two haploid generations followed by a diploid generation
d. a life cycle that limits the diploid state to a single cell produced by fertilization
e. a life cycle that limits the haploid state to a single cell which is immediately fertilized.

Answer: b
Difficulty: Moderate
Bloom's Taxonomy: Knowledge

Essay

74. Describe the two stages in meiosis where the most genetic recombination occurs.

Answer: The two stages of meiosis that are the most significant in terms of recombination are prophase I and anaphase I. Of the three events of sexual reproduction that contribute to genetic variability, two occur in meiosis itself. Prophase I is a significant stage due to crossing over events in the tetrad. This generates genetic variability at the chromosomal level. Anaphase I increases genetic variation via the randomness of segregation. This creates a new combination of paternal and maternal chromosomes.
Difficulty: Difficult
Bloom's Taxonomy: Analysis
Source: Section 11.1

75. Compare the functions of meiosis I and II in the overall generation of gametes

Answer: Meiosis I is important in providing opportunities for genetic variability to be introduced. This is accomplished via crossing over in prophase I and via random assortment in anaphase II. The primary purpose of meiosis II to create the haploid state after recombination and assortment created genetic variability.
Difficulty: Moderate
Bloom's Taxonomy: Synthesis
Source: Section 11.1

Labeling

Identify each numbered part of the following illustration.

76. _____ chiasmata
77. _____ nuclear envelope
78. _____ spindle microtubules
79. _____ tetrad
80. _____ centriole

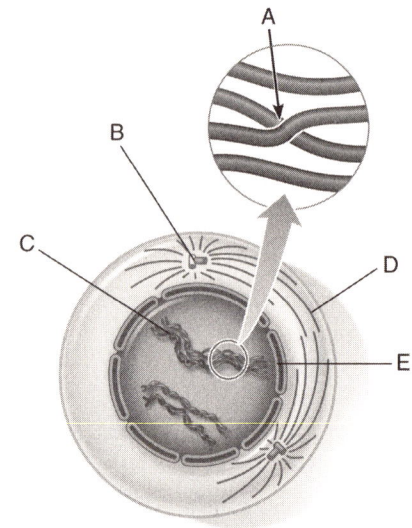

Answers: 76. A   77. E   78. D   79. C   80 B

Difficulty: Moderate
Bloom's Taxonomy: Knowledge
Source: Figure 11.3

12
MENDEL, GENES, AND INHERITANCE

Multiple-Choice

WHY IT MATTERS

1. Why does sickle-celled disease cause death?

a. The malformed red blood cells cannot transport oxygen.
b. The malformed red blood cells cannot enter capillaries and, in fact, block them.
c. The malformed blood cells assume a sickle-shape.
d. The patient has two copies of the mutant gene.
e. More cells assume the sickle shape as oxygen concentration in the tissues falls.

Answer: b
Difficulty: Easy
Bloom's Taxonomy: Knowledge

2. Why do biology textbooks (and biologists in general) always place such importance on an understanding of genetics?

a. Understanding how genes impact cellular function can give us insights to how cells themselves work.
b. Understanding patterns of inheritance allows us to predict which individuals are at higher risk for inherited disorders.
c. Understanding genetics helps explain the difference in individual personalities.
d. a, b, and c only
e. none of the above

Answer: d
Difficulty: Moderate
Bloom's Taxonomy: Knowledge

12.1 THE BEGINNINGS OF GENETICS: MENDEL'S GARDEN PEAS

3. What is the blending theory of inheritance that predominated before 1900?

a. It essentially means we inherit traits from both of our parents.
b. It states that traits are inherited via a mixing of parental blood.
c. It basically means that you're a perfect blend of traits from both of your biological parents.
d. It is a theory to explain the inheritance patterns of traits that skip a generation.
e. It explains why children of one tall parent and one short parent generally have an adult height between their parents' heights.

Answer: b
Difficulty: Easy
Bloom's Taxonomy: Knowledge

4. When did Gregor Mendel perform his experiments with garden peas?

a. in the 1860s, the same decade as the United States' Civil War
b. in the 1760s, predating the formation of the United States of America
c. in the 1620s, the decade in which William Shakespeare died
d. in the 1720s, the decade in which Bach composed the *Brandenburg Concertos*
e. in the 1820s, the decade in which Ecuador became independent from Spain

Answer: a
Difficulty: Easy
Bloom's Taxonomy: Knowledge

5. Mendel studied what he called characters and traits. What is the relationship between these terms?

a. Characters were heritable characteristics; traits were variations of characters.
b. Traits were heritable characteristics; characters were variations of traits.
c. Characters were the unknown package of transfer to the next generation; traits resulted from this transfer.
d. Characters were passed to the next generation; traits were never passed to the next generation.
e Characters and traits were synonymous in Mendel's writings.

Answer: a
Difficulty: Moderate
Bloom's Taxonomy: Knowledge

6. If purple is dominant in pea plants, a cross between P generation purple and white plants will result in

a. all purple flowers in the F1 generation.
b. all white flowers in the F1 generation.
c. all purple flowers in the F1 generation but a lighter purple than in the parents.
d. mostly purple flowers in the P1 generation, with an occasional white flower.
e. half of the plants having purple flowers, the other half having white flowers.

Answer: a
Difficulty: Easy
Bloom's Taxonomy: Knowledge

7.   Mendel crossed true-breeding plants having yellow peas with plants having green peas. The resulting plant all had yellow peas. An F_1 cross resulted in ¾ of the plants having yellow peas and ¼ of the plants having green peas. What does this tell you about the alleles for color?

a.   yellow is usually the dominant color, but sometimes green can be dominant
b.   green is the dominant color
c.   yellow is the dominant color
d.   yellow is the recessive color
e.   the F_1 plants must have had some green peas that went unnoticed.

Answer: c
Difficulty: Easy
Bloom's Taxonomy: Knowledge

8.   Use the product rule to determine the probability (P) of events X and Y both occurring.

a. $P = X^2 + Y^2$
b. $P = X^2Y^2$
c. $P = XY$
d. $P = X + Y$
e. $P = (X + Y)^2$

Answer: c
Difficulty: Easy
Bloom's Taxonomy: Knowledge

9.   If the probability of X occurring is 1 in 4, and the probability of Y occurring is 1 in 5, the probability of *both* occurring is

a. $(1/4)^2 + (1/5)^2 = (1/16) + (1/25) = (25/400) + (16/400) = 41/400.$
b. $(1/4) + (1/5) = (5/20) + (4/20) = 9/20.$
c. $(1/4)^2(1/5)^2 = (1/16)(1/25) = 1/400.$
d. $(1/4)(1/5) = 1/20.$
e. $(1/4 + 1/5)^2 = (5/20 + 4/20)^2 = (9/20)^2 = 81/400.$

Answer: d
Difficulty: Moderate
Bloom's Taxonomy: Knowledge

10.   Use the sum rule to determine the probability of either event X or event Y occurring if they cannot occur simultaneously. (In the equations below, P_x means probability of event X; P_y means probability of event Y)

a. $= P_x + P_y$
b. $= P_xP_y$
c. $= (P_x)^2 + (P_y)^2$
d. $= (P_x + P_y)^2$
e. $= P_x^2P_y^2$

Answer: a
Difficulty: Easy
Bloom's Taxonomy: Knowledge

11.   A testcross is used to

a.   determine if a parent with a dominant trait is heterozygous or homozygous.
b.   determine which allele is dominant.
c.   determine if the progeny of an experimental cross will get a random assortment of alleles.
d.   prove that an organism is true-breeding.
e.   cross an individual with a dominant phenotype with a homozygous dominant individual to prove the alleles are indeed dominant.

Answer: a
Difficulty: Easy
Bloom's Taxonomy: Knowledge

12.   Which of the following is a dihybrid cross?

a.   RrMM x Rrmm
b.   RRMM x rrmm
c.   RrMm x RrMm
d.   rrMM x RRmm
e.   RrMm x rrmm

Answer: c
Difficulty: Easy
Bloom's Taxonomy: Knowledge

13.   Which of the following is a test cross? (y means the allele's identity is unknown)

a.   RyMM x Rymm
b.   RRMM x rrmm
c.   RyMy x RyMy
d.   rrMM x RRmm
e.   RyMy x rrmm

Answer: e
Difficulty: Easy
Bloom's Taxonomy: Knowledge

14.   Which of the following shows an F_1 monohybrid cross?

a.   RrMM x Rrmm
b.   RRMM x rrmm
c.   Rr x Rr
d.   rr x RR
e.   Rr x rr

Answer: c
Difficulty: Easy
Bloom's Taxonomy: Knowledge

15. The principle of independent assortment states that

a. alleles on different homologous chromosomes are randomly sorted to individual gametes.
b. genes on the same chromosome are not randomly sorted to different gametes.
c. two alleles for a single trait are randomly sorted to individual gametes; there is no mention of chromosomes.
d. genes for different traits are always sorted to gametes independently of each other.
e. Both a and c

Answer: e
Difficulty: Easy
Bloom's Taxonomy: Knowledge

16. A parent has a genotype of RrYy. What is the probability of having a gamete with the RY genotype?

a. 1/2
b. 1/4
c. 3/4
d. 1/8
e. 0

Answer: b
Difficulty: Moderate
Bloom's Taxonomy: Application

17. Mendel selected 7 traits in pea plants to study. What was lucky about this choice?

a. Peas are easy to raise and have short life cycles.
b. All 7 traits were easy to characterize.
c. The 7 alleles segregated independently from each other.
d. He knew enough mathematics to apply rules of probability to his results.
e. The 7 traits were all located on the same plant chromosome.

Answer: c
Difficulty: Easy
Bloom's Taxonomy: Knowledge

18. Which individual established the connection between genes, meiosis, and fertilization?

a. Gregor Mendel
b. Hugo de Vries
c. Carl Correns
d. Erich von Tschermak
e. Walter Sutton

Answer: e
Difficulty: Easy
Bloom's Taxonomy: Knowledge

19. Identify the disorder caused by a *dominant* allele.

a. achondroplasia
b. cystic fibrosis
c. albinism
d. sickle cell disease
e. all of the above are caused by dominant alleles

Answer: a
Difficulty: Easy
Bloom's Taxonomy: Knowledge

20. An F_1 dihybrid cross with alleles that are either dominant or recessive

a. results in four phenotypes.
b. results in 1/16 of the progeny being homozygous recessive.
c. results in a 9:3:3:1 phenotypic ratio.
d. results in 1/16 of the progeny being homozygous dominant.
e. all of the above

Answer: c
Difficulty: Difficult
Bloom's Taxonomy: Knowledge, Application

21. Your father is heterozygous for a recessive disorder. You know your mom has two "good" alleles. What is the probability that you will have the disorder?

a. 0%
b. 25%
c. 50%
d. 75%
e 100%

Answer: a
Difficulty: Moderate
Bloom's Taxonomy: Application

22. Your father is heterozygous for sickle cell disease. You know your mom has two "good" alleles. What is the probability that you will have the disorder?

a. 0% but you will have a 50% chance of passing the "bad" gene on to your children
b. 0% and you don't need to worry about passing the "bad" allele on to your children
c. 25% for you and 25% for your children
d. 50% for you and 50% for your children
e. 50% for you and 25% that you'll pass the "bad" allele on to your children

Answer: a
Difficulty: Hard
Bloom's Taxonomy: Application

23.   Your parents are both heterozygous for sickle cell disease. What is the probability that you will have the disorder?

a.   25% regardless of the health of your 3 siblings
b.   50% no matter whether or not you have siblings
c.   75% no matter whether or not you have siblings
d.   100% even if you are an only child
e.   25% but only if your three siblings are healthy.

Answer: a
Difficulty: Moderate
Bloom's Taxonomy: Application

24.   R is the dominant allele for a round pea, r is the recessive allele for a wrinkled pea. If you cross plants having round peas with plants having wrinkled peas,

a.   you will be able to determine if the round pea plant is homozygous or heterozygous.
b.   you are conducting a test cross.
c.   you are crossing two parental pea plants.
d.   the progeny plants will all have wrinkled peas.
e.   a and b only

Answer: e
Difficulty: Difficult
Bloom's Taxonomy: Application

25.   If your mother has Huntington's disease, which is caused by a dominant allele, the odds of you inheriting the disorder are

a.   1/4.
b.   1/2.
c.   3/4.
d.   1.
e.   0.

Answer: b
Difficulty: Moderate
Bloom's Taxonomy: Application

26.   If your mother and father both have Huntington's disease, which is caused by a dominant allele, the odds of you having the disorder are

a.   1/4.
b.   1/2.
c.   3/4.
d.   1.
e.   0.

Answer: c
Difficulty: Moderate
Bloom's Taxonomy: Application

27.   If your mother has cystic fibrosis, which is caused by a recessive allele, the odds of you inheriting the disease are

a.   1/4.
b.   1/2.
c.   3/4.
d.   1.
e.   More information is needed to determine the odds.

Answer: e
Difficulty: Moderate
Bloom's Taxonomy: Application

28.   If your mother and father *both* have cystic fibrosis, which is caused by a recessive allele, the odds of you having cystic fibrosis are

a.   1/4.
b.   1/2.
c.   3/4.
d.   1.
e.   0.

Answer: d
Difficulty: Moderate
Bloom's Taxonomy: Application

29.   If your mother and father both are healthy but carry the allele for cystic fibrosis, which is caused by a recessive allele, the odds of you having *at least* one allele for the disorder are

a.   1/4.
b.   1/2.
c.   3/4.
d.   1.
e.   0.

Answer: c
Difficulty: Moderate
Bloom's Taxonomy: Application

30.   If your mother and father both have cystic fibrosis, which is caused by a recessive allele, the odds of you having *only* one allele for the disorder are

a.   1/4.
b.   1/2.
c.   3/4.
d.   1.
e.   0.

Answer: e
Difficulty: Moderate
Bloom's Taxonomy: Application

31. Round is the dominant allele for a pea; the recessive allele produces a wrinkled pea. How can you obtain true-breeding pea plants having round peas with the *least* amount of work?

a. Cross plants having round peas with plants having wrinkled peas. Select round pea plants from the progeny because they are now true-breeding.
b. Cross plants having round peas with other plants having round peas. Do this for multiple generations.
c. Cross plants having round peas with plants having wrinkled peas. Select round pea plants from the progeny and do a test cross to determine which parental plants were homozygous dominant. Use these homozygous pea plants as your true-breeding plants.
d. Cross plants having round peas with plants having wrinkled peas. This will tell you which round-pea plants are homozygous dominant and are thus true-breeding.
e. It's not possible. You can only get true-breeding plants that have wrinkled peas.

Answer: d
Difficulty: Difficult
Bloom's Taxonomy: Application

32. Your mother has albinism, which is a recessive trait. You learn that your spouse's mother also has albinism. What are the odds that your first child will also have albinism?

a. 1/4
b. 1/2
c. 3/4
d. 1
e. not enough information

Answer: e
Difficulty: Moderate
Bloom's Taxonomy: Application

33. Your mother has albinism, which is a recessive trait. Your father has cystic fibrosis, also caused by a recessive trait. You just got married and discover that your new father-in-law has albinism and cystic fibrosis. What are the odds that your first child will have *both* albinism and cystic fibrosis?

a. 1/16.
b. 1/8.
c. 1/4.
d. 1/2.
e. 0.

Answer: a
Difficulty: Difficult
Bloom's Taxonomy: Application

34. Your mother has albinism, which is a recessive trait. Your father has cystic fibrosis, also caused by a recessive trait. You just got married and discover that your new father-in-law has albinism and cystic fibrosis. What are the odds that your first child will have *either* albinism *or* cystic fibrosis, but not both?

a. 1/16
b. 3/8
c. 1/8
d. 1/4
e. 0

Answer: b
Difficulty: Difficult
Bloom's Taxonomy: Application

35. Your mother has albinism, which is a recessive trait. Your father has cystic fibrosis, also caused by a recessive trait. You just got married and discover that your new father-in-law has albinism and cystic fibrosis. What are the odds that your first child will have *neither* albinism *nor* cystic fibrosis?

a. 1/16
b. 3/8
c. 1/8
d. 2/5
e. 0

Answer: d
Difficulty: Difficult
Bloom's Taxonomy: Application

INSIGHTS FROM THE MOLECULAR REVOLUTION: WHY MENDEL'S DWARF PEA PLANTS WERE SO SHORT

36. The reason why dwarf pea plants are shorter is because

a. the enzymes involved in gibberellin synthesis are missing.
b. they have an altered gibberellin amino acid sequence.
c. they cannot make any gibberellin.
d. they produce very little gibberellin hormone.
e. they produce an excess of gibberellin.

Answer: d
Difficulty: Easy
Bloom's Taxonomy: Knowledge

12.2 LATER MODIFICATIONS AND ADDITIONS TO MENDEL'S HYPOTHESIS

37. The key difference between incomplete dominance and codominance is

a. with incomplete dominance, the recessive allele cannot be detected; in codominance, the expression of the recessive allele is apparent.
b. with incomplete dominance it is possible to detect the expression of a recessive allele; in codominance, both alleles contribute equally to the phenotype.
c. with codominance it is possible to detect the expression of a recessive allele; in incomplete dominance, both alleles contribute equally to the phenotype.
d. with incomplete dominance, it is possible to detect the expression of the dominant allele; in codominance, both alleles contribute equally to the phenotype.
e. the two terms mean the same thing

Answer: b
Difficulty: Moderate
Bloom's Taxonomy: Knowledge

38. You cross a pink snapdragon (C^RC^W) with a white snapdragon (C^WC^W). What percentage of the progeny will be red?

a. 0%
b. 25%
c. 50%
d. 75%
e. 100%

Answer: a
Difficulty: Moderate
Bloom's Taxonomy: Application

39. Snapdragons have incomplete dominance of the red and white alleles. What will be the phenotypes and ratios of the F_2 generation?

a. 100% pink
b. 100% red
c. 50% white, 50% red
d. 25% red, 50% pink, 25% white
e. 25% pink, 50% white, 25% red

Answer: d
Difficulty: Difficult
Bloom's Taxonomy: Knowledge, Application

40. Your father has B blood. Your mother has O blood. You get tested and learn that your blood is also type O. What does this tell you?

a. You were adopted and your parents didn't tell you.
b. Your mother had a secret affair.
c. Your father's genotype is I^Bi and your mother's genotype is ii.
d. Your father's genotype is I^BI^B and your mother's genotype is ii.
e. d. Your father's genotype is I^AI^B and your mother's genotype is ii.

Answer: c
Difficulty: Difficult
Bloom's Taxonomy: Knowledge, Application

41. You have type O blood (genotype ii). Who can you donate blood to in an emergency?

a. type O only
b. type B only
c. type A only
d. types A and B
e. types A, B, and O

Answer: e
Difficulty: Difficult
Bloom's Taxonomy: Application

42. You have type A blood (genotype I^Ai). Who can you donate blood to in an emergency?

a. type O only
b. type AB only
c. type A only
d. types A and B, not O
e. types A and AB

Answer: e
Difficulty: Difficult
Bloom's Taxonomy: Application

43. The different alleles in human blood type are a demonstration of

a. incomplete dominance.
b. codominance.
c. dominance and codominance.
d. dominance and incomplete dominance.
e. dominance, codominance, and incomplete dominance.

Answer: c
Difficulty: Moderate
Bloom's Taxonomy: Knowledge

44. An F_1 dihybrid cross results in a phenotypic ratio of 9:3:4. This tells you that

a. epistasis is occurring.
b. there is codominance at the second locus.
c. there is incomplete dominance at the second locus.
d. the second locus is dominant over the first locus.
e. none of these choices

Answer: a
Difficulty: Moderate
Bloom's Taxonomy: Knowledge

45. In snapdragons the red allele C^R is incompletely dominant over the white allele C^W. Which two plants would you cross to produce a true-breeding pink snapdragon?

a. pink with pink
b. pink with red
c. red with white
d. pink with white
e. it's impossible to accomplish

Answer: e
Difficulty: Moderate
Bloom's Taxonomy: Application

46. Mouse pigmentation is subject to epistasis of the B alleles by the d alleles. B (black) is dominant over b (brown). D is dominant over d. Homozygous d is epistatic to the black and brown genes. Given this information, what genotypes give you a white mouse?

a. BBdd
b. Bbdd
c. bbDD
d. bbDd
e. a and b both

Answer: e
Difficulty: Moderate
Bloom's Taxonomy: Application

47. Mouse pigmentation is subject to epistasis of the B alleles by the d alleles. B (black) is dominant over b (brown). D is dominant over d. Homozygous d is epistatic to the black and brown genes. Given this information, What will result from a F_1 cross between two mice?

a. 9/16 black, 3/16 brown, 4/16 white
b. 9/16 white, 3/16 brown, 4/16 black
c. 9/16 black, 6/16 brown, 1/16 white
d. 9/16 white, 6/16 brown, 1/16 black
e. All black mice

Answer: a
Difficulty: Moderate
Bloom's Taxonomy: Knowledge, Application

UNANSWERED QUESTIONS

48. We now have evidence that which of the following conditions is subject to epistasis?

a. Cystic fibrosis
b. Sickle cell disease
c. Bardet-Biedl syndrome
d. Albinism
e. Achondroplasia

Answer: c
Difficulty: Moderate
Bloom's Taxonomy: Knowledge

49. If you have a mutation *only* in the genes associated with Bardet-Biedl syndrome, your symptoms will be _____ severe compared to someone having the same mutation *and* a mutation in a different DNA sequence also associated with the disease because

_____.

a. more; Bardet-Biedl syndrome is lessened by epistasis
b. less; Bardet-Biedl syndrome is intensified by epistasis
c. the same; Bardet-Biedl syndrome is unaffected by epistasis
d. insufficient is provided to determine your severity of the disorder

Answer: b
Difficulty: Moderate
Bloom's Taxonomy: Knowledge, Comprehension

Integrative Multiple-Choice

50. We now know that some of the 7 alleles Mendel studied are on the same chromosome in pea plants. Despite this, the law of independent assortment still applies. How can you explain this?

a. There is recombination via the synaptonemal complex during mitosis.
b. There is recombination via the synaptonemal complex during meiosis.
c. The law of independent assortment applies to all alleles regardless of their arrangement on chromosomes.
d. Mendel was a good enough mathematician to design experiments that would result in the predicted ratios.
e. Mendel was incredibly lucky.

Answer: b
Difficulty: Moderate
Bloom's Taxonomy: Knowledge
Source: Section 11.2,12.1

51. A patient presents the following symptoms: anemia, dilation of heart, lung damage and pneumonia, rheumatism, abdominal pain, and kidney failure. After learning about their family history, you run a genetic test for which disorder?

a. Cystic fibrosis
b. Albinism
c. Sickle cell disease
d. Achondroplasia
e. Pleiotropy

Answer: c
Difficulty: Easy
Bloom's Taxonomy: Knowledge
Source: Section 12.1,12.2

Matching

Match each of the following terms with its correct definition.

52. _____ F_2 generation

53. _____ F_1 generation

54. _____ dominant

55. _____ homozygous

56. _____ recessive

57. _____ phenotype

58. _____ dihybrid

59. _____ heterozygous

60. _____ pleiotropy

61. _____ P generation

62. _____ probability

63. _____ genotype

64. _____ true breeding

65. _____ locus

66. _____ polygenic inheritance

67. _____ epistasis

68. _____ monohybrid

69. _____ incomplete dominance

A. The displayed traits are unchanged over multiple generations

B. The first generation of offspring from a parental cross

C. True-breeding plants used in the initial cross

D. A result of a cross between two first generation organisms

E. The allele that is expressed no matter what other allele is present

F. The allele that is only expressed if two identical copies are present.

G. Any organism with 2 identical copies of an allele

H. Any organism with 2 different forms of an allele

I. An F_1 heterozygote (for a single trait)

J. The genetic makeup of an organism

K. The physical traits of an organism

L. The likelihood of something occurring if the occurrence is a matter of chance

M. An organism that is heterozygous for two different traits

N. Where an allele is found on a chromosome

O. When one allele cannot completely mask the effects of another allele

P. Alleles at one locus mask the expression of alleles at a different locus

Q. Several different genes contribute to a particular phenotype

R. A single allele has multiple phenotypic effects.

Answers: **52. D**   **53. B**   **54. E**   **55. G**   **56. F**
               57. K   **58. M**   **59. H**   **60. R**   **61. C**
               62. L   **63. J**   **64. A**   **65. N**   **66. Q**
               67. P   **68. I**   **69. O**

Difficulty: Moderate
Bloom's Taxonomy: Knowledge
Source: Sections 12.1, 12.2

Classification

Use the five types of allele interactions listed below for questions 70–74.

a. Dominance
b. Incomplete dominance
c. Codominance
d. Epistasis
e. Polygenic inheritance

70. _____ Snapdragons have two alleles and three colors due to this.

70. _____ Human blood type AB is an example of this allele interaction.

72. _____ Mice have three colors of fur due to this allele interaction.

73. _____ Human height is an example of this.

74. _____ Peas being wrinkled or round is an example of this.

Answers: **70. b**   **71. c**   **72. d**   **73. e**   **74. a.**

Difficulty: Easy
Bloom's Taxonomy: Knowledge
Source: Section 12.1, 12.2

True/False

If the statement is true, write a "T" in the blank. If the statement is false, make it correct by changing the underlined word(s) and writing the correct word(s) in the answer blank.

75. _____ <u>Cross</u>-pollination is within a single plant while <u>self</u>-pollination is between two plants.

76. _____ An F_1 organism is always <u>homozygous.</u>

77. _____ A dihybrid test cross should always result in at least <u>50%</u> of the progeny being recessive for both traits.

78. _____ A test cross always uses a <u>homozygous recessive</u> organism as one parent.

79. _____ Dwarf pea plants have <u>deficient</u> levels of the hormone gibberellin.

80. _____ Mendel knew about and understood <u>incomplete dominance.</u>

81. _____ There no limit to the number of alleles that exist for a gene.

82. _____ The environment <u>can</u> impact the phenotype encoded by polygenic inheritance.

83. _____ Polygenic inheritance is proof of parental traits blending in the next generation according to the blending theory of inheritance.

Answers: *75. F, self, cross*   *76. F, heterozygous*   *77. F, 25%*   *78. T*   *79. T*   *80. F, complete dominance*   *81. T*   *82. T, 83. F*

Difficulty: Difficult
Bloom's Taxonomy: Knowledge
Source: Sections 12.1, 12.2

Select the Exception

84. Which of the following is not one of Mendel's three hypothesis to explain the results of his P, F_1 and F_2 crossed?

a. Adult plants carry a pair of genes that will determine the inheritance of each allele.
b. If an adult plant has two different alleles, one is dominant over the other.
c. Adult plants always have traits that are governed by a single pair of alleles, but the number of possible alleles varies with each trait.
d. The pairs of alleles separate in the formation of gametes so that each gamete gets one allele of the pair.

Answer: c
Difficulty: Moderate
Bloom's Taxonomy: Knowledge
Source: Section 12.1

85. Which of the following is NOT subject to polygenic inheritance?

a. human height
b. human skin color
c. flower color in snapdragons
d. seed color in wheat
e. ear length in corn

Answer: c
Difficulty: Easy
Bloom's Taxonomy: Knowledge
Source: 12.2

Short Answer

86. Define epistasis and give an example

Answer: Epistasis is where one set of genes will affect the expression of a different set of genes. An example would be coat color in Labrador retrievers. In labs, the genes for black and brown fur can be silenced by the expression of a homozygous recessive allele that blocks pigment production.
Difficulty: Moderate
Bloom's Taxonomy: Knowledge
Source: Section 12.1

87. Explain why human height appears to be a mixture of parental phenotypes when in fact height is genetically based.

Answer: Human height is a result of polygenic inheritance where there are multiple genes that contribute to this single phenotypic trait. When genes are inherited from two parents of vastly different heights, this results in a reduced number of alleles for tall height compared to the tall parent and an increased number of alleles for tall height compared to the shorter parent.
Difficulty: Moderate
Bloom's Taxonomy: Knowledge, Comprehension
Source: Section 12.2

Essay

88. Explain why Mendel's' work was so groundbreaking.

Answer: Mendel applied both the scientific method and mathematics to understanding the patterns of inheritance in pea plants. His work explained the rules of genetics even before chromosomes had been discovered, and while these rules have been refined, the fundamental rules of genetics that he established are still considered to be valid today. He was ahead of his time in his thinking, and his work wasn't well known or appreciated until long after his death.
Difficulty: Moderate
Bloom's Taxonomy: Analysis
Source: Section 12.1

89. Would meiosis, when it was discovered, have been understood without Mendel's work? Explain why or why not, using Mendel's three key findings about inheritance.

Answer: No, meiosis would not have been understood without Mendel's work to explain why the chromosome movements and interactions were taking place. Mendel discovered the fundamental patterns of inheritance, such as an individual having two alleles for a single trait, each derived from a parent. This was followed by the patterns of dominance which would explain why phenotypes could change over generations even though individuals were closely related to each other. And his principle of segregation explained why meiosis, and not mitosis, was necessary for the formation of gametes. The importance of Mendel's work cannot be overstated as it is the foundation upon which all of modern genetics rests.
Difficulty: Moderate
Bloom's Taxonomy: Comprehension, Synthesis
Source: Section 11.2, 12.1, 12.2

13
GENES, CHROMOSOMES, AND HUMAN GENETICS

Multiple-Choice

WHY IT MATTERS

1. Progeria is a genetic condition in some humans that causes

a. pattern baldness as early as age twenty.
b. socially inappropriate vocal outbursts and muscular twitches.
c. a high incidence of cancer with any exposure to sunlight.
d. muscular and mental deterioration shortly after puberty.
e. premature aging which typically leads to death in the early teens.

Answer: e
Difficulty: Easy
Bloom's Taxonomy: Comprehension

13.1 GENETIC LINKAGE AND RECOMBINATION

2. Linked genes are

a. genes whose effects combine to affect a single characteristic.
b. different alleles of the same gene.
c. genes that affect two different traits and that lead to a 9:3:3:1 phenotype ratio in a dihybrid cross.
d. genes that do not sort independently due to their being physically near each other on the same chromosome.
e. all of these

Answer: d
Difficulty: Moderate
Bloom's Taxonomy: Knowledge

3. Exceptions to the principle of independent assortment were discovered and explained by _____ as resulting from genes being physically associated with each other on the same chromosome.

a. Watson
b. Sturtevant
c. Morgan
d. Mendel
e. Crick

Answer: c
Difficulty: Moderate
Bloom's Taxonomy: Comprehension

4. Suppose that in studies of genes on the same chromosome you find the following recombination frequencies:

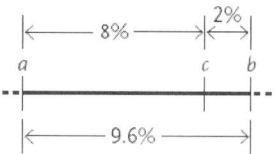

In this case it would be proper to say that *a*, *c*, and *b* are

a. linked genes.
b. different alleles of the same gene.
c. alternative alleles that are not physically possible since the numbers do not add up.
d. linked genes that are not physically possible since the numbers do not add up.
e. three of the possible alleles for determining a particular trait.

Answer: a
Difficulty: Moderate
Bloom's Taxonomy: Application

5. Genetic studies of an animal show that eye color is controlled by an autosomal gene with the dominant allele (*R*) for red eye color and the recessive allele (*r*) for yellow eye color. Also, another autosomal gene has the dominant allele (*T*) leading to paws with thumbs while the recessive allele (*t*) codes for paws without thumbs. The genetic cross *RRTT* x *rrtt* creates offspring with genotype *RrTt*. One of those dihybrids is mated in a testcross (*RrTt* x *rrtt*). Based on the principle of independent assortment the testcross should produce offspring with the phenotype ratio

a. 3 red-eyed with thumbs: 1 yellow-eyed without thumbs.
b. 1 red-eyed with thumbs: 1 yellow-eyed with thumbs: 1 red-eyed without thumbs: 1 yellow-eyed without thumbs.
c. 1 red-eyed with thumbs: 1 yellow-eyed without thumbs.
d. 9 red-eyed with thumbs: 3 yellow-eyed with thumbs: 3 red-eyed without thumbs: 1 yellow-eyed without thumbs.
e. 3 yellow-eyed with thumbs: 1 red-eyed without thumbs.

Answer: b
Difficulty: Moderate
Bloom's Taxonomy: Application

6. Genetic studies of an animal show that eye color is controlled by an autosomal gene with the dominant allcle (*R*) for red eye color and the recessive allele (*r*) for yellow eye color. Also, another autosomal gene has the dominant allele (*T*) leading to paws with thumbs while the recessive allele (*t*) codes for paws without thumbs. The genetic cross *RRTT* x *rrtt* creates offspring with genotype *RrTt*. One of those dihybrids is mated in a testcross (*RrTt* x *rrtt*). If the two genes are completely linked (no recombination occurs between them), then the testcross should produce offspring with the phenotype ratio

a. 3 red-eyed with thumbs: 1 yellow-eyed without thumbs.
b. 1 red-eyed with thumbs: 1 yellow-eyed with thumbs: 1 red-eyed without thumbs: 1 yellow-eyed without thumbs.
c. 1 red-eyed with thumbs: 1 yellow-eyed without thumbs.
d. 9 red-eyed with thumbs: 3 yellow-eyed with thumbs: 3 red-eyed without thumbs: 1 yellow-eyed without thumbs.
e. 1 yellow-eyed with thumbs: 1 red-eyed without thumbs.

Answer: c
Difficulty: Moderate
Bloom's Taxonomy: Application

7. When two genes on the same chromosome are located 10 map units from each other, _____ of the offspring from a test cross for genetic linkage should have a recombinant phenotype.

a. over 75%
b. about 50%
c. about 25%
d. about 10%
e. about 5%

Answer: d
Difficulty: Moderate
Bloom's Taxonomy: Application

8. When two genes are located on different chromosomes, _____ of the offspring from a test cross for genetic linkage should have a recombinant phenotype.

a. over 75%
b. about 50%
c. about 25%
d. about 10%
e. about 5%

Answer: b
Difficulty: Moderate
Bloom's Taxonomy: Application

9. Recombination frequency for two genes is a function of

a. the distance between the two genes on a single chromosome.
b. the overall size of the chromosome where the genes are located.
c. the distance between the two chromosomes that each have one of the two genes.
d. the relative sizes of the two chromosomes that each have one of the two genes.
e. the sex of the parent that supplies each gene.

Answer: a
Difficulty: Easy
Bloom's Taxonomy: Comprehension

10. In *Drosophila melanogaster* the allele for red eyes is dominant over the allele for purple eyes, and the allele for a gray body is dominant over the allele for a black body. A testcross was done to check for genetic linkage between the genes for these traits, with the following results for the offspring:

478 flies with red eyes and a black body
27 flies with red eyes and a gray body
462 flies with purple eyes and a gray body
33 flies with purple eyes and a black body

Which of the choices below best represents the map distance between the genes for these two traits?

a. 50.0 map units
b. 30.0 map units
c. 47.0 map units
d. 27.0 map units
e. 6.0 map units

Answer: e
Difficulty: Difficult
Bloom's Taxonomy: Application

11. In *Drosophila melanogaster* the allele for long wings is dominant over the allele for vestigial wings, and the allele for a gray body is dominant over the allele for a black body. A testcross was done to check for genetic linkage between the genes for these traits, with the following results for the offspring:

410 flies with long wings and a black body
105 flies with long wings and a gray body
390 flies with vestigial wings and a gray body
95 flies with vestigial wings and a black body

Which of the choices below best represents the map distance between the genes for these two traits?

a. 20.0 map units
b. 40.0 map units
c. 10.0 map units
d. 100.0 map units
e. 9.5 map units

Answer: a
Difficulty: Difficult
Bloom's Taxonomy: Application

12. The concept of mapping genes on chromosomes was developed by _____ when he was an undergraduate student.

a. Watson
b. Sturtevant
c. Morgan
d. Mendel
e. Crick

Answer: b
Difficulty: Moderate
Bloom's Taxonomy: Analysis

13. You are given a genetic mapping project as an undergraduate research assistant. You arc told that genes *a* and *b* are 7.4 map units from each other and that genes *b* and *c* are 5.7 map units from each other. You do a cross to determine the map distance between *a* and *c*. Which of the following results would indicate that *c* lies between *a* and *b*?

a. 11.1 map units
b. 15.0 map units
c. 2.0 map units
d. 14.8 map units
e. 13.1 map units

Answer: c
Difficulty: Moderate
Bloom's Taxonomy: Application

14. Genetic map units represent

a. absolute physical distances between genes.
b. locations of different alleles of a gene.
c. the chromosome on which a given gene is located.
d. the actual DNA sequence of a gene.
e. relative positions of genes with respect to each other.

Answer: e
Difficulty: Easy
Bloom's Taxonomy: Comprehension

15. Which process generally is the cause for production of recombinant offspring for two genes on the same chromosome?

a. movement of transposable elements
b. pairing of nonhomologous chromosomes
c. gene duplication
d. exon shuffling
e. crossing-over between homologous chromosomes

Answer: e
Difficulty: Moderate
Bloom's Taxonomy: Knowledge

FOCUS ON RESEARCH: **THE MARVELOUS FRUIT FLY,** *Drosophila melanogaster*

16. The discovery of sex-linked genes and production of the first chromosome map came from studies of

a. pea plants.
b. corn.
c. humans.
d. fruit flies.
e. mice.

Answer: d
Difficulty: Easy
Bloom's Taxonomy: Knowledge

17. For an individual *Drosophila melanogaster* fly the time from when it is laid as an egg to when it is an adult capable of breeding is about _____.

a. 3 months
b. 2 days
c. 6 weeks
d. 10 days
e. 1 month

Answer: d
Difficulty: Moderate
Bloom's Taxonomy: Knowledge

18. The genome of *Drosophila melanogaster* contains about _____ genes.

a. 1,000
b. 14,000
c. 165 million
d. 3 billion
e. 20,000

Answer: b
Difficulty: Difficult
Bloom's Taxonomy: Knowledge

13.2 SEX-LINKED GENES

19. Sex-linked genes are genes that are

a. expressed differently based on the sex of an individual.
b. linked to the sex determining gene.
c. located on sex chromosomes.
d. determinants of the sex of an individual.
e. found exclusively in one sex or the other.

Answer: c
Difficulty: Easy
Bloom's Taxonomy: Comprehension

20. In humans, sex determination generally depends upon

a. whether or not the X chromosome is present.
b. the number of X chromosomes present.
c. environment in the womb.
d. whether or not the Y chromosome is present.
e. all of these

Answer: d
Difficulty: Easy
Bloom's Taxonomy: Comprehension

21. An autosome is

a. a chromosome that is part of a pair that is the same in both males and females.
b. a structure for transport of genetic material to the cytoplasm.
c. a chromosome with a group of genes involved in autoimmune disorders.
d. a region of the DNA that is self-replicating.
e. a cellular body involved in autoimmune disorders.

Answer: a
Difficulty: Easy
Bloom's Taxonomy: Comprehension

22. In birds and butterflies _____ for sex chromosome inheritance.

a. males are XY and females are XX
b. males are ZZ and females are ZW
c. males are ZY and females are XW
d. males are XX and females are XY
e. males are ZW and females are ZZ

Answer: b
Difficulty: Moderate
Bloom's Taxonomy: Application

23. In humans, normally if _____ then the *SRY* gene switches development toward _____ at an early point in embryonic development.

a. a Y chromosome is present; maleness
b. an X chromosome is present; femaleness
c. two X chromosomes are present; femaleness
d. no X chromosome is present; maleness
e. all of the above

Answer: a
Difficulty: Moderate
Bloom's Taxonomy: Comprehension

24. In *Drosophila melanogaster* there is a sex-linked gene for eye color that is found only on the X chromosome. The allele for red eye color (X^{w+}) is dominant over the allele for white eye color (X^w). You examine a vial of 100 flies that are all offspring from a single genetic cross. You see only red-eyed females present, but you see both red-eyed and white-eyed males present. The genotypes of the parents were

a. $X^{w+}X^{w+}$; $X^{w+}Y$.
b. $X^{w+}X^w$; X^wY.
c. X^wX^w; $X^{w+}Y$.
d. $X^{w+}X^w$; X^wY.
e. $X^{w+}X^w$; $X^{w+}Y$.

Answer: e
Difficulty: Difficult
Bloom's Taxonomy: Application

25. In *Drosophila melanogaster* there is a sex-linked gene for eye color that is found only on the X chromosome. The allele for red eye color (X^{w+}) is dominant over the allele for white eye color (X^w). You examine a vial of 100 flies that are all offspring from a single genetic cross. You find both red-eyed females and white-eyed females as well as both red-eyed males and white-eyed males. The genotypes of the parents were

a. $X^{w+}X^{w+}$; X^wY.
b. $X^{w+}X^w$; X^wY.
c. X^wX^w; $X^{w+}Y$.
d. $X^{w+}X^w$; X^wY.
e. $X^{w+}X^w$; $X^{w+}Y$.

Answer: b
Difficulty: Moderate
Bloom's Taxonomy: Application

26. In *Drosophila melanogaster* there is a sex-linked gene for eye color that is found only on the X chromosome. The allele for red eye color (X^{w+}) is dominant over the allele for white eye color (X^w). You examine a vial of 100 flies that are all offspring from a single genetic cross. You see only red-eyed females and white-eyed males present. The genotypes of the parents were

a. $X^{w+}X^{w+}$; X^wY.
b. $X^{w+}X^w$; X^wY.
c. X^wX^w; $X^{w+}Y$.
d. $X^{w+}X^w$; X^wY.
e. $X^{w+}X^w$; $X^{w+}Y$.

Answer: c
Difficulty: Moderate
Bloom's Taxonomy: Application

27.   In *Drosophila melanogaster* there is a sex-linked gene for eye color that is found only on the X chromosome. The allele for red eye color (X^{w+}) is dominant over the allele for white eye color (X^w). You examine a vial of 100 flies that are all offspring from a single genetic cross. You find only red-eyed females and red-eyed males present. Flies from the vial were allowed to interbreed, and in the next generation you find only red-eyed females, but you find both red-eyed and white-eyed males. The genotypes of the original parents were

a. $X^{w+}X^{w+}$; X^wY.
b. $X^{w+}X^w$, X^wY.
c. X^wX^w; $X^{w+}Y$.
d. $X^{w+}X^w$; X^wY.
e. $X^{w+}X^w$; $X^{w+}Y$.

Answer: a
Difficulty: Difficult
Bloom's Taxonomy: Application

28.   An individual who is a carrier of a genetically inherited disease

a.   has the disease, and any of their offspring will have the disease.
b.   does not have the disease but must have a parent with the disease.
c.   does not have the disease but may have offspring with the disease.
d.   has the disease and must have a parent with the disease.
e.   does not have the disease but must have a parent with the disease and may have offspring with the disease.

Answer: c
Difficulty: Moderate
Bloom's Taxonomy: Knowledge

29.   A woman with normal blood clotting mates with a man who has hemophilia. Their first child is a boy who has hemophilia. Tests show that the father and son both have the same form of hemophilia, that it is X-linked, and that the boy has normal genetic inheritance. You can predict that if the couple produces more children together, then the odds are that

a.   all of the children will have hemophilia.
b.   half of the boys and none of the girls will have hemophilia.
c.   none of the rest of children should have hemophilia.
d.   all of the boys and half of the girls will have hemophilia.
e.   half of the boys and half of the girls will have hemophilia.

Answer: e
Difficulty: Moderate
Bloom's Taxonomy: Application

30.   A woman with normal blood clotting mates with a man who also has normal blood clotting. Their first child is a boy who has hemophilia. Tests show that the child's hemophilia is X-linked and that he has normal genetic inheritance. You can predict that if the couple produces more children together, then the odds are that

a.   all of the children will have hemophilia.
b.   half of the boys and none of the girls will have hemophilia.
c.   none of the rest of children should have hemophilia.
d.   all of the boys and half of the girls will have hemophilia.
e.   half of the boys and half of the girls will have hemophilia.

Answer: b
Difficulty: Moderate
Bloom's Taxonomy: Application

31.   In placental mammals such as humans the dosage compensation mechanism to essentially equalize expression of sex-linked genes in males and females is

a.   doubling the gene expression for most genes on the X chromosome in males.
b.   doubling the number of X chromosomes in males early in embryonic development in all cells except for precursors of sex cells.
c.   having all major genes on the X chromosome also present on the Y chromosome.
d.   halving the gene expression for most genes on each X chromosome in females.
e.   inactivation of one of the two X chromosomes in most body cells of females.

Answer: e
Difficulty: Moderate
Bloom's Taxonomy: Comprehension

32.   If you see a male calico cat, you can be fairly certain that his diploid cells have a sex chromosome combination of _____.

a.   XYY
b.   XY
c.   XX
d.   XXY
e.   XO

Answer: d
Difficulty: Moderate
Bloom's Taxonomy: Synthesis

13.3 CHROMOSOMAL ALTERATIONS THAT AFFECT INHERITANCE

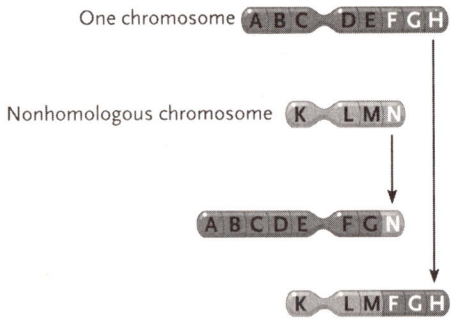

33. The change in the chromosomes depicted between the top and the bottom in the figure above represents a(n) _____.

a. inversion
b. duplication
c. reciprocal translocation
d. deletion
e. none of these

Answer: c
Difficulty: Difficult
Bloom's Taxonomy: Application
Source: Figure 13.11

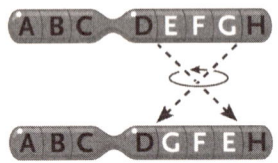

34. The change in the chromosome depicted between the top and the bottom in the figure above represents a(n) _____.

a. inversion
b. duplication
c. reciprocal translocation
d. deletion
e. none of these

Answer: a
Difficulty: Moderate
Bloom's Taxonomy: Application
Source: Figure 13.11

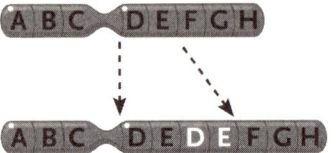

35. The change in the chromosome depicted between the top and the bottom in the figure above represents a(n) _____.

a. inversion
b. duplication
c. reciprocal translocation
d. deletion
e. none of these

Answer: b
Difficulty: Moderate
Bloom's Taxonomy: Application
Source: Figure 13.11

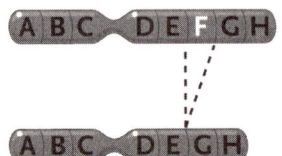

36. The change in the chromosome depicted between the top and the bottom in the figure above represents a(n) _____.

a. inversion
b. duplication
c. reciprocal translocation
d. deletion
e. none of these

Answer: d
Difficulty: Moderate
Bloom's Taxonomy: Application
Source: Figure 13.11

37. Which of the following is likely to be found in sperm produced when crossing over during meiosis occurs within a region where one of the homologous chromosomes has an inversion with respect to the other chromosome?

a. duplications
b. translocations
c. deletions
d. balanced translocations
e. both duplications and deletions

Answer: e
Difficulty: Moderate
Bloom's Taxonomy: Analysis

38. In humans a deletion from chromosome 5 typically leads to severe mental retardation and a malformed larynx; this disorder is known as _____.

a. Turner syndrome
b. Down syndrome
c. *cri-du-chat*
d. Triple-X syndrome
e. Klinefelter syndrome

Answer: c
Difficulty: Difficult
Bloom's Taxonomy: Knowledge

39. The presences of different genes for several types of hemoglobin in mammals, but not in sharks and many other vertebrates, is evidence of _____ of genetic material.

a. an inversion
b. duplication
c. reciprocal translocation
d. deletion
e. both duplication and deletion

Answer: b
Difficulty: Moderate
Bloom's Taxonomy: Application

40. Nondisjuction refers to

a. failure of homologous pairs to separate during meiosis.
b. improper pairing of nonhomologous chromosomes during meiosis.
c. failure of sister chromatids to pair during mitosis.
d. failure of homologous pairs or sister chromatids to separate during meiosis.
e. failure of homologous pairs to separate during mitosis.

Answer: d
Difficulty: Moderate
Bloom's Taxonomy: Knowledge

41. Individuals with extra or missing copies of some of their chromosomes are called _____.

a. polyploids
b. haploids
c. aneuploids
d. euploids
e. diploids

Answer: c
Difficulty: Moderate
Bloom's Taxonomy: Knowledge

42. Individuals with three or more copies of each of their chromosomes are called _____.

a. polyploids
b. haploids
c. aneuploids
d. euploids
e. diploids

Answer: a
Difficulty: Easy
Bloom's Taxonomy: Knowledge

43. Which is most likely to happen to a human polyploid?

a. natural abortion
b. death around the age of 1 year
c. survival until the early teens
d. survival until the mid-thirties
e. death around the age of 1 month

Answer: a
Difficulty: Moderate
Bloom's Taxonomy: Synthesis

44. Examinations of miscarried human embryos show that about 70% of such embryos are _____.

a. polyploids
b. haploids
c. aneuploids
d. euploids
e. diploids

Answer: c
Difficulty: Moderate
Bloom's Taxonomy: Knowledge

45. In humans an extra copy of chromosome 21 typically leads to moderate to severe mental retardation and sterility, as well as a greater likelihood of heart defects and other problems. This disorder is known as _____.

a. Turner syndrome
b. Down syndrome
c. *cri-du-chat*
d. Triple-X syndrome
e. Klinefelter syndrome

Answer: b
Difficulty: Easy
Bloom's Taxonomy: Knowledge

46. Which of the following conditions would most likely lead to an apparently normal human female?

a. Turner syndrome
b. Down syndrome
c. *cri-du-chat*
d. Triple-X syndrome
e. Klinefelter syndrome

Answer: d
Difficulty: Moderate
Bloom's Taxonomy: Knowledge

47. Which genetic condition is revealed in the karyotype display shown in the figure above?

a. Turner syndrome
b. Down syndrome
c. *cri-du-chat*
d. Triple-X syndrome
e. Klinefelter syndrome

Answer: b
Difficulty: Moderate
Bloom's Taxonomy: Application
Source: Figure 13.13

48. About half of all flowering plant species, including many important crop plants, are _____.

a. polyploids
b. haploids
c. aneuploids
d. euploids
e. diploids

Answer: a
Difficulty: Moderate
Bloom's Taxonomy: Knowledge

13.4 HUMAN GENETICS AND GENETIC COUNSELING

49. About 10–15% of African Americans are carriers for _____, an autosomal recessive genetic disorder where a defective version of hemoglobin is produced.

a. Duchenne muscular dystrophy
b. phenylketonuria
c. achondroplasia
d. sickle-cell anemia
e. cystic fibrosis

Answer: d
Difficulty: Easy
Bloom's Taxonomy: Knowledge

50. Carriers for sickle-cell anemia have a genetic advantage in some situations over those who do not have any copies of the sickle-cell allele because carriers have

a. lower blood pressure.
b. the ability to carry more oxygen in their blood.
c. increased resistance to malaria.
d. hyperactive immune systems.
e. extra red blood cells.

Answer: c
Difficulty: Moderate
Bloom's Taxonomy: Knowledge

51. About 4% of those of Northern European descent are carriers for _____, an autosomal recessive genetic disorder where a defective membrane transport protein leads to abnormal chloride levels in extracellular fluids.

a. Duchenne muscular dystrophy
b. phenylketonuria
c. achondroplasia
d. sickle-cell anemia
e. cystic fibrosis

Answer: e
Difficulty: Moderate
Bloom's Taxonomy: Knowledge

52. Most hospitals in the United States routinely test all newborns for this autosomal recessive disorder where an enzyme in amino acid metabolism is not produced, leading to a buildup of compounds that damage brain tissue and can lead to mental retardation unless a restricted diet is followed.

a. Duchenne muscular dystrophy
b. phenylketonuria
c. achondroplasia
d. sickle-cell anemia
e. cystic fibrosis

Answer: b
Difficulty: Moderate
Bloom's Taxonomy: Knowledge

53. This autosomal dominant genetic trait associated with a gene on human chromosome 4 leads to a type of dwarfing due to defective cartilage formation.

a. Duchenne muscular dystrophy
b. phenylketonuria
c. achondroplasia
d. sickle-cell anemia
e. cystic fibrosis

Answer: c
Difficulty: Moderate
Bloom's Taxonomy: Knowledge

54. Which of the following has X-linked recessive inheritance?

a. Duchenne muscular dystrophy
b. phenylketonuria
c. achondroplasia
d. sickle-cell anemia
e. cystic fibrosis

Answer: a
Difficulty: Moderate
Bloom's Taxonomy: Knowledge

55. The percentage of children's hospital patients being treated for problems arising from inherited disorders is about

a. 1%.
b. 1% to 3%.
c. 5% to 10%.
d. 10% to 25%.
e. 70%.

Answer: d
Difficulty: Difficult
Bloom's Taxonomy: Knowledge

56. As part of his or her job, a genetic counselor may

a. analyze family pedigrees.
b. advise a couple to have direct testing for altered proteins.
c. advise a couple to have DNA testing.
d. predict the chances of a couple having a child with a specific trait.
e. all of these

Answer: e
Difficulty: Moderate
Bloom's Taxonomy: Comprehension

57. You are genetic counselor, and a couple comes to you with concerns that if they have a child together it could have cystic fibrosis. Genetic tests reveal that the woman is a carrier for cystic fibrosis but the man is not. Which of the following would be a correct thing to tell them?

a. Any child that they produce has a 50% chance of having cystic fibrosis, and the rest will be carriers.
b. All of their offspring will have cystic fibrosis.
c. They should not have any concerns, as no child that they produce should wind up with cystic fibrosis.
d. No child that they produce should wind up with cystic fibrosis, but each of their offspring will have a 50% chance of being a carrier.
e. Any child that they produce will have a 25% chance of having cystic fibrosis.

Answer: d
Difficulty: Moderate
Bloom's Taxonomy: Application

58. You are genetic counselor, and a couple comes to you with concerns that if they have a child together it could have sickle-cell anemia. Genetic tests reveal that the man and woman are both carriers for sickle-cell anemia. Which of the following would be a correct thing to tell them?

a. Any child that they produce has a 50% chance of having sickle-cell anemia, and the rest will be carriers.
b. All of their offspring will have sickle-cell anemia.
c. They should not have any concerns, as no child that they produce should wind up with sickle-cell anemia.
d. No child that they produce should wind up with sickle-cell anemia, but each of their offspring will have a 50% chance of being a carrier.
e. Any child that they produce will have a 25% chance of having sickle-cell anemia.

Answer: e
Difficulty: Moderate
Bloom's Taxonomy: Application

59. Prenatal diagnosis techniques such as amniocentesis and chorionic villus sampling remove cells that were produced by an embryo. These cells are then used

a. to treat diseases.
b. for implantation to produce pregnancy.
c. to develop embryonic stem cells.
d. to test for the presence of mutant alleles or chromosomal alterations.
e. for cloning.

Answer: d
Difficulty: Moderate
Bloom's Taxonomy: Comprehension

INSIGHTS FROM THE MOLECULAR REVOLUTION: ACHONDROPLASTIC DWARFING BY A SINGLE AMINO ACID CHANGE

60. A mutation in the human gene for a fibroblast growth factor receptor (FGFR) appears to be responsible for _____.

a. Duchenne muscular dystrophy
b. phenylketonuria
c. achondroplasia
d. sickle-cell anemia
e. cystic fibrosis

Answer: c
Difficulty: Moderate
Bloom's Taxonomy: Knowledge

13.5 NONTRADITIONAL PATTERNS OF INHERITANCE

61. Cytoplasmic inheritance refers to genes found in

a. sex chromosomes.
b. mitochondria and chloroplasts.
c. viruses.
d. bacteria.
e. only one parent.

Answer: b
Difficulty: Moderate
Bloom's Taxonomy: Comprehension

62. For most multicellular eukaryotes, including humans, mitochondria are inherited

a. only from the father.
b. from both the father and the mother.
c. from the mother for females and from the father for males.
d. randomly from either the father or the mother.
e. only from the mother.

Answer: e
Difficulty: Moderate
Bloom's Taxonomy: Comprehension

63. The expression of only one allele of a gene and silencing of the other allele, all based on which parent contributed each allele, is

a. sex-linked inheritance.
b. uniparental inheritance.
c. cytoplasmic inheritance.
d. genomic imprinting.
e. maternal inheritance.

Answer: d
Difficulty: Moderate
Bloom's Taxonomy: Knowledge

64. Prader-Willi syndrome and Angelman syndrome are both caused by the same genetic deletion on human chromosome 15, but they have very different phenotypes. Which disorder occurs depends upon which parent provided the chromosome with the deletion. This is an example of

a. sex-linked inheritance.
b. uniparental inheritance.
c. cytoplasmic inheritance.
d. genomic imprinting.
e. maternal inheritance.

Answer: d
Difficulty: Moderate
Bloom's Taxonomy: Comprehension

65. The molecular mechanism that produces genomic imprinting by inactivating an allele involves _____ of DNA in the allele that will not be expressed.

a. methylation
b. mutation
c. ligation
d. deletion
e. inversion

Answer: a
Difficulty: Moderate
Bloom's Taxonomy: Knowledge

66. Cancers associated with mammalian insulin growth factor 2 (*IGf2*) typically result from

a. translocations.
b. X-inactivation.
c. inversions.
d. duplications.
e. loss of imprinting.

Answer: e
Difficulty: Difficult
Bloom's Taxonomy: Knowledge

UNANSWERED QUESTIONS

67. Genetic recombination in somatic cells is used to

a. produce variability in offspring.
b. determine cell types during development.
c. repair damaged or broken chromosomes.
d. produce specialized tissues in organs.
e. establish the sex of the individual.

Answer: c
Difficulty: Moderate
Bloom's Taxonomy: Knowledge

68. Research findings indicate that the *SRY* gene probably evolved from

a. a fibroblast growth factor receptor gene.
b. a testosterone receptor gene on the Y chromosome.
c. a chromosomal translocation between an autosome and the X chromosome.
d. a gene on the Y chromosome that determines ear hair length.
e. a brain-determining gene on the X chromosome.

Answer: e
Difficulty: Difficult
Bloom's Taxonomy: Knowledge

69. In cells of human individuals with three or more X chromosomes

a. one X chromosome remains active and the rest are inactivated.
b. all of the X chromosomes remain active.
c. two X chromosomes remain active, and any more are inactivated.
d. two X chromosomes remain active in females and one remains active in males; the rest are inactivated.
e. only one X chromosome is inactivated.

Answer: a
Difficulty: Moderate
Bloom's Taxonomy: Knowledge

Integrative Multiple-Choice

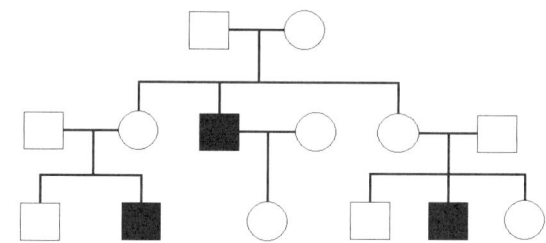

70. Examine the pedigree in the figure above, where individuals that have the genetic condition being tested are marked with filled squares or circles. Which of the following inheritance patterns is most likely correct for this condition?

a. autosomal dominant
b. X-linked recessive
c. X-linked dominant
d. cytoplasmic inheritance
e. autosomal recessive

Answer: b
Difficulty: Moderate
Bloom's Taxonomy: Application
Source: Sections 13.2 and 13.4

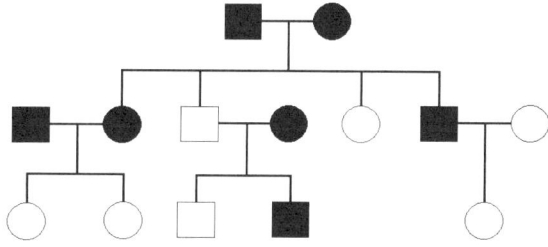

71. Examine the pedigree in the figure above, where individuals that have the genetic condition being tested are marked with filled squares or circles. Which of the following inheritance patterns is most likely correct for this condition?

a. autosomal dominant
b. X-linked recessive
c. X-linked dominant
d. cytoplasmic inheritance
e. autosomal recessive

Answer: a
Difficulty: Moderate
Bloom's Taxonomy: Application
Source: Sections 13.2 and 13.4

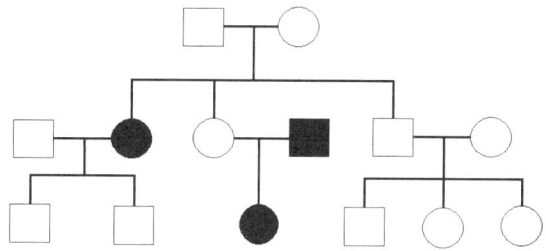

72. Examine the pedigree in the figure above, where individuals that have the genetic condition being tested are marked with filled squares or circles. Which of the following inheritance patterns is most likely correct for this condition?

a. autosomal dominant
b. X-linked recessive
c. X-linked dominant
d. cytoplasmic inheritance
e. autosomal recessive

Answer: e
Difficulty: Moderate
Bloom's Taxonomy: Application
Source: Sections 13.2 and 13.4

73. You are genetic counselor, and a couple comes to you with concerns that if they have a child together it could have hemophilia. The man has an X-linked recessive form of hemophilia, but the woman does not. Genetic tests reveal the woman is not a carrier for hemophilia. Which of the following would be a correct thing to tell them?

a. Any child that they produce will have a 50% chance of having hemophilia.
b. All of their male offspring will have hemophilia, and each of their female offspring will have a 50% chance of having hemophilia.
c. None of their offspring should have hemophilia, but all of their female offspring will be carriers for hemophilia.
d. Each of their male offspring will have a 50% chance of having hemophilia, and while their female offspring should not have hemophilia, they will each have a 50% chance of being carriers.
e. All of their male offspring will have hemophilia, and while their female offspring should not have hemophilia, they will all be carriers.

Answer: c
Difficulty: Moderate
Bloom's Taxonomy: Application
Source: Sections 13.2 and 13.4

74 You are genetic counselor, and a couple comes to you with concerns that if they have a child together it could have hemophilia. The woman has an X-linked recessive form of hemophilia, but the man does not. Which of the following would be a correct thing to tell them?

a. Any child that they produce will have a 50% chance of having hemophilia.
b. All of their male offspring will have hemophilia, and each of their female offspring will have a 50% chance of having hemophilia.
c. None of their offspring should have hemophilia, but all of their female offspring will be carriers for hemophilia.
d. Each of their male offspring will have a 50% chance of having hemophilia, and while their female offspring should not have hemophilia they will each have a 50% chance of being carriers.
e. All of their male offspring will have hemophilia, and while their female offspring should not have hemophilia, they will all be carriers.

Answer: e
Difficulty: Moderate
Bloom's Taxonomy: Application
Source: Sections 13.2 and 13.4

75. You are genetic counselor, and a couple comes to you with concerns that if they have a child together it could have hemophilia. Neither one of them has hemophilia, but the woman's biological father did have an X-linked recessive form of hemophilia. Which of the following would be a correct thing to tell them?

a. Any child that they produce will have a 50% chance of having hemophilia.
b. All of their male offspring will have hemophilia, and each of their female offspring will have a 50% chance of having hemophilia.
c. None of their offspring should have hemophilia, but all of their female offspring will be carriers for hemophilia.
d. Each of their male offspring will have a 50% chance of having hemophilia, and while their female offspring should not have hemophilia, they will each have a 50% chance of being carriers.
e. All of their male offspring will have hemophilia, and while their female offspring should not have hemophilia, they will all be carriers.

Answer: d
Difficulty: Difficult
Bloom's Taxonomy: Application
Source: Sections 13.2 and 13.4

Choice

Choose the mode of inheritance that has been determined for each of the following genetic conditions in humans.

a. autosomal recessive
b. autosomal dominant
c. X-linked recessive
d. aneuploidy
e. genomic imprinting

76. _____ Down syndrome

77. _____ achondroplasia

78. _____ sickle cell anemia

79. _____ Angelman syndrome

80. _____ phenylketonuria

81. _____ hemophilia A

82. _____ cystic fibrosis

83. _____ red-green colorblindness

84. _____ Huntington disease

Answers: **76. d** **77. b** **78. a** **79. e** **80. a**
81. c **82. a** **83. c** **84. b**

Difficulty: Moderate
Bloom's Taxonomy: Knowledge
Source: Sections 13.2–13.5

Short Answer

85. Suppose that you have discovered a new mutant in *Drosophila melanogaster.* What kinds of genetic crosses should you do to determine if the allele causing the mutation is inherited as an autosomal dominant trait, an autosomal recessive trait, or a sex-linked trait?

Answer: You should do at least the following crosses:
male with the mutant phenotype X female with the mutant phenotype
male with the mutant phenotype X female without the mutant phenotype
male without the mutant phenotype X female with the mutant phenotype
After examining the offspring of these crosses, you may need to do even more crosses to determine the inheritance pattern. Crosses within some of the offspring sets may be very useful in such cases.
Difficulty: Moderate
Bloom's Taxonomy: Evaluation
Source: Section 13.2

Essay

86. You are working in a research lab on the organism *Drosophila melanogaster* and you have been given the job of determining the relative positions of three linked genes for your undergraduate research project. A graduate student in the lab tells you that gene *a* is 5.0 centimorgans from gene *b*, and that gene *c* is 2.7 centimorgans from gene *b*. What should you do to finish mapping these three genes? What are the likely results that may occur, and how would you interpret such results?

Answer: To complete this task, one should perform crosses that will allow determination of the recombination frequency between gene a and gene c. The recombination frequency will give the map distance between genes a and c. There are two likely outcomes: an a-c map distance of about 2.3 centimorgans and an a-c map distance of about 7.7 centimorgans. The results may not be exactly these values, but they should be close to one or the other of them. A result of close to 2.3 centimorgans would indicate that gene c is between genes a and b, while a result of close to 7.7 centimorgans would indicate that gene b is between genes a and c.
Difficulty: Moderate
Bloom's Taxonomy: Evaluation
Source: Section 13.1

87. Your cousin finds out that she is a carrier for phenylketonuria. Also, her husband's biological mother has phenylketonuria. She asks you to explain what this could mean if she and her husband have children and asks what, if anything, could be done to best protect any child that they have. What would you tell her?

Answer: First, I would tell her that phenylketonuria is easily treated by dietary adjustments, and that with proper guidance and behavior people with phenylketonuria are able to lead full, productive, and essentially normal lives. I would tell her that the way phenylketonuria is inherited guarantees that her husband is also a carrier, since his mother has it. He got a normal copy of the gene involved in phenylketonuria from his father, and he got the phenylketonuria form of the gene from his mother. The normal copy from his father prevents him from having phenylketonuria himself. I would tell her that, as a carrier, she is in the same situation—she has one normal copy of the gene and one abnormal copy, and the normal copy keeps her from having phenylketonuria. Since the odds of passing on any gene copy are essentially the same as flipping a coin, any child that she has will have a 50% chance of getting a normal copy of the gene from her and a 50% chance of getting an abnormal copy. The same will apply for her husband. Since having phenylketonuria requires having an abnormal gene copy from both parents, each child that they produce will have a 25% chance of having phenylketonuria. Then, I would tell her that nearly all hospitals in the United States screen newborn infants for phenylketonuria, and that she should ensure that such a test is done for any child that she has. If the results of the test indicate that the child has phenylketonuria, she should follow the dietary instructions for the child closely to ensure that phenylketonuria does not affect the mental abilities of the child.
Difficulty: Difficult
Bloom's Taxonomy: Evaluation
Source: Section 13.4

14

DNA STRUCTURE, REPLICATION, AND ORGANIZATION

Multiple-Choice

WHY IT MATTERS

1. The genetic material of all living organisms is
    a. protein.
    b. deoxyribonucleic acid.
    c. Ribonucleic acid.
    d. glycoprotein.
    e. polypeptide.

Answer: b
Difficulty: Easy
Bloom's Taxonomy: Knowledge

2. Watson and Crick are famous for determining
    a. the presence of nuclein (DNA) in pus cells.
    b. the role of DNA in cells.
    c. what chemical components are in DNA.
    d. the three-dimensional structure of DNA.
    e. the location of DNA in cells.

Answer: d
Difficulty: Easy
Bloom's Taxonomy: Comprehension

14.1 ESTABLISHING DNA AS THE HEREDITARY MOLECULE

3. Prior to the 1940s, many biologists thought that _____, being made up of 20 types of _____, was the best candidate for genetic material.
    a. protein; amino acids
    b. DNA; amino acids
    c. protein; nucleotides
    d. DNA; nucleotides
    e. RNA; nucleotides

Answer: a
Difficulty: Easy
Bloom's Taxonomy: Knowledge

4. What happens when living *R* strain *Streptococcus pneumoniae* bacteria are mixed with heat-killed *S* strain *Streptococcus pneumoniae* bacteria?
    a. The *S* strain bacteria come back to life.
    b. The *R* strain bacteria are killed, and the *S* strain bacteria remain dead.
    c. The *R* strain bacteria are transformed into *S* strain bacteria.
    d. The *S* strain bacteria are transformed into *R* strain bacteria.
    e. The *R* strain bacteria are killed, and the *S* strain bacteria come back to life.

Answer: c
Difficulty: Moderate
Bloom's Taxonomy: Comprehension

5. The *transforming principle* that Griffith described from his work with *Streptococcus pneumoniae* bacteria turned out to be _____.
    a. a polysaccharide capsule
    b. a phospholipid
    c. protein
    d. RNA
    e. DNA

Answer: e
Difficulty: Easy
Bloom's Taxonomy: Knowledge

6. In their experiments to determine the transforming principle, Avery, MacLeod, and McCarty used enzymes that break down _____.
    a. protein, DNA, and RNA
    b. protein and RNA
    c. DNA
    d. protein, lipids, and RNA
    e. lipids and DNA

Answer: a
Difficulty: Difficult
Bloom's Taxonomy: Knowledge

7. In the Hershey and Chase experiment, the pellet was radioactive after bacteria had been infected with ^{32}P-labeled viruses and centrifuged. Why?
    a. Bacteria were centrifuged to form the pellet, and they had incorporated radioactive proteins into their DNA.
    b. Bacteria were centrifuged to form the pellet, and they had incorporated radioactive DNA.
    c. Viruses were centrifuged to form the pellet, and they had incorporated radioactive proteins from the bacterial DNA.
    d. Bacteria were centrifuged to form the pellet, and they had incorporated radioactive proteins into their cell membranes.
    e. Viruses were centrifuged to form the pellet, and they had incorporated radioactive DNA.

Answer: b
Difficulty: Moderate
Bloom's Taxonomy: Comprehension, Synthesis

8. The T2 bacteriophages used in the Hershey and Chase experiment are made of what?
    a. DNA
    b. RNA and protein
    c. protein, phospholipid, and DNA
    d. DNA and protein
    e. protein, phospholipid, DNA, and RNA

Answer: d
Difficulty: Moderate
Bloom's Taxonomy: Analysis

9. In the Hershey and Chase experiment, ^{32}P was used to label _____ and ^{35}S was used to label _____.

a. RNA; protein
b. protein; DNA
c. phospholipids; protein
d. protein; phospholipids
e. DNA; protein

Answer: e
Difficulty: Moderate
Bloom's Taxonomy: Comprehension

14.2 DNA STRUCTURE

10. Each DNA nucleotide is made up of

a. a six-carbon sugar, a phosphate group, and one of twenty amino acids.
b. a five-carbon sugar, a nitrogenous base, and one of twenty amino acids.
c. a five-carbon sugar, a phosphate group, and one of four nitrogenous bases.
d. a six-carbon sugar, a nitrogenous base, and onc of four amino acids.
e. a five-carbon sugar, a phosphate group, and one of four amino acids.

Answer: c
Difficulty: Moderate
Bloom's Taxonomy: Knowledge

11. Adjacent nucleotides on a strand of DNA are connected to each other by a(n) _____.

a. hydrophobic interaction
b. phosphodiester bond
c. hydrogen bond
d. peptide bond
e. ionic bond

Answer: b
Difficulty: Moderate
Bloom's Taxonomy: Knowledge

12. The DNA of an organism is studied and found to contain 14% guanine. This organism should have _____% thymine and _____% cytosine in its DNA.

a. 86; 14
b. 14; 36
c. 36; 36
d. 14; 86
e. 36; 14

Answer: e
Difficulty: Difficult
Bloom's Taxonomy: Application

13. The DNA of an organism is studied and found to contain 30% adenine. Based on this you would predict that the DNA of this organism also contains 30% _____.

a. thymine
b. cytosine
c. each of cytosine and guanine
d. each of thymine and guanine
e. guanine

Answer: a
Difficulty: Moderate
Bloom's Taxonomy: Application

14. In DNA the purines are

a. thymine and cytosine.
b. adenine and cytosine.
c. adenine and guanine.
d. thymine and adenine.
e. guanine and thymine.

Answer: c
Difficulty: Difficult
Bloom's Taxonomy: Knowledge

15. In DNA the pyrimidines are

a. thymine and cytosine.
b. adenine and cytosine.
c. adenine and guanine.
d. thymine and adenine.
e. guanine and thymine.

Answer: a
Difficulty: Difficult
Bloom's Taxonomy: Knowledge

16. Which of the following nitrogenous bases is NOT normally found in DNA?

a. thymine
b. cytosine
c. adenine
d. uracil
e. guanine

Answer: d
Difficulty: Easy
Bloom's Taxonomy: Knowledge

17. Which of the following nucleotide sequences represents the complementary sequence that would bind to the DNA strand 5'–GACGTT–3'?

a. 5'–TCATGG–3'
b. 3'–TCATGG–5'
c. 3'–CTGCAA–5'
d. 3'–AGTACC–5'
e. 5'–TTGCAG–3'

Answer: c
Difficulty: Easy
Bloom's Taxonomy: Application

18. Wilkins and Franklin studied the structure of DNA using _____.

a. molecular scale models of nucleotides
b. X-ray diffraction
c. computer-assisted graphics
d. electron microscopy
e. light microscopy

Answer: b
Difficulty: Moderate
Bloom's Taxonomy: Knowledge

19. A DNA double helix has two strands that are held to each other by _____.

a. hydrogen bonds
b. ionic bonds
c. hydrophobic interactions
d. phosphodiester bonds
e. covalent bonds

Answer: a
Difficulty: Easy
Bloom's Taxonomy: Knowledge

20. The width of a DNA double helix

a. is constant.
b. is narrower where adenine is present than where cytosine is present.
c. is wider where purines are present than where pyrimidines are present.
d. varies randomly.
e. is wider where pyrimidines are present than where purines are present.

Answer: a
Difficulty: Easy
Bloom's Taxonomy: Comprehension

21. The two strands of a DNA double helix are said to be antiparallel. This means that

a. the 5' end of one strand is directly paired with the 5' end of the other strand.
b. since the double helix twists, it is not perfectly parallel.
c. one strand has a negative charge and the other strand has a positive charge.
d. the 5' end of one strand is directly paired with the 3' end of the other strand.
e. both strands have a negative charge.

Answer: d
Difficulty: Easy
Bloom's Taxonomy: Comprehension

14.3 DNA REPLICATION

22. DNA replication is said to be semiconservative because

a. the number of nucleotides within genes remains constant.
b. half of the DNA in a cell comes from one parent and the other half from the other parent.
c. the same process of DNA replication is used by all organisms.
d. the total amount of DNA within an individual remains the same.
e. each new DNA molecule is composed of one old strand and one new strand.

Answer: e
Difficulty: Easy
Bloom's Taxonomy: Comprehension

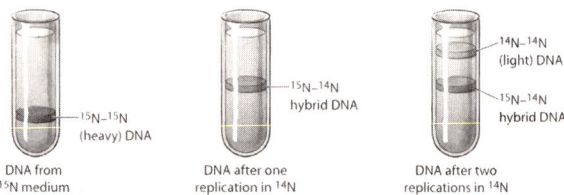

23. The figure above depicts the result of an experiment to determine how DNA replication occurs. Based on these results, it appears that after replication each DNA molecule is made of

a. either entirely old DNA strands or entirely new DNA strands.
b. one old DNA strand and one new DNA strand.
c. entirely new DNA.
d. some DNA helix regions that are old DNA, alternating with some DNA regions that are new DNA.
e. two strands that are each a mix of old and new DNA.

Answer: b
Difficulty: Difficult
Bloom's Taxonomy: Evaluation
Source: Fig. 14.9

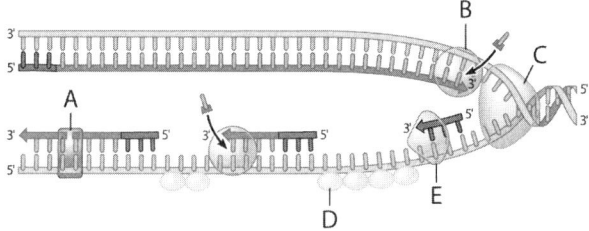

Use the figure above for questions 24–28.

24. The structure labeled "A" in this figure of a replication fork represents _____.

a. single-stranded binding protein
b. primase
c. DNA ligase
d. helicase
e. DNA polymerase

Answer: c
Difficulty: Moderate
Bloom's Taxonomy: Analysis
Source: Fig. 14.12

25. The structure labeled "B" in this figure of a replication fork represents _____.

a. single-stranded binding protein
b. primase
c. DNA ligase
d. helicase
e. DNA polymerase

Answer: e
Difficulty: Moderate
Bloom's Taxonomy: Analysis
Source: Fig. 14.12

26. The structure labeled "C" in this figure of a replication fork represents _____.

a. single-stranded binding protein
b. primase
c. DNA ligase
d. helicase
e. DNA polymerase

Answer: d
Difficulty: Moderate
Bloom's Taxonomy: Analysis
Source: Fig. 14.12

27. The structure labeled "D" in this figure of a replication fork represents _____.

a. single-stranded binding protein
b. primase
c. DNA ligase
d. helicase
e. DNA polymerase

Answer: a
Difficulty: Difficult
Bloom's Taxonomy: Analysis
Source: Fig. 14.12

28. The structure labeled "E" in this figure of a replication fork represents _____.

a. single-stranded binding protein
b. primase
c. DNA ligase
d. helicase
e. DNA polymerase

Answer: b
Difficulty: Moderate
Bloom's Taxonomy: Analysis
Source: Fig. 14.12

29. Which of the following adds individual nucleotides to the 3' end of an existing strand to build a new DNA strand during DNA replication?

a. topoisomerase
b. primase
c. DNA polymerase
d. helicase
e. DNA ligase

Answer: c
Difficulty: Moderate
Bloom's Taxonomy: Knowledge

30. Which of the following acts to remove overtwisting and strain ahead of the replication fork during DNA replication?

a. topoisomerase
b. primase
c. DNA polymerase
d. helicase
e. DNA ligase

Answer: a
Difficulty: Difficult
Bloom's Taxonomy: Knowledge

31. Which of the following closes "nicks" between DNA fragments, forming a covalent bond that ties or joins the fragments together?

a. topoisomerase
b. primase
c. DNA polymerase
d. helicase
e. DNA ligase

Answer: e
Difficulty: Easy
Bloom's Taxonomy: Knowledge

32. Which of the following catalyzes the unwinding of the DNA double helix during DNA replication?

a. topoisomerase
b. primase
c. DNA polymerase
d. helicase
e. DNA ligase

Answer: d
Difficulty: Easy
Bloom's Taxonomy: Knowledge

33. Which of the following assembles a short RNA chain as the first nucleotides in a new DNA strand?

a. topoisomerase
b. primase
c. DNA polymerase
d. helicase
e. DNA ligase

Answer: b
Difficulty: Moderate
Bloom's Taxonomy: Knowledge

34. During DNA replication, DNA ligase is most active on the lagging strand. Why?

a. The lagging strands contain more short DNA segments than the leading strand, and these short segments are joined together by DNA ligase.
b. The lagging strand is synthesized more slowly, and DNA ligase speeds up the DNA polymerase.
c. The lagging strand synthesizes DNA in the 3' → 5' direction.
d. The lagging strand requires DNA ligase to couple the RNA primer to the Okazaki fragments.
e. The lagging strand has no RNA primase activity; it is replaced by DNA ligase.

Answer: a
Difficulty: Difficult
Bloom's Taxonomy: Synthesis

35. Adding nucleotides onto a growing DNA strand during DNA replication in cells occurs in

a. the 5' → 3' direction for the leading strand and the 3' → 5' direction on the lagging strand.
b. either the 5' → 3' direction or the 3' → 5' direction on both strands, depending on where replication begins.
c. the 5' → 3' direction only.
d. the 3' → 5' direction for the leading strand and the 5' → 3' direction on the lagging strand.
e. the 3' → 5' direction only.

Answer: c
Difficulty: Moderate
Bloom's Taxonomy: Knowledge

36. Reiji Okazaki discovered that what are now called "Okazaki fragments" are produced during DNA replication. These are

a. short lengths of new DNA on the leading strand.
b. RNA primers on the lagging strand.
c. RNA primers on both the lagging and leading strand.
d. short lengths of new DNA on the lagging strand.
e. RNA primers on the leading strand.

Answer: d
Difficulty: Easy
Bloom's Taxonomy: Knowledge

37. During DNA replication, the _____ strand is assembled in the _____ direction that the DNA double helix unwinds and is produced by _____ replication.

a. leading; opposite; continuous
b. lagging; same; discontinuous
c. leading; same; discontinuous
d. lagging; opposite; continuous
e. leading; same; continuous

Answer: e
Difficulty: Moderate
Bloom's Taxonomy: Knowledge

38. DNA replication must start at a replication origin. In eukaryotes the DNA molecule that makes up a chromosome is typically _____ and usually has _____ replication origin.

a. circular; one
b. circular; more than one
c. linear; more than one
d. linear; one

Answer: c
Difficulty: Easy
Bloom's Taxonomy: Knowledge

39. The energy to form the new bond when a nucleotide is added to a growing DNA strand is provided primarily by _____.

a. unwinding of the DNA double helix
b. hydrolysis of pyrophosphate
c. breaking hydrogen bonds between base pairs
d. DNA polymerase
e. forming hydrogen bonds between base pairs

Answer: b
Difficulty: Difficult
Bloom's Taxonomy: Knowledge, Synthesis

40. Imagine that a cell has a genetic mutation so that the primase enzyme is unable to make RNA strands. Assuming that all of the other enzymes directly involved in DNA replication are still functional in these cells, how much of the process of DNA replication would you expect to see in these cells?

a. The leading strand would be synthesized, but not the lagging strand.
b. None of it at all, no part of the DNA replication process could occur.
c. The DNA helix would be unwound by helicase, but no new strands will be produced at all.
d. Both the leading and lagging strand would be synthesized, but pieces of discontinuous strands would not be joined together.
e. DNA replication would still proceed completely, since RNA strands are not part of the final product of DNA replication.

Answer: c
Difficulty: Difficult
Bloom's Taxonomy: Synthesis

41. Telomeres are found

a. in the middle of chromosomes.
b. at replication origins.
c. where DNA strands are joined together.
d. within genes.
e. at the ends of a chromosome.

Answer: e
Difficulty: Easy
Bloom's Taxonomy: Knowledge

42. In humans telomerase functions to

a. add telomere repeats in some human cells.
b. remove telomere repeats in all human cells.
c. add telomere repeats in all human cells.
d. remove telomere repeats in some human cells.

Answer: a
Difficulty: Easy
Bloom's Taxonomy: Knowledge

43. During normal DNA replication, part of the DNA at the ends of linear chromosomes is not copied into the new DNA strands because

a. DNA ligase cannot join pieces at the end of a chromosome.
b. RNA primers at the beginning of a new strand cannot be replaced with DNA.
c. those ends are Okazaki fragments that are lost.
d. cells do not need the DNA at the ends of chromosomes.
e. the ends of chromosomes are made of protein, not DNA.

Answer: b
Difficulty: Moderate
Bloom's Taxonomy: Knowledge

44. Suppose you take a cell from an adult cow and attempt to use it to produce a clone of that cow. If for some reason telomerase is not functioning in that cell or in any cell that comes from it, what will you expect to happen with your clone?

a. The cell will never be able to divide at all.
b. When the clone grows up, it will most likely have cancer.
c. The cell may divide, but after a certain number of divisions, cell division will stop.
d. The lack of telomerase should have no effect on the clone.
e. When the clone grows up, it will most likely be sterile.

Answer: c
Difficulty: Difficult
Bloom's Taxonomy: Synthesis

INSIGHTS FROM THE MOLECULAR REVOLUTION: A FRAGILE CONNECTION BETWEEN DNA REPLICATION AND MENTAL RETARDATION

45. Fragile X syndrome is one of the most common sources of inherited _____.

a. sterility
b. mental retardation
c. blindness
d. cancer
e. deafness

Answer: b
Difficulty: Easy
Bloom's Taxonomy: Knowledge

46. Fragile X syndrome is caused by _____ on the X chromosome in humans.

a. the presence of CCG repeats
b. mutations within CCG repeats
c. deletion of CCG repeats
d. overreplication of CCG repeats
e. the absence of CCG repeats

Answer: d
Difficulty: Moderate
Bloom's Taxonomy: Knowledge

14.4 MECHANISMS THAT CORRECT REPLICATION ERRORS

47. Many errors made during DNA replication are corrected during DNA replication through proofreading by _____.

a. DNA polymerase
b. primase
c. telomerase
d. DNA ligase
e. helicase

Answer: a
Difficulty: Easy
Bloom's Taxonomy: Knowledge

48. Without proofreading, the rate of errors in DNA replication in bacteria is as high as about one for every _____ nucleotides assembled.

a. 10 to 100
b. 100,000 to 1,000,000
c. 1000 to 10,000
d. 10,000,000 to 100,000,000
e. 100,000,000 to 1,000,000,000

Answer: c
Difficulty: Difficult
Bloom's Taxonomy: Knowledge

49. DNA repair enzymes typically find mismatched base pairs by scanning for _____.

a. missing hydrogen bonds
b. unsealed nicks in the DNA strands
c. Okazaki fragments
d. extra hydrogen bonds
e. distortions in the DNA double helix

Answer: e
Difficulty: Difficult
Bloom's Taxonomy: Knowledge

50. After DNA repair enzymes cut out part of a DNA strand that had an incorrect nucleotide, _____ is/are needed to complete the repair.

a. primase, DNA polymerase, and DNA ligase
b. DNA polymerase
c. DNA polymerase and DNA ligase
d. primase and DNA ligase
e. primase and DNA polymerase

Answer: c
Difficulty: Moderate
Bloom's Taxonomy: Comprehension

51. Individuals with *xeroderma pigmentosum* inherit a faulty DNA repair mechanism. As a consequence, they

a. are sterile.
b. have no proofreading during DNA replication.
c. cannot join the Okazaki fragments produced during DNA replication.
d. easily develop skin cancer when exposed to sunlight.
e. lose part of the DNA on the ends of chromosomes during each round of DNA replication.

Answer: d
Difficulty: Moderate
Bloom's Taxonomy: Knowledge

52. _____ are the ultimate source of variability in offspring.

a. Mutations
b. Nucleosomes
c. Okazaki fragments
d. DNA repairs
e. RNA primers

Answer: a
Difficulty: Easy
Bloom's Taxonomy: Knowledge

14.5 DNA ORGANIZATION IN EUKARYOTES AND PROKARYOTES

53. Chromatin is made up of

a. DNA only.
b. DNA, RNA, and protein.
c. DNA and RNA.
d. RNA and protein.
e. DNA and protein.

Answer: e
Difficulty: Easy
Bloom's Taxonomy: Knowledge

54. Nucleosomes are best described as

a. prokaryotic DNA associated with nonhistone proteins.
b. eukaryotic DNA associated with histone proteins.
c. prokaryotic DNA associated with histone proteins.
d. eukaryotic DNA associated with nonhistone proteins.

Answer: b
Difficulty: Easy
Bloom's Taxonomy: Application

55. Histones are _____ and _____ charged proteins.

a. large; negatively
b. large; positively
c. small; positively
d. small; negatively

Answer: c
Difficulty: Moderate
Bloom's Taxonomy: Knowledge

56. The DNA in the nucleus of a typical human cell nucleus would be about _____ long if fully stretched out.

a. 2 m
b. 10 mm
c. 100 m
d. 10 μm
e. 2 km

Answer: a
Difficulty: Difficult
Bloom's Taxonomy: Knowledge

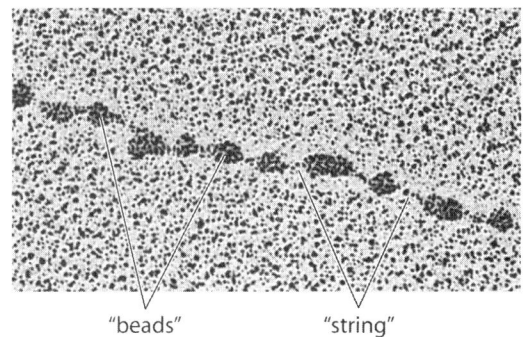

"beads"   "string"

57. The electron micrograph above shows chromatin organized like "beads on a string." At this level of DNA compaction the "beads" are _____ and the "string" is _____.

a. nucleosomes; DNA
b. 30-nm chromatin fibers; 10-nm chromatin fiber
c. DNA; 10-nm chromatin fiber
d. nucleosomes; histone H1
e. DNA; histone H1

Answer: a
Difficulty: Moderate
Bloom's Taxonomy: Application
Source: Figure 14.18

58. Association of histone H1 with linker DNA is required to form _____.

a. heterochromatin
b. the 10-nm chromatin fiber
c. nucleosomes
d. the solenoid
e. euchromatin

Answer: d
Difficulty: Difficult
Bloom's Taxonomy: Knowledge

59. The nucleosome core particle is composed of _____.

a. DNA
b. helicase
c. single-stranded binding proteins
d. RNA primer
e. histones

Answer: e
Difficulty: Moderate
Bloom's Taxonomy: Knowledge, Analysis

60. Which of these items associated with DNA packaging is the thickest?

a. a DNA double helix
b. a solenoid
c. a nucleosome
d. an H2A histone protein
e. an H1 histone protein

Answer: b
Difficulty: Moderate
Bloom's Taxonomy: Application

61. Histones bond to DNA by _____ of DNA.

a. an ionic attraction to the phosphate groups
b. hydrogen bonding to the deoxyribose sugars
c. an ionic attraction to the nitrogenous bases
d. hydrogen bonding with the nitrogenous bases
e. covalent bonds to the deoxyribose sugars

Answer: a
Difficulty: Moderate
Bloom's Taxonomy: Synthesis

62. The Barr body found in mammalian females is an X chromosome compacted to the level of _____.

a. euchromatin
b. 10-nm chromatin fibers
c. heterochromatin
d. nucleosomes
e. solenoids

Answer: c
Difficulty: Moderate
Bloom's Taxonomy: Knowledge

63. DNA replication must start at a replication origin. In prokaryotes the DNA molecule that makes up a chromosome is typically _____ and usually has _____ replication origin.

a. circular; one
b. circular; more than one
c. linear; more than one
d. linear; one

Answer: a
Difficulty: Easy
Bloom's Taxonomy: Knowledge

64. Studies of human egg formation have shown that at ovulation the egg cell is stuck in metaphase II and only completes meiosis if a sperm enters the egg. In metaphase II, chromosomes are tightly condensed, so you should expect to find _____ in human egg cells that have not been entered by sperm.

a. very high levels of gene expression
b. mostly euchromatin
c. very little gene expression
d. no nucleosomes
e. no solenoids

Answer: c
Difficulty: Moderate
Bloom's Taxonomy: Application

UNANSWERED QUESTIONS

65. Senescing cells are kept from entering _____ because they cannot _____.

a. G_1 phase; divide
b. S phase; initiate DNA replication
c. M phase; form compacted chromosomes
d. G_2 phase; finish DNA replication
e. M phase; regenerate telomere repeats

Answer: b
Difficulty: Difficult
Bloom's Taxonomy: Knowledge

Integrative Multiple-Choice

66. Suppose that you performed a version of the Hershey and Chase experiment, this time using ^{32}P-labeled viruses that insert their DNA into the DNA of the cells that they infect. The viral DNA is then treated as part of the cell's own DNA and is replicated during DNA replication and passed on to daughter cells when the cell divides. You infect a population of cells with the ^{32}P-labeled viruses, then let the infected cells go through two generations of cell divisions. If you then examine the cells, you should find ^{32}P-labeled DNA in

a. none of them.
b. about ¼ of them.
c. about ½ of them.
d. about ¾ of them.
e. all of them.

Answer: c
Difficulty: Difficult
Bloom's Taxonomy: Synthesis
Source: Sections 14.1 and 14.3

67. Suppose that a mistake made during DNA replication in a cell is not corrected, but instead the mutation remains in the single strand where it occurred. Then suppose that after the cell divides both daughter cells survive, and those cells then go on to have DNA replication and ultimately cell division. There are now four cells where there once was one cell. Assuming that the mistake was never corrected and that no new uncorrected mistakes occurred, how many of the four cells will have the mutation in their DNA?

a. none
b. one
c. two
d. three
e. four

Answer: b
Difficulty: Difficult
Bloom's Taxonomy: Application, Synthesis
Source: Sections 14.3 and 14.4

Classification

For questions 68–72 identify the researcher or researchers associated with the discovery or experiment from the list below.

a. Watson and Crick
b. Avery, MacLeod, and McCarty
c. Griffith
d. Hershey and Chase
e. Meselson and Stahl

68. _____ Showed that DNA is the transforming principle from heat-killed *S* strain *Streptococcus pneumoniae* that can make the *R* strain virulent

69. _____ Showed that DNA replication in *Escherichia coli* is semiconservative

70. _____ Showed that a transforming principle from heat-killed *S* strain *Streptococcus pneumoniae* could be used to make the *R* strain virulent

71. _____ Worked out the double helix model for DNA structure

72. _____ Showed that the genetic material of bacteriophage T2 is DNA

Answers: 68. b 69. e 70. c 71. a 72. d
Difficulty: Moderate
Bloom's Taxonomy: Knowledge
Source: Sections 14.1, 14.2, and 14.3

Short Answer

73. More than 90% of cancer cells have fully active telomerase enzymes. Explain how that might play a role in enabling cancer cells to keep rapidly dividing.

Answer: Cancer cells are characterized by their ability to divide out of control and with no apparent limit to the number of generations that they can produce. Active telomerase would appear to play a key role in allowing cancer cells to do just that. During each round of DNA replication, part of the ends of chromosomes cannot be replicated. Thus, with each cell generation, a cell loses some of its DNA. Active telomerase places new telomere repeats at the ends of chromosomes so that the DNA that is lost will be from telomere repeats and not critical genes. For most cells, telomerase is not active, and this places a natural limit on how many replications can be performed before the cell loses critical genetic information and is able to replicate and divide no more. Cancer cells with active telomerase are able to override this natural limit on cell replication.
Difficulty: Moderate
Bloom's Taxonomy: Synthesis
Source: Section 14.3 and Unanswered Questions

74. Acyclovir is a chemical analog of a DNA nucleoside that is used to treat people who are infected by herpes simplex viruses (HSV). How does acyclovir work as a treatment without harming the patients?

Answer: In infected cells, a virus-encoded enzyme is able to use acyclovir to make acyclovir monophosate (uninfected cells cannot do this effectively). Acyclovir monophosphate is then converted to acyclovir triphosphate, which then interferes with viral DNA replication by viral DNA polymerase. Normal cellular DNA polymerase function is not very sensitive to interference by acyclovir triphosphate when compared to viral DNA polymerase, and uninfected cells cannot effectively produce acyclovir triphosphate from acyclovir anyway. Thus, acyclovir is fairly specific in targeting the virus and not the patient.

Difficulty: Difficult
Bloom's Taxonomy: Comprehension
Source: Unanswered Questions

15

FROM DNA TO PROTEIN

Multiple-Choice

WHY IT MATTERS

1. Byssus is

a. an enzyme involved in transcription.
b. a tough, adhesive material produced by some mussels.
c. a flexible fabric produced from some animal hairs.
d. an enzyme involved in translation.
e. an antibiotic produced by some yeast.

Answer: b
Difficulty: Easy
Bloom's Taxonomy: Knowledge

2. Every protein is assembled on _____ according to instructions that are copied from _____.

a. mRNAs; tRNAs
b. tRNAs; mRNAs
c. ribosomes; mRNA
d. tRNAs; DNA
e. ribosomes; DNA

Answer: e
Difficulty: Easy
Bloom's Taxonomy: Knowledge

15.1 THE CONNECTION BETWEEN DNA, RNA, AND PROTEIN

3. The first evidence of a connection between genes and metabolism came from studies of alkaptonuria conducted by _____.

a. Beadle and Tatum
b. Griffith
c. Watson and Crick
d. Garrod
e. Nirenberg and Leder

Answer: d
Difficulty: Moderate
Bloom's Taxonomy: Knowledge

4. Evidence of a direct relationship between genes and enzymes, including development of the one gene-one enzyme hypothesis, came from studies conducted by _____ of arginine biosynthesis in mutant stains of *Neurospora*.

a. Beadle and Tatum
b. Griffith
c. Watson and Crick
d. Garrod
e. Nirenberg and Leder

Answer: a
Difficulty: Moderate
Bloom's Taxonomy: Knowledge

5. Consider a mutant organism that is unable to make the amino acid arginine. Knowing that the metabolic pathway to production of arginine is ornithine → citrulline → arginosuccinate → arginine, you test the ability of the mutant to grow in the presence of each one of these compounds, providing just one of the compounds in each of your tests. You find that the mutant can grow in the presence of arginosuccinate or arginine, but not in the presence of citrulline or ornithine. From this you can conclude that the product of the gene mutated in the mutant is most directly involved in production of _____.

a. arginine from ornithine
b. arginosuccinate from citrulline
c. citrulline from ornithine
d. arginine from arginosuccinate
e. arginosuccinate from ornithine

Answer: b
Difficulty: Moderate
Bloom's Taxonomy: Analysis

6. The central dogma describes the flow of information of gene expression as

a. DNA → RNA → protein.
b. RNA → DNA.
c. RNA → DNA → protein.
d. protein → DNA → RNA.
e. DNA → protein → RNA.

Answer: a
Difficulty: Easy
Bloom's Taxonomy: Comprehension

7. The process of transcription refers to the use of information encoded in _____ to make _____.

a. RNA; a DNA strand
b. DNA; a polypeptide
c. DNA; a complementary RNA copy
d. a polypeptide; RNA
e. RNA; a polypeptide

Answer: c
Difficulty: Easy
Bloom's Taxonomy: Comprehension

8. The process of translation refers to the use of information encoded in _____ to make _____.

a. RNA; a DNA strand
b. DNA; a polypeptide
c. DNA; a complementary RNA copy
d. a polypeptide; RNA
e. RNA; a polypeptide

Answer: e
Difficulty: Easy
Bloom's Taxonomy: Comprehension

9. The RNA "alphabet" used in the genetic code has _____ letters.

a. 5
b. 64
c. 4
d. 20
e. 3

Answer: c
Difficulty: Easy
Bloom's Taxonomy: Knowledge

10. Except for the stop codons, the codons in the genetic code specify which of _____ amino acids to use.

a. 5
b. 64
c. 4
d. 20
e. 3

Answer: d
Difficulty: Easy
Bloom's Taxonomy: Knowledge, Application

11. In both prokaryotes and eukaryotes the start codon (or initiator codon) is _____, which codes for the amino acid _____.

a. UGA; proline
b. UUU; phenylalanine
c. AAA; lysine
d. ACG; threonine
e. AUG; methionine

Answer: e
Difficulty: Moderate
Bloom's Taxonomy: Knowledge

12. The degeneracy of the genetic code refers to the fact that

a. the code has some codons that do not specify an amino acid.
b. the code has most amino acids represented by more than one codon.
c. the code varies considerably between different organisms.
d. the code is commaless, with no indicators of spaces between codons.
e. the code varies considerably between different cell types within a multicellular organism.

Answer: b
Difficulty: Easy
Bloom's Taxonomy: Comprehension

13. With minor exceptions, the genetic code

a. is the same for all living organisms and viruses.
b. is specialized so that viruses have a different code from all living organisms.
c. differs between different organisms and viruses.
d. is the same for all viruses and for single-celled organisms, but is more complex in multicellular organisms.
e. has three versions: one for viruses, one for single-celled organisms, and one for multicellular organisms.

Answer: a
Difficulty: Easy
Bloom's Taxonomy: Comprehension, Analysis

15.2 TRANSCRIPTION: DNA-DIRECTED RNA SYNTHESIS

14. During transcription

a. double-stranded RNA chains are produced.
b. the entire DNA molecule is used.
c. primase creates an RNA primer to start the RNA strand.
d. only one of the two DNA strands acts as a template.
e. all of these

Answer: d
Difficulty: Moderate
Bloom's Taxonomy: Evaluation

15. The place where RNA polymerase first associates with DNA so that transcription can begin is called the _____ and is located _____ of the transcribed region.

a. promoter; upstream
b. initiator; downstream
c. initiator; upstream
d. promoter; downstream
e. intron; downstream

Answer: a
Difficulty: Moderate
Bloom's Taxonomy: Comprehension

16. The TATA box is a key element of the _____ of most eukaryotic protein-coding genes.

a. terminator
b. coding region
c. promoter
d. transcription start point
e. introns

Answer: c
Difficulty: Easy
Bloom's Taxonomy: Knowledge

17. Proteins called transcription factors are involved in _____ of transcription.

a. both initiation and termination stages
b. both initiation of and elongation stages
c. the termination stage
d. the initiation stage
e. the initiation, elongation, and termination stages

Answer: d
Difficulty: Moderate
Bloom's Taxonomy: Synthesis

18. During the elongation stage of transcription

a. RNA nucleotides are added to the transcript.
b. the transcript grows in the 5'→3' direction.
c. RNA polymerase unwinds the DNA.
d. the double helix reforms behind RNA polymerase.
e. all of these

Answer: e
Difficulty: Moderate
Bloom's Taxonomy: Analysis

19. Stopping transcription in prokaryotes requires

a. generation of a stop codon.
b. copying a terminator sequence in mRNA.
c. splicing introns out and pasting exons together.
d. activation of gene repressors.
e. generation of a poly-A tail.

Answer: b
Difficulty: Difficult
Bloom's Taxonomy: Synthesis

20. Protein-encoding genes in eukaryotes are transcribed by

a. RNA polymerase I.
b. RNA polymerase II.
c. RNA polymerase III.
d. either RNA polymerase I or III.
e. either RNA polymerase II or III.

Answer: b
Difficulty: Moderate
Bloom's Taxonomy: Knowledge

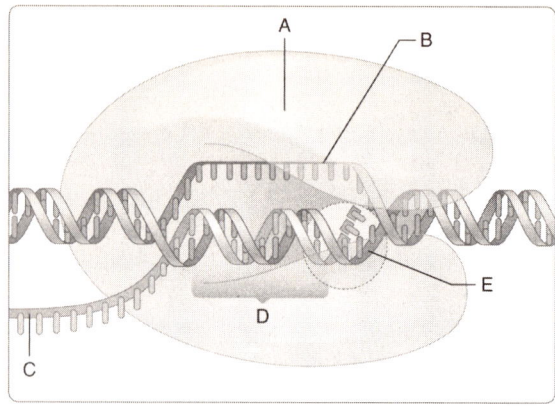

21. In the figure above, representing transcription, the item labeled A is

a. reverse transcriptase.
b. primase.
c. the spliceosome.
d. RNA polymerase.
e. DNA helicase.

Answer: d
Difficulty: Easy
Bloom's Taxonomy: Analysis
Source: Fig. 15.7

22. In the figure above, representing transcription, the item labeled B is

a. the nontemplate strand.
b. RNA-DNA hybrid.
c. DNA double helix.
d. the RNA copy.
e. the template strand.

Answer: a
Difficulty: Moderate
Bloom's Taxonomy: Analysis
Source: Fig. 15.7

23. In the figure above, representing transcription, the item labeled C is

a. the nontemplate strand.
b. RNA-DNA hybrid.
c. the template strand.
d. DNA double helix.
e. the RNA copy.

Answer: e
Difficulty: Moderate
Bloom's Taxonomy: Analysis
Source: Fig. 15.7

24. In the figure above, representing transcription, the item labeled D is

a. the nontemplate strand.
b. RNA-DNA hybrid.
c. the template strand.
d. DNA double helix.
e. the RNA copy.

Answer: b
Difficulty: Moderate
Bloom's Taxonomy: Analysis
Source: Fig. 15.7

25. In the figure above, representing transcription, the item labeled E is

a. the nontemplate strand.
b. RNA-DNA hybrid.
c. the template strand.
d. DNA double helix.
e. the RNA copy.

Answer: c
Difficulty: Moderate
Bloom's Taxonomy: Analysis
Source: Fig. 15.7

15.3 PRODUCTION OF mRNAs IN EUKARYOTES

26. In a messenger RNA the 3' UTR refers to the region of the mRNA that is

a. before the start codon.
b. before the site for initiation of transcription.
c. after the stop codon.
d. the coding region.
e. after the site for termination of transcription.

Answer: c
Difficulty: Difficult
Bloom's Taxonomy: Application

27. Precursor-mRNA (pre-mRNA) typically exists

a. only in prokaryotes.
b. in both the nucleus and cytoplasm of eukaryotes.
c. only in the cytoplasm of eukaryotes.
d. in both prokaryotes and eukaryotes.
e. only in the nucleus of eukaryotes.

Answer: e
Difficulty: Easy
Bloom's Taxonomy: Synthesis

28. While a eukaryotic pre-mRNA is still being synthesized it is modified on its 5' end when

a. exon shuffling mixes mRNA pieces.
b. enzymes recognize a polyadenylation signal in the trailing RNA sequence.
c. the RNA introns are cut out by the spliceosome.
d. an enzyme adds a guanine-containing cap.
e. the RNA exons are spliced together.

Answer: d
Difficulty: Moderate
Bloom's Taxonomy: Synthesis

29. The cap on an mRNA is the site where

a. ribosomes attach at the start of translation.
b. the start codon is located.
c. the process of translation ends.
d. the stop codon is covered until needed.
e. the mRNA is protected from attack by RNA-digesting enzymes.

Answer: a
Difficulty: Moderate
Bloom's Taxonomy: Knowledge

30. The process of transcription in eukaryotes ends when

a. exon shuffling mixes mRNA pieces.
b. enzymes recognize a polyadenylation signal in the trailing RNA sequence.
c. the RNA introns are cut out by the spliceosome.
d. an enzyme adds a guanine-containing cap.
e. the RNA exons are spliced together.

Answer: b
Difficulty: Difficult
Bloom's Taxonomy: Synthesis

31. The poly(A) tail of an mRNA

a. is where the start codon is located.
b. covers the stop codon until it is needed.
c. protects the mRNA from attack by RNA-digesting enzymes.
d. is where ribosomes attach at the start of translation.
e. is where the process of translation ends.

Answer: c
Difficulty: Moderate
Bloom's Taxonomy: Knowledge

32. After processing of pre-mRNAs, the regions that are retained in finished mRNAs are called _____.

a. UTRs
b. snRNPs
c. exons
d. domains
e. introns

Answer: c
Difficulty: Moderate
Bloom's Taxonomy: Knowledge, Comprehension

33. The process of removing introns from mRNA and putting the remaining exons together occurs in a complex called the _____.

a. ribosome
b. anticodon
c. lariat
d. polysome
e. spliceosome

Answer: e
Difficulty: Easy
Bloom's Taxonomy: Knowledge, Comprehension

34. Small ribonucleoprotein particles (snRNPs) are involved in _____.

a. mRNA splicing
b. initiation of transcription
c. aminoacylation of tRNA
d. initiation of translation
e. termination of translation

Answer: a
Difficulty: Moderate
Bloom's Taxonomy: Analysis

35. In the process of mRNA splicing, the lariat structure is

a. the splicing complex.
b. the region where two exons are pasted together.
c. the enzyme that cuts the pre-mRNA.
d. the released intron.
e. the region where two introns are pasted together.

Answer: d
Difficulty: Moderate
Bloom's Taxonomy: Knowledge

36. Human cells produce perhaps as many as 100,000 different proteins, and yet the human genome has only about 25,000 genes. This is best explained by

a. aminoacylation.
b. exon shuffling.
c. the wobble hypothesis.
d. degeneracy.
e. alternative splicing.

Answer: e
Difficulty: Moderate
Bloom's Taxonomy: Application

37. Smooth muscle and striated muscle have distinct mRNA forms due to _____.

a. aminoacylation
b. alternative splicing
c. polyadenylation
d. degeneracy
e. exon shuffling

Answer: b
Difficulty: Moderate
Bloom's Taxonomy: Knowledge

38. The human protein TPA has domains that are genetically similar to domains from other proteins, including epidermal growth factor and fibronectin. This is best explained by _____.

a. alternative splicing
b. degeneracy
c. exon shuffling
d. the wobble hypothesis
e. aminoacylation

Answer: c
Difficulty: Difficult
Bloom's Taxonomy: Application

15.4 TRANSLATION: mRNA-DIRECTED POLYPEPTIDE SYNTHESIS

39. The process of translation of nuclear protein-coding genes in eukaryotic cells occurs in the _____.

a. cytoplasm
b. Golgi apparatus
c. nucleolus
d. nucleus
e. mitochondria

Answer: a
Difficulty: Easy
Bloom's Taxonomy: Application

40. This type of RNA is typically found as a strand about 75–90 bases long that folds into a structure with four double-helical segments.

a. snRNA
b. pre-mRNA
c. rRNA
d. tRNA
e. mRNA

Answer: d
Difficulty: Difficult
Bloom's Taxonomy: Knowledge

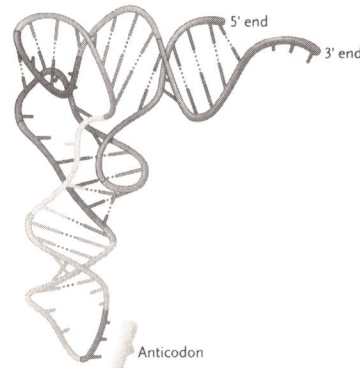

41. The structure in the figure above (Figure 15.2) represents a molecule of _____.

a. pre-mRNA
b. tRNA
c. snRNA
d. mRNA
e. rRNA

Answer: b
Difficulty: Moderate
Bloom's Taxonomy: Analysis
Source: Fig. 15.12

42. The region in a tRNA that bonds with mRNA during translation is the _____.

a. anticodon
b. aminoacylation site
c. TATA box
d. cloverleaf
e. reading frame

Answer: a
Difficulty: Easy
Bloom's Taxonomy: Knowledge

43. The ability of some tRNAs to pair with different codons is described according to _____ as being due to imprecise basepairing with the third base in the codon.

a. alternative splicing
b. degeneracy
c. exon shuffling
d. the wobble hypothesis
e. aminoacylation

Answer: d
Difficulty: Moderate
Bloom's Taxonomy: Knowledge

44. If the anticodon sequence of a(n) _____ is correct, a special enzyme will bind to it and "charge" it by adding the correct amino acid to its 3' end.

a. pre-mRNA
b. tRNA
c. snRNA
d. mRNA
e. rRNA

Answer: b
Difficulty: Moderate
Bloom's Taxonomy: Application

45. The process of adding the correct amino acid onto a tRNA molecule is catalyzed by _____.

a. the tRNA itself
b. RNA polymerase
c. an mRNA
d. the ribosome
e. an aminoacyl-tRNA synthetase

Answer: e
Difficulty: Easy
Bloom's Taxonomy: Knowledge

46. The formation of peptide bond during translation is catalyzed by peptidyl transferase, an enzyme that is

a. a protein in the small ribosomal subunit.
b. part of an rRNA in the small ribosomal subunit.
c. part of an rRNA in the large ribosomal subunit.
d. a protein in the large ribosomal subunit.

Answer: c
Difficulty: Moderate
Bloom's Taxonomy: Knowledge

47. The site of translation in all living cells is the ribosome, which is

a. made of large and small subunits in both prokaryotes and eukaryotes.
b. made of large and small subunits in eukaryotes, but is a single unit in prokaryotes.
c. a single unit in both prokaryotes and eukaryotes.
d. made of large and small subunits in prokaryotes but is a single unit in eukaryotes.
e. none of these

Answer: a
Difficulty: Easy
Bloom's Taxonomy: Application

48. The antibiotics streptomycin and erythromycin function by inhibiting the function of _____ in prokaryotes but not eukaryotes.

a. RNA polymerases
b. aminoacyl-tRNA synthetases
c. spliceosomes
d. DNA polymerases
e. ribosomes

Answer: e
Difficulty: Difficult
Bloom's Taxonomy: Knowledge

49. At the start of translation the initiator tRNA is basepaired with the start codon at _____ in the ribosome.

a. the A site (aminoacyl site)
b. first the A site (aminoacyl site) and then the E site (exit site)
c. the P site (peptidyl site)
d. first the A site (aminoacyl site) and then the P site (peptidyl site)
e. the E site (exit site)

Answer: c
Difficulty: Difficult
Bloom's Taxonomy: Synthesis

50. Which of the regions of the ribosome accepts charged tRNA molecules during the elongation phase of translation?

a. the A site (aminoacyl site)
b. first the A site (aminoacyl site) and then the E site (exit site)
c. the P site (peptidyl site)
d. first the A site (aminoacyl site) and then the P site (peptidyl site)
e. the E site (exit site)

Answer: a
Difficulty: Difficult
Bloom's Taxonomy: Knowledge

51. Which of these events in translation in eukaryotes does NOT directly involve hydrolysis of GTP?

a. release of initiation factors
b. binding of a proper aminoacyl tRNA with a codon at the A site
c. formation of the peptide bond
d. translocation of the ribosome along the mRNA
e. all of these events directly involve GTP hydrolysis

Answer: c
Difficulty: Difficult
Bloom's Taxonomy: Synthesis

52. Initiation factors are _____ that assist in the initiation of transcription.

a. mRNAs
b. proteins
c. tRNAs
d. amino acids
e. snRNPs

Answer: b
Difficulty: Easy
Bloom's Taxonomy: Knowledge

53. The ribosome binding site is located _____ and is where _____.

a. at the start codon; the large ribosome subunit binds the mRNA-tRNA complex in prokaryotes
b. just upstream of the start codon; the large ribosome subunit binds the mRNA in eukaryotes
c. on initiator tRNA; the large ribosome subunit binds the mRNA-tRNA complex in eukaryotes
d. on aminoacyl-tRNA; the large ribosome subunit binds the tRNA in prokaryotes
e. just upstream of the start codon; the small ribosome subunit binds the mRNA in prokaryotes

Answer: e
Difficulty: Moderate
Bloom's Taxonomy: Knowledge

54. In translation the reading frame is established by

a. removal of the cap from mRNA.
b. pairing of initiator tRNA with the start codon.
c. the first base in the mRNA molecule.
d. the first tRNA to bind a codon at the A site.
e. aligning a stop codon with the A site.

Answer: b
Difficulty: Moderate
Bloom's Taxonomy: Comprehension

55. Which of the regions of the ribosome accepts charged tRNA molecules during the elongation phase of translation?

a. the A site (aminoacyl site)
b. either the A site (aminoacyl site) or the E site (exit site)
c. the P site (peptidyl site)
d. either the P site (peptidyl site) or the E site (exit site)
e. the E site (exit site)

Answer: a
Difficulty: Moderate
Bloom's Taxonomy: Knowledge

56. In translation the joining of amino acids is catalyzed by the enzyme _____.

a. RNA polymerase
b. reverse transcriptase
c. peptidyl transferase
d. polyadenylase
e. aminoacyl tRNA synthetase

Answer: c
Difficulty: Moderate
Bloom's Taxonomy: Knowledge

57. Relative to the mRNA, which of the following moves during translocation?

a. attached tRNAs and the ribosome
b. attached tRNAs and the polypeptide chain
c. the ribosome only
d. the ribosome and the polypeptide chain
e. the polypeptide chain only

Answer: c
Difficulty: Difficult
Bloom's Taxonomy: Synthesis

58. The energy for translocation during translation comes from

a. formation of the peptide bond.
b. breaking of bonds between tRNA and mRNA.
c. breaking of bonds between mRNA and the ribosome.
d. removal of a tRNA from the E site.
e. hydrolysis of GTP.

Answer: e
Difficulty: Moderate
Bloom's Taxonomy: Comprehension

59. During translation, tRNAs bond to mRNAs using

a. ionic bonds.
b. van der Waals forces.
c. nonpolar covalent bonds.
d. hydrogen bonds.
e. polar covalent bonds.

Answer: d
Difficulty: Moderate
Bloom's Taxonomy: Application

60. Translation ends when a stop codon is at the _____ site, allowing a _____ to bind there.

a. P; release factor
b. A; terminator tRNA
c. E; terminator tRNA
d. A; release factor
e. P; terminator tRNA

Answer: d
Difficulty: Moderate
Bloom's Taxonomy: Comprehension

61. A ribosome is made of

a. rRNA and proteins.
b. tRNA and rRNA.
c. tRNA, rRNA, and mRNA.
d. rRNA and mRNA.
e. tRNA, rRNA, and proteins.

Answer: a
Difficulty: Easy
Bloom's Taxonomy: Knowledge

62. A polysome is

a. the combination of a large and a small ribosome subunit.
b. the complex where mRNA splicing occurs.
c. the assembly at a promoter at the start of transcription.
d. an mRNA with multiple ribosomes attached.
e. the complex that puts a poly(A) tail on an mRNA.

Answer: d
Difficulty: Moderate
Bloom's Taxonomy: Knowledge

63. Suppose that the first part of a new polypeptide chain being produced in a eukaryotic cell has a signal peptide. Which of the following would you associate with the location of translation to form that polypeptide?

a. cytoplasm only
b. first cytoplasm, then rough ER
c. Golgi apparatus only
d. first rough ER, then Golgi apparatus
e. rough ER only

Answer: b
Difficulty: Moderate
Bloom's Taxonomy: Application

64. Suppose that the first part of a new polypeptide chain being produced in a eukaryotic cell does not have a signal peptide. Which of the following would you associate with the location of translation to form that polypeptide?

a. cytoplasm only
b. first cytoplasm, then rough ER
c. Golgi apparatus only
d. first rough ER, then Golgi apparatus
e. rough ER only

Answer: a
Difficulty: Moderate
Bloom's Taxonomy: Application

65. Which of the following best describes the role of the signal recognition particle (SRP)?

a. assists in mRNA splicing
b. activates gene expression
c. terminates translation
d. assists in polyadenylation of mRNA
e. temporarily blocks translation

Answer: e
Difficulty: Moderate
Bloom's Taxonomy: Knowledge

66. The routing of proteins to their final destinations in eukaryotic cells is

a. essentially random.
b. based on transcription factors.
c. controlled by the type of ribosome used.
d. directed by signals that are part of the proteins.
e. primarily determined by mRNA splicing.

Answer: d
Difficulty: Easy
Bloom's Taxonomy: Comprehension

67. Substitution of one base pair for another in a coding region of a gene can result in a _____ mutation where the changed codon still codes for the same amino acid that was previously coded for.

a. missense
b. chromosomal
c. frameshift
d. silent
e. nonsense

Answer: d
Difficulty: Easy
Bloom's Taxonomy: Comprehension

68. Substitution of one base pair for another in a coding region of a gene can result in a _____ mutation where the changed codon codes for a different amino acid than was previously coded for.

a. missense
b. chromosomal
c. frameshift
d. silent
e. nonsense

Answer: a
Difficulty: Easy
Bloom's Taxonomy: Comprehension

69. Substitution of one base pair for another in a coding region of a gene can result in a _____ mutation where the changed codon codes for a stop codon where an amino acid was previously coded for.

a. missense
b. chromosomal
c. frameshift
d. silent
e. nonsense

Answer: e
Difficulty: Moderate
Bloom's Taxonomy: Comprehension

70. Insertion of two bases into the coding region of a gene just after the start codon of a gene will result in a _____ mutation.

a. missense
b. chromosomal
c. frameshift
d. silent
e. nonsense

Answer: c
Difficulty: Moderate
Bloom's Taxonomy: Comprehension

71. Individuals with the genetic condition sickle cell anemia have hemoglobin β chains with the amino acid valine where normal hemoglobin β chain has glutamic acid. This is the result of a _____ mutation.

a. missense
b. chromosomal
c. frameshift
d. silent
e. nonsense

Answer: a
Difficulty: Moderate
Bloom's Taxonomy: Application

INSIGHTS FROM THE MOLECULAR REVOLUTION: MEASURING RIBOSOMES WITH A MOLECULAR RULER

72. The molecular scissors that Noller and his coworkers use to study ribosomes use _____ to break the sugar-phosphate backbone of an RNA molecule.

a. snRNPs
b. restriction enzymes
c. a modified mRNA
d. an iron atom
e. spliceosomes

Answer: d
Difficulty: Moderate
Bloom's Taxonomy: Knowledge

73. The molecular scissors that Noller and his coworkers use to study ribosomes showed that the A site is formed by _____ and that the P site is formed by _____.

a. both the large and small ribosome subunits; only the large ribosome subunit
b. both the large and small ribosome subunits; both the large and small ribosome subunits
c. only the small ribosome subunit; only the large ribosome subunit
d. only the large ribosome subunit; both the large and small ribosome subunits
e. both the large and small ribosome subunits; only the small ribosome subunit

Answer: b
Difficulty: Difficult
Bloom's Taxonomy: Knowledge

74. In yeast the transcription machinery at a promoter includes _____ polypeptides.

a. less than ten
b. over three hundred
c. nearly one hundred
d. about a dozen
e. fifteen

Answer: c
Difficulty: Moderate
Bloom's Taxonomy: Knowledge

75. Research by Noller and his coworkers indicates that translocation is a property of _____.

a. tRNA
b. mRNA
c. protein factors
d. GTP
e. the ribosome

Answer: e
Difficulty: Moderate
Bloom's Taxonomy: Knowledge

Integrative Multiple-Choice

76. Which of the following is NOT unique to eukaryotes?

a. snRNPs
b. sorting signals on proteins
c. capping the 5' end of mRNA
d. adding a poly(A) tail to mRNA
e. RNA polymerase II

Answer: b
Difficulty: Moderate
Bloom's Taxonomy: Synthesis
Source: Sections 15.2–15.4

77. Which of the following is unique to prokaryotes?

a. UTRs in mRNA
b. introns
c. polysomes connected to DNA
d. use of aminoacyl-tRNA synthetases
e. translocation during translation

Answer: c
Difficulty: Moderate
Bloom's Taxonomy: Synthesis
Source: Sections 15.2–15.4

78. The following are major events in expression of a gene for insulin, which is secreted outside of certain mammalian cells. Place these events in order, from earliest to latest.

    1: binding of a release factor
    2: binding of signal recognition particle (SRP)
    3: binding of transcription factors
    4: removal of introns
    5: scanning of mRNA

a. $4 \rightarrow 2 \rightarrow 3 \rightarrow 1 \rightarrow 5$
b. $1 \rightarrow 4 \rightarrow 3 \rightarrow 2 \rightarrow 5$
c. $5 \rightarrow 2 \rightarrow 1 \rightarrow 4 \rightarrow 3$
d. $3 \rightarrow 4 \rightarrow 5 \rightarrow 2 \rightarrow 1$
e. $3 \rightarrow 2 \rightarrow 1 \rightarrow 5 \rightarrow 4$

Answer: d
Difficulty: Difficult
Bloom's Taxonomy: Synthesis
Source: Sections 15.2–15.4

Second base of codon

79. Assume that an mRNA molecule is made beginning complementary to this DNA sequence:
    3'-CTTACATGGCATCC-5'.

See the genetic code table above. The second codon (counting the start codon as the first codon) directs incorporation of which amino acid in the polypeptide?

a. asparagine
b. tyrosine
c. arginine
d. proline
e. cysteine

Answer: b
Difficulty: Moderate
Bloom's Taxonomy: Application
Source: Figure 15.5; Sections 15.2 and 15.4

80. Assume that an mRNA molecule is made beginning complementary to this DNA sequence:
    3'-CTTACATGGCATCC-5'.

See the genetic code table above. The third codon (counting the start codon as the first codon) directs incorporation of which amino acid in the polypeptide?

a. cysteine
b. methionine
c. threonine
d. histidine
e. arginine

Answer: e
Difficulty: Moderate
Bloom's Taxonomy: Application
Source: Figure 15.5; Sections 15.2 and 15.4

81. See the genetic code table above. The start codon for translation directs incorporation of which amino acid?

a. lysine
b. phenylalanine
c. methionine
d. glycine
e. none, it just indicates the start

Answer: c
Difficulty: Easy
Bloom's Taxonomy: Application
Source: Figure 15.5; Sections 15.2 and 15.4

82. Which of the following processes occurs in the cytoplasm of a eukaryotic cell?

a. translation
b. DNA replication and transcription
c. transcription
d. DNA replication
e. translation and transcription

Answer: a
Difficulty: Moderate
Bloom's Taxonomy: Synthesis
Sections 15.2 and 15.4

Matching

Match each of the following terms with its correct definition.

83. _____ promoter

84. _____ exon

85. _____ stop codon

86. _____ terminator

87. _____ start codon

A. Indicates the end for transcription

B. First codon read in translation

C. Amino-acid coding sequence retained in finished mRNA

D. Indicates the end for translation

E. Control sequence ahead of a transcription unit

Answers: 83. E   84. C   85. D   86. A   87. B

Difficulty: Easy
Bloom's Taxonomy: Knowledge
Source: Sections 15.2-15.4

Match each of the following types of RNA with the correct description.

88. _____   snRNA

89. _____   pre-mRNA

90. _____   tRNA

91. _____   rRNA

92. _____   mRNA

A. RNA transcribed from a protein-coding gene that is ready to be used in translation

B. RNA that binds with proteins to make a particle that is involved in removal of introns and joining of exons

C. RNA that forms part of the ribosome

D. RNA with an anticodon and a linkage site for a specific amino acid

E. RNA in eukaryotes that must be processed in the nucleus before it is ready to be translated

Answers: 88. B   89. E   90. D   91. C   92. A

Difficulty: Easy
Bloom's Taxonomy: Knowledge
Source: Sections 15.2–15.4

Choice

For each of these items or events, indicate which of the following processes it is associated with.

a. processing of pre-mRNA
b. translation
c. sorting proteins in cells
d. transcription

93. _____   promoter

94. _____   ribosome

95. _____   spliceosome

96. _____   intron

97. _____   ER membrane

98. _____   tRNA

99. _____   RNA polymerase

100. _____   SRP receptor

101. _____   capping enzyme

102. _____   RNA-DNA double helix

103. _____   release factor

104. _____   signal peptidase

105. _____   TATA box

106. _____   peptidyl transferase

107. _____   snRNP

108. _____   signal recognition particle

Answers: 93. d   94. b   95. a   96. a   97. c
98. b   99. d   100. c   101. a.   102. d
103. b   104. c   105. d   106. b   107. a
108. c

Difficulty: Moderate
Bloom's Taxonomy: Knowledge, Comprehension
Source: Sections 15.2–15.4

Short Answer

109. Briefly describe the places where tRNA would be found in a ribosome.

Answer: A ribosome has three locations where tRNAs can be found, the E, P, and A sites. The E site is where an uncharged tRNA will leave the ribosome. The P site is where a tRNA with an attached polypeptide chain is located after translocation and before peptide bond formation. The A site is where a new, charged tRNA will first bind to the translation assembly.
Difficulty: Moderate
Bloom's Taxonomy: Synthesis
Source: Section 15.4

110. Describe how exon shuffling could lead to the formation of novel proteins.

Answer: Exon-intron junctions often fall at points that divide major functional regions, or domains, in encoded proteins. Genetic duplications and rearrangements can occasionally cut and paste parts of different genes together. When the cutting and pasting happens within introns, this will bring together novel combinations of exons that will produce a new protein. The individual domains generally retain their specific functionality in the new protein. This mechanism can produce changes more quickly and more efficiently than random point mutations that only affect individual amino acids in a protein.
Difficulty: Difficult
Bloom's Taxonomy: Application
Source: Section 15.3

Essay

111. Describe the process of transcription in prokaryotes.

Answer: First, RNA polymerase binds to the promoter, immediately upstream of where production of the mRNA will begin. RNA polymerase unwinds the DNA and starts synthesis of a new RNA molecule just past the promoter region. Only one of the DNA strands is used as a template, and the RNA strand is synthesized in the 5'→3' direction antiparallel to the template DNA strand. RNA polymerase progresses away from the promoter, growing the RNA strand. As the RNA strand elongates, it is displaced from the DNA immediately behind the RNA polymerase as the DNA strands reform the DNA double helix. Eventually the RNA polymerase reaches a region of specific DNA terminator sequence. A protein binds to the terminator, triggering the complete release of RNA polymerase and RNA from the DNA and each other. This ends transcription in prokaryotes.
Difficulty: Moderate
Bloom's Taxonomy: Synthesis
Source: Section 15.2

112. Describe the process of mRNA splicing.

Answer: First, snRNPs bind with sequences at the junctions of each intron and exon. These snRNPs at the borders of an intron associate with each other and other form a complex that loops out the intron. Other snRNPs are also recruited to the complex. The active spliceosome then cuts the beginning of the intron, and the intron bonds to itself. The spliceosome then cuts intron at its end and joins together the exons that were on either side of the intron. The intron is released and later degraded. Meanwhile, the spliceosome disassembles as the snRNPs are released. When all introns are removed and all exons joined together the mRNA is finished and ready for translation.
Difficulty: Difficult
Bloom's Taxonomy: Synthesis
Source: Section 15.3

16
CONTROL OF GENE EXPRESSION

Multiple-Choice

WHY IT MATTERS

1. The primary difference between different nucleated cells in a developing zygote is

a. the genetic code of each cell.
b. the expression of housekeeping genes.
c. the expression of a gene or group of genes.
d. the number of chromosomes present in the nucleus.
e. the presence or absence of gene regulators.

Answer: c
Difficulty: Easy
Bloom's Taxonomy: Knowledge

2. What is the primary difference between eukaryotes and prokaryotes in terms of gene expression?

a. Prokaryotes express different genes as a response their environment; eukaryotic cells do not do this.
b. Prokaryotes express different genes as a response to selective pressures; eukaryotes express the same genes most of their lives.
c. Prokaryotes generally alter gene expression in response to immediate environmental changes; eukaryotes generally only respond to selective pressures.
d. Prokaryotes generally alter gene expression in response to immediate environmental changes; eukaryotic cells do that as well as alter gene expression for the purpose of cell differentiation.
e. There is really no difference between gene expression in eukaryotes and prokaryotes.

Answer: d
Difficulty: Easy
Bloom's Taxonomy: Knowledge, Comprehension

16.1 REGULATION OF GENE EXPRESSION IN PROKARYOTES

3. In prokaryotes, the genes for metabolic pathways are

a. always on for early steps in the pathway, but genes for later steps are generally off.
b. always on so the bacteria can respond rapidly to changing conditions.
c. turned on and off as conditions change.
d. independently regulated.
e. expressed in low levels all of the time, but will be expressed in higher levels when conditions warrant.

Answer: c
Difficulty: Easy
Bloom's Taxonomy: Knowledge

4. The genes in an operon are

a. each transcribed into their own mRNA molecule.
b. all transcribed into a single mRNA molecule.
c. transcribed in subsets.
d. transcribed as a single unit in prokaryotes, but as individual genes in eukaryotes.
e. located at different locations on the chromosome and so can never be transcribed to the same mRNA molecule.

Answer: b
Difficulty: Easy
Bloom's Taxonomy: Knowledge

5. What is the relationship between operons and transcription units?

a. They are two terms for the same thing.
b. An operon is made up of a transcription unit and associated regulatory DNA sequences.
c. A transcription unit is made up of an operon and associated regulatory DNA sequences.
d. An operon is comprised of multiple transcription units.
e. A transcription unit is comprised of multiple operons.

Answer: b
Difficulty: Moderate
Bloom's Taxonomy: Knowledge

6. What is the relationship between operators, transcription units, and operons?

a. Operators are comprised of transcription units and operons.
b. Operons are comprised of transcription units and operators.
c. Operators are comprised of multiple transcription units associated with a single operon.
d. Operons are comprised of multiple operators associated with a single transcription unit.
e. Operons are comprised of multiple transcription units associated with a single operator.

Answer: b
Difficulty: Moderate
Bloom's Taxonomy: Knowledge

7. Where does RNA polymerase first bind to on the *E. coli* lac operon ?

a. the lac I repressor
b. the promoter
c. the operator
d. the transcription initiation site
e. the first codon of the lacZ gene

Answer: b
Difficulty: Easy
Bloom's Taxonomy: Knowledge

8. Which statement makes a correct comparison between the regulation of an operon and controlling the motion of a car?

a. The removal of a repressor is like stepping on the gas pedal of a car.
b. The removal of a repressor is like removing your foot from the brake pedal of a car.
c. The removal of a repressor is like moving your foot from the brake pedal to the gas pedal of a car.
d. The removal of a repressor is like moving your foot from the gas pedal to the brake pedal of a car.
e. The removal of a repressor is like removing your left foot from the brake pedal of car while your right foot remains on the gas pedal.

Answer: b
Difficulty: Moderate
Bloom's Taxonomy: Comprehension

9. The *E. coli* lac operon encodes three genes. These genes

a. cleave lactose into glucose and galactose, produce transacetylase, and actively transport lactose into the cell.
b. encode for an enzyme that cleaves glucose, a transport protein for active transport. and a transacetylase enzyme for solubilizing lactose in the cytosol.
c. encode for an enzyme that cleaves lactose, a transport protein for active transport of lactose, and a transacetylase enzyme whose function is still unknown.
d. catalyze the synthesis of lactose from glucose and galactose, produce transacetylase, and actively export lactose from the cell.
e. directly interact with the lac repressor and initiate their own transcription.

Answer: c
Difficulty: Moderate
Bloom's Taxonomy: Knowledge, Comprehension

10. Suppose the lacI repressor gene were permanently silenced by an alteration to the DNA sequence. What would be the impact on the function of the lac operon?

a. The lac operon would be transcribed but at a low level.
b. The lac operon would be transcribed at a high level.
c. There would be no real impact on lac operon expression.
d. The lac operon would only be expressed when lactose was present.
e. The lac operon would not be expressed even when lactose was present.

Answer: b
Difficulty: Difficult
Bloom's Taxonomy: Application

11. How is lac operon transcription regulated by the presence or absence of lactose?

a. When lactose is not available, the repressor will bind to the promoter and block RNA polymerase from transcribing the genes of the lac operon.
b. When lactose is available, the repressor will bind to the lactose instead of blocking RNA polymerase.
c. When lactose is not available, the repressor will bind to allolactose.
d. When lactose if available, allolactose is produced and binds to the operator, blocking RNA polymerase.
e. In the absence of lactose, transcription occurs at a low level resulting in the production of allolactose which in turn stimulates the cell to take up lactose.

Answer: b
Difficulty: Easy
Bloom's Taxonomy: Knowledge, Comprehension

12. Put the following steps in order:
    1. allolactose binds to the lac repressor protein
    2. RNA polymerase binds to the promoter
    3. the lac repressor is expressed
    4. beta-galactosidase cleaves lactose and generates allolactose
    5. mRNA is transcribed from the lac operon at low levels
    6. beta-galactosidase, permease, and transacetylase are produced at high levels

a. 1, 3, 2, 5, 4, 6
b. 4, 1, 2, 3, 5, 6
c. 3, 5, 4, 1, 2, 6
d. 3, 4, 1, 6, 2, 5
e. 5, 3, 1, 2, 4, 6

Answer: c
Difficulty: Moderate
Bloom's Taxonomy: Knowledge, Comprehension

13. What turns off the lac operon?

a. the short lifespan of mRNA molecules
b. the short lifespan of the lac operon gene products
c. the depletion of lactose from the cytosol
d. the lack of allolactose to bind to the repressor
e. the combination of all of the above

Answer: e
Difficulty: Easy
Bloom's Taxonomy: Knowledge

14. Where is the trp repressor gene located?

a. just upstream of the trp operon
b. on a region of the chromosome quite distant from the trp operon
c. adjacent to the lac repressor gene
d. just downstream of the trp operon
e. directly adjacent to the promoter region

Answer: b
Difficulty: Easy
Bloom's Taxonomy: Knowledge

15. Why does *E. coli* shut down the trp operon if tryptophan is available in the environment?

a. Synthesizing an amino acid takes energy so it is a waste of energy to make something that is already available.
b. Environmental tryptophan is of higher quality than what the *E. coli* can make itself.
c. The trp operon encodes genes that export tryptophan from the cell; if there is already tryptophan in the environment, further export isn't necessary.
d. It doesn't shut down the trp operon but only lowers the level of trp operon activity.
e. *E. coli* actually turns the trp operon *on* when tryptophan is present; tryptophan is an energy source for the cell, and there is no reason to make enzymes to degrade something that is lacking in the environment.

Answer: a
Difficulty: Moderate
Bloom's Taxonomy: Knowledge, Comprehension

16.2 REGULATION OF TRANSCRIPTION IN EUKARYOTES

16. At which level(s) is eukaryotic gene expression regulated?

a. transcription
b. translation
c. posttranslation
d. a and b only
e. a, b, and c

Answer: e
Difficulty: Easy
Bloom's Taxonomy: Knowledge

17. If the histones are rearranged to control access to a gene, this is called

a. transcription.
b. chromatin remodeling.
c. DNA condensation.
d. remodeling complex formation.
e. acetylation.

Answer: b
Difficulty: Easy
Bloom's Taxonomy: Knowledge

18. Acetylation

a. adds an acetyl group (CH_3CO-) to the cyotsine nucleotides of DNA.
b. adds an acetyl group (CH_3CH_2-) to the DNA of a promoter sequence.
c. adds an acetyl group (CH_3CO-) to the histone protecting the transcription unit of a gene.
d. adds an acetyl group (CH_3CO-) to the histone protecting the promoter region of a gene.
e. adds an acetyl group (CH_3CH_2-) to the RNA polymerase that will initiate transcription at the promoter region.

Answer: d
Difficulty: Difficult
Bloom's Taxonomy: Knowledge

19. Which of the following is the single most important stage when regulating gene expression?

a. translational regulation
b. degradation of mRNA
c. initiation of transcription
d. RNA interference
e. removal of masking segments

Answer: c
Difficulty: Easy
Bloom's Taxonomy: Knowledge

20. How does RNA polymerase II get involved in transcription?

a. It binds directly to the eukaryotic promoter sequence.
b. It binds directly to the TATA box.
c. It binds to transcription factors in the nucleus, and the complex then binds to the promoter.
d. Transcription factors bind to the promoter, and they in turn bind to RNA polymerase II.
e. Some transcription factors bind to the promoter, others bind to RNA polymerase II, and then the two groups of proteins bind to each other.

Answer: d
Difficulty: Easy
Bloom's Taxonomy: Knowledge

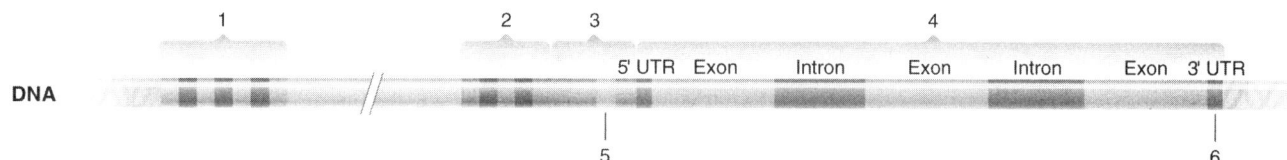

DNA

1   2   3   4

5' UTR   Exon   Intron   Exon   Intron   Exon   3' UTR

5   6

Use figure shown above for questions 21–28.

21. Which number identifies a sequence that is transcribed but not translated?

a. 2
b. 3
c. 4
d. 5
e. 6

Answer: e
Difficulty: Moderate
Bloom's Taxonomy: Knowledge

22. Which number identifies the promoter?

a. 1
b. 2
c. 3
d. 4
e. 5

Answer: c
Difficulty: Easy
Bloom's Taxonomy: Knowledge

23. Which number identifies the proximal promoter?

a. 1
b. 2
c. 3
d. 4
e. 5

Answer: b
Difficulty: Easy
Bloom's Taxonomy: Knowledge

24. Which number identifies the site where the transcriptional complex begins to form?

a. 1
b. 2
c. 3
d. 4
e. 5

Answer: d
Difficulty: Moderate
Bloom's Taxonomy: Knowledge

25. Which number identifies a region near the promoter that increases the rate of transcription?

a. 2
b. 3
c. 4
d. 5
e. 6

Answer: a
Difficulty: Moderate
Bloom's Taxonomy: Knowledge

26. Which number identifies the site where activators bind?

a. 2
b. 3
c. 4
d. 5
e. 6

Answer: d
Difficulty: Easy
Bloom's Taxonomy: Knowledge

27. Which number identifies a regulatory sequence that can increase the rate of transcription?

a. 1
b. 3
c. 4
d. 5
e. 6

Answer: a
Difficulty: Easy
Bloom's Taxonomy: Knowledge

28. Which number identifies the sequence most directly impacted by chromatin remodeling?

a. 1
b. 2
c. 3
d. 4
e. 5

Answer: c
Difficulty: Moderate
Bloom's Taxonomy: Knowledge

29. Other than the proteins that they encode, what is a major difference between cell-specific genes and housekeeping genes? (PPE= promoter proximal element)

a. Cell-specific genes are expressed only in a *single* cell type; housekeeping genes are expressed in *all* cell types.
b. Cell-specific genes have PPEs and enhancers; housekeeping genes have only PPEs.
c. Cell-specific genes have PPEs that use specific activators; housekeeping genes have "universal" activators for their PPEs.
d. Cell-specific genes are only found in certain cell types; housekeeping genes are found in all cell types.
e. Cell-specific genes have PPEs that use "universal" activators; housekeeping genes have PPEs that respond only to specific activators.

Answer: c
Difficulty: Moderate
Bloom's Taxonomy: Knowledge, Comprehension

30.	What is/are the function(s) of a coactivator?

a.	form a bridge between the enhancer and the promoter proximal region
b.	cause the DNA to form a loop
c.	bind to RNA polymerase
d.	increase the rate of transcription
e.	all of the above

Answer: e
Difficulty: Difficult
Bloom's Taxonomy: Knowledge

31.	How do transcription repressors work?

a.	They bind to the same sequence where activators normally bind.
b.	They bind to and disable the activator.
c.	They recruit histone acetylation enzymes and thus interfere with chromatin remodeling.
d.	They bind to RNA polymerase II and prevent it from binding to the transcription factor complex
e.	all of the above

Answer: a
Difficulty: Moderate
Bloom's Taxonomy: Knowledge, Comprehension

32.	The way a eukaryotic cell efficiently regulates the expression multiple genes is to

a.	have individual activators for individual genes.
b.	have a few activators that can interact with small subsets of genes.
c.	have activators that serve as catalysts for further activator protein synthesis.
d.	have a few activators that work in different combinations to activate different genes.
e.	have a single activator that undergoes modifications to enable interaction with different promoters.

Answer: d
Difficulty: Moderate
Bloom's Taxonomy: Knowledge, Comprehension

33.	Steroid hormones can trigger gene expression in a select number of cells because

a.	only target cells allow the steroid hormone to cross the plasma membrane.
b.	only the target cells have a steroid hormone response element encoded in their DNA.
c.	only the target cells have the correct receptor in their cytosol to bind to the hormone.
d.	nontarget cells lack the genes found in target cells.
e.	b and c.

Answer: e
Difficulty: Moderate
Bloom's Taxonomy: Knowledge, Comprehension

34.	Methylation regulates transcription

a.	via the addition of a methyl group to cytosine bases of DNA.
b.	via the addition of a methyl group to cysteine residues on RNA polymerase II.
c.	via the addition of a methyl group to cysteine bases of DNA.
d.	via the addition of a methyl group to cytosine residues on RNA polymerase II.
e.	by interfering with the chromatin remodeling process.

Answer: a
Difficulty: Easy
Bloom's Taxonomy: Knowledge

35.	When a DNA promoter sequence is methylated,

a.	it is permanently silenced because the methyl group can never be removed.
b.	it is temporarily silenced because the methyl group can be removed.
c.	it is perpetually transcribed because the methyl group can never be removed.
d.	it is temporarily transcribed because the methyl group can be removed.
e.	the impact on expression is dependent on and specific to a particular gene.

Answer: b
Difficulty: Easy
Bloom's Taxonomy: Knowledge, Comprehension

36.	When a DNA sequence is permanently rendered inaccessible to transcription via methylation and chromatin modifications, we call this

a.	silencing.
b.	imprinting.
c.	a Barr body.
d.	inactivation.
e.	quiescence.

Answer: b
Difficulty: Easy
Bloom's Taxonomy: Knowledge

16.3 POSTTRANSCRIPTIONAL, TRANSLATIONAL, AND POSTTRANSLATIONAL REGULATION

37. You are asked to locate some mRNA masking proteins and handed two rabbits (one male and one female) to use as your cell sources. You decide to do some needle biopsies of one particular tissue and extract the masking proteins. What tissues or organ samples do you collect for analysis?

a. skin cells
b. ovaries
c. testes
d. liver
e. uterus

Answer: b
Difficulty: Moderate
Bloom's Taxonomy: Application.

38. One function of the 5' UTR (untranslated region) of mRNA is to

a. control the half-life of mRNA.
b. extend the half-life of mRNA.
c. decrease the half-life of mRNA.
d. stabilize the mRNA structure.
e. bind ribosomes for initiating translation.

Answer: a
Difficulty: Moderate
Bloom's Taxonomy: Knowledge

39. Micro RNAs (miRNAs)

a. are short sequences only 22 or so bases long.
b. are noncoding RNA sequences that bind to a protein complex.
c. can regulate mRNA translation by cleaving the mRNA.
d. can prevent translation of its target mRNA by blocking the ribosome.
e. are correctly characterized by all of the above statements.

Answer: e
Difficulty: Moderate
Bloom's Taxonomy: Knowledge

40. Which statement accurately describes the relationship between RNA interference, small interfering RNA, and micro RNAs?

a. They all describe different methods of post-translational gene regulation.
b. Micro RNAs can be broken down into two types: RNA interference molecules and small interfering RNAs.
c. RNA interference occurs in two ways: by micro RNA or by small interfering RNAs.
d. Small interfering RNAs bind to dicer protein, microRNAs do not; both are types of RNA interference.
e. MicroRNAs are transcribed from viruses, small interfering RNAs are transcribed from nuclear DNA; both are types of RNA interference.

Answer: c
Difficulty: Difficult
Bloom's Taxonomy: Knowledge

41. A small molecule of RNA is transcribed in the nucleus. It is folded, cleaved by dicer protein, and then binds to a target molecule of mRNA. This molecule of RNA must be

a. miRNA.
b. siRNA.
c. an mRNA that was not properly capped.
d. an mRNA that never received its poly-A tail.
e. a transcript for a masking protein.

Answer: a
Difficulty: Easy
Bloom's Taxonomy: Knowledge

42. The vast majority of proteins expressed early in an animal's development

a. are translated from preexisting mRNAs present in the unfertilized egg.
b. are translated from preexisting mRNAs delivered by the sperm that fertilized an egg.
c. are translated from mRNAs that were transcribed from the zygote's DNA in the first three rounds of cell division.
d. are transcribed and translated from the unfertilized egg's DNA postfertilization.
e. were present as inactive proteins in the unfertilized egg.

Answer: a
Difficulty: Easy
Bloom's Taxonomy: Knowledge

43. Posttranslational modification includes

a. cleavage of poly-A tails from mRNA.
b. the binding of miRNAs to mRNA.
c. the chemical modification, processing, and degradation of proteins.
d. chromatin remodeling.
e. a, b, and c only.

Answer: c
Difficulty: Easy
Bloom's Taxonomy: Knowledge

44. Pepsin, a digestive enzyme that degrades proteins in the stomach, is synthesized as pepsinogen and converted to active pepsin in the stomach by the removal of several amino acids. The activation of pepsin is an example of

a. transcriptional regulation.
b. translational regulation.
c. posttranscriptional regulation.
d. posttranslational regulation.
e. RNA interference.

Answer: d
Difficulty: Easy
Bloom's Taxonomy: Knowledge

45. Why ubiquitin nicknamed the "doom tag"?

a. When ubiquitin binds to a target protein, the ubiquitin enzyme degrades he protein and destroys it.
b. When ubiquitin binds to a target protein, the complex is sent to the proteasome and the complex is hydrolyzed to amino acids.
c. If a cell is deficient in ubiquitin, the cell will die.
d. High levels of ubiquitin in a cell cause the cell to destroy vital proteins, leading to premature cell death.
e. Ubiquitin catalyzes the conversion of pepsinogen to active pepsin, an enzyme vital to protein digestion in the stomach of mammals. A lack of ubiquitin in the stomach causes an organism to starve.

Answer: b
Difficulty: Moderate
Bloom's Taxonomy: Knowledge, Comprehension.

16.4 THE LOSS OF REGULATORY CONTROLS IN CANCER

46. These genes encode proteins that generally regulate the cell cycle in normal cells.

a. tumor suppressor genes
b. proto-oncogenes
c. oncogenes
d. tumor suppressor genes and proto-oncogenes
e. tumor suppressor genes, proto-oncogenes, and oncogenes

Answer: b
Difficulty: Easy
Bloom's Taxonomy: Knowledge

47. Many proto-oncogenes encode for

a. extracellular receptors.
b. intracellular receptors.
c. tumor suppressor proteins.
d. protein kinases.
e. extracelluar receptors and protein kinases.

Answer: e
Difficulty: Easy
Bloom's Taxonomy: Knowledge

48. The role of a tumor suppressor protein in a cell is to

a. promote cell division of healthy cells.
b. halt cell division in healthy cells experiencing difficulties in DNA replication.
c. promote cell division of abnormal cells.
d. trigger DNA replication in preparation for cell division.
e. activate cyclin-dependant protein kinases.

Answer: b
Difficulty: Easy
Bloom's Taxonomy: Knowledge

INSIGHTS FROM THE MOLECULAR REVOLUTION: A VIRAL TAX ON TRANSCRIPTIONAL REGULATION

49. What is the role of CREB in a white blood cell?

a. It is a tumor suppressor protein.
b. It stimulates the activity of p53.
c. It binds to CRE and activates cell division.
d. It increases cAMP production during an infection.
e. It decreases cAMP production during an infection.

Answer: c
Difficulty: Easy
Bloom's Taxonomy: Knowledge, Comprehension

50. How does the Tax protein contribute to the increased growth rate of white blood cells?

a. It stabilizes the CREB dimers that bind to the CRE DNA sequence.
b. It inhibits the ability of CREB to bind to DNA.
c. It phosphorylates the CREB activator protein.
d. It allows CREB to bind the CRE-like viral enhancer.
e. It stimulates the production of cAMP in a cell.

Answer: a
Difficulty: Moderate
Bloom's Taxonomy: Knowledge, Comprehension

UNANSWERED QUESTIONS

51. Chromatin remodeling

a. is subject to well-characterized stages of regulation by the cell.
b. blocks access to genes that should not be expressed in a particular cell.
c. is unaffected by alterations to the DNA sequences of a cell.
d. results in specific patterns of gene expression.
e. is generally independent of histone protein arrangements.

Answer: d
Difficulty: Moderate
Bloom's Taxonomy: Knowledge, Comprehension

52. RNAi could be used to treat macular degeneration by

a. preventing the development of excess blood vessels.
b. increasing the turnover rate of blood vessels in the retina.
c. preventing the replication of the virus that causes macular degeneration.
d. blocking expression of the viral genes that allow infection of retinal cells to occur.
e. stimulating the capillaries to tighten their walls, preventing the plasma leakage that clouds vision.

Answer: a
Difficulty: Moderate
Bloom's Taxonomy: Knowledge, Comprehension

Integrative Multiple-Choice

53. A key difference between eukaryotic and prokaryotic genomes is

a. the lack of controls of gene expression in prokaryotes.
b. the physical arrangement of genes on the chromosome(s).
c. the near simultaneous transcription and translation that occurs in eukaryotes and is strictly sequential in prokaryotes.
d. the physical separation of transcription and translation in prokaryotes compared to no separation of the processes in eukaryotes.
e. the regulation of the cell cycle by eukaryotes which is not observed in prokaryotes.

Answer: b
Difficulty: Moderate
Bloom's Taxonomy: Knowledge
Source: Sections 16.1, 16.2

54. In prokaryotes, which of the following types of regulation will *not* occur?

a. chromatin remodeling
b. variations of mRNA turnover
c. variations in protein turnover
d. activation of promoters
e. regulation of transcription

Answer: a
Difficulty: Easy
Bloom's Taxonomy: Knowledge
Source: Section 16.1, 16.2, 16.3

Matching

Match each of the following terms with its correct definition.

55. _____ miRNA
56. _____ repressor
57. _____ malignant
58. _____ activator
59. _____ chromatin remodeling
60. _____ transcriptional regulation
61. _____ silencing
62. _____ inducer
63. _____ siRNA
64. _____ operon
65. _____ benign
66. _____ metastasis

A. A regulatory protein that activates the expression of an operon's genes

B. Interfering RNA molecules originating encoded by nuclear DNA

C. A process of controlling the expression of genes that takes place at the DNA level

D. A regulatory protein that prevents an operon's genes from being expressed

E. The inhibition of transcription via DNA methylation

F. Interfering RNA molecules associated with some viral life cycles

G. A tumor that is comprised of cells that invade and disrupt the surrounding tissues

H. The presence of this molecule, such as allolactose, will increase expression of an operon's genes

I. A cluster of prokaryotic genes and their associated DNA regulatory sequences

J. A tumor that is comprised of undifferentiated cells that stay together in a single mass

K. The spread of malignant cells through the blood or lymphatic system

L. One example of this would be removing a nucleosome from DNA to expose a promoter

Answers: *55. B*   *56. D*   *57. G*   *58. A*   *59. L*
                60. C   *61. E*   *62. H*   *63. F*   *64. I*
                65. J   *66. K*

Difficulty: Moderate
Bloom's Taxonomy: Knowledge
Source: Sections 16.1, 16.2, 16.3, 16.4

Short Answer

67. Explain the difference between benign and malignant tumors.

Answer: In benign tumors, the cells have dedifferentiated into a less specialized state. The cells remain in a single mass and do not invade surrounding tissues. For this reason, they are generally not life threatening. Malignant tumors, in contrast, are disruptive to the surrounding tissues. Like benign tumors, they are comprised of dedifferentiated cells. Due to their disruptive nature and ability to metastasize, they are life threatening.
Difficulty: Moderate
Bloom's Taxonomy: Knowledge
Source: Section 16.4

68. Compare DNA methylation to histone acetylation.

Answer: DNA methylation and histone acetylation are two types of transcriptional regulation. Histone acetylation is a highly reversible way of granting transcription machinery access to a promoter. DNA methylation by itself is reversible but can be permanent when paired with additional chromatin modifications. Unlike histone acetylation, DNA methylation can be preserved from one generation to the next.
Difficulty: Moderate
Bloom's Taxonomy: Knowledge, Comprehension
Source: Section 16.3

Sequence

Imagine you are tracing the regulatory journey of a single protein being expressed in a single cell. There are multiple stages of regulation, and many of them can only occur in a particular order. Write the letter of the first regulatory step next to 69. The letter of the last stage of regulation is written next to 76.

69. _____   A.   Regulation of transcription initiation
70. _____   B.   Protein breakdown
71. _____   C.   Pre-mRNA processing
72. _____   D.   Protein modification
73. _____   E.   Chromatin remodeling
74. _____   F.   RNA interference
75. _____   G.   Initiation of translation
76. _____   H.   mRNA breakdown

Answers: 69. E   70. A   71. C   72. F   73. G
74. H   75. D   76. B

Difficulty: Moderate
Bloom's Taxonomy: Knowledge
Source: Section 16.3

True/False

If the statement is true, write a "T" in the blank. If the statement is false, make it correct by changing the underlined word(s) and writing the correct word(s) in the answer blank.

77. _____   An operator and transcription unit are called an <u>operon.</u>

78. _____   Many <u>eukaryotic operons</u> are subject to numerous regulatory mechanisms.

79. _____   <u>Eukaryotic</u> genes are organized into operons.

80. _____   <u>Eukaryotic</u> genes consist of protein-coding sequences and adjacent regulatory sequences.

81. _____   Acetylation is <u>irreversible.</u>

82. _____   Each steroid hormone can bind to its own <u>steroid hormone response unit</u> (SHRU) but not to the SHRU associated with other steroid hormones.

83. _____   When DNA is imprinted, the methylation pattern is <u>preserved.</u>

84. _____   RNAi is a key regulator or <u>transcription.</u>

85. _____   The DNA encoding siRNA <u>is not</u> normally found in the nucleus.

86. _____   There is a direct correlation between the level of translation and the length of a poly-A tail.

87. _____   A <u>benign</u> tumor is usually not life-threatening.

88. _____   RNAi could be used to treat macular degeneration in the future.

Answers: 77. T   78. F, prokaryotic operons
79. F, prokaryotic   80. T   81. F   reversible
82. T   83. T   84. F,   translation   85. T
86. T   87. T   88. T

Difficulty: Difficult
Bloom's Taxonomy: Knowledge
Source: Sections 16.1, 16.2, 16.3, 16.4

Select the Exception

89.   Which of the following is *not* a form of regulation of gene expression?

a.   transcriptional regulation
b.   translational regulation
c.   pretranscriptional regulation
d.   posttranscriptional regulation
e.   posttranslational regulation

Answer: c
Difficulty: Easy
Bloom's Taxonomy: Knowledge
Source: Why it matters

90.   Which of the following is *not* a reason why eukaryotic gene regulation is more complicated than prokaryotic gene regulation?

a.   Eukaryotic DNA is sequestered in the nucleus.
b.   Eukaryotic DNA is complexed with histones.
c.   Eukaryotic transcription is completed before translation begins.
d.   Multicellular eukaryotic organisms produce many different types of cells.
e.   Eukaryotic regulatory sequences are scattered throughout the genome.

Answer: e
Difficulty: Moderate
Bloom's Taxonomy: Knowledge
Source: Section 16.2

91.   Which of the following will *not* result in a proto-oncogene's conversion into an oncogene?

a.   A mutation in the proto-oncogene's promoter that increases transcription of the gene.
b.   A mutation in the coding sequence of a protein that renders a protein abnormally active.
c.   An alteration to the p53 coding sequence rendering the p53 protein inactive.
d.   An introduction of new DNA sequences to a chromosome that alters the regulation of the cell cycle.
e.   The translocation of a proto-oncogene to a location in the cell and under the control of a more active promoter than present in the original location.

Answer: c
Difficulty: Moderate
Bloom's Taxonomy: Knowledge
Source: Section 16.4

Essay

92.   The default state for the lac operon is off, while the default state for the trp operon is on. Explain this contradiction and the mechanism of control via the repressor proteins.

Answer: The lac operon's function is to allow a cell to utilize lactose as an energy source. It wouldn't make sense to express these genes unless lactose were available in the environment. Because of this, the presence of lactose serves to block the repressor from binding to the operator. This results in the lac operon only being active when lactose is available.

In contrast, the trp operon's function is to allow the bacterial cell to synthesize tryptophan, which is an amino acid used as a building block in protein synthesis. The cell must make it unless it is available in the environment. For this reason, the trp operon is on unless there is excess tryptophan available (as would be the case when it is in the environment) to bind to the trp repressor, activate it, and cause the repressor to bind to the operator.
Difficulty: Moderate
Bloom's Taxonomy: Comprehension
Source: Section 16.1

93.   When comparing gene regulation in prokaryotes to that in eukaryotes, which is better? Justify your answer.

Answer: Neither is superior to the other; instead, each reflects the nature of the organisms using that system. In prokaryotes, the entire organism is a single cell so differentiation is not necessary or desirable. Rather, the genome is organized and regulated to maximize efficiency of a cell during what might be a very short lifespan. In contrast, multicelluar organisms must have cell differentiation for efficiency. This requires a method of regulation that is not needed or possible in the prokaryotic cell. By regulating gene expression at multiple stages, the organism can coordinate cell division, differentiation, and body development. In summary, each type of gene regulation is tailored to the life strategy of the organism in which it is found.
Difficulty: Moderate
Bloom's Taxonomy: Evaluation
Source: Section 16.1, 16.2, 16.3

Labeling

Identify each labeled part of the following illustration of eukaryotic DNA.

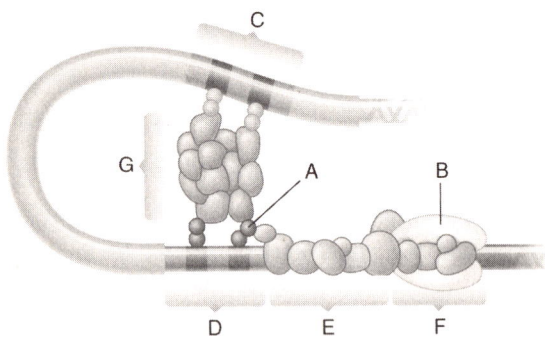

94. _____ activators
95. _____ coactivator multiprotein complex
96. _____ promoter
97. _____ transcription initiation site
98. _____ enhancer
99. _____ proximal promoter region
100. _____ RNA polymerase

Answer: 94. A   95. G   96. E   97. F   98. C
99. D   100. B

Difficulty: Moderate
Bloom's Taxonomy: Knowledge
Source: Figure 16.10

Identify each labeled part of the following illustration.

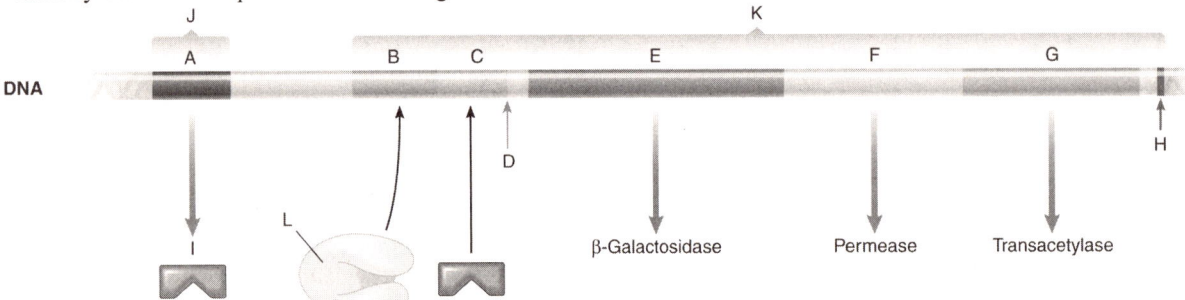

101. _____ lacZ
102. _____ promoter
103. _____ lacY
104. _____ transcription termination site
105. _____ regulatory gene
106. _____ lac operon
107. _____ lac repressor
108. _____ lacI
109. _____ lacA
110. _____ operator
111. _____ RNA polymerase
112. _____ transcription initiation site

Answer: 101. E   102. B   103. F   104. H   105. J
106. K   107. I   108. A   109. G   110. C   111. L
112. D

Difficulty: Moderate
Bloom's Taxonomy: Knowledge
Source: Figure 16.2

17

BACTERIAL AND VIRAL GENETICS

Multiple-Choice

WHY IT MATTERS

1. *Escherichia coli* is used in scientific investigations because it:

a. grows quickly in huge numbers in nutrient solutions that are easy to prepare.
b. is affected by bacteriophages.
c. can be used in genetic studies.
d. allows researchers to detect events that occur once within millions of offspring.
e. all of the above

Answer: e
Difficulty: Easy
Russell's Biology: Knowledge

2. *E. coli* biochemical studies have added immeasurably to:

a. the definition of genes.
b. the definition of gene activities.
c. the identification of biochemical pathways.
d. all of the above
e. b and c only

Answer: d
Difficulty: Moderate
Russell's Biology: Knowledge

3. *Neurospora* and *Aspergillus*

a. are fungi.
b. are used in a similar manner as *E. coli* to elucidate biochemical pathways.
c. are eukaryotes.
d. can be grown and analyzed biochemically in large numbers.
e. all of the above

Answer: e
Difficulty: Moderate
Russell's Biology: Synthesis, Knowledge

4. Eukaryotes that can be used in molecular studies include:

a. Arabidopsis.
b. yeasts.
c. Drosophila.
d. all of the above
e. none of the above

Answer: d
Difficulty: Moderate
Russell's Biology: Knowledge

5. Transposable elements are:

a. peptide sequences that move from place to place in the nucleus.
b. sequences that can move from place to place in bacterial DNA.
c. sequences that do not move from place to place in bacterial DNA.
d. all of the above
e. none of the above

Answer: b
Difficulty: Moderate
Russell's Biology: Knowledge

17.1 Gene Transfer and Genetic Recombination in Bacteria

6. Bacterial genes can be transferred from one bacterium to another by different mechanisms. The newly introduced DNA can:

a. recombine with DNA already present.
b. generate genetic variability.
c. exchange alleles between homologous regions from two different individuals.
d. none of the above
e. a, b, and c only

Answer: e
Difficulty: Moderate
Russell's Biology: Knowledge

7. As early as the 1940s, researchers knew that bacteria could be grown in a minimal medium. In this minimal medium the most common source of organic carbon is:

a. peptidase.
b. glucose.
c. rubisco.
d. catalase.
e. ammonium chloride.

Answer: b
Difficulty: Easy
Russell's Biology: Knowledge

8. Ammonium chloride is a commonly used salt in minimal media. The reason ammonium chloride is used is because:

a. it is an organic source of carbon.
b. it provides nitrogen.
c. it jellifies the media.
d. it liquefies the media.
e. a and b only

Answer: b
Difficulty: Moderate
Russell's Biology: Synthesis, Knowledge

9. Bacterial cultures with a large number of genetically identical cells are called:

a. clone cultures.
b. equal cultures.
c. cultures.
d. identical cultures.
e. a, b, and d only

Answer: a
Difficulty: Easy
Russell's Biology: Knowledge

10. Auxotroph bacteria

a. are mutant bacteria.
b. cannot grow on minimal media.
c. feed on proteins
d. all of the above
e. a and b only

Answer: e
Difficulty: Easy
Russell's Biology: Knowledge

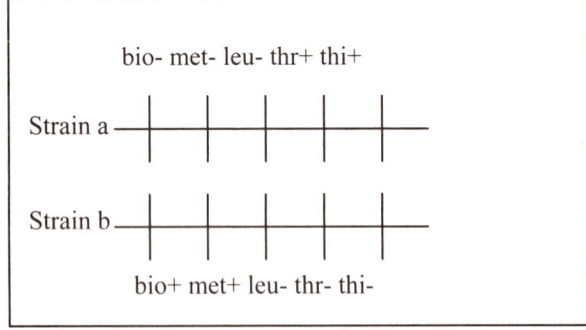

Use this figure to answer questions 11 through 15.

11. bio refers to a gene that governs a cell's ability to synthesize biotin from inorganic precursors. The designation bio+ indicates:

a. that the allele is normal.
b. that the allele is the mutant allele.
c. that the bacterium is biologically active.
d. that the bacterium is biologically inactive.
e. none of the above

Answer: a
Difficulty: Easy
Russell's Biology: Knowledge, Application

12. met refers to a gene that governs a cells ability to synthesize methionine from inorganic precursors. The designation met- indicates:

a. that the allele is normal.
b. that the allele is the mutant allele.
c. that the bacterium is biologically active.
d. that the bacterium is biologically inactive.
e. none of the above

Answer: b
Difficulty: Easy
Russell's Biology: Knowledge, Application

13. leu refers to a gene that governs a cells ability to synthesize leucine from inorganic precursors. The designation leu- indicates:

a. that the allele is normal.
b. that the allele is the mutant allele.
c. that the bacterium is biologically active.
d. that the bacterium is biologically inactive.
e. none of the above

Answer: b
Difficulty: Easy
Russell's Biology: Knowledge, Application

14. thr refers to a gene that governs a cells ability to synthesize threonine from inorganic precursors. The designation thr+ indicates:

a. that the allele is normal.
b. that the allele is the mutant allele.
c. that the bacterium is biologically active.
d. that the bacterium is biologically inactive.
e. none of the above

Answer: a
Difficulty: Easy
Russell's Biology: Knowledge, Application

15. If strain a and strain b conjugate, the most likely outcome of their offspring would be:

a. bio- met- leu- thr- thi+.
b. bio+ met+ leu+ thr+ thi+.
c. bio+ met- leu- thr- thi+.
d. bio+ met+ leu- thr- thi+.
e. none of the above.

Answer: b
Difficulty: Difficult
Russell's Biology: Knowledge, Application

16. During bacterial conjugation, bacterial DNA with different alleles get together. Bacteria use a long tubular structure called a _____ to obtain differing alleles.

a. metabolic pilus
b. asexual pilus
c. flagellum
d. cilia
e. sex pilus

Answer: e
Difficulty: Easy
Russell's Biology: Knowledge

17. Conjugation is a process by which the bacterial cells

a. contact each other using a sex pilus to obtain new alleles.
b. contact each other for gratification purposes.
c. disrupt each other
d. a and b only
e. a, b, and c

Answer: a
Difficulty: Easy
Russell's Biology: Knowledge

18. The ability to conjugate depends on the presence within the donor cell of a plasmid called the _____.

a. X factor
b. F factor
c. C factor
d. sex pilus
e. S factor

Answer: b
Difficulty: Easy
Russell's Biology: Knowledge

19. Small circles of DNA that occur in bacteria in addition to the main circular chromosomal DNA molecule are called:

a. transformants.
b. plastids.
c. plasmids.
d. plasmitrons.
e. b, c and d only

Answer: c
Difficulty: Easy
Russell's Biology: Knowledge

20. Donor cells in conjugation are labeled _____, whereas recipient cells are labeled _____.

a. F+ cells , F- cells
b. F+ cells, F+ cells
c. F- cells, F+ cells
d. F1 cells, F2 cells
e. F+ cells, F2 cells

Answer: a
Difficulty: Moderate
Russell's Biology: Knowledge, Application

21. The purpose of the sex pilus is to:

a. form a cytoplasmic bridge and conjugate.
b. form a gap junction and conjugate.
c. facilitate DNA exchange.
d. a and c only
e. a, b, and c

Answer: d
Difficulty: Moderate
Russell's Biology: Knowledge, Application

22. F+ X F- mating does not result in

a. plasmid genetic recombination.
b. chromosomal genetic recombination.
c. circularization of DNA.
d. conjugation.
e. all of the above

Answer: b
Difficulty: Moderate
Russell's Biology: Knowledge, Application

23. Hfr cells are cells that can integrate the f factor and some genes into the bacterial chromosome. Hfr cells can do this through:

a. mitosis.
b. meiosis.
c. prophase.
d. crossing-over.
e. none of the above

Answer: d
Difficulty: Moderate
Russell's Biology: Knowledge, Application

24. Genetic recombination by conjugation was discovered by:

a. Francois Jacob.
b. Elie L. Wollman.
c. Gregor Mendel.
d. a and b only
e. b and c only

Answer: d
Difficulty: Moderate
Russell's Biology: Knowledge, Application

25. The full transfer of an entire DNA molecule to an F- cell would take about _____ minutes.

a. 10–20
b. 20–30
c. 300–400
d. 4–7
e. 90–100

Answer: e
Difficulty: Moderate
Russell's Biology: Knowledge, Application

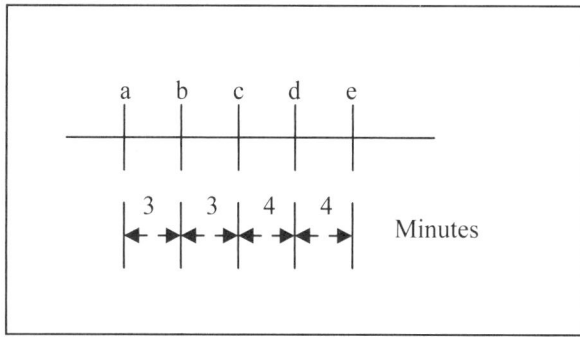

Use this figure to answer questions 26 through 29.

26. In order to transfer gene a, b, and c to an F- cell, how long must the sex pilus be in place?

a. Less than 4 minutes.
b. 10 minutes.
c. 14 minutes.
d. More than 8 minutes but less than 20.
e. none of the above

Answer: b
Difficulty: Difficult
Russell's Biology: Knowledge, Application

27. In order to transfer gene a to an F- cell, the sex pilus must be in place for _____ minutes.

a. 1
b. 2
c. 3
d. 4
e. 5

Answer: c
Difficulty: Difficult
Bloom's Taxonomy: Knowledge, Application

28. In order to transfer all the genes to an F- cell we would need how long?

a. 1 minute
b. 20 minutes
c. 3 minutes
d. 14 minutes
e. 5 minutes

Answer: d
Difficulty: Difficult
Russell's Biology: Knowledge, Application

29. How long would it take to transfer genes a and b to an F- cell?

a. 1 minute
b. 7 minutes
c. 90 minutes
d. 4 minutes
e. 6 minutes

Answer: e
Difficulty: Difficult
Russell's Biology: Knowledge, Application

30. R plasmids

a. provide antibiotic resistance.
b. provide a fertility factor.
c. provide reliance to bacteria.
d. provide recombination to bacteria.
e. all of the above

Answer: a
Difficulty: Easy
Russell's Biology: Knowledge

31. In transformation, cells

a. take up pieces of DNA that are released as other cells disintegrate.
b. take up pieces of DNA through infection of a virus.
c. replicate DNA molecules.
d. make replicate copies of one another.
e. generate their own DNA.

Answer: a
Difficulty: Easy
Russell's Biology: Knowledge

32. Transformation was discovered by:

a. Francois Jacob.
b. Elie L. Wollman.
c. Gregor Mendel.
d. Joshua Lederberg.
e. Fred Griffith.

Answer: e
Difficulty: Easy
Russell's Biology: Knowledge

33. A colony of mice is treated with a noninfective form of the bacterium *Streptococcus pneumoniae*. A new researcher exposes this noninfective form to heat-killed cells of an infective strain. What would be the effect of this action in the mice population?

a. The mice population will remain healthy.
b. Some mice will die, but most will live.
c. The mice population will get ill with pneumonia.
d. b and c only.
e. a and b only.

Answer: c
Difficulty: Moderate
Russell's Biology: Knowledge, Application

34. A colony of mice is treated with a noninfective form of the bacterium *Streptococcus pneumoniae*. A researcher exposes this noninfective form to heat-killed cells of an infective strain. The bacterial cells are said to be:

a. transduced.
b. artificially transformed.
c. transformed.
d. noninfectious.
e. none of the above

Answer: c
Difficulty: Moderate
Russell's Biology: Knowledge, Application

35. The linear DNA fragments taken up from disrupted infective cells recombine with the chromosomal DNA of the noninfective cells by _____, much in the same way as genetic recombination takes place in conjugation.

a. meiosis
b. metaphase II
c. nuclear division
d. double crossovers
e. a, b, and c

Answer: d
Difficulty: Easy
Russell's Biology: Knowledge, Application

36. Bacterial cells that cannot readily pick up DNA molecules from their surroundings can be forced to do so by _____.

a. artificial transformation
b. double crossing over
c. crossing over
d. constriction
e. b and c only

Answer: a
Difficulty: Easy
Russell's Biology: Knowledge, Application

37. Exposing *E. coli* to calcium ions, then incubating the culture at low temperatures and finally heat shocking the culture will accomplish:

a. artificial transformation.
b. double crossing over.
c. crossing over.
d. constriction.
e. b and c only

Answer: a
Difficulty: Easy
Russell's Biology: Knowledge, Application

38. A technique that is used to artificially transform a bacterium is called electroporation. Electroporation:

a. exposes cells briefly to rapid pulses of electrical current.
b. exposes cells to calcium ions.
c. exposes cells to stomes.
d. inserts the foreign DNA into the DNA in the chloroplast.
e. inserts the foreign DNA into the plant's mitochondrial DNA.

Answer: a
Difficulty: Moderate
Russell's Biology: Knowledge, Application

39. In transduction, bacterial cells

a. take up pieces of DNA that are released as other cells disintegrate.
b. take up pieces of DNA through the use of a virus.
c. replicate DNA molecules.
d. make replicate copies of one another.
e. generate their own DNA.

Answer: b
Difficulty: Easy
Russell's Biology: Knowledge, Application

40. Transduction was discovered in 1952 by this Nobel Prize winner.

a. Francois Jacob
b. Elie L. Wollman
c. Gregor Mendel
d. Joshua Lederberg
e. Fred Griffith

Answer: d
Difficulty: Easy
Russell's Biology: Knowledge

41. Norton Zinder and Joshua Lederberg used this microorganism to understand transduction.

a. *Escherichia coli*
b. *Neurospora*
c. *Aspergillus*
d. *Salmonella typhimurium*
e. *Streptococcus pneumoniae*

Answer: d
Difficulty: Moderate
Russell's Biology: Knowledge

42. A plate of solid growth medium with colonies on it is pressed gently onto sterile velveteen. The velveteen "stamp" is used to make identical plates. This technique is called:

a. replication.
b. replica plating.
c. plating.
d. plate making.
e. none of the above

Answer: b
Difficulty: Easy
Russell's Biology: Knowledge

43. Media that contain a full complement of nutrient substances, including amino acids, and other chemicals that normal strains use for themselves, is called:

a. Agar media.
b. minimal medium.
c. nutritious media.
d. complete medium.
e. none of the above

Answer: d
Difficulty: Easy
Russell's Biology: Knowledge

44. A researcher is looking for met- strains. He/she compares two plates. One plate has colonies growing on complete medium. The other plate has colonies growing on methionine deficient medium. How can the researcher determine which colonies are met-?

a. The colonies growing on the methionine plate are met-.
b. By comparing the colonies in both plates we can identify met- recombinants because they do not grow on the plate lacking methionine.
c. We cannot determine which colonies are met- because we do not have enough information.
d. Only the colonies growing on the complete medium are met-.
e. The complete medium is the answer.

Answer: b
Difficulty: Difficult
Russell's Biology: Knowledge, Application

17.2 Viruses and Viral Recombination

45. Viruses can undergo genetic recombination when

a. the DNA of one virus infects a single cell.
b. bacterial DNA enters a bacterium.
c. a and b only
d. the DNA of two viruses infect a single cell.
e. none of the above

Answer: d
Difficulty: Moderate
Russell's Biology: Knowledge

46. Viruses are composed of only

a. proteins.
b. carbohydrates and proteins.
c. proteins and nucleic acids.
d. nucleic acids.
e. carbohydrates, proteins, lipids, and nucleic acids.

Answer: c
Difficulty: Easy
Russell's Biology: Knowledge

47. The core of a virus is

a. a protein.
b. DNA or RNA.
c. a nucleic acid.
d. b and c only
e. a, b, and c

Answer: d
Difficulty: Moderate
Russell's Biology: Knowledge

48. When viruses are not inside a cell, they are carried about passively by random molecular movements and perform none of the metabolic activities of _____.

a. homeostasis.
b. life.
c. eukaryotes.
d. prokaryotes.
e. autotrophs.

Answer: b
Difficulty: Easy
Russell's Biology: Knowledge

49. All viruses have genes encoding at least:

a. their coat proteins.
b. enzymes required for nucleic acid replication.
c. a and b only
d. a plasma membrane.
e. pili.

Answer: c
Difficulty: Easy
Russell's Biology: Knowledge, Application

50. Viruses that kill their host cells during each cycle of infection are called

a. virulent bacteriophages.
b. bacteriophages.
c. temperate bacteriophages.
d. phages.
e. all of the above

Answer: a
Difficulty: Moderate
Russell's Biology: Knowledge

51. Viruses that enter an inactive phase in which the host cell replicates and passes on a bacteriophage DNA for generations before the phage becomes active and kills the host is called

a. virulent bacteriophages.
b. bacteriophages.
c. temperate bacteriophages.
d. phages.
e. all of the above

Answer: b
Difficulty: Moderate
Russell's Biology: Knowledge

52. T-even bacteriophages infect

a. *Escherichia coli.*
b. *Neurospora.*
c. *Aspergillus.*
d. *Salmonella typhimurium.*
e. *Streptococcus pneumoniae.*

Answer: a
Difficulty: Easy
Russell's Biology: Knowledge

53. The series of events from infection of one cell through the release of progeny phages from broken, open, or lysed cells is called

a. lytic cycle.
b. lysogenic cycle.
c. Krebs cycle.
d. citric acid cycle.
e. a and b only

Answer: a
Difficulty: Moderate
Russell's Biology: Knowledge

54. For some virulent phages, fragments of the host DNA may be included in the heads as the viral particles assemble, providing the basis for transduction of bacterial genes during the next cycle of infection. This mechanism is termed:

a. transformation.
b. transmutation.
c. metabolism.
d. specific transduction.
e. generalized transduction.

Answer: e
Difficulty: Easy
Russell's Biology: Knowledge

55. Bacteriophage lambda

a. is a temperate phage.
b. is a virulent phage.
c. is used in plants.
d. is used in animals.
e. all of these

Answer: a
Difficulty: Easy
Russell's Biology: Knowledge

56. This cycle begins when the lambda circular chromosome integrates into the host cell's DNA by crossing over. Once integrated, the lambda genes are mostly inactive. This cycle is referred as

a. lytic cycle.
b. lysogenic cycle.
c. Krebs cycle.
d. citric acid cycle.
e. a and b only

Answer: b
Difficulty: Moderate
Russell's Biology: Knowledge

57. A virus in the lysogenic cycle is referred to as a

a. prophage.
b. virulent phage.
c. tempered phage.
d. none of the above.
e. all of these

Answer: a
Difficulty: Moderate
Russell's Biology: Knowledge

Matching

Match each of the following terms with its correct definition.

58. _____ lytic pathway
59. _____ lysogenic pathway
60. _____ transduction
61. _____ phage
62. _____ virulent phage
63. _____ temperate phage
64. _____ generalized transduction
65. _____ specialized transduction
66. _____ transposable elements
67. _____ insertion sequences
68. _____ transposase
69. _____ transposon
70. _____ retrotransposons
71. _____ reverse transcriptase
72. _____ TE-instigated
73. _____ bacteriophage lambda

A. For some virulent phages, fragments of the host DNA may be included in the heads as the viral particles assemble, providing the basis for transduction of bacterial genes during the next cycle of infection.

B. A virus

C. This cycle begins when the lambda circular chromosome integrates into the host cell's DNA by crossing over. Once integrated, the lambda genes are mostly inactive.

D. Type of virus that does not kill a host cell in every replication cycle

E. Type of virus that kills a host cell in every replication cycle

F. When only genes that are adjacent to the integration site of a temperate phage can be cut out with the viral DNA and included in phage particles during the lytic stage.

G. Particular segments of DNA that move from one place to another.

H. Simplest TEs.

I. The series of events from infection of one cell through the release of progeny phages from broken, open, or lysed cells.

J. Enzyme that catalyses reactions to insert or remove TEs

K. Has an inverted repeat sequence at each end.

L.   Transpose by a copy-and-paste mechanism using RNA.

M.   Certain kinds of cancer have been linked to these genes.

N.   Enzyme that uses and RNA template to make a DNA copy of the retrotransposon.

O.   Transferring DNA using a virus.

P.   *E. coli* phage.

Answers: *58. I*   *59. C*   *60. O*   *61. B*   *62. E*
            63. D   *64. A*   *65. F*   *66. G*   *67. H*   *68. J*
            69. K   *70. L*   *71. N*   *72. M*   *73. P*

Difficulty: Moderate
Russell's Biology: Knowledge
Source: Section 17.1, 17.2, 17.3

18
DNA TECHNOLOGIES AND GENOMICS

Multiple-Choice

WHY IT MATTERS

1. DNA technology can
a. link a criminal to a crime scene.
b. match tissue samples with the probable source animal.
c. help investigators identify a body.
d. identify which individual is *not* the rapist in a particular case.
e. all of the above

Answer: e
Difficulty: Easy
Bloom's Taxonomy: Knowledge

2. Which of the following is **not** an example of biotechnology?
a. genetic engineering
b. yogurt manufacturing
c. cheese production
d. crop fertilization
e. DNA fingerprinting

Answer: d
Difficulty: Easy
Bloom's Taxonomy: Knowledge

18.1 DNA Cloning

3. When cloning DNA into bacteria,
a. the DNA sequence is inserted into a plasmid.
b. the DNA sequence is inserted into the bacterial chromosome.
c. the linear DNA sequence is circularized, thus generating a plasmid.
d. the new DNA is directly introduced into the bacterial cell.
e. either the approach in b or c is used; the choice depends on the type of bacteria.

Answer: a
Difficulty: Easy
Bloom's Taxonomy: Knowledge

4. What is the *natural* function of restriction endonucleases?
a. research applications
b. DNA manipulation *in vitro*
c. defense against viruses that infect bacteria
d. regulating bacterial gene expression
e. breaking phosphodiester bonds in mammalian DNA

Answer: c
Difficulty: Easy
Bloom's Taxonomy: Knowledge

5. Some restriction endonuclease cuts the DNA in such a way that short, single stranded regions are created. We call these regions
a. hydrogen-bonding ends
b. sticky ends
c. tacky ends
d. blunt ends
e. jagged ends

Answer: b
Difficulty: Easy
Bloom's Taxonomy: Knowledge

6. What do restriction endonucleases do?
a. break phosphodiester bonds
b. break ester bonds
c. break glycosidic bonds
d. break peptide bonds
e. all of the above

Answer: a
Difficulty: Easy
Bloom's Taxonomy: Knowledge

7. The DNA sequence recognized by restriction endonucleases is generally how long?
a. 1–2 nucleotides
b. 4–8 nucleotides
c. 10–15 nucleotides
d. 24–32 nucleotides
e. 36–47 nucleotides

Answer: b
Difficulty: Easy
Bloom's Taxonomy: Knowledge

8. Cloning plasmids generally contain genes for which two traits?
a. lacZ and amp^R
b. beta-galactosidase enzyme production and ampicillin resistance
c. DNA ligase production and sticky ends
d. DNA circulase and DNA recombinase enzyme expression
e. DNA replication ability and antibiotic resistance

Answer: b
Difficulty: Moderate
Bloom's Taxonomy: Knowledge

Enzyme	Recognition sequence and cut site (vertical line). Only the top strand of the double stranded DNA sequence is shown.
Eco RI	G|AATTC
Hind III	A|AGCTT
Bam HI	G|GATCC
Cla I	AT|CGAT
Pvu I	CGAT|CGC

Use the table above for questions 9–11.

9. Which enzyme did I use if I cut a DNA fragment and end up with a double stranded sequence that looks like this:

```
AATTC--------G
    G--------CTTAA
```

a. Eco RI
b. Hind III
c. Bam HI
d. Cla I
e. Pvu I

Answer: a
Difficulty: Difficult
Bloom's Taxonomy: Application

10. Which enzymes did I use if I cut a DNA fragment and end up with a double stranded sequence that looks like this:

```
GATCC--------AT
    G--------TAGC
```

a. Eco RI and Hind III
b. Hind III amd Bam HI
c. Bam HI and Cla I
d. Cla I and Pvu I
e. Pvu I and Bam HI

Answer: c
Difficulty: Difficult
Bloom's Taxonomy: Application

11. Which enzymes should I use to generate a double stranded sequence that looks like this:

```
AGCTT--------G
    A--------CCTAG
```

a. Eco RI and Hind III
b. Hind III amd Bam HI
c. Bam HI and Cla I
d. Cla I and Pvu I
e. Pvu I and Bam HI

Answer: b
Difficulty: Difficult
Bloom's Taxonomy: Application

12. Why is it important to have an antibiotic resistance gene in the cloning plasmid?

a. It makes the bacteria resistant to the antibiotic.
b. It provides a way for researchers to sort the bacteria that have the cloning plasmid from the bacteria that don't have the cloning plasmid.
c. It aids scientists in developing new antibiotic treatments.
d. It provides a way for sorting bacteria that have the gene of interest from the bacteria that just have an "empty" cloning plasmid.
e. It makes the bacteria produce more clones.

Answer: b
Difficulty: Moderate
Bloom's Taxonomy: Comprehension

13. Why do scientists insert their DNA of interest into the middle of the lacZ coding sequence?

a. It makes the bacteria resistant to the antibiotic.
b. It provides a way for researchers to sort the bacteria that have the cloning plasmid from the bacteria that don't have the cloning plasmid.
c. It aids scientists in developing new antibiotic treatments.
d. It provides a way for sorting bacteria that have the gene of interest from the bacteria that just have an "empty" cloning plasmid.
e. It makes the bacteria produce more clones.

Answer: d
Difficulty: Moderate
Bloom's Taxonomy: Comprehension

14. What is the substrate for beta-galactosidase in bacterial screening assays that check for colonies having recombinant DNA?

a. lactose
b. X-gal
c. sucrose
d. blue azure
e. bromophenol blue

Answer: b
Difficulty: Easy
Bloom's Taxonomy: Knowledge

15. If multiple fragments from a genome have been cloned into bacteria, how can researcher identify the bacteria that contains the particular sequence they are interested in?

a. The bacteria containing the gene of interest are antibiotic resistant and unable to digest X-gal.
b. The DNA purified from individual colonies of bacteria is probed with a short, single stranded DNA fragment that is complementary to the gene of interest.
c. The scientists have to sequence all of the DNA from all of the white bacterial colonies; fortunately computers let us automate this process.
d. The scientists have to sequence all of the DNA from all of the blue bacterial colonies; fortunately computers let us automate this process.
e. We have to isolate the DNA of interest before transforming it into the bacteria. That is the only way to know which bacteria have the DNA of interest.

Answer: b
Difficulty: Moderate
Bloom's Taxonomy: Comprehension

16. Why would researchers want to have both genomic libraries and cDNA libraries from a particular organisim?

a. The genomic library is generally too big to be helpful; a cDNA library, being smaller, is much easier to study.
b. A cDNA library will be the same in every cell; the genomic library tells researchers which genes are being expressed in individual cell types.
c. The cDNA library helps us understand how retroviruses work; the genomic library does not.
d. The cDNA library is a lot easier to clone than a genomic library.
e. A genomic library will be the same in every cell; the cDNA library tells researchers which genes are being expressed in individual cell types.

Answer: e
Difficulty: Easy
Bloom's Taxonomy: Knowledge

17. What is/are the advantage(s) of a polymerase chain reaction?

a. PCR allows for the mass-production of a specific DNA sequence without cloning.
b. PCR permits scientists to target a single copy of a single gene among millions of other sequences.
c. PCR allows scientists to study DNA from samples of finite amounts, such as from fossils, or small tissue samples left at a crime scene.
d. a and b only
e. a, b, and c.

Answer: e
Difficulty: Moderate
Bloom's Taxonomy: Knowledge

18. I need to determine if the gene for a particular protein is the same length in frogs, humans, and trees. Which method or methods will be most helpful to me?

a. PCR and agarose gel electrophoresis
b. DNA cloning and DNA hybridization
c. cDNA library construction and agarose gel electrophoresis
d. genomic library construction
e. agarose gel electrophoresis alone

Answer: a
Difficulty: Easy
Bloom's Taxonomy: Knowledge

***FOCUS ON RESEARCH:* RESEARCH METHODS: CLONING A GENE OF INTEREST IN A PLASMID CLONING VECTOR**

19. Identify the correct steps and the correct order for cloning a gene of interest.

1. Transform the bacteria.
2. Mix the postrestriction endonuclease gene of interest with the postrestriction endonuclease cloning plasmid.
3. Incubate the combined DNA fragments with the liagase enzyme.
4. Use restriction endonucleases to cut the gene of interest and the target DNA.
5. Spread bacteria on medium containing lactose and ampicillin.
6. Spread bacteria on medium containing X-gal and ampicillin.

a. 4, 2, 3, 1, 6
b. 4, 3, 2, 1, 5
c. 4, 3, 2, 1, 6
d. 4, 2, 3, 1, 5
e. 2, 4, 1, 5, 3

Answer: c
Difficulty: Moderate
Bloom's Taxonomy: Knowledge

FOCUS ON RESEARCH: RESEARCH METHODS: DNA HYBRIDIZATION TO IDENTIFY A DNA SEQUENCE OF INTEREST

20.   I've cloned a genomic library and know that my bacteria have foreign DNA fragments inserted into the cloning vector. I now need to identify the specific bacteria that have the gene coding for protein X. How can I locate these bacteria?

a.   DNA hybridization of the different colonies
b.   PCR of the same bacterial mixture that was spread on agarose plates
c.   restriction endonuclease treatment of DNA from some of the bacterial colonies
d.   agarose gel electorphoresis of DNA isolated from the bacterial colonies
e.   transfer the bacterial colonies to filter paper

Answer: a
Difficulty: Easy
Bloom's Taxonomy: Knowledge

FOCUS ON RESEARCH: RESEARCH METHODS: THE POLYMERASE CHAIN REACTION (PCR)

21.   Which of the following do I need to amplify the DNA for gene X and not for gene Y?

a.   Some prior knowledge of the sequence of gene X so appropriate primers can be designed.
b.   a heat-stable DNA polymerase
c.   restriction endonucleases that cut gene X and not gene Y
d.   plasmid vectors that have multiple antibiotic resistance genes.
e.   A cDNA library and access to probes for gene X

Answer: a
Difficulty: Easy
Bloom's Taxonomy: Knowledge

22.   Why does the polymerase chain reaction require the samples to undergo numerous cycles, each containing three different temperatures?

a.   The initial cycles don't produce many copies of the original gene since the amplification is exponential in nature.
b.   The three temperatures are for different processes; the highest temperature separates the double stranded DNA into single strands, the lower temperatures allow a target primer to bind and a DNA polymerase to generate new sequences.
c.   The enzyme used is from an organism that thrives in high temperatures, so high temperatures must be used.
d.   a and b only
e.   a, b, and c.

Answer: a
Difficulty: Easy
Bloom's Taxonomy: Knowledge

FOCUS ON RESEARCH: RESEARCH METHODS: SEPARATION OF DNA FRAGMENTS BY AGAROSE GEL ELECTROPHORESIS

23.   In figure below, which DNA fragment is the smallest?

a.   A
b.   B
c.   C
d.   D
e.   E

Answer: e
Difficulty: Moderate
Bloom's Taxonomy: Application

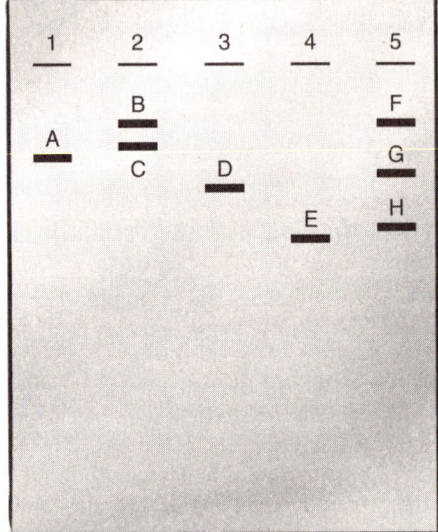

24.   In the figure above, which DNA fragment is closest to the positive electrode?

a.   A
b.   B
c.   E
d.   F
e.   H

Answer: b
Difficulty: Moderate
Bloom's Taxonomy: Application

25.   In figure above, which lane shows a very large, linear DNA fragment that was cut by one enzyme?

a.   1
b.   2
c.   3
d.   4
e.   5

Answer: b
Difficulty: Moderate
Bloom's Taxonomy: Application

26. In figure above, which lane shows a very large, linear DNA fragment that was cut by two enzymes?

a. 1
b. 2
c. 3
d. 4
e. 5

Answer: e
Difficulty: Moderate
Bloom's Taxonomy: Application

18.2 Applications of DNA Technologies

27. Restriction fragment length polymorphisms are used to

a. test the effectiveness of different restriction enzymes on a sequence of DNA.
b. compare the DNA sequences between individuals by looking for changes in restriction enzyme digest patterns.
c. compare DNA sequences by determining the full DNA sequence.
d. prepare DNA for further sequence analysis.
e. treat sickle-cell anemia.

Answer: b
Difficulty: Moderate
Bloom's Taxonomy: Knowledge
Figure 18.2

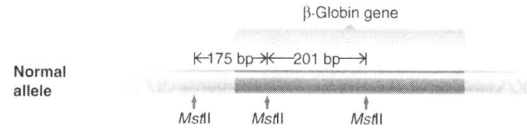

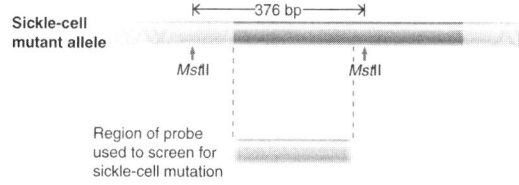

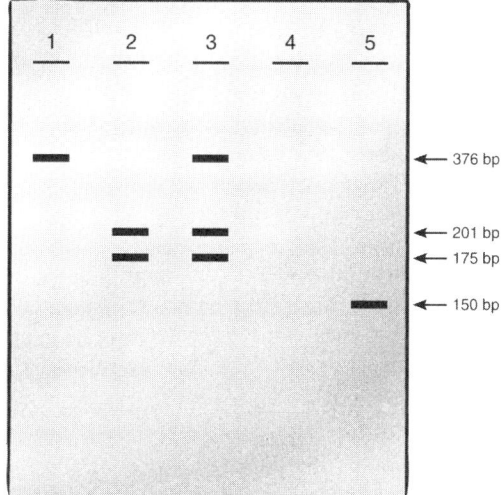

28. Using the information presented in figure above, where nuclear DNA was digested with MstII, subjected to gel electrophoresis, and probed in a Southern blot with the indicated probe, which lane shows an individual that is heterozygous for sickle cell?

a. 1
b. 2
c. 3
d. 4
e. 5

Answer: c
Difficulty: Difficult
Bloom's Taxonomy: Knowledge, Application

29. Using the information presented in figure above, where nuclear DNA was digested with MstII, subjected to gel electrophoresis, and probed in a Southern blot with the indicated probe, which lane had a DNA sample from a palm tree?

a. 1
b. 2
c. 3
d. 4
e. 5

Answer: d
Difficulty: Difficult
Bloom's Taxonomy: Knowledge, Application

30. Using the information presented in figure above (where nuclear DNA was digested with MstII, subjected to gel electrophoresis, and probed in a Southern blot with the indicated probe), which lane probably had a DNA sample from a nonhuman source, at least partial homology to the beta-globin gene, and a different arrangement of MstII restriction enzyme sites than found in humans?

a. 1
b. 2
c. 3
d. 4
e. 5

Answer: e
Difficulty: Difficult
Bloom's Taxonomy: Knowledge, Application

31. DNA fingerprinting is

a. a way to identify the species a sample came from.
b. a way to distinguish between two individuals of the same species.
c. a method that examines the DNA sequences encoding for the tiny ridges and valleys that provide texture to the pads of human fingers.
d. a way of extracting DNA samples from human fingerprints.
e. a way of determining sequencing the human genome.

Answer: b
Difficulty: Easy
Bloom's Taxonomy: Knowledge

32. In standardized DNA fingerprinting used in the United States,

a. 13 different noncoding loci are examined.
b. 13 different coding loci are examined.
c. only loci with short tandem repeats are tested; the number varies with individuals.
d. the loci with long tandem repeats are tested; the number varies with individuals.
e. the sequence of short tandem repeats is determined.

Answer: a
Difficulty: Easy
Bloom's Taxonomy: Knowledge

33. Genetic engineering has been used to

a. induce bacteria to mass-produce human insulin.
b. produce a vaccine against hoof-and-mouth disease.
c. create mice with altered genomes so we can model certain human diseases.
d. create plants that are more resistant to certain pests or herbicides.
e. all of the above

Answer: e
Difficulty: Easy
Bloom's Taxonomy: Knowledge

34. The main difference between germ-line gene therapy and somatic gene therapy is

a. only somatic gene therapy can result in the modified genes being passed to the next generation.
b. only germ-line gene therapy will have results limited to the current generation.
c. in germ-line gene therapy, the gametes are altered; in somatic gene therapy, that doesn't happen.
d. in somatic gene therapy, the gametes are altered; in germ-line gene therapy, only the body cells are altered.
e. only germ-line gene therapy is ethical for research purposes in rodents.

Answer: c
Difficulty: Easy
Bloom's Taxonomy: Knowledge

35. Which of the following mammals has *not* been successfully cloned?

a. sheep
b. pigs
c. humans
d. nonhuman primates
e. dogs

Answer: c
Difficulty: Easy
Bloom's Taxonomy: Knowledge

36. Why might we want to introduce new genes into plants?

a. to improve their nutritional value to primary consumers
b. to reduce their reliance on photosynthesis.
c. to create plants that can produce pharmaceutical drugs.
d. a and c only
e. a, b, and c.

Answer: d
Difficulty: Easy
Bloom's Taxonomy: Knowledge

37. Which of the following is *not* a method of introducing genes into plants?

a. using Ti plasmid and *Rhizobium radiobacter* bacteria to infect cells with the new genes
b. creating transgenic plant cells in cultures to create a callus which then grows into a transgenic plant
c. crossing plants with desirable traits with other plants with the same trait for multiple generations
d. actually all of the above are used; the question is wrong

Answer: c
Difficulty: Easy
Bloom's Taxonomy: Knowledge

FOCUS ON RESEARCH: RESEARCH METHODS: SOUTHERN BLOT ANALYSIS

38. Why is Southern blotting used?

a. to identify samples with specific DNA sequences
b. to transfer DNA samples to a special membrane or filter paper
c. to separate DNA based on size
d. to amplify a genomic region of interest
e. to prepare DNA for cloning

Answer: a
Difficulty: Easy
Bloom's Taxonomy: Knowledge

FOCUS ON RESEARCH: RESEARCH METHODS: INTRODUCTION OF GENES INTO MOUSE EMBRYOS USING EMBRYONIC GERM-LINE CELLS

39. At which stage is the foreign gene introduced into the developing embryo?

a. blastocyst, via a transenic cell
b. egg, via direct microinjection of the new DNA
c. sperm, via direct microinjection of the new DNA
d. in cell culture, prior to fertilization
e. it all depends on the method the particular researcher prefers.

Answer: a
Difficulty: Easy
Bloom's Taxonomy: Knowledge

FOCUS ON RESEARCH: **RESEARCH METHODS: USING THE TI PLASMID OF** *RHIZOBIUM RADIOBACTER* **TO PRODUCE TRANSGENIC PLANTS**

40. How is the Ti plasmid used to create transgenic plants?

a. The Ti plamsid allows all of the plant cells to host the bacteria *Rhizobium radiobacter* and thus express the foreign gene that was inserted into the bacterial chromosome.
b. The Ti plasmid inserts the foreign gene directly into the plant's nuclear DNA.
c. Once the *Rhizobium radiobacter* forms a tumor in the plant, the tumor cells can take up the foreign DNA and incorporate it into their chromosomes.
d. The Ti plasmid inserts the foreign DNA into the DNA in the chloroplast.
e. The Ti plasmid inserts the foreign DNA into the plant's mitochondrial DNA.

Answer: b
Difficulty: Moderate
Bloom's Taxonomy: Knowledge

18.3 GENOME ANALYSIS

41. The human genome has a size of approximately

a. 1 billion base pairs
b. 3 billion base pairs
c. 7 billion base pairs
d. 10 billion base pairs
e. 14 billion base pairs

Answer: b
Difficulty: Easy
Bloom's Taxonomy: Knowledge

42. What is the difference between dideoxyribonucleotides (ddNTP) used in Sanger sequencing and the deoxyribonucleotides (dNTP) normally found in DNA?

a. ddNTPs have a 3' H rather than OH
b. dNTPs have a 3' H rather than OH
c. ddNTPs have a 5' OH rather than H
d. dNTPs have a 5' OH rather than H
e. ddNTPs have a 3' OOH rather than OH

Answer: a
Difficulty: Moderate
Bloom's Taxonomy: Knowledge

43. Why do normal nucleotides need to be included in the Sanger sequencing reaction mix?

a. so the DNA polymerase has a substrate to work with
b. to enhance the action of the dideoxynucleotides
c. so a longer sequence (up to 300 additional base pairs) can be determined than without (normally 200 bp).
d. so multiple sequences of different lengths will be generated
e. to prevent the DNA polymerase from making too many errors

Answer: d
Difficulty: Moderate
Bloom's Taxonomy: Knowledge

44. What is an open reading frame?

a. the protein coding sequence in a genome and the associated regulatory sequences
b. the nucleotides between and including a start codon and an end codon in all chromosomes
c. the nucleotides between and including a start codon and an end codon, minus the introns
d. the nucleotides between and including a start codon and an end codon, minus the exons
e. the nucleotides between a start codon and end codon in prokaryotes

Answer: b
Difficulty: Easy
Bloom's Taxonomy: Knowledge

45. What percentage of a typical eukaryotic genome is noncoding and often has no known function?

a. 10–20%
b. 1–2%
c. 20–30%
d. 25–50%
e. 50–75%

Answer: d
Difficulty: Easy
Bloom's Taxonomy: Knowledge

46. Now that the human genome has been sequenced, we know that there are fewer than expected protein coding genes (approximately 25,000). Yet, the total number of proteins produced in humans approaches 100,000. What accounts for this discrepancy in numbers?

a. We have not yet identified all of the open reading frames in the human genome.
b. We have not yet fully sequenced the human genome.
c. The various mechanisms of mRNA processing lead to more proteins being produced than the DNA directly encodes.
d. Some proteins are converted into completely new proteins after translation.
e. There has been a gross over-estimation of the number of proteins produced in humans.

Answer: c
Difficulty: Moderate
Bloom's Taxonomy: Knowledge, Comprehension

47. Now that we have sequenced the human genome, why can't we cure all genetic diseases?

a. We do not yet understand how all of the parts of the genome work together.
b. We have not identified all of the specific genes that cause certain genetic disorders.
c. Knowing how a gene causes a disorder is not the same as being able to fix the DNA.
d. We don't have the technology to safely correct a genetic defect in all individuals.
e. all of the above.

Answer: e
Difficulty: Moderate
Bloom's Taxonomy: Knowledge, Comprehension

48. How can microarrays help us understand cellular functions?

a. Microarrays let us study different cells under different conditions.
b. Microarrays let us directly measure protein expression in individual cells.
c. Microarrays, by hybridizing cDNAs to DNA sequences already present on a chip, let us identify which portions of a genome were being expressing a cell at a particular time.
d. Microarrays let us identify which DNA sequences are present in a particular cell type under certain conditions.
e. Microarrays, by hybridizing DNA to cDNA sequences already present on a chip, let us identify which portions of a genome were being expressing a cell at a particular time.

Answer: c
Difficulty: Easy
Bloom's Taxonomy: Knowledge

49. What is the benefit of a protein microarray over a DNA microarray?

a. Protein microarrays allow researchers to measure the relative concentrations of different proteins.
b. DNA microarrays are not as sensitive as protein microarrays.
c. Protein microarrays allow modified proteins to be studied directly.
d. a and c only
e. a, b, and c

Answer: d
Difficulty: Moderate
Bloom's Taxonomy: Knowledge

INSIGHTS FROM THE MOLECULAR REVOLUTION: ENGINEERING RICE FOR BLIGHT RESISTANCE

50. Why did scientists turn to genetic engineering for rice to create resistance to blight?

a. The bacteria causing blight cannot be identified.
b. Efforts to cross-breed crop rice with blight-resistant rice produced rice unsuitable as a food source.
c. It is the cheapest and easiest way to produce blight-resistant rice.
d. Almost ¾ of the crop rice planted is lost each year to blight.
e. It provided the funding necessary to sequence the rice genome.

Answer: b
Difficulty: Easy
Bloom's Taxonomy: Knowledge

51. The blight-resistant rice are resistant because

a. they code a receptor protein for their plasma membranes that somehow alters the biochemistry of the cell and inhibits bacterial growth.
b. they don't allow the bacteria to enter their cells.
c. they produce an antibacterial compound that is toxic to the blight-causing bacteria.
d. they have extra membrane receptors compared to normal rice plants.
e. they lack a membrane receptor found in other rice plants.

Answer: a
Difficulty: Easy
Bloom's Taxonomy: Knowledge

FOCUS ON RESEARCH: RESEARCH METHODS: DIDEOXY (SANGER) METHOD OF SEQUENCING DNA

52.   What is it about the mixture of ddNTPs and dNTPs that allows a DNA sequence to be determined using the Sanger method?

a.   By using fluorescently labeled dNTPs in the mixture of nucleotides, the sequencing reaction is randomly terminated with a labeled nucleotide. We can sort the fragments based on size and determine the sequence using the fluorescent tags.

b.   By using fluorescently labeled ddNTPs in the mixture of nucleotides, the sequencing reaction is randomly terminated with a labeled nucleotide. We can sort the fragments based on size and determine the sequence using the fluorescent tags.

c.   By using fluorescently labeled ddNTPs, we ensure that all of the nucleotides are fluorescent. This lets us sort the different fragments by light wavelengths and thus determine the sequence.

d.   By using fluorescently labeled dNTPs, we ensure that all of the nucleotides are fluorescent. This lets us sort the different fragments by light wavelengths and thus determine the sequence.

e.   The combination of fluorescent dNTPs vs. the non-fluorescent ddNTPs changes the emission spectrum of the DNA fragments. We then sort the DNA fragments based on size and measure the degree of fluorescence to determine the nucleotide sequence.

Answer: b
Difficulty: Moderate
Bloom's Taxonomy: Knowledge
Figure 18.3

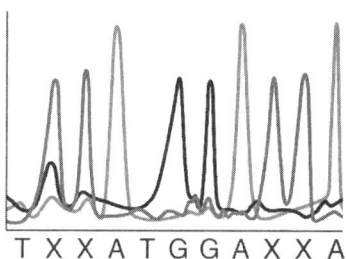

T X X A T G G A X X A

53.   I tried to sequence a DNA fragment and ended up with the trace pattern shown in figure above. What is the most likely cause of this result?

a.   I forgot to add the ddCTPs to the sequencing reaction.

b.   I forgot to add the normal dCTPs to the sequencing reaction.

c.   I forgot to add DNA polymerase to the sequencing reaction mix.

d.   The fluorescent tags for the ddGTPs are degraded.

e.   The polyacrylamide gel didn't polymerize properly and is altering the migration patterns of the DNA fragments.

Answer: a
Difficulty: Difficult
Bloom's Taxonomy: Application

FOCUS ON RESEARCH: RESEARCH METHODS: WHOLE-GENOME SHOTGUN SEQUENCING

54.   In the whole-genome shotgun method, the genome is fragmented and individual fragments are sequenced. How do we determine the order of the nucleotides in the intact chromosome?

a.   The fragmentation is done in a very systematic way such that the physical arrangement of fragments is readily apparent.

b.   The ends of the fragments overlap with the ends of other fragments. Via analysis of sequence overlap, we can determine the sequence of the intact chromosome.

c.   We can sequence all of the fragments with this method but must supplement our information with data from a different technique to make a determination of the final chromosome arrangement.

d.   After the sequences of fragments are determined, DNA hybridization assays are conducted to determine the physical arrangement of the genes on the chromosome.

e.   After sequencing, the fragments are labeled and used as probes in a Southern blot, telling us where the individual fragments are positioned on a chromsome that was subjected to gel electrophoresis.

Answer: b
Difficulty: Moderate
Bloom's Taxonomy: Comprehension

FOCUS ON RESEARCH: RESEARCH METHODS: DNA MICROARRAY ANALYSIS OF GENE EXPRESSION LEVELS

55.   In the microarray shown in your book, the cDNAs were labeled with red and green fluorescent tags. How then, do you end up with spots on the microarray emitting yellow light?

a.   In light, red and green are two of the primary colors, and the combination of the two will produce yellow light.

b.   The yellow light comes from the laser, and a yellow spot indicates that neither cDNA hybridized to that spot on the microarray.

c.   The over expression of one cDNA relative to the other will skew the color pattern of the spot on the microarray. This result in the yellow color.

d.   The color choice was an arbitrary decision by the artist and doesn't reflect how the process actually works.

e.   none of the above

Answer: a
Difficulty: Moderate
Bloom's Taxonomy: Comprehension

56.   Pharmacogenomics

a.   seeks to customize drugs to the cellular proteomes of a patient.
b.   might target specific disease mechanisms and thus minimize side effects.
c.   would account for an individual's metabolism.
d.   would require a person's genetic profile to be determined and subsequently stored in some database.
e.   all of the above

Answer: e
Difficulty: Easy
Bloom's Taxonomy: Knowledge

57.   Microarrays have been used to

a.   distinguish between two types of cancer.
b.   determine the rate of cancer progression.
c.   design a drug to specifically treat an individual's particular form of cancer
d.   replace cellular biopsies to diagnose cancer.
e.   all of these

Answer: a
Difficulty: Easy
Bloom's Taxonomy: Knowledge

Integrative Multiple-Choice

58.   I generate cDNA libraries for the same cell line under different conditions. When I compare the libraries, I find some cDNAs in one library that are missing from the other. What is the best explanation for this result?

a.   One of the the cDNA libraries is really a genomic library.
b.   Even the same cell type will change gene expression as conditions change. The difference in cDNAs tells me which genes are expressed in each set of conditions.
c.   There was an error in generating one of the cDNA libraries and some of the sequences were not cloned properly.
d.   The probe I used in the Southern blot of the libraries must have been poorly constructed.
e.   There is no explanation other than experimental error.

Answer: b
Difficulty: Moderate
Bloom's Taxonomy: Knowledge
Source: Section 18.1, 18.2

59.   Which of the following is *not* an example of a *current* use of bioinformatics?

a.   locating individual genes in a genomic sequence
b.   aligning sequences in databases to determine similarities between organisms
c.   predicting the structure and function of gene products
d.   identifying possible evolutionary relationships between organisms
e.   determining the final protein structure of novel proteins based on just the nucleotide sequence

Answer: e
Difficulty: Moderate
Bloom's Taxonomy: Knowledge
Source: Section 3.5, 18.3

Matching

Match each of the following terms with its correct definition.

60.   _____   genetic engineering

61.   _____   recombinant DNA

62.   _____   restriction endonucleases

63.   _____   restriction fragments

64.   _____   genomic library

65.   _____   DNA hybridization

66.   _____   cDNA library

67.   _____   PCR

68.   _____   agarose gel elecrophoresis

69.   _____   Southern blot

70.   _____   DNA fingerprinting

71.   _____   transgenic

72.   _____   structural genomics

73.   _____   functional genomics

74.   _____   proteomics

75.   _____   systems biology

A.   This uses DNA technologies to alter genes in a cell or organism.

B.   A branch of biology that tries to combine proteomics and genomics at an organismal level.

C.   The determination and analysis of the nucleotide sequences in the chromosomes in an organism.

D.   The products of a restriction endonuclease's action.

E.   A collection of clones that together contain every DNA sequence in a genome, including non-coding sequences.

F. The use of a short ssDNA sequence, labeled with some sort of tag, to bind to complementary DNA and thus identify it.

G. A collection of clones that together contain a DNA version of every mRNA sequence expressed in a particular cell type.

H. This procedure can be described as "a photocopy machine for specific DNA sequences."

I. DNA from two different sources that have been joined together into a single molecule.

J. This term is used to describe organisms that have been subject to genetic engineering.

K. The study of all of the proteins produced from an organism's genome.

L. The study of the functions of genes and non-coding portions of the genome, including their interactions with each other.

M. Bacterial enzymes that recognize and cut specific DNA sequences.

N. A method of hybridizing labeled DNA to DNA fragments that were previously subjected to gel electrophoresis.

O. This technique is used to separate DNA molecules based on their relative sizes.

P. A method of identifying individuals based on their individual patterns of short tandem repeats located on certain loci of their genome.

Answers: **60. A**   **61. I**   **62. M**   **63. D**   **64. E**
   65. F   **66. G**   **67. H**   **68. O**   **69. N**   **70. P**
   71. J   **72. C**   **73. L**   **74. K**   **75. B**

Difficulty: Moderate
Bloom's Taxonomy: Knowledge
Source: Section 18.1, 18.2, 18.3

Classification

Use the processes listed below for questions 76–82.

   a. DNA library
   b. Restriction enzyme digest
   c. Gene cloning
   d. Southern blotting
   e. PCR

76. _____ You'd use this tool if you wanted to check and see if a particular gene was contained in an organism's genome.

77. _____ This technique will let you move a gene into a vector for expression in different cells.

78. _____ You could use this technique to see if a DNA sequence is present in an individual's genome without having to use Sanger sequencing.

79. _____ This technique is used to amplify specific sequences of DNA.

80. _____ You'd use this to cut DNA at a selected site.

81. _____ How can you mass-produce a gene in a bacterial cell that doesn't normally have this gene?

82. _____ You use this method determine the presence or absence of particular 20 bp nucleotide pattern in a noncoding region of a genome.

Use the processes listed below for questions 83–88.

   a. DNA fingerprinting
   b. Trangenics
   c. DNA sequencing
   d. Microarray analysis
   e. Proteomics

83. _____ You want to study the interaction of all of the proteins in a cell. What methodology is appropriate?

84. _____ You're a forensic scientist trying to determine if a suspect is the actual rapist. Which technique will you use?

85. _____ You want to generate mice that express a fluorescent protein in every cell of their body. What methodology are you considering?

86. _____ You use this to determine the nucleotide sequence of a particular gene in an organism.

87. _____ This methodology lets you compare the gene expression patterns of identical cells under different conditions.

Answers: **76. a**   **77. c**   **78. d**   **79. e**   **80. b**
   81. c   **82. d**   **83. e**   **84. a**   **85. b**
   86. c   **87. d**

Difficulty: Moderate
Bloom's Taxonomy: Comprehension
Source: Section 18.1, 18.2, 18.3

Choice

For each of the following statements, choose the most appropriate macromolecule being studied or manipulated from the list below.

a. DNA   b. mRNA/cDNA
c. protein

88. _____ Southern Blotting

89. _____ PCR

90. _____ Sequencing

91. _____ Microarray

92. _____ Genomics

93. _____ Cellular proteomics

94. _____ Transgenics

95. _____ Shotgun library

96. _____ Fingerprinting

97. _____ Proteomics

98. _____ Cloning

99. _____ Agarose Gel Electrophoresis

100. _____ Polyacrylamide Gel Electrophoresis

101. _____ Sanger method

Answers: **88. a   89. a   90. a   91. b   92. a**
93. c   94. a   95. a   96. a   97. c   98. a
99. a   100. a   101. a

Difficulty: Easy
Bloom's Taxonomy: Knowledge
Source: Section 18.1, 18.2, 18.3

Short Answer

102. Define the purpose of cloning a gene into a plasmid cloning vector.

Answer: By cloning a gene into a vector, you can generate hundreds of copies of the gene very easily. This allows the DNA to be produced in amounts that are then usable in other applications, such as restriction enzyme analysis, Southern blotting, or Sanger sequencing. It also lets one mass-produce a protein if the sequence that is cloned is a protein coding region.
Difficulty: Moderate
Bloom's Taxonomy: Knowledge, Comprehension
Source: Section 18.1

103. Explain the ethical concerns with germ-line gene therapy in humans.

Answer: If gene therapy were applied to germ-line cells in humans rather than being limited to somatic cells, we would be doing an experiment on the next generation without their consent. We would also be creating changes to the human genome that were being passed to all future generations, and we don't know what the consequences of those changes will be. For these reasons, we limit gene therapy in humans to somatic cells only.
Difficulty: Moderate
Bloom's Taxonomy: Knowledge, Comprehension
Source: Section 18.2

True/False

If the statement is true, write a "T" in the blank. If the statement is false, make it correct by changing the underlined word(s) and writing the correct word(s) in the answer blank.

104. _____ DNA fingerprinting is limited to <u>human studies</u>.

105. _____ There are often different methodologies that can be used to obtain the same information about a DNA sequence.

106. _____ Gene therapy is a safe, effective, and commercially available treatment for numerous genetic disorders.

107. _____ Gene therapy has been successful in curing some individuals of <u>autoimmune disorder and sickle cell disease.</u>

108. _____ Genetically modified crops have been <u>banned</u> in the United States.

109. _____ Some genetically modified <u>tobacco plants </u>can express the light-producing chemicals present in fireflies.

110. _____ When using embryonic germ line-cells to create transgenic mice, the offspring of the genetically modified animal will have <u>all transgenic</u> cells in their bodies.

111. _____ Blight-resistant crop rice <u>has been tested and shown to be nutritionally equivalent</u> to the non-genetically modified crop rice.

112. _____ <u>Shotgun sequencing </u>is the method of choice for generating whole genome sequences.

113. _____ <u>Open Reading Frames</u> include the start and stop codons for a protein.

Answers: 104. F, any organism can be analyzed   105. T   106. F, we're not there yet.   107. T   108. F approved 109. T   110. T   111. F, is being tested; results are not yet available.   112. T   113. T

Difficulty: Difficult
Bloom's Taxonomy: Knowledge
Source: Section 18.1, 18.2

Essay

114. Make an argument for which molecular biology technique discussed in this chapter is the single most important technique for forensic scientists and justify your answer with the benefits of this method compared to the others that are available to you.

Answer: In forensic science, sample sizes are very small and finite. For this reason, the most important tool is PCR. This allows even a minute sample to be copied exponentially, generating relatively large amounts of DNA that are then usable in other assays using the techniques described. Many other research methods (Southern blotting, fingerprinting, sequencing, microarray analysis) require a substantial sample size in order to do the assay. In the case of forensic science, the available materials are very limited in quantity, and thus PCR is the method of choice.
Difficulty: Moderate
Bloom's Taxonomy: Analysis
Source: Section 18.1, 18.2, 18.3

115. Explain how proteomics is considered by some to be even more important than genome sequencing.

Answer: There are limits to what genome sequencing can tell us about an organism. It has been said by some that the genome sequences are like a dictionary; the dictionary can tell you words, but it doesn't tell you how a language is used. The proteomics field seeks to determine how all of the proteins expressed in a cell work together. This will help us understand how the cells actually function and tell us which genes are being expressed and for what purpose. If the genome is a dictionary, then the proteome is a novel.
Difficulty: Moderate
Bloom's Taxonomy: Synthesis
Source: Section 18.3

19

THE DEVELOPMENT OF EVOLUTIONARY THOUGHT

Multiple-Choice

WHY IT MATTERS

1. In *The Origin of Species by Natural Selection* _____ proposed that natural mechanisms produce and transform the diversity of life on Earth.

a. Alfred Russel Wallace
b. Charles Lyell
c. Charles Darwin
d. Joseph Hooker
e. Charles Darwin and Alfred Russel Wallace

Answer: c
Difficulty: Easy
Bloom's Taxonomy: Knowledge

2. Biological evolution occurs in _____ when specific processes cause genomes of organisms to differ from their ancestors

a. individuals
b. populations
c. communities
d. phenotypes
e. sub-species

Answer: b
Difficulty: Easy
Bloom's Taxonomy: Knowledge

3. All biological research is undertaken with the explicit or implicit recognition that _____.

a. Charles Darwin's view of evolution put forth in *The Origin* is without error
b. all forms of life are related and have evolved from ancestral forms
c. all species share the same common ancestor
d. the products of evolution are easily predicted
e. the theory of evolution can be ignored unless the research is explicitly examining evolutionary process

Answer: b
Difficulty: Moderate
Bloom's Taxonomy: Knowledge, Comprehension

19.1 RECOGNITION OF EVOLUTIONARY CHANGE

4. Who developed the science of taxonomy?

a. Charles Darwin
b. Aristotle
c. Alfred Russel Wallace
d. Sir Isaac Newton
e. Carolus Linnaeus

Answer: e
Difficulty: Easy
Bloom's Taxonomy: Knowledge

5. Which of the following branches of biology examines the form and variety of organisms in their natural environment?

a. Taxonomy
b. Natural history
c. Comparative morphology
d. Biogeography
e. Geology

Answer: b
Difficulty: Easy
Bloom's Taxonomy: Knowledge

6. The method that Carolus Linnaeus used for classifying organisms was

a. geology
b. biogeography
c. comparative morphology
d. homology
e. evolutionary biology

Answer: c
Difficulty: Moderate
Bloom's Taxonomy: Comprehension

7. Which of the following contributed to an understanding that species were probably not immutable?

a. homologous bones
b. vestigial structures
c. global exploration of fauna and flora
d. fossils
e. all of these

Answer: e
Difficulty: Moderate
Bloom's Taxonomy: Comprehension

8. Homologous structures were "discovered" by findings that indicated that _____.

a. species on different continents were not similar in form
b. different geological layers held different kinds of fossils
c. structures of similar form and position occurred in animals of markedly different size, shape, and lifestyle
d. some structures found in animals are apparently useless
e. some kinds of animals had apparently gone extinct

Answer: c
Difficulty: Moderate
Bloom's Taxonomy: Knowledge, Comprehension

9. The proposition that species change through time was proposed by which of the following?

a. Charles Darwin
b. Jean Baptiste de Lamarck
c. Alfred Russel Wallace
d. a and b
e. a, b, and c

Answer: d
Difficulty: Easy
Bloom's Taxonomy: Knowledge

10. According to the *principle of use and disuse* the form of body parts in offspring _____.

a. are inherited based on phenotypic changes that occur in parents during their lifetime
b. are the result of how much the offspring uses a particular body part
c. are the result of natural selection
d. are immutable
e. are a result of biogeographic location

Answer: b
Difficulty: Easy
Bloom's Taxonomy: Knowledge

11. Lamarck's theory of evolution proposed which of the following?

a. Species change through time.
b. New characteristics are passed from one generation to the next.
c. Organisms change in response to their environment.
d. Changes that an animal acquires during its lifetime are inherited by its offspring.
e. All of these.

Answer: b
Difficulty: Easy
Bloom's Taxonomy: Knowledge

12. Darwin was probably influenced most by which of the following when developing his theory of evolution?

a. *Scala Naturea*
b. Catastrophism
c. Uniformitarianism
d. Taxonomy
e. Vestigial structures

Answer: c
Difficulty: Moderate
Bloom's Taxonomy: Analysis

19.2 DARWIN'S JOURNEYS

13. As a young child, you preferred to collects shells and insects, dig through the mud around a pond, and watch birds rather than play video games. What did you have in common with Charles Darwin?

a. an interest in history
b. an interest in evolution
c. an interest in natural history
d. an interest in geology
e. an interest in biogeography

Answer: c
Difficulty: Easy
Bloom's Taxonomy: Knowledge, Comprehension

14. The reason Darwin sailed aboard the H.M.S. *Beagle* was because he _____.

a. wanted to develop his theory of evolution
b. was invited to be the ship's naturalist
c. was invited to be the ship's geologist
d. was invited to be the captain's dining companion
e. was invited to be the captain's personal physician

Answer: d
Difficulty: Easy
Bloom's Taxonomy: Knowledge

15. Which of the following observations did not influence Darwin's later thoughts about evolution?

a. The discovery of fossils that resembled extant organisms in the same region.
b. The observation that animals from similar habitats in South America and Europe were very different.
c. A great variety of species exist across the globe.
d. Species on islands differed from species on the mainland closest to the island.
e. North American and South American species were more similar to each other than European and African species.

Answer: e
Difficulty: Easy
Bloom's Taxonomy: Knowledge

16. Darwin's voyage took place between _____.

a. 1730 and 1735
b. 1830 and 1835
c. 1853 and 1858
d. 1930 and 1938
e. none of the choices

Answer: b
Difficulty: Easy
Bloom's Taxonomy: Knowledge

17. Morphological differences in European beaver and _____ contributed to Darwin's thoughts on reasons for species differences.

a. South American nutria
b. South American iguana
c. extinct glyptodont
d. nine banded armadillo
e. South American beaver

Answer: a
Difficulty: Moderate
Bloom's Taxonomy: Knowledge

18. Darwin's observations of finches from the Galápagos included which of the following?

a. The 13 finch species were descended from a common ancestor.
b. The 13 finch species each had differently-shaped bills.
c. The 13 finch species had migrated from South America.
d. Finches were catalogued according to island of origin and captured for later analysis.
e. All of these.

Answer: d
Difficulty: Moderate
Bloom's Taxonomy: Knowledge

19. A basic truth of inheritance that had been known well before Darwin's time was that _____.

a. offspring frequently resemble their parents
b. offspring inherit genes from their parents
c. selective breeding improves domesticated plants and animals
d. a and c
e. a, b, and c

Answer: d
Difficulty: Easy
Bloom's Taxonomy: Knowledge

20. Thomas Malthus' influence on Darwin can best be characterized by which of the following?

a. Darwin was convinced artificial selection could work in nature.
b. Darwin realized species typically produce many more offspring than are needed to replace the parent generation.
c. Darwin was convinced the fate of humanity was dependent on artificial selection.
d. Darwin was convinced the fate of humanity was dependent on natural selection.
e. Darwin was convinced to publish his ideas about natural selection

Answer: b
Difficulty: Moderate
Bloom's Taxonomy: Knowledge, Comprehension

21. Darwin calculated that if reproduction was not limited, a single pair of elephants could leave _____ descendants after 750 years.

a. 750
b. 15,000
c. 150,000
d. 19 million
e. 190 million

Answer: d
Difficulty: Moderate
Bloom's Taxonomy: Knowledge

22. Darwin's inference that individuals within a population compete for limited resources was based on which observation(s)?

a. Most organisms produce more than one or two offspring.
b. Populations do not increase in size indefinitely.
c. Food and other resources are limited for most populations.
d. a and b.
e. a, b, and c.

Answer: e
Difficulty: Easy
Bloom's Taxonomy: Knowledge

23. An example of evolutionary divergence is best characterized by Darwin's observation of _____.

a. fossils along the coast of Argentina
b. body armor similarities between armadillos and fossilized glyptodonts
c. differences in appearance of nutria and beaver
d. differences in bill shape and food habits of finches
e. the great variety of form of species

Answer: d
Difficulty: Difficult
Bloom's Taxonomy: Comprehension

24. What was the major stumbling block for the acceptance of natural selection as a mechanism for evolution when proposed by Darwin?

a. strong evidence for inheritance of acquired traits
b. lack of a fossil record
c. lack of a plausible theory of heredity
d. lack of evidence for artificial selection
e. lack of observational and experimental data

Answer: c
Difficulty: Easy
Bloom's Taxonomy: Knowledge

FOCUS ON RESEARCH: BASIC RESEARCH: CHARLES DARWIN'S LIFE AS A SCIENTIST

25. Charles Darwin's approach to science can best be characterized as _____.

a. reductionism
b. focused
c. haphazard
d. holism
e. publish as often as possible

Answer: d
Difficulty: Difficult
Bloom's Taxonomy: Comprehension

26. Upon Darwin's return from his voyage he spent time doing which of the following?

a. breeding dogs
b. breeding pigeons
c. studying barnacles for eight years
d. a and b
e. a and c

Answer: e
Difficulty: Easy
Bloom's Taxonomy: Knowledge

27. On what basis might a biologist suggest that biological knowledge gained after Darwin is just "filling in the last 2 percent."

a. Darwin's concept of evolution still forms the unifying intellectual paradigm within which all biological research is undertaken.
b. Darwin is considered the father of many biological sciences, from soil biology to pollination biology.
c. Darwin's studies and theory touched on every field of biology.
d. Darwin revolutionized the study of biology, with all advances since filling in the details.
e. All of these.

Answer: e
Difficulty: Moderate
Bloom's Taxonomy: Comprehension

INSIGHTS FROM THE MOLECULAR REVOLUTION: ARTIFICIAL SELECTION IN THE TEST TUBE

28. How did John Toole and his colleagues mimic the evolutionary process of selection on the molecular scale?

a. They used an experimental process to select specific DNA molecules from many random nucleotide sequences.
b. They discovered a way to cure unwanted blood clotting.
c. They discovered the DNA sequence for thrombin.
d. They proved evolution could occur in a test tube.
e. They were able to get DNA to mutate into RNA.

Answer: a
Difficulty: Moderate
Bloom's Taxonomy: Knowledge

29. Darwin would probably be most fascinated by which aspect of experiments that can generate artificially selected DNA molecules?

a. their utility to cure diseases
b. that DNA is the herditary material
c. that anticlotting molecules that might work in monkeys would also work in humans
d. that DNA can be experimentally replicated
e. that they were able to get DNA to mutate into RNA

Answer: b
Difficulty: Moderate
Bloom's Taxonomy: Comprehension

19.3 EVOLUTIONARY BIOLOGY SINCE DARWIN

30. The perceived fundamental conflict between Darwin's and Mendel's theories was that _____.

a. Mendel's theory was based on experimentation while Darwin's was based on observation
b. Mendel's experiments were based on simple traits while Darwin's evidence was based on complex characteristics
c. Mendel's experiments were unbiased, while Darwin set out to prove evolution by natural selection
d. Mendel's experiments were based on peas while Darwin used a variety of examples
e. Darwin was a scientist while Mendel was a monk

Answer: b
Difficulty: Moderate
Bloom's Taxonomy: Knowledge

31. The field of study that linked Mendel's and Darwin's work was _____.

a. population biology
b. paleontology
c. population genetics
d. biogeography
e. modern synthesis

Answer: c
Difficulty: Easy
Bloom's Taxonomy: Knowledge

32. Which of the following matches between evidence for evolution and biological disciplines is <u>not</u> correct?

a. Pesticide resistance:Historical biogeography
b. Forelimbs or all four legged vertebrates are homologous:Comparative morphology
c. Short tails of African and Asian monkeys:Historical biogeography
d. Gill pouches in embryos of four-limbed vertebrates:Comparative embryology
e. Species on oceanic islands often closely resemble species on the nearby mainland:Comparative embryology

Answer: a
Difficulty: Easy
Bloom's Taxonomy: Knowledge

33. Extinction is an example of a _____.

a. microevolutionary change
b. macroevolutionary change
c. natural selection
d. Mendelian genetics
e. "hopeful monster"

Answer: b
Difficulty: Easy
Bloom's Taxonomy: Knowledge

UNANSWERED QUESTIONS

34. What percentage of the species that have ever lived have become extinct?

a. 1 percent
b. 9 percent
c. 50 percent
d. 75 percent
e. 99 percent

Answer: e
Difficulty: Easy
Bloom's Taxonomy: Knowledge

35. New genetic variations sometimes become more common within populations because _____.

a. most parts of the genome are non-coding and are available for selection
b. proteins can be "recruited" for advantageous function
c. proteins for which they code are advantageous and selected
d. adaptation occurs rapidly in the few species that have survived through time
e. none of these

Answer: c
Difficulty: Moderate
Bloom's Taxonomy: Knowledge

36. Evolution probably occurs most commonly though _____.

a. macroevolutionary mechanisms
b. microevolutionary mechanisms
c. protein recruitment
d. extinction
e. mutations during embryonic development

Answer: c
Difficulty: Moderate
Bloom's Taxonomy: Comprehension

37. Based on what we do not know about evolution, which of the following can be inferred from what we do know?

a. Most evolutionary changes in species are probably due to the invention of new genes and protein products.
b. Most evolutionary changes in species are probably due to the reorganization of existing genes and protein products.
c. The most well adapted species never go extinct.
d. Evolutionary change is due mostly to mutation within coding regions of the genome.
e. Evolution proceeds by weeding out genes and protein products so only those that have advantageous function remain.

Answer: b
Difficulty: Difficult
Bloom's Taxonomy: Comprehension

Integrative Multiple-Choice

38. Evolution by natural selection was not conceivable prior to _____.

a. Darwin conceiving the idea
b. Darwin and Wallace presenting the idea to the Linnean Society of London
c. Mendel's experiments with peas
d. the modern synthesis
e. None of these

Answer: e
Difficulty: Moderate
Bloom's Taxonomy: Comprehension
Source: Sections 19.1, 19.2, and 19.3

39. For evolution to become the theory that underpins all of the biological sciences, which of the following had to occur?

a. Darwin had to write *The Origin*.
b. Lamarck's theory of biological evolution had to be disproved.
c. Darwin had to prove that finches in the Galápagos were descended from a common ancestor.
d. Mendel had to discover genes.
e. Research in many fields had to provide evidence for evolution.

Answer: e
Difficulty: Moderate
Bloom's Taxonomy: Knowledge, Comprehension
Source: Sections 19.1, 19.2, and 19.3

40. To know that the 13 finch species that inhabit the Galápagos are descended from a common ancestor, biologist had to rely on _____.

a. Darwin's notebooks describing his observations
b. comparative molecular biology
c. comparative embryology
d. Mendelian genetics
e. modern historical biogeography and Darwin's observations

Answer: b
Difficulty: Moderate
Bloom's Taxonomy: Knowledge, Comprehension
Source: Sections 19.1 and 19.3

Matching

Match each of the following person with the correct concept.

41. _____ Charles Darwin

42. _____ Charles Lyell

43. _____ Jean Baptiste de Lamarck

44. _____ Alfred Russel Wallace

45. _____ George-Louis de Buffon

46. _____ James Hutton

A. Vestigial structures

B. Evolution by natural selection

C. Inheritance of acquired characteristics

D. Gradualism

E. Uniformitarianism

Answers: 41. B   42. E   43. C   44. B   45. A 46. D

Difficulty: Easy
Bloom's Taxonomy: Knowledge
Source: Why it Matters, Section 19.1

Classification

Use the five disciplines listed below for questions 47–53.

a. population genetics
b. taxonomy
c. historical biogeography
d. comparative morphology
e. comparative embryology
f. comparative molecular biology
g. paleobiology

47. _____ Using fossil evidence is important in this discipline.

48. _____ Classifying organisms is done by practitioners of this discipline.

49. _____ In this discipline genetics and mathematics are combined to predict how selection and other processes such as non-random reproduction affect a population's genetic changes through time.

50. _____ A practitioner of this discipline might observe that species on oceanic islands often closely resemble species on the nearest mainland.

51. _____ If you observe that the arms of humans and wings of birds are constructed similarly, you would be practicing this discipline.

52. _____ Practitioners of this discipline are responsible for the finding that cytochrome *c* is found within the mitochondria of all eukaryotic organisms.

53. _____ Evidence that fishes, amphibians, reptiles, and birds go through similar stages of development was found by practitioners of this discipline.

Answers: 47. g   48. b   49. a   50. c   51. d 52. f   53. e

Difficulty: Moderate
Bloom's Taxonomy: Knowledge
Source: Sections 19.1, 19.3

Choice

Based on the figure, place the appropriate species with the number of amino acids that differ from the human sequence a–g.

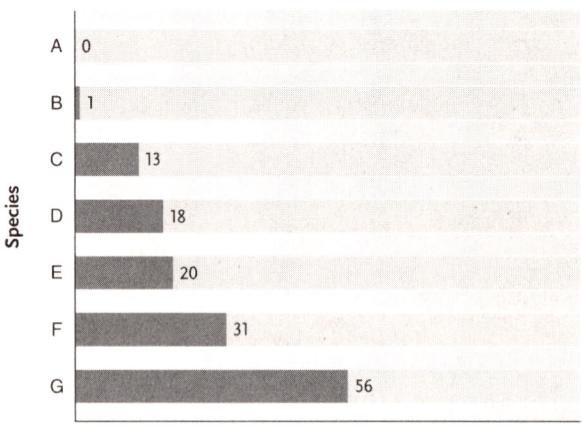

Number of amino acids that differ from the human sequence

54. _____   Domestic dog

55. _____   Rhesus monkey

56. _____   Turtle

57. _____   Rattlesnake

58. _____   Chimpanzee

59. _____   Yeast

60. _____   Domestic chicken

*Answers: 54. C   55. B   56. F   57. E   58. A
59. G   60. D*

Difficulty: Difficult
Bloom's Taxonomy: Knowledge
Source: Section 19.3

Short Answer

61. What are the products of evolution?

Answer: Genetic changes (and genetic variability) in populations that result in changes in phenotypes.
Difficulty: Moderate
Bloom's Taxonomy: Knowledge
Source: Section 19.1

62. Explain how similarity in amino acid sequences among species indicates shared ancestry.

Answer: Species that share a more common ancestor should have more similar amino acid sequences (coding for specific protein/s).
Difficulty: Moderate
Bloom's Taxonomy: Knowledge, Comprehension
Source: Section 19.3

63. Explain how observing artificial selection provides an inference for natural selection.

Answer: The mechanism of passing on genes selectively (i.e., non-random mating), if it occurs in nature, would mean both types of selection share a common mechanism.
Difficulty: Difficult
Bloom's Taxonomy: Comprehension
Source: Section 19.2

Sequence

Imagine you are tracing the history of biological evolution. Arrange the following historical figures in order of when their contribution influenced thoughts on the subject (directly or indirectly). Write the letter of the earliest next to 64, the letter of the most recent to 71.

64. _____   A.   Lamarck
65. _____   B.   Wallace
66. _____   C.   Futuyma
67. _____   D.   Mendel
68. _____   E.   Buffon
69. _____   F.   Aristotle
70. _____   G.   Lyell
71. _____   H.   Malthus

*Answers: 64. F   65. E   66. A   67. H   68. G
69. B   70. D   71. C*

Difficulty: Moderate
Bloom's Taxonomy: Knowledge
Source: Sections 19.1, 19.2, 19.3

True/False

If the statement is true, write a "T" in the blank. If the statement is false, make it correct by changing the underlined word(s) and writing the correct word(s) in the answer blank.

72. _____   Darwin's finches illustrate evolution by <u>speciation</u>.

73. _____   "Mutationism" suggested that evolution occurred <u>gradually through time</u>.

74. _____   <u>Macroevolution</u> describes large scale patterns in the history of life.

75. _____   Insecticide resistance is an example of <u>artificial selection</u>.

76. _____ The concept of orthogenesis proposed evolution was <u>goal oriented</u>.

77. _____ Evolution has been shown <u>to be</u> goal oriented.

78. _____ The process by which Darwin came to the idea of evolution by natural selection was <u>quick</u> and based on <u>little</u> evidence.

79. _____ There are still <u>many</u> unanswered questions about evolution.

Answers: *72. F, natural selection*   *73. F, in spurts*
       74. T   *75. F, natural selection*   *76. T*
   77. F, to be not   *78. F, slow, enormous*
   79. T

Difficulty: Difficult
Bloom's Taxonomy: Knowledge
Source: Sections 19.

Select the Exception

80. Four of the five people below influenced Darwin's writing of the book *On the Origin of Species by Means of Natural Selection*. Select the exception.

a. Mendel
b. Wallace
c. Lyell
d. Malthus
e. Lamarck

Answer: a
Difficulty: Easy
Bloom's Taxonomy: Knowledge
Source: Sections 19.1, 19.2, 19.3

81. Four of the five indicate that evolution by natural selection is a plausible mechanism leading to the diversity of species that have and do exist. Select the exception.

a. artificial selection
b. Darwin's finches
c. insect resistance to pesticides
d. a parent's traits developed through life are passed on to offspring via natural selection
e. biogeographical patterns in species

Answer: d
Difficulty: Moderate
Bloom's Taxonomy: Knowledge
Source: Sections 19.1, 19.2, 19.3

Essay

82. You are a farmer that grows corn. You are offered a genetically modified variety of corn that is 100 percent resistant to corn borers. You have to decide what proportion of your field you should plant in resistant corn versus a variety that is susceptible to corn borers. Explain why you might choose one of the following proportions of resistant to susceptible: 100:0, 90:10, 75:25, or 50:50.

Answer: The ultimate answer depends on rates of selection pressure associated with the resistant variety of corn. The question should be studied using methods of population genetics. Some variables to consider are: Is the yield of resistant and susceptible corn the same in the absence of corn borers, and in the presence of corn borers? What is the probability the corn borer will evolve resistance to the new variety of corn, which in part depends on the population genetics of the system? In effect, if the corn borer population contains a lot of genetic variability (highly heterozygous) then planting at the higher proportions of resistance will reduce rates of evolution.
Difficulty: Difficult
Bloom's Taxonomy: Analysis
Source: Section 19.3

83. Discuss the links between Darwin's observations and inferences. Are some inferences more logical than others? Is there evidence for all of the inferences?

Answer: Any number of examples can be used for each observation—both from the text and just general knowledge (e.g., the elephant example, finches, Fig. 19.12, Fig. 19.14). Given the advances in molecular biology, it is probably most obvious that Darwin's inference about heredity is the most logical today. However, there is ample evidence for all the observations. For example, competition has clearly been shown to influence natural selection (this will be examined in detail in a later chapter).
Difficulty: Moderate
Bloom's Taxonomy: Synthesis
Source: Section 19.2

20
MICROEVOLUTION: GENETIC CHANGES WITHIN POPULATIONS

Multiple-Choice

WHY IT MATTERS

1. A group of interbreeding individuals in a specific geographical location is called a

a. gene pool.
b. species.
c. population.
d. community.
e. ecosystem.

Answer: c
Difficulty: Easy
Bloom's Taxonomy: Knowledge

20.1 VARIATION IN NATURAL POPULATIONS

20.2 POPULATION GENETICS

2. The Hardy-Weinberg formula is valuable for the calculation of changes in

a. population size.
b. speciation.
c. allele frequencies.
d. mutation.
e. dimorphism.

Answer: c
Difficulty: Easy
Bloom's Taxonomy: Knowledge

3. Natural selection acting on a population may favor

a. individuals at the left end of the range of variation.
b. individuals at the right end of the range.
c. extreme individuals at both ends of the range.
d. the intermediate individuals within the range.
e. all of the above

Answer: e
Difficulty: Easy
Bloom's Taxonomy: Knowledge

4. Microevolution can be said to have taken place when

a. a population experiences a shift in allele frequencies.
b. a mutation occurs in a population.
c. several mutations occur in a population.
d. a feature of an individual animal changes through use or disuse.
e. when a population has different forms of the same gene.

Answer: a
Difficulty: Moderate
Bloom's Taxonomy: Knowledge, Comprehension

5. In an isolated population of fruit flies, 4 percent of the individuals have "pink eyes", a homozygous recessive condition, and 96 percent have the dominant black eye phenotype. What percentage of the population are heterozygotes?

a. 32 percent
b. 16 percent
c. 48 percent
d. 88 percent
e. 4 percent

Answer: a.
Difficulty: Difficult
Bloom's Taxonomy: Analysis

6. The Hardy-Weinberg principle of genetic equilibrium tells us what to expect when a sexually reproducing population is

a. decreasing with each generation.
b. increasing with each generation.
c. migrating.
d. evolving.
e. not evolving.

Answer: e
Difficulty: Easy
Bloom's Taxonomy: Knowledge

7. Which of the following will disrupt Hardy-Weinberg equilibrium the least?

a. directional selection
b. stabilizing selection
c. migration
d. reduction to a small population size
e. non-random mating

Answer: b
Difficulty: Easy
Bloom's Taxonomy: Knowledge

8. Which of the following is a correct expression of the Hardy-Weinberg equation?

a. $p+q = p^2+q^2$
b. $p^2+2pq+q^2 = 0$
c. $p^2+pq-q^2 = 1$
d. $p^2+2pq+q^2 = 1$
e. $p+q = 0$

Answer: d
Difficulty: Easy
Bloom's Taxonomy: Knowledge

9. If there are two alleles for a gene in a population, and the frequency of the dominant allele (p) is .5, then the frequency of the recessive allele (q) is

a. .5.
b. .025.
c. .25.
d. .75.
e. .05.

Answer: a
Difficulty: Easy
Bloom's Taxonomy: Knowledge

10. In the Hardy-Weinberg equation "2pq" represents

a. the homozygous recessive genotypes.
b. the heterozygous genotypes.
c. the homozygous dominant genotypes.
d. the total pool of alleles for the dominant phenotypes.

Answer: b
Difficulty: Easy
Bloom's Taxonomy: Knowledge

11. If the frequency of the recessive allele (q) for a particular gene in a population is .2, what percentage of the individuals in the population will be heterozygotes for that gene (only two alleles present for that gene)?

a. .8
b. .2
c. .32
d. .16
e. .64

Answer: c
Difficulty: Moderate
Bloom's Taxonomy: Analysis

12. In a colony of 100 guinea pigs, 16 show the recessive trait of bristly hair. What is the frequency of the recessive allele in the population?

a. .4
b. .16
c. .04
d. .016
e. .2

Answer: a
Difficulty: Moderate
Bloom's Taxonomy: Analysis

13. In the same example as above, what is the percentage of heterozygotes in the population?

a. 48 percent
b. 52 percent
c. 36 percent
d. 24 percent
e. 84 percent

Answer: a
Difficulty: Difficult
Bloom's Taxonomy: Analysis

14. If we want to know the percentage of particular genotypes within an actual population, assuming complete dominance and two alleles, the one measurement we have to actually make is of the frequency of the

a. heterozygous phenotypes.
b. heterozygous genotypes.
c. homozygous dominant genotypes.
d. dominant phenotypes.
e. recessive phenotypes.

Answer: e
Difficulty: Moderate
Bloom's Taxonomy: Comprehension

20.3 THE AGENTS OF MICROEVOLUTION

15. The contribution an individual makes to the gene pool of the next generation relative to the contributions of other individuals is called

a. competition.
b. genetic drift.
c. relative fitness.
d. gene flow.
e. mutation.

Answer: c
Difficulty: Easy
Bloom's Taxonomy: Knowledge

16. Microevolution is

a. the appearance of a life form adapted to a new adaptive zone.
b. the formation of a new species.
c. the occurrence of a new mutation.
d. a change in allele frequencies within a population.
e. the gradual change in the form of fossils over geological time.

Answer: d
Difficulty: Easy
Bloom's Taxonomy: Knowledge

17. A sudden reduction in population size generally results in

a. extinction.
b. mutation.
c. speciation.
d. genetic drift.
e. natural selection.

Answer: d
Difficulty: Easy
Bloom's Taxonomy: Knowledge

18. The source of new alleles in a population is

a. natural selection.
b. mutation.
c. microevolution.
d. adaptation.
e. genetic drift.

Answer: b
Difficulty: Easy
Bloom's Taxonomy: Knowledge

19. The only survivors of a colony on Venus are a man and a woman, who happen to be both originally from the southern Ukraine. Their descendants will show the effect of

a. genetic drift.
b. punctuated equilibrium.
c. excessive mutation.
d. heterozygote advantage.
e. frequency-dependent selection.

Answer: a
Difficulty: Easy
Bloom's Taxonomy: Knowledge

20. We are sending a small number of dogs to colonize a new planet. Through a random selection process, we end up only with dachshunds and chihuahuas. The subsequent colony of dogs are all small with short legs. This is an example of

a. adaptive radiation.
b. punctuated equilibrium.
c. genetic drift.
d. microevolution.
e. mutation.

Answer: c
Difficulty: Easy
Bloom's Taxonomy: Knowledge

21. Because of fluctuations in the environment, such as depletion of food supply or an outbreak of disease, a population may periodically experience a rapid and marked decrease in number of individuals, leading to a form of genetic drift called

a. a founders effect.
b. migration.
c. gene flow.
d. natural selection.
e. a genetic bottleneck.

Answer: e
Difficulty: Easy
Bloom's Taxonomy: Knowledge

22. The production of random evolutionary changes in small breeding populations is known as

a. gene flow.
b. genetic drift.
c. disruptive selection.
d. mutation.
e. natural selection.

Answer: b
Difficulty: Easy
Bloom's Taxonomy: Knowledge

23. Which is the least likely result of a genetic bottleneck?

a. complete elimination of the less common allele
b. increase in the frequency of the less common allele
c. allele frequencies identical to the previous generation
d. decrease in the frequency of an uncommon allele
e. decrease in the frequency of the most common allele

Answer: c
Difficulty: Difficult
Bloom's Taxonomy: Comprehension

24. Directional selection favors

a. intermediate phenotypes.
b. phenotypes at one end of the distribution .
c. phenotypic extremes at both ends.
d. heterozygotes.
e. homozygotes.

Answer: b
Difficulty: Easy
Bloom's Taxonomy: Knowledge

25. Natural selection that shifts the adaptation of an entire populations is

a. stabilizing selection.
b. directional selection.
c. disruptive selection.
d. artificial selection.
e. sexual selection.

Answer: b
Difficulty: Easy
Bloom's Taxonomy: Knowledge

26. Natural selection results in

a. adaptation.
b. a mutation.
c. a new species.
d. genetic drift.
e. stabilizing selection.

Answer: a
Difficulty: Moderate
Bloom's Taxonomy: Comprehension

27. Changing environmental conditions would most likely cause an existing species to undergo

a. disruptive selection.
b. mutation.
c. directional selection.
d. stabilizing selection.
e. paedomorphosis.

Answer: c
Difficulty: Moderate
Bloom's Taxonomy: Comprehension

28. The founder effect is a type of

a. natural selection.
b. mutation.
c. genetic drift.
d. gene flow.
e. adaptation.

Answer: c
Difficulty: Easy
Bloom's Taxonomy: Knowledge

29. Which process results in microevolution without any form of natural selection?

a. genetic drift
b. mutation
c. development of new characteristics during an individual's lifetime
d. genetic variation
e. none

Answer: a
Difficulty: Moderate
Bloom's Taxonomy: Knowledge

30. Evolution can be said to have taken place when

a. a population experiences a shift in allele frequencies.
b. a mutation occurs in a population.
c. several mutations occur in a population.
d. a feature of an individual animal changes through use or disuse.
e. a population has different forms of the same gene.

Answer: a
Difficulty: Easy
Bloom's Taxonomy: Knowledge

31. If individuals move from one population to another, it may cause a shift in allele frequencies due to

a. genetic drift.
b. directional selection.
c. natural selection.
d. mutation.
e. gene flow.

Answer: e
Difficulty level: Moderate
Bloom's Taxonomy: Knowledge

32. Genetic drift will have a progressively larger impact on allele frequencies in a population as

a. gene flow increases.
b. population size decreases.
c. mutation rate decreases.
d. the number of heterozygous loci increases.
e. random mating increases.

Answer: b
Difficulty: Moderate
Bloom's Taxonomy: Comprehension

33. A baby born with wrinkled, poorly-formed skin and blood vessels prone to rupturing can be said to have inherited a

a. lethal mutation.
b. deleterious mutation.
c. neutral mutation.
d. advantageous mutation.
e. partial mutation.

Answer: b
Difficulty: Moderate
Bloom's Taxonomy: Knowledge

20.4 MAINTAINING GENETIC AND PHENOTYPIC ADVANTAGE

34. Balanced polymorphism in a population is likely the result of

a. disruptive selection.
b. sexual selection.
c. directional selection.
d. reproductive isolation.
e. stabilizing selection.

Answer: a
Difficulty: Moderate
Bloom's Taxonomy: Knowledge, Comprehension

35. The stable presence of more than one allele for a particular gene in a population is called

a. balanced polymorphism.
b. mixed gene pool.
c. mutation pressure.
d. genetic drift.
e. competition.

Answer: a
Difficulty: Easy
Bloom's Taxonomy: Knowledge

36. Sometimes the most abundant color form of an animal is preyed upon more extensively than less common forms. This illustrates the phenomenon of

a. stabilizing selection.
b. directional selection.
c. disruptive selection.
d. frequency dependent selection.
e. genetic drift.

Answer: d
Difficulty: Easy
Bloom's Taxonomy: Knowledge

37. People with the disease known as sickle-cell anemia are often better able to survive malaria because of

a. balanced polymorphism.
b. heterozygote advantage.
c. frequency dependent selection.
d. disruptive selection.
e. stabilizing selection.

Answer: b
Difficulty: Easy
Bloom's Taxonomy: Knowledge

38. Mutations that confer no apparent selective advantage or disadvantage in a particular environment are said to be selectively

a. random.
b. deleterious.
c. neutral.
d. benign.
e. stabilizing.

Answer: c
Difficulty: Easy
Bloom's Taxonomy: Knowledge

39. The distribution of the genetic disorder called sickle-cell anemia is correlated with the distribution of malaria in tropical countries. This is an example of

a. heterozygote advantage.
b. frequency-dependent selection.
c. a genetic bottleneck.
d. genetic drift.
e. evolutionary convergence.

Answer: a
Difficulty: Easy
Bloom's Taxonomy: Knowledge

40. The decline in the occurrence of sickle cell anemia in the American population is most likely the result of

a. a lower mutation rate in the United States than in Africa.
b. the advantage of both homozygous forms over the heterozygous form.
c. the development of appropriate medical treatment for the sickle cell condition in the United States.
d. a decline in the occurrence of malaria in the United States.
e. the absence of mosquitoes.

Answer: d
Difficulty: Moderate
Bloom's Taxonomy: Knowledge, Comprehension

41. The difference in the appearance of the male and the female of the same species is called

a. polymorphism.
b. sexual dimorphism.
c. the dioecious condition.
d. sexual selection.
e. a primary sexual characteristic.

Answer: b
Difficulty: Easy
Bloom's Taxonomy: Knowledge

42. In frequency-dependent selection, highest mortality can be expected in

a. the most abundant genotype.
b. the least abundant genotype.
c. the least well-adapted genotype.
d. heterozygous genotypes.
e. homozygous recessive genotypes.

Answer: a

Difficulty: Moderate
Bloom's Taxonomy: Knowledge, Comprehension

20.5 ADAPTATION AND EVOLUTIONARY CONSTRAINTS

43. According to the modern synthetic theory of evolution, the accumulation of mutations over time results in

a. microevolution.
b. adaptation.
c. an increase in genetic variation.
d. speciation.
e. macroevolution.

Answer: c
Difficulty: Moderate
Bloom's Taxonomy: Knowledge, Comprehension

44. Which of the following evolution-related events is in the correct cause-and-effect sequence?

a. mutation => variation => natural selection => adaptation => speciation
b. variation => adaptation => mutation => natural selection => speciation
c. speciation => adaptation => variation => mutation => natural selection
d. mutation => speciation =>adaptation => variation => natural selection
e. natural selection => variation => mutation =>adaptation => speciation

Answer: a
Difficulty: Moderate
Bloom's Taxonomy: Comprehension

45. Any product of natural selection that increases the relative fitness of an organism in its environment is called

a. mutation.
b. adaptive trait.
c. adaptation.
d. speciation.
e. variation.

Answer: b
Difficulty: Easy
Bloom's Taxonomy: Knowledge

Matching

Match each of the following definitions with the correct agent.

46. _____ A heritable change in DNA

47. _____ Change in allele frequencies as individuals join a population and reproduce

48. _____ Random changes in allele frequencies caused by chance events

49. _____ Differential survivorship or reproduction of individuals with different genotypes

50. _____ Choice of mates based on their phenotypes and genotypes

A. Gene flow

B. Nonrandom mating

C. Genetic drift

D. Mutation

E. Natural selection

Answers: 46. D   47. A   48. C   49. E   50. B

Difficulty: Moderate
Bloom's Taxonomy: Knowledge
Source: Section 20.2

Match the Hardy Weinberg factor with the correct genetic designation.

51 _____ p

52. _____ q

53. _____ q^2

54. _____ p^2

55. _____ 2pq

56. _____ p+q

A. Frequency of heterozygotes

B. Frequency of the dominant allele

C. Frequency of the recessive allele

D. The total gene pool for a gene with two alleles

E. Frequency of homozygous recessive genotypes

F. Frequency of homozygous dominant genotypes

Answers: 51. B   52. C   53. E   54. F   55. A   56. D

Difficulty: Moderate
Bloom's Taxonomy: Knowledge
Source: Section 20.2

Short Answer

57. If you wanted to calculate the frequencies of the two different alleles for a particular gene in a population, what data would you have to gather and how would you use it?

Answer: You would have to sample the population to determine the proportion of individuals displaying the homozygous recessive phenotypes, which would be equivalent to q^2. You would take the square root of that to determine the frequency of the recessive allele q, and then subtract q from 1.0 to obtain the frequency of the dominant allele, p.
Difficulty: Moderate
Bloom's Taxonomy: Analysis
Source: Section 20.2

58. How can an allele that confers no selective advantage spread through a population?

Answer: If two alleles are situated close together on a chromosome, and one is strongly selected for, the other will be spread with it, even if it confers no advantage. This spread of alleles adjacent to selected alleles is called a selective sweep.
Difficulty: Difficult
Bloom's Taxonomy: Comprehension
Source: Unanswered Questions

True/False

If the statement is true, write a "T" in the blank. If the statement is false, make it correct by changing the underlined word(s) and writing the correct word(s) in the answer blank.

59. _____ Natural selection results in organisms <u>perfectly adapted</u> to their environments.

60. _____ Natural selection <u>quickly eliminates</u> harmful alleles from the population.

61. _____ Genetic drift can eliminate alleles from a population <u>more quickly than</u> natural selection

62. _____ When multiple color forms of an organism exist in a population, the least abundant form is likely to <u>increase</u> in frequency.

63. _____ Bacteria readily <u>develop immunity</u> to many forms of antibiotics.

64. _____ In an unchanging environment, a well-adapted population is likely to experience <u>no natural selection</u> of any kind.

65. _____ Genetic drift is non-selective <u>and therefore does not contribute to microevolution</u>.

66. _____ The reason why some specialty breeds of dogs are more prone to genetic disease is because of genetic drift.

67. _____ Small populations on islands can be said to have experienced a form of genetic drift called the <u>founders effect</u>.

68. _____ Phenotypic variation <u>cannot</u> be passed on from one generation to the next.

69. _____ Intermediate forms are often <u>absent</u> in characters showing qualitative variation.

70. _____ <u>Genetic variation</u> can originate both from the production of new alleles and from the arrangement of existing alleles.

Answers: 59. F, well adapted, never perfect 60. F, may retain (especially if recessive) 61. T   62. T   63. T   64. F, stabilizing selection   65. F, but does contribute to microevolution by changing allele frequencies 66. T   67. T   68. F, is 69. T   70. T

Difficulty: Easy
Bloom's Taxonomy: Knowledge
Source: Throughout chapter

Select the Exception

71.   Four of the five processes listed below will disrupt Hardy-Weinberg equilibrium. Select the exception.

a.   non-random mating
b.   natural selection
c.   migration
d.   punctuated equilibrium
e.   genetic drift

Answer: d
Difficulty: Moderate
Bloom's Taxonomy: Comprehension
Source: Section 20.3

72.   Four of the five population attributes can be calculated with the Hardy-Weinberg equation. Select the exception.

a.   the frequency of heterozygotes
b.   the frequency of homozygous dominant genotypes
c.   the frequency of a dominant allele
d.   the frequency of a recessive allele
e.   the frequency of mutation

Answer: e
Difficulty: Moderate
Bloom's Taxonomy: Comprehension
Source: Section 20.3

73.   Four of the five processes below are agents of microevolutionary change. Select the exception.

a.   mutation
b.   genetic drift
c.   natural selection
d.   random mating
e.   gene flow

Answer: d
Difficulty: Moderate
Bloom's Taxonomy: Knowledge, Comprehension
Source: Section 20.3

74.   Four of the five processes below are components of natural selection. Select the exception.

a.   genetic drift
b.   overproduction of offspring
c.   limitations in vital resources
d.   genetic variation
e.   differential reproductive success

Answer: a
Difficulty: Easy
Bloom's Taxonomy: Knowledge
Source: Section 20.3

Essay

75. If you were planning the colonization of a distant planet, what would you do to insure that the population remained genetically healthy for a long time?

Answer: To avoid genetic drift and the occurrence of genetic disease, you would want to begin with as large a population as possible. You would also want that population to be as genetically diverse as possible, and so you would include as many different racial and ethnic groups as possible. You would, of course, also screen applicants for their genetic background to avoid the introduction of alleles concerned with genetic disease in the first place. You would encourage random mating among the colonists, urging them to avoid cliques of similar racial and genetic makeup. You might also monitor levels of cosmic and other radiation on the planet, and take steps to shield the colonists from potential mutagenic or carcinogenic radiation.
Difficulty: Moderate
Bloom's Taxonomy: Analysis
Source: Section 20.3

21
SPECIATION

Multiple Choice

WHY IT MATTERS

1.   What mechanism seemed to separate the "birds of paradise" in New Guinea?

a.   chromosomal differences
b.   different mating rituals
c.   geographic isolation
d.   interference from humans
e.   mechanical differences between species

Answer: c
Difficulty: Moderate
Bloom's Taxonomy: Application

2.   The habitats of "birds of paradise" in New Guinea are on mountains separated by

a.   areas of unsuitable land.
b.   fierce predators.
c.   man-made borders.
d.   the ocean.
e.   rivers and lakes.

Answer: a
Difficulty: Moderate
Bloom's Taxonomy: Knowledge

3.   Ernst Mayr, the scientist from the New Guinea study on "birds of paradise", knew that the natives understood speciation because they

a.   have a museum with many catalogued specimens.
b.   have individual names for many of the species.
c.   have many biologists in New Guinea that studied the speciation.
d.   helped to achieve the speciation.
e.   teach speciation theories in school.

Answer: b
Difficulty: Easy
Bloom's Taxonomy: Knowledge

21.1 WHAT IS A SPECIES?

4.   An example of the morphological species concept would be achieved by

a.   doing a genetic analysis on the two organisms.
b.   looking at the appearance of two organisms.
c.   mating the two organisms to see if viable offspring result.
d.   observing the mating rituals of the two organisms.
e.   tracing the common ancestry of the two organisms.

Answer: b
Difficulty: Moderate
Bloom's taxonomy: Application

5.   The morphological approach to speciation is prone to misinterpretation, as shown by the example of

a.   the many species of "birds of paradise" that look alike.
b.   a chihuahua and a great dane that do not look similar.
c.   snails that resemble each other, but cannot mate.
d.   the sea shells that do not resemble each other, but house only one species of snail.
e.   all of the above

Answer: e
Difficulty: Moderate
Bloom's Taxonomy: Application

6.   The biological species concept relies primarily on the species' ability to

a.   adapt to the environment.
b.   have greater longevity.
c.   intermingle with other species.
d.   live in a variety of habitats.
e.   produce fertile offspring.

Answer: e
Difficulty: Moderate
Bloom's Taxonomy: Knowledge

7. The phylogenetic species concept looks at both the

a. environmental adaptations and morphology.
b. genetic data and environmental adaptations.
c. morphology and environmental adaptations.
d. morphology and genetic data.
e. species response to adverse conditions and genetic data.

Answer: d
Difficulty: Difficult
Bloom's Taxonomy: Knowledge

8. The concept for examining speciation by reconstructing the evolutionary tree is the

a. biological species concept.
b. examination of ring species.
c. examination of clinal variation.
d. morphological species concept.
e. phylogenetic species concept.

Answer: e
Difficulty: Moderate
Bloom's Taxonomy: Knowledge

9. A ring species is one where

a. a clinal variation occurs.
b. all of the various populations can successfully mate in nature.
c. an area's climate interferes with speciation.
d. none of the various populations can successfully mate in nature.
e. only intermediary populations can mate successfully.

Answer: e
Difficulty: Moderate
Bloom's Taxonomy: Knowledge

10. Birds living in a cold environment tend to have

a. long legs and larger bodies.
b. long legs and smaller bodies.
c. short legs and larger bodies.
d. short legs and smaller bodies.
e. the leg length and body size of birds does not relate to the environment.

Answer: c
Difficulty: Moderate
Bloom's Taxonomy: Knowledge

11. In the biological species concept, it is specified that two organisms are of the same species if they breed and

a. are unsuccessful in producing offspring.
b. produce any offspring.
c. produce fertile offspring under laboratory conditions only.
d. produce fertile offspring under normal circumstances.
e. produce infertile offspring.

Answer: d
Difficulty: Easy
Bloom's Taxonomy: Knowledge

12. When geographic variation occurs, the organisms

a. may be mistaken for separate species.
b. can mate successfully.
c. can produce offspring with intermediate characteristics.
d. may be present in different locations.
e. all of the above

Answer: e
Difficulty: Moderate
Bloom's Taxonomy: Knowledge

13. Gene flow between organisms of ring species occurs

a. between any of the organisms if they are placed in the same environment.
b. freely.
c. never.
d. only between intermediary populations.
e. only under laboratory conditions.

Answer: d
Difficulty: Moderate
Bloom's Taxonomy: Application

14. In a clinal variation, the organisms may

a. look identical.
b. not have a normal lifespan.
c. never be able to mate successfully.
d. vary in appearance due to adaptations to different environments.
e. vary in appearance due to human interference.

Answer: d
Difficulty: Moderate
Bloom's Taxonomy: Knowledge

21.2 MAINTAINING REPRODUCTIVE ISOLATION

15.   When two organisms do not mate due to the fact that they mate at different times of the year, it is known as _____.

a.   behavioral isolation
b.   ecological isolation
c.   gametic isolation
d.   mechanical isolation
e.   temporal isolation

Answer: e
Difficulty: Easy
Bloom's Taxonomy: Knowledge

16.   An example of mechanical isolation would be

a.   flowers that bloom months apart.
b.   two organisms that are pollinated by different insects.
c.   two organisms that have different habitats.
d.   two organisms that have incompatible gametes.
e.   two organisms that have different mating rituals.

Answer: b
Difficulty: Difficult
Bloom's Taxonomy: Application

17.   Which of the following combinations represent prezygotic mechanisms of isolation?

a.   hybrid breakdown and temporal isolation
b.   hybrid breakdown and gametic isolation
c.   hybrid sterility and temporal isolation
d.   mechanical and temporal isolation
e.   mechanical and hybrid breakdown

Answer: d
Difficulty: Moderate
Bloom's Taxonomy: Knowledge

18.   Postzygotic isolating mechanisms occur when the offspring

a.   do not survive.
b.   have poor health.
c.   inherit different amounts of chromosomes from each parent.
d.   inherit different sets of instructions for development from each parent.
e.   all of the above

Answer: e
Difficulty: Moderate
Bloom's Taxonomy: Application

19.   If the F_2 generation illustrates a lack of fitness, this is an example of

a.   behavioral isolation.
b.   gametic isolation.
c.   hybrid breakdown.
d.   crossing over.
e.   hybrid sterility.

Answer: c
Difficulty: Moderate
Bloom's Taxonomy: Knowledge

20.   A mule is an example of

a.   hybrid breakdown.
b.   hybrid inviability.
c.   hybrid sterility.
d.   mechanical isolation.
e.   temporal isolation.

Answer: c
Difficulty: Easy
Bloom's Taxonomy: Knowledge

21.   If in a laboratory situation, I cross an organism with 64 chromosomes with a closely related organism with 62 chromosomes,

a.   many unhealthy offspring will result due to hybrid breakdown.
b.   many unhealthy offspring will result due to hybrid inviability.
c.   no offspring will result due to hybrid sterility.
d.   sterile offspring will result due to hybrid inviability.
e.   sterile offspring will result due to hybrid sterility.

Answer: e
Difficulty: Moderate
Bloom's Taxonomy: Application

22.   If two species of lizards do not mate because their mating rituals differ greatly, this is known as

a.   behavioral isolation.
b.   ecological isolation.
c.   gametic isolation.
d.   mechanical isolation.
e.   temporal isolation.

Answer: a
Difficulty: Easy
Bloom's Taxonomy: Knowledge

23. If one plant has a very small tubular type of flower and cannot be pollinated by a large honeybee, it may develop which type of prezygotic isolation?

a. behavioral isolation
b. ecological isolation
c. gametic isolation
d. mechanical isolation
e. temporal isolation

Answer: d
Difficulty: Moderate
Bloom's Taxonomy: Application

24. Some *Drosophila* reject the sperm of another species, which is known as

a. gametic isolation.
b. hybrid breakdown.
c. hybrid inviability.
d. mechanical isolation.
e. temporal isolation.

Answer: a
Difficulty: Moderate
Bloom's Taxonomy: Application

25. Crocuses that bloom in the early spring and marigolds that bloom in the summer would definitely experience which type of isolation?

a. behavior
b. ecological
c. gametic
d. temporal
e. should not exhibit any type of isolation

Answer: d
Difficulty: Moderate
Bloom's Taxonomy: Application

26. The fact that eggs have the ability to recognize surface proteins on sperm of their own species illustrates

a. behavioral isolation.
b. gametic isolation.
c. hybrid breakdown.
d. hybrid inviability.
e. hybrid sterility.

Answer: b
Difficulty: Moderate
Bloom's Taxonomy: Application

21.3 THE GEOGRAPHY OF SPECIATION

27. A host race develops when an organism's

a. genes for the host plant choice change.
b. genes for mating preferences change.
c. geographical barriers exist.
d. both a and b
e. both b and c

Answer: d
Difficulty: Difficult
Bloom's Taxonomy: Knowledge

28. What conclusions can be drawn from the studies on bent grass in a copper-polluted field and the neighboring population in a non-polluted field?

a. Flowers in the polluted field will not survive.
b. The one population now requires copper to survive.
c. They have different flowering schedules, so isolation may be developing.
d. They have different types of gametes, so isolation has taken place.
e. They show no differences.

Answer: c
Difficulty: Difficult
Bloom's Taxonomy: Comprehension

29. If a flood separated a large population into two populations, this would be an example of which process?

a. allopatric speciation
b. development of a host race
c. parapatric speciation
d. polyploidy
e. sympatric speciation

Answer: a
Difficulty: Moderate
Bloom's Taxonomy: Application

30. The study in the textbook of which of the following species gives an example of adaptive radiation?

a. apple maggots
b. Baltimore Orioles
c. bent grass
d. fish near the isthmus of Panama
e. Hawaiian fruit flies

Answer: e
Difficulty: Difficult
Bloom's Taxonomy: Knowledge

31. What is the definition of a species cluster?

a. a group of closely related species with a common ancestor
b. a group of species that developed host races
c. a group of species that exhibit polyploidy
d. a group of totally unrelated species living in a common habitat
e. when allopatric and parapatric speciation occur together

Answer: a
Difficulty: Moderate
Bloom's Taxonomy: Knowledge

32. Polyploidy refers to _____ and occurs most often in _____.

a. an individual getting one or more additional sets of chromosomes; plants
b. an individual getting one or more additional sets of chromosomes; animals
c. an individual lacking one or more sets of chromosomes; plants
d. an individual lacking one or more sets of chromosomes; animals
e. an individual lacking one or more sets of chromosomes and receiving additional sets of other chromosomes; animals

Answer: a
Difficulty: Moderate
Bloom's Taxonomy: Knowledge

33. What is a hybrid zone?

a. an area where no hybrids exist
b. an area where two populations may breed and produce inviable, fertile offspring
c. an area where two populations may breed and produce inviable, infertile offspring
d. an area where two populations may breed and produce viable, fertile offspring
e. an area where two populations may breed and produce viable, infertile offspring

Answer: d
Difficulty: Moderate
Bloom's Taxonomy: Knowledge

34. If a hybrid organism has a shorter lifespan, which situation may result?

a. allopatric speciation
b. hybrid zones
c. parapatric speciation
d. species clusters
e. sympatric speciation

Answer: c
Difficulty: Moderate
Bloom's Taxonomy: Knowledge

21.4 GENETIC MECHANISMS OF SPECIATION

35. From the studies on chromosome similarities, it looks as though humans are most closely related to

a. apes
b. apes and gorillas
c. chimpanzees
d. gorillas
e. lemurs

Answer: c
Difficulty: Easy
Bloom's Taxonomy: Knowledge

36. In the research on monkey-flower speciation, researchers found that mutations in as few as _____ genes could result in speciation.

a. 2
b. 4
c. 8
d. 16
e. 25

Answer: c
Difficulty: Moderate
Bloom's Taxonomy: Knowledge

37. If _____ mutation(s) occur(s) in snails, the direction of coiling may reverse, which _____ mating.

a. 1; encourages
b. 1; prohibits
c. 2; encourages
d. 2; prohibits
e. 3; encourages

Answer: b
Difficulty: Moderate
Bloom's: knowledge

38. The definition of allopolyploidy is

a. a genetic divergence that results in nonviable offspring.
b. a decrease in chromosome number due to hybridization of different species.
c. a decrease in chromosome number within a single species.
d. an increase in chromosome number due to hybridization of different species.
e. an increase in chromosome number within a single species.

Answer: d
Difficulty: Difficult
Bloom's Taxonomy: Knowledge

39. When a gamete receives the same amount of chromosomes as a somatic cell,

a. a reduced gamete is formed and allopolyploidy is present.
b. a reduced gamete is formed and autopolyploidy is present.
c. an unreduced gamete is formed and allopolyploidy is present.
d. an unreduced gamete is formed and autopolyploidy is present.
e. an unreduced gamete is formed and allopolyploidy and autopolyploidy are present.

Answer: d
Difficulty: Difficult
Bloom's Taxonomy: Comprehension

40. A triploid organism is

a. not viable.
b. usually sterile because of improper chromosome segregation.
c. usually sterile because of proper chromosome segregation.
d. usually fertile because of improper chromosome segregation.
e. usually fertile because of proper chromosome segregation.

Answer: b
Difficulty: Difficult
Bloom's Taxonomy: Knowledge

UNANSWERED QUESTIONS

41. If speciation occurs with the production of two or more species in a small area, _____ is (are) probably responsible.

a. both sympatric and parapatric speciation
b. either sympatric or parapatric speciation
c. neither sympatric or parapatric speciation
d. only sympatric speciation
e. only parapatric speciation

Answer: b
Difficulty: Difficult
Bloom's Taxonomy: Knowledge

42. What avenues could be explored to determine if asexual organisms form species?

a. determine if a cluster of bacteria constitutes a species
b. explore various niches in which different populations reside
c. look into previously undiscovered types of bacteria
d. genomic research
e. all of the above

Answer: e
Difficulty: Easy
Bloom's Taxonomy: Application

Integrative Multiple Choice

43. The morphological species concept would relate more to

a. prezygotic isolation.
b. postzygotic isolation.
c. both prezygotic and postzygotic isolation
d. neither prezygotic nor postzygotic isolation
e. it would depend on the species used for the study

Answer: b
Difficulty: Difficult
Bloom's Taxonomy: Application
Sources: Sections 21.1, 21.2

44. Some of the work with Hawaiian fruit flies illustrated the presence of a

a. prezygotic isolating mechanism, which involved behavioral isolation.
b. prezygotic isolating mechanism, which involved gametic isolation.
c. prezygotic isolating mechanism, which involved temporal isolation.
d. postzygotic isolating mechanism, which involved behavioral isolation.
e. postzygotic isolating mechanism, which involved gametic isolation.

Answer: a
Difficulty: Difficult
Bloom's Taxonomy: Comprehension
Sources: Sections 21.2, 21.3

Matching

Match the appropriate definitions to the scientific terms:

45. _____ reproductive isolation due to mating at different times of the year

46. _____ determining different species by observing appearance

47. _____ the second generation of the mating shows poor fitness

48. _____ determining different species by checking to see if fertile offspring are produced

49. _____ a "race"

50. _____ closely related species form polyploidy offspring by hybridization

51. _____ an error in mitosis or meiosis forms polyploidy offspring

52. _____ an individual has extra copies of the haploid complement of chromosomes

53. _____ reproductive isolation between subgroups of a single population

54. _____ speciation occurs between adjacent populations

55. _____ a physical barrier divides a population into separate areas, and speciation results

56. _____ speciation occurs due to different mating rituals

57. _____ there is a physical reason why the two organisms cannot mate, and speciation occurs

A. Allopolyploidy

B. Allopatric speciation

C. Autopolyploidy

D. Behavioral isolation

E. Biological species concept

F. Hybrid breakdown

G. Mechanical isolation

H. Morphological species concept

I. Parapatric speciation

J. Polyploidy

K. Subspecies

L. Sympatric speciation

M. Temporal isolation

Answers: *45. M   46. H   47. F   48. E   49. K*
              50. C   51. C   52. J   53. L   54.
              I   55. B   56. D   57. G

Difficulty: Moderate
Bloom's Taxonomy: Knowledge
Sources: Sections 21.1, 21.2, 21.3, 21.4

Short Answer

58. Based on the morphological species concept, why would you assume that there are many different species of rat snakes?

Answer: The rat snake is present in many diverse areas of the country. Individual snakes from these different locations have different banding patterns and do not resemble each other. Based on a morphological approach, one would assume that they are different species. In reality, they are subspecies that could successfully mate if brought to a common location.
Difficulty: Difficult
Bloom's Taxonomy: Application
Source: Section 21.1

59. How do animals show clinal variation?

Answer: Clinal variation results from gene flow between nearby populations when each population is adapting to slightly different conditions. The sparrows mentioned in the text developed changes in leg length and body size based on different environmental conditions.
Difficulty: Moderate
Bloom's Taxonomy: Knowledge
Source: Section 21.1

60. How did researchers establish the connection between chimpanzees and humans?

Answer: Researchers examined the chromosomes of humans, apes, gorillas, and chimpanzees. They looked for similar banding patterns among the species studied and established that those of chimpanzees most closely related to the human chromosomes.
Difficulty: Moderate
Bloom's Taxonomy: Knowledge
Source: Section 21.4

Sequence

Place the following steps of allopatric speciation in their correct order:

A. A flood causes a physical barrier down the middle of a population.
B. A population covers a large area.
C. The two separate populations evolve differently and form new species.
D. When the water recedes, the two new species do not mate.

61. ____ first event
62. ____ second event
63. ____ third event
64. ____ fourth event

Answers: 61. B   62. A   63. C   64. D

Difficulty: Easy
Bloom's Taxonomy: Application
Source: Section 21.3

Labeling

Reproductive Isolating Mechanisms		
Timing Relative to Fertilization	Mechanism	Mode of Action
Prezygotic ("premating") mechanisms	65	Species live in different habitats
	66	Species breed at different times
	67	Species cannot communicate
	68	Species cannot physically mate
	69	Species have nonmatching receptors on gametes
Postzygotic ("postmating") mechanisms	70	Hybrid offspring do not complete development
	71	Hybrid offspring cannot produce gametes
	72	Hybrid offspring have reduced survival or fertility

Place the following terms appropriately in the accompanying table:

65. _____
66. _____
67. _____
68. _____
69. _____
70. _____
71. _____
72. _____

Answers (in order from top to bottom): 65. ecological isolation   66. temporal isolation   67. behavioral isolation   68. mechanical isolation 69. gametic isolation 70. hybrid inviability   71. hybrid sterility   72. hybrid breakdown

Difficulty: Moderate
Bloom's Taxonomy: Knowledge
Source: Section 21.2

True/False Questions

If the statement is true, write a "T" in the blank. If the statement is false, make it correct by changing the underlined word(s) and writing the correct word(s) in the answer blank.

73. _____ Unreduced gametes are those that contain additional chromosome numbers.

74. _____ Polyploid plants are often unhealthy as compared to other plants.

75. _____ The apple maggot was used as an example of allopatric speciation.

76. _____ If two organisms cannot mate due to physical differences, it is known as temporal isolation.

77. _____ An example of a prezygotic isolating mechanism is behavioral isolation.

Answer: 73. T   74. F, robust or larger   75. F, sympatric 76. F, mechanical   77. T

Difficulty: Moderate
Bloom's Taxonomy: Knowledge
Source: Sections 21.2, 21.3, 21.4

Select the Exception

78. All of the following are common polyploidy crops except:

a. broccoli
b. coffee
c. cotton
d. potatoes
e. wheat

Answer: a
Difficulty: Easy
Bloom's Taxonomy: Knowledge
Source: Section 21.4

79. Which of the following is not an example of a prezygotic isolating mechanism?

a. An iris blooms at a different time of the year than a daffodil.
b. A peacock puts on a different mating display than a sparrow.
c. A snail with a shell coiling in one direction cannot mate with a snail whose shell curls the other way.
d. A zebroid is sterile.
e. One type of roundworm would reject the sperm from another species of roundworm.

Answer: d
Difficulty: Moderate
Bloom's Taxonomy: Application
Source: Section 21.2

80. Which of the following categories of organisms is not known to produce viable polyploids?

a. amphibians
b. fish
c. mammals
d. fish
e. reptiles

Answer: c
Difficulty: Easy
Bloom's Taxonomy: Knowledge
Source: Section 21.4

Essays

81. How does temporal isolation differ in plants and animals?

Answer: Plants detect the length of day by the use of plant hormones. This process tells the plant when it should flower. If two plants flower at different times of the year, they will not cross-pollinate naturally. Many animals breed in the springtime, so that the offspring will be hardy by the time the harsh winter conditions arrive. They may mate at slightly different times of the year, or even times of the day, that will prohibit mating between those individuals.
Difficulty: Difficult
Bloom's Taxonomy: Comprehension
Source: Section 21.2

82. Do you feel that viruses show speciation, and do you consider viruses like the smallpox virus to be an "endangered species"?

Answer: This is a very thought provoking question. It really boils down to whether you consider viruses as living or non-living. Many scientists categorize them as an intermediate category between the living and non-living. If you consider viruses as alive, they mutate frequently so you could consider that as speciation. Also if you consider viruses as living things, I suppose the smallpox virus is "endangered" since it is not present in the environment currently.
Difficulty: Difficult
Bloom's Taxonomy: Application
Source: Unanswered Questions

22
PALEOBIOLOGY AND MACROEVOLUTION

Multiple-Choice

WHY IT MATTERS

1.   Cuvier shocked the scientific world in 1796 by suggesting that

a.   species evolved.
b.   species could become extinct.
c.   fossils were dead organisms.
d.   the earth was several billion years old.
e.   organisms changed through blind natural selection.

Answer: b
Difficulty: Easy
Bloom's Taxonomy: Knowledge

2.   If we are analyzing the origin of new families, orders, etc. over time, we are studying

a.   microevolution
b.   mutation
c.   macroevolution
d.   natural selection
e.   anagenesis

Answer: c
Difficulty: Easy
Bloom's Taxonomy: Knowledge, Comprehension

22.1 THE FOSSIL RECORD

3.   The era during which the earliest signs of life appeared was the

a.   Archaean
b.   Paleozoic
c.   Mesozoic
d.   Cambrian
e.   Cenozoic

Answer: a
Difficulty: Easy
Bloom's Taxonomy: Knowledge

4.   Dinosaurs flourished during the

a.   Archaean
b.   Paleozoic
c.   Mesozoic
d.   Cambrian
e.   Cenozoic

Answer: c
Difficulty: Easy
Bloom's Taxonomy: Knowledge

5.   The radioactive decay of carbon-14 can be used to determine

a.   relative dating.
b.   absolute dating.
c.   mutation rates.
d.   plate tectonics.
e.   extinction rates.

Answer: b
Difficulty: Easy
Bloom's Taxonomy: Knowledge

6.   The Cretaceous is the period in which

a.   mammals diversified and flourished.
b.   life appeared on this planet.
c.   the first animals with shells and skeletons appeared.
d.   the dinosaurs reached their peak and then became extinct.
e.   the first human-like apes appeared.

Answer: d
Difficulty: Moderate
Bloom's Taxonomy: Knowledge

7.   Fossils are found only in

a.   metamorphic rocks.
b.   igneous rocks.
c.   sedimentary rocks.
d.   crystalline rocks.
e.   radioactive rocks.

Answer: c
Difficulty: Easy
Bloom's Taxonomy: Knowledge

8.   A paleontologist estimates that when a particular rock formed, it contained 12 mg of the radioactive isotope potassium-40, which has a half-life of 1.3 billion years. The rock now contains 3 mg of the isotope. About how old is the rock?

a.   2.6 billion years
b.   5.2 billion years
c.   1.3 billion years
d.   0.3 billion years
e.   0.4 billion years

Answer: b
Difficulty: Moderate
Bloom's Taxonomy: Analysis

9. The oldest of the Eras that make up the geological time scale is the

a. Archaean.
b. Paleozoic.
c. Mesozoic.
d. Cenozoic.
e. Jurassic.

Answer: a
Difficulty: Easy
Bloom's Taxonomy: Knowledge

10. The time interval used in measuring the age of rocks is called

a. radioisotope unit.
b. radiometric unit.
c. radiometric life.
d. radioisotope life.
e. half-life.

Answer: e
Difficulty: Easy
Bloom's Taxonomy: Knowledge

11. Most dinosaurs became extinct

a. 10,000 years ago.
b. 1 million years ago.
c. 500 million years ago.
d. 65 million years ago.
e. 1 billion years ago.

Answer: d
Difficulty: Easy
Bloom's Taxonomy: Knowledge

12. Mammals underwent an evolutionary "explosion" or major adaptive radiation in the

a. Mesozoic.
b. Cambrian.
c. Triassic.
d. Devonian.
e. Cenozoic.

Answer: e
Difficulty: Easy
Bloom's Taxonomy: Knowledge

13. Most dinosaurs died out at the end of the

a. Triassic.
b. Jurassic.
c. Cretaceous.
d. Tertiary.
e. Permian.

Answer: c
Difficulty: Easy
Bloom's Taxonomy: Knowledge

14. Which group of animals represents the surviving dinosaurs?

a. birds
b. crocodiles and alligators
c. snakes and lizards
d. turtles
e. mammals

Answer: a
Difficulty: Moderate
Bloom's Taxonomy: Comprehension

15. The age that saw a great diversification of land plants, as well as the first amphibians, was the

a. Cambrian.
b. Cretaceous.
c. Devonian.
d. Jurassic.
e. Ordovician.

Answer: c
Difficulty: Easy
Bloom's Taxonomy: Knowledge

16. Over time, the carbon isotope ^{14}C decays into

a. ^{12}C.
b. ^{14}N.
c. ^{12}N.
d. ^{13}C.
e. ^{12}B.

Answer: b
Difficulty: Easy
Bloom's Taxonomy: Knowledge

17. Radiometric dating works best with

a. sedimentary rocks.
b. limestone.
c. metamorphic rock.
d. volcanic rocks.
e. meteorites.

Answer: d
Difficulty: Easy
Bloom's Taxonomy: Knowledge

18. Wood and bone fragments found intact in sediments can be accurately aged using

a. Thorium-232.
b. Uranium-238.
c. Uranium-235.
d. Potassium-40.
e. Carbon-14.

Answer: e
Difficulty: Moderate
Bloom's Taxonomy: Knowledge

22.2 EARTH HISTORY, BIOGEOGRAPHY, AND CONVERGENT EVOLUTION

19.   Because of continental drift, dinosaurs during the Jurassic Period were able to migrate freely from

a.   Africa to Europe.
b.   Antarctica to Asia.
c.   Australia to Asia.
d.   South America to North America.
e.   South America to Africa.

Answer: e
Difficulty: Moderate
Bloom's Taxonomy: Comprehension

20.   When all continents were fused into one during the Permian Period, they were called

a.   Laurasia.
b.   Pangaea.
c.   Gondwanaland.
d.   Eurasia.
e.   Amerasia.

Answer: b
Difficulty: Easy
Bloom's Taxonomy: Knowledge

21.   Australia, Antarctica, and South America represent fragments of

a.   Australasia.
b.   Laurasia.
c.   Gondwanaland.
d.   Eurasia.
e.   Amerasia.

Answer: c
Difficulty: Easy
Bloom's Taxonomy: Knowledge

22.   Oceanic ridges form

a.   along oceanic trenches.
b.   where shallow seas dry up.
c.   in zones of rising magma.
d.   where old oceanic crust sinks below continental crust.
e.   where two continents collide.

Answer: c
Difficulty: Easy
Bloom's Taxonomy: Knowledge

23.   According to Wallace's biogeographical realms, North America belongs primarily to the

a.   Neotropical.
b.   Ethiopian.
c.   Palearctic.
d.   Nearctic.
e.   Oriental.

Answer: d
Difficulty: Easy
Bloom's Taxonomy: Knowledge

24.   If we were to discover beings on a distant planet that looked very much like us, it would probably be cited as an example of

a.   coincidence.
b.   convergent evolution.
c.   homology.
d.   adaptive radiation.
e.   genetic drift.

Answer: b
Difficulty: Moderate
Bloom's Taxonomy: Comprehension

25.   Australia and Eurasia each have a mouse-like mammal, one a marsupial, the other a placental. This is an example of

a.   long-distance migration.
b.   homology.
c.   adaptive radiation.
d.   convergent evolution.
e.   coincidence.

Answer: d
Difficulty: Moderate
Bloom's Taxonomy: Comprehension

26.   A species that is confined to a specific, relatively small geographic area is called

a.   endemic.
b.   allopatric.
c.   sympatric.
d.   autopolyploid.
e.   polymorphic.

Answer: a
Difficulty: Easy
Bloom's Taxonomy: Knowledge

22.3 INTERPRETING EVOLUTIONARY LINEAGES

27.   The evolution of horses is best characterized as

a.   a linear sequence of intermediate forms leading to modern horses.
b.   an example of evolutionary convergence.
c.   poorly documented and speculative.
d.   a gradual increase in body size.
e.   a complex branched history with many extinct lineages.

Answer: e
Difficulty: Moderate
Bloom's Taxonomy: Comprehension

28.   The apparent lack of intermediate forms in the fossil record might best be explained as due to

a.   disruptive selection.
b.   frequency dependent selection.
c.   allopatric speciation.
d.   genetic drift.
e.   punctuated equilibrium.

Answer: e
Difficulty: Easy
Bloom's Taxonomy: Knowledge

29. The evolutionary mode in which species evolve rapidly at first and then remain stable for very long periods of time is called

a. catastrophism.
b. uniformitarianism.
c. disruptive selection.
d. punctuated equilibrium.
e. anagenesis.

Answer: d
Difficulty: Easy
Bloom's Taxonomy: Knowledge

30. A view of Earth's history that attributes profound change to the cumulative product of slow but continuous processes is

a. gradualism.
b. homology.
c. punctuated equilibrium.
d. descent with modification.
e. cladogenesis.

Answer: a
Difficulty: Easy
Bloom's Taxonomy: Knowledge

31. A series of changes in a single lineage of species, without branching to create additional species, is called

a. cladogenesis.
b. paedomorphosis.
c. punctuated equilibrium.
d. adaptive radiation.
e. anagenesis.

Answer: e
Difficulty: Moderate
Bloom's Taxonomy: Knowledge

22.4 MACROEVOLUTIONARY TRENDS IN MORPHOLOGY

32. The changing size ratio between an animal's head and the rest of its body during development is an example of

a. allometric growth.
b. punctuated equilibrium.
c. anagenesis.
d. preadaptation.
e. paedomorphosis.

Answer: a
Difficulty: Easy
Bloom's Taxonomy: Knowledge

33. Some animals have evolved sexual maturity in juvenile body forms. This is called

a. allometric growth.
b. preadaptation.
c. paedomorphosis.
d. punctuated equilibrium.
e. anagenesis.

Answer: c
Difficulty: Easy
Bloom's Taxonomy: Knowledge

34. A trait that is adaptive in one context and later turns out to have adaptive value in a rather different context is called

a. paedomorphosis.
b. anagenesis.
c. cladogenesis.
d. preadaptation.
e. endemic.

Answer: d
Difficulty: Moderate
Bloom's Taxonomy: Knowledge

22.5 MACROEVOLUTIONARY TRENDS IN BIODIVERSITY

35. Adaptive radiation is

a. the basis of radiometric dating.
b. development of similar body form in unrelated organisms.
c. formation of beneficial alleles through exposure to radiation.
d. origin of many species from a single common ancestor.
e. spread of successful alleles from one population to another.

Answer: d
Difficulty: Easy
Bloom's Taxonomy: Knowledge

36. A single seed lands on one of a group of isolated islands. Several million years later, its descendants have evolved into a number of new species on the different islands, representing a variety of trees, shrubs, and vines. At that point in time, we have an example of

a. adaptive radiation.
b. punctuated equilibrium.
c. genetic drift.
d. microevolution.
e. mutation.

Answer: a
Difficulty: Moderate
Bloom's Taxonomy: Comprehension

22.6 EVOLUTIONARY DEVELOPMENTAL BIOLOGY

37.   A gene that determines the structure of body parts during embryonic development is a

a.   homeotic gene.
b.   homeostatic gene.
c.   homologous gene.
d.   homeobox.
e.   homogenous gene.

Answer: a
Difficulty: Easy
Bloom's Taxonomy: Knowledge

38.   The overall body plan of animals is controlled by

a.   Hox genes.
b.   Pax genes.
c.   homeobox genes.
d.   homeodomain genes.
e.   homeostatic genes.

Answer: a
Difficulty: Moderate
Bloom's Taxonomy: Knowledge, Comprehension

Matching

Match each of the following events with the correct time period.

39.   _____   diversification of terrestrial vascular plants; first amphibians and insects

40.   _____   modern birds and mammals diversify

41.   _____   origin of mammals; Pangaea begins to break up

42.   _____   large swamp forests occur; first seed plants

43.   _____   birds appear; dinosaurs diversify and dominate

44.   _____   initial radiation of animal phyla

45.   _____   asteroid at the end of this period causes extinction of dinosaurs

46.   _____   origin of life; evolution of prokaryotes

47.   _____   first vascular plants and jawed fishes appear

48.   _____   continents coalesce into Pangaea; mass extinction at the end of this period destroys 85 percent of life

A.   Archaean

B.   Cambrian

C.   Ordovician

D.   Silurian

E.   Devonian

F.   Carboniferous

G.   Permian

H.   Triassic

I.   Jurassic

J.   Cretaceous

K.   Cenozoic – Tertiary

Answers: 39. E   40. K   41. H   42. F   43. I
44. B   45. J   46 A   47. D   48. G

Difficulty: Moderate
Bloom's Taxonomy: Knowledge
Source: Section 22.1

Match the following radioisotopes with their half-lives

49.   _____   Samarium-147

50.   _____   Rubidium-87

51.   _____   Thorium-232

52.   _____   Uranium-238

53.   _____   Uranium-235

54.   _____   Potassium-40

55.   _____   Carbon-14

A.   1.25 billion years

B.   106 billion years

C.   4.5 billion years

D.   14 billion years

E.   700 million years

F.   5730 years

G.   48 billion years

Answers: 49. B   50. G   51. D   52. C   53. E
54. A   55. F

Difficulty: Moderate
Bloom's Taxonomy: Knowledge
Source: Section 22.1

Short Answer

56.   Why would migration to an island by a single pair of birds be likely to lead to adaptive radiation in their descendents?

Answer: Islands are often poor in species, with many unoccupied ecological niches, because relatively small numbers of animals are capable of crossing large expanses of ocean.
Difficulty: Moderate
Bloom's Taxonomy: Comprehension
Source: Section 22.5

57.   Why is it difficult to find intermediate fossils in organisms that have evolved under the punctuated equilibrium model?

Answer: Evolutionary change in these organisms has been so rapid that relatively few intermediate individuals lived, and so were unlikely to have been fossilized.
Difficulty: Moderate
Bloom's Taxonomy: Comprehension
Source: Section 22.3

Sequence

Arrange the following levels of geological eras in the correct order. Write the letter of the oldest period next to 58, the letter of the newest period next to 67.

58. _____   A.   Triassic
59. _____   B.   Permian
60. _____   C.   Archaean
61. _____   D.   Cretaceous
62. _____   E.   Devonian
63. _____   F.   Cambrian
64. _____   G.   Silurian
65. _____   H.   Jurassic
66. _____   I.   Ordovician
67. _____   J.   Carboniferous

Answers: 58. C   59. F   60. I   61. G   62. E
63. J   64. B   65. A   66. H   67. D

Difficulty: Moderate
Bloom's Taxonomy: Knowledge
Source: Section 22.1

True/False

If the statement is true, write a "T" in the blank. If the statement is false, make it correct by changing the underlined word(s) and writing the correct word(s) in the answer blank.

68. _____   The evolution of horses involves primarily <u>anagenesis.</u>

69. _____   Oceanic trenches form primarily where one crustal plate <u>sinks below another.</u>

70. _____   As animals develop from embryos to adults, the relative sizes of body organs <u>remain constant.</u>

71. _____   Although they have changed shape through erosion and mountain building, the positions of the major continents relative to one another has <u>remained fixed</u> since the Earth was formed.

72. _____   Adaptive radiation represents an extensive episode of <u>cladogenesis.</u>

73. _____   Primitive feathers in certain dinosaurs evolved <u>in anticipation of flight in their descendents, the birds.</u>

74. _____   Most likely, about <u>10 percent</u> of all species go extinct every million years.

75. _____   The most severe mass extinction occurred at the end of the <u>Permian</u> Period.

76. _____   The extinction of dinosaurs was probably due to <u>the breakup of Pangaea and the resulting cooling of the world's climate.</u>

77. _____   The extinction of the dinosaurs occurred <u>rapidly as a result of a single cataclysmic event.</u>

Answers: 68. F, cladogenesis 69. T   70. F, changes 71. F, shifted extensively through continental drift 72. T   73. F, as adaptations for other factors 74. T   75. T 76. F, an asteroid impact 77. F, over a 40,000 year period after an asteroid impact

Difficulty: Moderate
Bloom's Taxonomy: Knowledge
Source: Section 22.1

Select the Exception

78. Four of the following were innovations that led to new adaptive zones. Select the exception.

a.   flight-capable wings in birds
b.   amniotic eggs in early reptiles
c.   Galapagos finches evolved to eat differently sized seeds
d.   sperm cells carried in pollen grains in seed plants
e.   silk glands in spiders

Answer: c
Difficulty: Difficult
Bloom's Taxonomy: Comprehension
Source: Section 22.5

Essay

79.   What are the factors that make the fossil record biased and incomplete?

Answer: Fossilization is more likely to occur in lowland areas where sediments are being deposited, and so species living in mountainous or xeric areas are less likely to fossilize. In addition, organisms with hard body parts are more likely to fossilize that soft-bodied organisms. Hence there is a far more complete fossil record of vertebrate evolution than of the evolution of flowers, for example. Finally, abundant, widespread organisms are more likely to have been fossilized than rare organisms, or those restricted to very specialized habitats.
Difficulty: Moderate
Bloom's Taxonomy: Analysis
Source: Section 22.1

23
SYSTEMATIC BIOLOGY: PHYLOGENY AND CLASSIFICATION

Multiple Choice

WHY IT MATTERS

1.   The classification of species often leads to

a.   understanding evolutionary relationships that directly and positively influence human health.
b.   new knowledge .
c.   a list of names with no significance to biological function.
d.   a and b
e.   a, b, and c

Answer: d
Difficulty: Easy
Bloom's Taxonomy: Knowledge, Comprehension

2.   What is a reason for the eradication of malaria in Europe owing a debt to systematics?

a.   Systematists developed insecticides to fight malaria.
b.   Systematists re-categorized mosquitoes into six species.
c.   Systematists discovered that mosquitoes breed in standing water.
d.   All of the above.
e.   None of the above.

Answer: b
Difficulty: Easy
Bloom's Taxonomy: Knowledge

3.   The eradication of malaria in Europe clearly illustrates the process of

a.   systematics.
b.   science in action.
c.   evolution in action.
d.   a and b
e.   a, b, and c

Answer: d
Difficulty: Moderate
Bloom's Taxonomy: Comprehension

23.1 SYSTEMATIC BIOLOGY: AN OVERVIEW

4.   A plant systematist aims to

a.   identify and name plant species.
b.   classify plant species within larger groups.
c.   understand the evolutionary history and relationships of plant species.
d.   all of the above
e.   none of the above

Answer: d
Difficulty: Easy
Bloom's Taxonomy: Knowledge

5.   A diagram that hypothesizes a set of organisms' evolutionary relationships to one another is a

a.   comparison tree.
b.   phylogenetic tree.
c.   classification tree.
d.   taxonomic tree.
e.   none of the above

Answer: b
Difficulty: Easy
Bloom's Taxonomy: Knowledge

6.   In what other areas of biological study is systematics important?

a.   ecology
b.   physiology
c.   anatomy
d.   genetics
e.   all of the above

Answer: e
Difficulty: Easy
Bloom's Taxonomy: Knowledge

23.2 THE LINNAEAN SYSTEM OF TAXONOMY

7.   Who was responsible for establishing the classification system that uses binomial nomenclature?

a.   Charles Darwin
b.   Louis Pasteur
c.   Carolus Linnaeus
d.   Jean-Baptiste Lamarck
e.   Gregor Mendel

Answer: c
Difficulty: Easy
Bloom's Taxonomy: Knowledge

8. According to the convention for properly writing Latin binomials, the scientific name for modern human beings should be written as

a. *Homo sapiens.*
b. Homo Sapiens.
c. Homo *sapiens.*
d. homo sapiens.
e. homo sapiens.

Answer: a
Difficulty: Easy
Bloom's Taxonomy: Knowledge

9. Given the plant species' name *Claytonia virginica*, what might you surmise about it?

a. It was named by somebody named Clayton who found it in Virginia.
b. It grows on clay soils in virgin forests.
c. It is made of a ton of clay and is virile and used for making gin.
d. It is related to *Ostrya virginiana.*
e. Absolutely nothing.

Answer: a
Difficulty: Easy
Bloom's Taxonomy: Comprehension

10. Currently, taxonomic domains include

a. Kingdom, Phylum, Class, and Species.
b. Plants, Animals, Fungi, Protists, and Monerans.
c. Archaea, Eubacteria, Eukarya.
d. Mammalia, Insecta, Aves, and Planta.
e. North America, South America, Eurasia, Africa, and Australia.

Answer: c
Difficulty: Easy
Bloom's Taxonomy: Knowledge

23.3 ORGANISMAL TRAITS AS SYSTEMATIC CHARACTERS

11. Which of the following kinds of traits was originally used for classifying organisms into taxonomic categories?

a. morphology and behavior
b. DNA and amino acid sequences
c. physiology and biochemical pathways
d. geography and ecology
e. all of the above

Answer: a
Difficulty: Easy
Bloom's Taxonomy: Knowledge

12. Morphological characteristics are widely used for taxonomic classifications because

a. they are the most likely indicators of evolutionary relationships.
b. they are the most likely to represent reproductive isolation among groups.
c. they are easy to assess in living, preserved, or fossilized specimens.
d. they are always the exact phenotypic expression of genotypes.
e. all of the above.

Answer: c
Difficulty: Moderate
Bloom's Taxonomy: Knowledge, Comprehension

13. Behavioral traits are useful in distinguishing among species because

a. behaviors are always genetically determined.
b. courtship behaviors and timing of mating may create prezygotic isolation.
c. behaviors are easy to assess in dead and fossilized specimens.
d. behaviors are independent of morphological characteristics.
e. all of the above

Answer: b
Difficulty: Easy
Bloom's Taxonomy: Knowledge, Comprehension

23.4 EVALUATING SYSTEMATIC CHARACTERS

14. Which of the following pairs of traits are homologous?

a. butterfly wings and bat wings
b. the surface areas of bird and bat wings
c. the bones of bird and bat wings
d. butterfly wings and bird wings
e. all of the above

Answer: c
Difficulty: Moderate
Bloom's Taxonomy: Comprehension

15. Fossils of plants found in deeper geologic layers do not show vascular tissues, while fossils in shallower geologic layers include some specimens with and some without vascular tissues. Thus, vascular tissue in plants is considered a _____ character.

a. ancestral
b. mosaic
c. primitive
d. derived
e. physiologically essential

Answer: ***b***
Difficulty: Moderate
Bloom's Taxonomy: Comprehension

16. Of the following choices, the best outgroup to use to resolve a phylogeny for great apes would be

a. lemurs.
b. butterflies.
c. angiosperms.
d. slime molds.
e. Archaea.

Answer: ***a***
Difficulty: Moderate
Bloom's Taxonomy: Comprehension, Application

23.5 PHYLOGENETIC INFERENCE AND CLASSIFICATION

17. The Class Reptilia is considered to be _____ because the group contains species that share a common ancestor, but does not include all of that ancestor's descendants.

a. monophyletic
b. polyphyletic
c. paraphyletic
d. an outgroup
e. a mosaic

Answer: ***c***
Difficulty: Easy
Bloom's Taxonomy: Knowledge, Comprehension

18. The most parsimonious explanation for the existence of beaks in birds is that

a. they arose once in a common ancestor of beaked groups.
b. they arose multiple times in different bird lineages.
c. they are a product of convergent evolution among birds.
d. they are not a genetically determined trait, and thus independent characters.
e. all of the above.

Answer: ***a***
Difficulty: Moderate
Bloom's Taxonomy: Comprehension, Application

19. On a cladogram, the tips of the branches represent

a. common ancestors.
b. groups of organisms that share a common ancestor.
c. distinguishing characters.
d. common characters.
e. extinct species.

Answer: ***b***
Difficulty: Moderate
Bloom's Taxonomy: Knowledge, Comprehension

23.6 MOLECULAR PHYLOGENETICS

20. Molecular phylogenies are constructed with data from sequences of

a. DNA and RNA.
b. proteins.
c. amino acids.
d. a and b only
e. all of the above

Answer: ***d***
Difficulty: Easy
Bloom's Taxonomy: Knowledge

21. Character changes in molecular sequences are generated by

a. insertions.
b. deletions.
c. substitutions.
d. base pairing.
e. a, b, and c

Answer: ***e***
Difficulty: Easy
Bloom's Taxonomy: Knowledge

22. Polymerase chain reaction (PCR) has advanced an understanding of systematics because it allows

a. for analysis of morphological traits.
b. for analysis of small amounts of DNA from dried museum specimens and some fossils.
c. for the reconstruction of entire individuals from a small amount of DNA.
d. for the reconstruction of entire individuals from a small amount of RNA.
e. for the analysis of an entire genome from a small fragment of the genome.

Answer: ***b***
Difficulty: Easy
Bloom's Taxonomy: Knowledge

23. Which of the following is the most important contribution of molecular phylogenetics?

a. the discovery of rRNA
b. the reanalysis of the morphology of *Amborella trichopoda*
c. the reorganization of living systems from five to six Kingdoms
d. the reorganization of living systems into three domains
e. the realization that morphological characters are not representative of phylogenetic relationships

Answer: d
Difficulty: Moderate
Bloom's Taxonomy: Knowledge, Comprehension

INSIGHTS FROM THE MOLECULAR REVOLUTION: WHALES WITH COW COUSINS?

24. SINEs in molecular phylogenetics are

a. Substituted Integrated Elements.
b. Short Interspersed Elements.
c. Single Insertion Elements.
d. Smart Innovative Entrepreneurs.
e. none of the above

Answer: b
Difficulty: Moderate
Bloom's Taxonomy: Knowledge

25. A recent study of SINEs in whales and other mammals generated results that show that whales are derived from a common ancestor with

a. pigs and camels.
b. dogs and cats.
c. ruminants and hippos.
d. sharks and rays.
e. none of the above

Answer: c
Difficulty: Moderate
Bloom's Taxonomy: Knowledge

UNANSWERED QUESTIONS

26. The systematics of today is very different than the systematics learned a generation ago. As a result we now know that

a. the biological species concept is not valid
b. the number of existing species is 1.56 million
c. the Linnaean system of classification is wrong most of the time
d. morphological traits are not reliable for determining species relationships
e. none of the above

Answer: e
Difficulty: Moderate
Bloom's Taxonomy: Comprehension

Integrative Multiple-Choice

27. Birds and mammals are considered polyphyletic when considering that

a. birds have wings and mammals do not.
b. birds have feathers and mammals do not.
c. birds and mammals do not share the same common reptilian ancestor.
d. birds and mammals cannot interbreed.
e. molecular clock data indicates birds and mammals diverged at the same time.

Answer: c
Difficulty: Difficult
Bloom's Taxonomy: Comprehension
Source: Sections 23.4, 23.5, 23.6

28. Which of the following is/are inaccurate about the Linnaean system of classification?

a. It does not rely on molecular data.
b. It does not contribute to an understanding of evolutionary relationships.
c. It does not help solve problems that are related to human health and welfare.
d. Both a and b are inaccurate.
e. Both b and c are inaccurate.

Answer: e
Difficulty: Moderate
Bloom's Taxonomy: Comprehension
Source: Sections Why It Matters, 23.2, 23.6

Matching

Match each of the following term with its correct definition.

29. _____ Phylogeny

30. _____ Taxonomy

31. _____ Domain

32. _____ Homology

33. _____ Homoplasy

34. _____ Mosaic Evolution

35. _____ Outgroup

36. _____ Monophyletic taxa

37. _____ Phylocode

38. _____ Molecular clock

39. _____ PCR

A. Cladistic system of naming groups based on evolutionary relationships

B. Distantly related group used as a comparison in a phylogeny

C. Similarities in characters that arose from shared ancestry

D. Technique to amplify DNA for use in sequencing

E. Group with one ancestral species and all of its descendants

F. Mutation rate that indexes time of divergence among groups

G. Highest level of taxonomic hierarchy

H. Variation in rates of evolution of characters within a group

I. Identifying, naming, and classifying species

J. Evolutionary history of a group of organisms

K. Phenotypic similarities that arose in different lineages

Answers: 29. J   30. I   31. G   32. C   33. K
34. H   35. B   36. E   37. A
38. F   39. D

Difficulty: Moderate
Bloom's Taxonomy: Knowledge
Source: Sections 23.1, 23.2, 23.3, 23.4, 23.5, 23.6

Sequence

Using the following list of taxonomic groups, place them in order from least inclusive to most inclusive, labeling the taxonomic hierarchy for each group.

Canis canis, Carnivora, Animalia, Mammalia, Chordata, *Canis*, Canidae, Eukarya

40. _____ (Least Inclusive)
41. _____
42. _____
43. _____
44. _____
45. _____
46. _____
47. _____ (Most Inclusive)

Answer: 40. Canis canis   Species
41. Canis   Genus
42. Canidae   Family
43. Carnivora   Order
44. Mammalia   Class
45. Chordata   Phylum
46. Animalia   Kingdom
47. Eukarya   Domain

Difficulty: Moderate
Bloom's Taxonomy: Application
Source: Section 23.2

True/False Questions

If the statement is true, write a "T" in the blank. If the statement is false, make it correct by changing the underlined word(s) and writing the correct word(s) in the answer blank.

48. _____ A plant systematist studies the <u>physiological systems</u> of plants.

49. _____ Common names for a species may vary widely among different groups of people who use those names, and thus are not reliable for <u>scientific</u> use.

50. _____ Carolus Linnaeus developed a classification system for organisms based on their <u>DNA sequences</u>.

51. _____ Homoplasies are the result of <u>convergent evolution</u>.

52. _____ <u>Homoplasies</u> are useful characters for phylogenetic analyses.

53. _____ All evolutionarily meaningful characters evolve at <u>the same rate</u>.

54. _____ Systematists strive to organize taxa that represent <u>monophyletic</u> groups.

55. _____ Cladograms allow for <u>paraphyletic and polyphyletic</u> groupings

56. _____ <u>Traditional evolutionary systematics</u> used Linnaean systems of classification.

57. _____ <u>Polymerase Chain Reaction</u> is a technique to produce multiple copies of a sequence of DNA.

58. _____ The traditional Kingdom <u>Monera</u> is currently divided between two Domains.

Answer: 48. F, evolutionary relationships   49: T
50. F, morphology   51. T   52. F, homologous
characters 53. F, different rates 54. T   55. F,
only monophyletic   56. T   57. T   58. T

Difficulty: Moderate
Bloom's Taxonomy: Knowledge, Comprehension
Source: Sections 23.1, 23.2, 23.3, 23.4, 23.5, 23.6

Alignment Question

For the following DNA sequences:
Segment A: A A T G C G G T A C A T G C A T T G
Segment B: A A T G G C G G T A C A T G C A T T G
Segment C: A A T G C G G A C A T G C T T G
Segment D: A A T G G G G T A C A T C C A T T G

Align the segments.

59. Segment A: _____

60. Segment B: _____

61. Segment C: _____

62. Segment D: _____

Answer: (must be lined up in spaces above) 59.
Segment A: A A T G C G G T A C A T G C A T
T G   60. Segment B: A A T G G C G G T A C
A T G C A T T G   61. Segment C: A A T
G C G G A C A T G C T T G   62.
Segment D: A A T G   G G G T A C A T C C A T
T G

Assuming segment A is the ancestral group, indicate where the following mutations have occurred in the derived groups.

63. insertions _____
64. deletions _____
65. substitutions _____

Answer: 63. Segment B: G insertion between
positions 4 and 5 64. Segment C: Deletion of T at
position 8 and deletion of A at position 15
65. Segment D: Substitution of G for C at position
5, substitution of C for G at position 13
Difficulty: Moderate
Bloom's Taxonomy: Comprehension, Application
Source: 23.6

Cladogram questions

Refer to the figure below for questions 66–76.

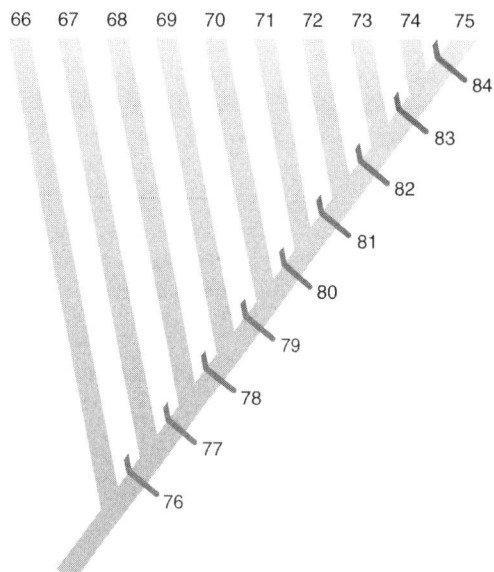

Using the information in the following table, fill in the cladogram with the appropriate groups (questions 66–75) and distinguishing characteristics (76–84).

	Vertebrae	Jaws	Swim bladder or lungs	Paired limbs	Extraembryonic membranes	Mammary glands	Dry, scaly skin	Two openings at back of skull	One opening in front of eye	Feathers
Lancelets	–	–	–	–	–	–	–	–	–	–
Amphibians	+	+	+	+	–	–	–	–	–	–
Birds	+	+	+	+	+	–	+	+	+	+
Bony fishes	+	+	+	–	–	–	–	–	–	–
Crocodilians	+	+	+	+	+	–	+	+	+	–
Lampreys	+	–	+	+	–	–	–	–	–	–
Lizards	+	+	+	+	+	–	+	+	–	–
Mammals	+	+	+	+	+	+	–	–	–	–
Sharks	+	+	–	–	–	–	–	–	–	–
Turtles	+	+	+	+	+	–	+	–	–	–

66. _____
67. _____
68. _____
69. _____
70. _____
71. _____
72. _____
73. _____
74. _____
75. _____
76. _____
77. _____
78. _____
79. _____
80. _____

81. _____
82. _____
83. _____
84. _____
85. _____
86. _____

Answers: 66. Lancelets 67. Lampreys 68.
Sharks 69. Bony fishes 70. Amphibians
71. Mammals 72. Turtles 73.
Lizards 74. Crocodilians 75. Birds
76. Vertebrae 77. Jaws 78.
Swim bladder or lungs 79. Paired limbs 80.
Extraembryonic membranes 81. Dry, scaly
skin 82. Two openings at back of skull 83. One
opening in front of eye 84. Feathers
Difficulty: Moderate
Bloom's Taxonomy: Comprehension, Application
Source: 23.5

Essay

87. Discuss the advantages and disadvantages of using molecular sequences for phylogenetic analysis. Describe the methods a molecular systematist would use to resolve the hypotheses generated by multiple possible phylogenies from molecular data. Then, explain how systematists might resolve a strongly supported molecular phylogeny with a different traditionally established morphological phylogeny.

Answer: Advantages include that molecular data provide independence of analysis for numerous characters, and are appropriate for distantly or closely related group analyses. Disadvantages include that lack of variation in characters may occur due to developmental or environmental factors; that DNA or RNA data provide a limited set of characters (just four) for evaluation; and that base pair changes can arise multiple times. To resolve the multiple phylogenetic trees that could arise from molecular data, systematists would use principles of parsimony and statistical tools such as maximum likelihood to determine the most strongly supported trees. To resolve discrepancies in hypothetical trees from molecular versus morphological data, a larger set of molecular sequences or of morphological characters would have to be compared. Also, a more comprehensive grouping of taxa would help provide resolution. The ongoing publishing and presenting of trees in primary literature and at meetings provides venues for systematists to compare and resolve trees through discussion, argument, and inclusion of novel information and interpretations.
Difficulty: Difficult
Bloom's Taxonomy: Synthesis
Source: 23.6

88. A conservation manager is planning to reintroduce collared lizards into an isolated glade in Missouri, where she assumes they have gone locally extinct. Several other isolated glades in the region have populations of collared lizards from which she can draw. Based on your understandings of phylogenetics, provide arguments for why she should draw individuals from a single extant glade population versus from multiple extant glade populations to repopulate the empty glade. What tools might she use to make this decision?

Answer: First, the manager needs to determine how closely related the other populations of collared lizards are to each other. If they have been isolated long enough to diverge into new taxa (subspecies or even new species) then mixing individuals from different populations would not be productive. Morphological observations (How different do they look?), behavioral observations (Do they still have the same feeding, courtship, and reproductive patterns?), and genetic assessments (How much have their genetic make-ups—either protein or DNA sequences—changed since isolation?) will provide information on the degree to which each population has diverged from the others. Finally, a test of putting a few individual males and females together in a controlled setting to see if they will successfully mate and produce viable and fertile offspring will confirm if these groups can be introduced together to a new location.
Difficulty: Difficult
Bloom's Taxonomy: Application and Synthesis
Source: Sections Why it Matters, 23.3, 23.6

89. Explain how the Linnaean system of classification, although developed without any concept of evolutionary theory, provided a reasonable, if not perfect, framework for later application of evolutionary systematics. What limitations of the Linnaean system have spurred cladists to argue for its abandonment and replacement by modern cladistic approaches?

Answer: The Linnaean system was based on morphological descriptions, which provide a reasonable basis for sketching out a logical hierarchy for grouping organisms. The hierarchical system is still applied to cladistic approaches, in that more ancestral forms link groups deeper in a cladogram, and thus represent a more inclusive (or higher on the hierarchical framework) grouping or clade. While Linnaeus did not envision a historical sequence of ancestral and descendant relationships, as did Darwin, the increasingly inclusive groupings of genus, family, order, class, phylum, and kingdom can capture the essence of shallower and deeper evolutionary divergences. The limitations of this approach are that the morphological data can provide numerous ways of grouping taxa, which may or may not accurately reflect the historical divergences, as can now be understood more fully with genetic data and with more complete fossil data (for example, Class Reptilia). The Linnaean classifications contained, in retrospect, many polyphyletic and paraphyletic taxa, which Cladists do not allow.

Difficulty: Difficult
Bloom's Taxonomy: Synthesis and Evaluation
Source: Sections 23.2, Unanswered Questions

24
THE ORIGIN OF LIFE

Multiple-Choice

WHY IT MATTERS

1. Modern cells are thought to have arisen from

a. protocells.
b. pre-cells.
c. gas and dust clouds.
d. prokaryotic cells that predated protocells.
e. prokaryotic cells that predated pre-cells.

Answer: a
Difficulty: Easy
Bloom's Taxonomy: Knowledge

2. Assuming the deposits found in Australia are actually fossils of prokaryotes, life on Earth may have originated

a. 14 billion years ago.
b. 4.6 billion years ago.
c. 3.5 billion years ago.
d. 2.1 billion years ago.
e. 1.2 billion years ago.

Answer: c
Difficulty: Easy
Bloom's Taxonomy: Knowledge

24.1 THE FORMATION OF MOLECULES NECESSARY FOR LIFE

3. When scientists design experiments to model how life may have arisen, they need to make some assumptions. Which of the following is an assumption that is commonly made by these researchers?

a. The chemical reactions of living things are only possible in living systems.
b. Living organisms are composed of elements commonly found in the universe and on Earth.
c. Conditions on Earth have not changed over time.
d. Life arose on Earth from non-living matter.
e. The development of living cells from nonliving matter was *very* rapid.

Answer: d
Difficulty: Moderate
Bloom's Taxonomy: Knowledge

4. Where did Earth's atmosphere originate?

a. Earth's gravity trapped gasses from passing comets.
b. Gases were released from the Earth itself as it cooled.
c. The atmosphere was derived from dust particles from the original gas cloud around the planet.
d. choices a, b and c
e. choices b and c only

Answer: e
Difficulty: Moderate
Bloom's Taxonomy: Knowledge

5. Why is the distance of the Earth from the sun so crucial for life as we know it?

a. The distance provides the optimal temperature for water to occur in a liquid form.
b. The distance allows the Earth's orbit to have a year of a reasonable length.
c. The distance allows for multiple seasons in temperate climates, which was crucial for development of different chemicals used to build the first protocells.
d. The distance provides ample opportunity for water to freeze into ice.
e. The distance ensures that ample water is available for all of the organisms on Earth.

Answer: a
Difficulty: Easy
Bloom's Taxonomy: Knowledge

6. Why is the absence of atmospheric oxygen (O_2) so important to the Oparin-Haldane hypothesis?

a. Oxygen is corrosive and thus would have degraded many of the metallic compounds on the surface, releasing toxic gasses into the new atmosphere.
b. Oxygen is more reduced than hydrogen gas, methane, or ammonia.
c. Oxygen would have supported microorganisms that help chemical decay proceed.
d. Oxygen is able to reverse reactions by removing electrons and hydrogen from organic molecules, thus destroying the first organic compounds as quickly as they developed.
e. Oxygen would have caused multiple explosions, converting chemical energy into heat energy and rendering that energy unavailable for driving the synthesis of organic molecules.

Answer: d
Difficulty: Easy
Bloom's Taxonomy: Knowledge

7. Which of the following statements about the Miller-Urey experiments highlights a *limitation* of its support for the Oparin-Haldane hypothesis?

a. The energy supplied in the Miller-Urey experiment was constant, but was only available intermittently from electrical storms on early Earth.
b. The building blocks of cells produced in the Miller-Urey experiments didn't include phospholipids.
c. The repetition of the Miller-Urey experiment with the inclusion of oxygen virtually eliminated the presence of organic molecules, meaning that early Earth's atmosphere must have contained little to no oxygen gas.
d. Water had to be present in both liquid and gaseous forms, which was probably not the case in Earth's early atmosphere.
e. Hydrogen cyanide (HCN) and formaldehyde (CH_2O) had to be provided in order to generate organic molecules in the Miller-Urey experiment. These compounds probably didn't exist in Earth's early atmosphere.

Answer: a
Difficulty: Difficult
Bloom's Taxonomy: Comprehension

8. If life on Earth didn't develop via organic compounds produced in a reducing atmosphere, where else could it have first occurred?

a. Life could have arisen from the organic compounds formed inside the hydrothermal vents far below the ocean floor.
b. Life could have developed on another planet and was then delivered by meteorites.
c. Life on Earth could have developed from organic compounds delivered to Earth by meteorites.
d. Life on Earth may have arisen inside land-based volcanoes.
e. Life on Earth could have developed in the reduced atmosphere located right above volcanoes.

Answer: c
Difficulty: Moderate
Bloom's Taxonomy: Knowledge

24.2 THE ORIGIN OF CELLS

9. After organic molecules have been formed in abundance, what is the *next* step that must occur in the progression toward a living organism?

a. evaporation of the water the organic compounds are dissolved in
b. dehydration reactions linking macromolecule monomers into larger molecules
c. increasing availability of organic compounds
d. the formation of a semi-permeable membrane
e. aggregation of the different organic compounds so all of the pre-cell building blocks are in the same localized region

Answer: e
Difficulty: Moderate
Bloom's Taxonomy: Knowledge

10. How might absorption of organic compounds into clays assist in their aggregation and organization into membrane-bound protocells?

a. Clays contain many minerals in extremely thin layers and readily repels ions, forcing multiple chemical interactions.
b. Clays are highly structured, contain water, and store potential energy that could drive the chemical reactions forward.
c. Clays lack water, forcing the concentration of the organic molecules and thus driving multiple interactions.
d. Clays are extremely resistant to temperature changes and thus would insulate the organic compounds from temperature changes in the atmosphere.
e. None of the above statements accurately explain clay's possible role in the formation of protocells.

Answer: b
Difficulty: Moderate
Bloom's Taxonomy: Knowledge, Comprehension

11. What happens when phospholipids are placed in an aqueous solution?

a. They disperse throughout the solution and form polar bonds with other organic molecules present in the solution.
b. They move to the water's surface and form a single-layered membrane between the water and the atmosphere.
c. They spontaneously form double membrane vesicles.
d. They spontaneously form bilayers.
e. The hydrophobic tales force the phospholipids to form "beads" at the surface, much like you see when vegetable oil is added to an aqueous solution.

Answer: d
Difficulty: Moderate
Bloom's Taxonomy: Knowledge

12. Which part of the metabolic pathways in cells probably originated first?

a. use of ATP as a coupling agent
b. single step redox reactions
c. cellular respiration
d. glycolysis
e. electron transfer systems

Answer: b
Difficulty: Easy
Bloom's Taxonomy: Knowledge

13. Which of the following was probably the first information storage molecule?

a. DNA
b. RNA
c. protein
d. carbohydrates
e. lipids

Answer: b
Difficulty: Easy
Bloom's Taxonomy: Knowledge

14. Why do some scientists believe that RNA was the first information storage molecule?

a. because some viruses use RNA as their information storage molecule
b. because RNA can be a catalyst as well as an information storage molecule
c. because amino acids cannot form peptide bonds without ribosomes to catalyze their formation
d. because RNA is a less stable molecule than DNA
e. because proteins are made of far more amino acid monomers (over 20) than RNA (4)

Answer: b
Difficulty: Easy
Bloom's Taxonomy: Knowledge

15. If RNA was the first information storage molecule, what macromolecule would have been the "next" molecule produced in the progression toward living cells?

a. DNA
b. lipids
c. carbohydrates
d. sugars
e. proteins

Answer: e
Difficulty: Easy
Bloom's Taxonomy: Knowledge

16. What evidence do we have for the development of photosynthetic cells that used water as an electron source at least 3 billion years ago?

a. fossils of the first eukaryotic cells
b. the discovery of the first plant fossils
c. the discovery of stromatolites
d. fossils of the first animals
e. fossils of the earliest chloroplasts

Answer: c
Difficulty: Easy
Bloom's Taxonomy: Knowledge

INSIGHTS FROM THE MOLECULAR REVOLUTION: REPLICATING THE RNA WORLD

17. The experiment showing that a ribozyme could generate sequences complementary to a template strand

a. proved that the first information storage molecules must have been comprised of RNA.
b. proved that RNA predates DNA as an information molecule.
c. proved for the first time that RNA can have catalytic activity.
d. provides evidence that RNA could have catalyzed RNA replication in the development of the first true cells.
e. refutes the theory that RNA was the likely information storage molecule used by protocells and the first true cells.

Answer: d
Difficulty: Easy
Bloom's Taxonomy: Knowledge

18. In the experiment with a ribozyme that catalyzed the synthesis of a complementary RNA strand,

a. the complementary strands had no errors despite 100 of them being sequenced.
b. all of the 100 complementary strands sequenced had multiple errors.
c. most of the 100 complementary strands sequenced were correct, but three of them contained a single error each.
d. approximately one error per strand was found in 11 of 100 complementary strands that were sequenced.
e. the error rate could not be determined.

Answer: d
Difficulty: Moderate
Bloom's Taxonomy: Knowledge

24.3 THE ORIGINS OF EUKARYOTIC CELLS

19. Who developed the modern endosymbiotic theory?

a. Charles Darwin
b. Lynn Margulis
c. Richard Dickerson
d. Wendy Johnson
e. Stanley Miller

Answer: b
Difficulty: Moderate
Bloom's Taxonomy: Knowledge

20. The hypothesis that mitochondria and chloroplasts originated from the phagocytic activity of prokaryotes that was not followed by digestion is called the

a. theory of natural selection.
b. endosymbiotic theory.
c. ribozymes-first theory.
d. Oparin-Haldane hypothesis.
e. stromatolite theory.

Answer: b
Difficulty: Easy
Bloom's Taxonomy: Knowledge

21. According to the endosymbiont hypothesis,

a. most of the genes from the endocytosized bacteria moved to the nucleus.
b. the host anaerobe became dependent on the endocytosized bacteria.
c. the endocytosized bacteria became dependent on the host cell.
d. the endocytosized bacteria were all photosynthetic.
e. a, b, and c only.

Answer: e
Difficulty: Easy
Bloom's Taxonomy: Knowledge

22. Even though the three domain taxonomic system used in this book is widely accepted, there is a key relationship between the domains that is unclear. That relationship is

a. between Archaea and Eukarya.
b. between Archaea and Bacteria.
c. between Bacteria and Eukarya.
d. between Archaea, Eukarya, and Bacteria.
e. whether the three domains all have a common ancestor.

Answer: d
Difficulty: Moderate
Bloom's Taxonomy: Knowledge

23. Which feature of Archaea is shared with Bacteria but not with Eukarya?

a. the sequestration of DNA into a nucleus
b. the common occurrence of introns
c. the chemical structure of the cell walls
d. the plasma membrane structure
e. a genome comprised of a single, circular molecule of DNA

Answer: e
Difficulty: Easy
Bloom's Taxonomy: Knowledge

24. The development of multicellular organisms likely originated

a. from colonies of unicellular organisms of the same species.
b. from endocytosis of other cells followed by mitosis.
c. from cooperation between different species of cells.
d. from colonies of numerous species of cells.

Answer: a
Difficulty: Easy
Bloom's Taxonomy: Knowledge

UNANSWERED QUESTIONS

25. Why is the protein-first hypothesis not accepted by most scientists?

a. Proteins cannot be catalysts.
b. There is no known mechanism of protein self-replication.
c. Of all of the macromolecules, only DNA can self-replicate.
d. There is no evidence that proteins can store information.
e. Proteins are too complex to have been the first "polymers of life."

Answer: b
Difficulty: Moderate
Bloom's Taxonomy: Knowledge

26. What is the composition of the "minimal RNA cell" scientists are trying to create in the laboratory?

a. a phospholipid membrane surrounding two ribozymes
b. a phospholipid membrane encasing a ribosome, a genome, and some amino acids
c. a protein shell surrounding nucleotides and two ribozymes
d. a single RNA sequence surrounded by a single phospholipid "sphere"
e. none of these

Answer: a
Difficulty: Easy
Bloom's Taxonomy: Knowledge

Integrative Multiple-Choice

27.   Which of the following does *not* describe modern, living cells?

a.   a semi-permeable membrane keeping the internal environment of the cell separate from the external environment
b.   nucleic acids linked in a sequence to form at least one large molecule
c.   a requirement for the nucleic acids to be contained behind a nuclear membrane
d.   multiple pathways for converting energy from one form into another
e.   a method of converting the nucleic acid sequences into proteins

Answer: c
Difficulty: Easy
Bloom's Taxonomy: Knowledge
Source: Fig. 1.1, 5.2, 24.1

28.   What do some scientists believe lead to the increase in the level of oxygen in the atmosphere?

a.   the first photosynthetic reactions
b.   photosynthetic cells that used water and not H_2S as a source of electrons
c.   The development of cellular respiration pathways in cells
d.   the use of glycolysis in cells
e.   photosynthetic pathways that stripped oxygen off of carbon dioxide and retained only the carbon in subsequent steps

Answer: b
Difficulty: Easy
Bloom's Taxonomy: Knowledge
Source: 9.2, 24.2

Short Answer

29.   Explain the endosymbiotic hypothesis.

Answer: According to the endosymbiotic hypothesis, chloroplasts and mitochondria in modern cells are derived from free-living bacteria that were incorporated into other cells as symbiotes. Over time, the relationship became obligatory and permanent as some of the DNA from the bacteria was transferred to the host cells' genomes.
Difficulty: Moderate
Bloom's Taxonomy: Knowledge
Source: Section 24.3

True/False

If the statement is true, write a "T" in the blank. If the statement is false, make it correct by changing the underlined word(s) and writing the correct word(s) in the answer blank.

30.   _____   Scientists believe life arose spontaneously from nonliving matter.

31.   _____   Scientists agree that life originated on Earth.

32.   _____   Proteins may have been the first information-storage molecules because they can self-replicate as well as catalyze reactions.

33.   _____   Clays do not facilitate the formation of phospholipid bilayers.

34.   _____   Phospholipid bilayers created in a laboratory can incorporate proteins into their surfaces.

35.   _____   Stromatolites are fossils of ancient, photosynthetic cyanobacteria.

36.   _____   Most scientists believe that prokaryotes evolved from eukaryotes.

37.   _____   Mitochondria probably developed before the atmosphere was oxygen rich.

38.   _____   The development of the Golgi, endoplasmic reticulum, and nucleus are part of the endosymbiotic theory.

39.   _____   Archaea genomes contain introns.

40.   _____   The first truly living cells must have been prokaryotic in structure.

41.   _____   Multi-cellularity is believed to have evolved multiple times on Earth.

Answers: 30. F, transitioned gradually   31. F, possibly arose on a different planet and traveled to Earth via a meteor. 32. F, RNA 33. T
34. T 35. T 36. F, eukaryotes, prokaryotes
37. F, after 38. F, chloroplast, mitochondria
39. T 40. T 41. T

Difficulty: Difficult
Bloom's Taxonomy: Knowledge
Source: Section 8.10

Select the Exception

42. Evidence for the endosymbiotic theory does *not* include which of the following?

a. DNA in prokaryotes, mitochondria, and chloroplasts is arranged in a single, circular genome.
b. Endocytosized cells have been directly observed to survive in the cytoplasm of the host cell.
c. Endocytosized organelles have been directly observed to survive in the cytoplasm of the host cell.
d. Fossils of prokaryotic cells in the process of endocytosizing cyanobacteria have been discovered.
e. There are some rRNA sequences that are similar between prokaryotes, mitochondria, and chloroplasts.

Answer: d
Difficulty: Easy
Bloom's Taxonomy: Knowledge

43. Which of the following was probably *not* a natural energy source that drove the chemical reactions leading to the first organic molecules?

a. sunlight
b. electrical storms (lightening)
c. photosynthesis
d. radioactive decay
e. heat energy released from Earth's core

Answer: c
Difficulty: Easy
Bloom's Taxonomy: Knowledge

44. According to the Oparin-Haldane hypothesis, which of the following was probably *not* a major component of Earth's early atmosphere?

a. O_2 (oxygen)
b. H_2 (hydrogen)
c. CH_4 (methane)
d. H_2O (in vapor form)
e. NH_3 (ammonia)

Answer: a
Difficulty: Easy
Bloom's Taxonomy: Knowledge

Essay

45. Describe the supporting evidence for the endosymbiotic theory.

Answer: There are two lines of evidence that directly support the endosymbiotic theory. Firstly, the DNA in both mitochondria and chloroplasts is arranged in a circular molecule, just as is observed in prokaryotic genomes. Secondly, phagocytized cells and organelles have been directly observed surviving inside the phagocytic cell. In addition to the direct evidence, the development of a symbiotic relationship would have benefited both participants. Over time, it would make sense for some of the DNA to be transferred to the host cell's genome so that is was under the control of the host cell. This process would have led to the development of the chloroplasts and mitochondria present in cells today.
Difficulty: Moderate
Bloom's Taxonomy: Comprehension
Source: Section 24.3

46. Define the Oparin-Haldane hypothesis and explain how it has been tested in a laboratory

Answer: The Oparin-Haldane hypothesis proposes that conditions on early Earth allowed the formation of organic compounds that were essential building blocks in the first cells. In a closed vessel, hydrogen, methane, ammonia, and water vapor were exposed to charged electrodes. Of the course of a week, urea, amino acids, lactic acid, formic acid, and acetic acids had all formed in the container. If the original atmosphere had the same composition as what was in the container, this experiment showed one way in which organic molecules could have formed.
Difficulty: Moderate
Bloom's Taxonomy: Knowledge, Comprehension
Source: Section 24.1

25
PROKARYOTES AND VIRUSES

Multiple-Choice

WHY IT MATTERS

1.   Which of the following statements about prokaryotes is *false*?

a.   They lack a true nucleus.
b.   They are generally smaller than eukaryotes.
c.   They are found only in certain niches.
d.   They have a wide range of metabolic activities.
e.   They may have a bigger biomass than plants.

Answer: c
Difficulty: Moderate
Bloom's Taxonomy: Knowledge

2.   The three domains of life are

a.   eukaryotes, prokaryotes, and viruses.
b.   animals, plants, and microorganisms.
c.   Prokaryota, Eukaryota, and Protoctista.
d.   Archaebacteria, Eubacteria, and Eukaryota.
e.   Archaea, Bacteria, and Eukarya.

Answer: e
Difficulty: Easy
Bloom's Taxonomy: Knowledge

3.   Biosphere 2 was an attempt by scientists to build a completely closed ecosystem in Arizona. It failed because of the

a.   decreased level of oxygen through respiration by soil microorganisms.
b.   contamination of food by soil microorganisms.
c.   disease outbreak caused by soil microorganisms.
d.   increased level of carbon dioxide produced by photosynthetic microorganisms.
e.   inability of soil microorganisms to recycle nitrogen.

Answer: a
Difficulty: Moderate
Bloom's Taxonomy: Comprehension

25.1 PROKARYOTIC STRUCTURE AND FUNCTION

4.   Which of the following does not contribute to prokaryotic genetic variability?

a.   mutation
b.   binary fission
c.   gene transfer by conjugation
d.   gene transfer by transduction
e.   gene transfer by transformation

Answer: b
Difficulty: Moderate
Bloom's Taxonomy: Knowledge

5.   DNA in prokaryotes is found in

a.   the nucleoid.
b.   the nucleolus.
c.   plasmids.
d.   the nucleoid and plasmids.
e.   the nucleolus and plasmids.

Answer: d
Difficulty: Moderate
Bloom's Taxonomy: Knowledge

6.   The genome of prokaryotes consists of

a.   a single linear DNA molecule.
b.   a single circular DNA molecule.
c.   many linear DNA molecules.
d.   many circular DNA molecules.
e.   a single DNA molecule that may be circular or linear, depending on the species.

Answer: e
Difficulty: Moderate
Bloom's Taxonomy: Comprehension

7.   Prokaryotes that are curved and commalike are called

a.   cocci.
b.   bacilli.
c.   vibrios.
d.   spirilla.
e.   sarcina.

Answer: c
Difficulty: Easy
Bloom's Taxonomy: Knowledge

8. The major structural component of bacterial cell walls is

a. cellulose.
b. chitin.
c. proteoglycan.
d. peptidoglycan.
e. arabinogalactan.

Answer: d
Difficulty: Easy
Bloom's Taxonomy: Knowledge

9. The correct sequence of reagents in the Gram stain technique is

a. iodine, alcohol, safranin, crystal violet.
b. safranin, crystal violet, alcohol, iodine.
c. crystal violet, iodine, alcohol, safranin.
d. crystal violet, alcohol, iodine, safranin.
e. alcohol, iodine, crystal violet, safranin.

Answer: c
Difficulty: Moderate
Bloom's Taxonomy: Knowledge

10. After performing a Gram stain of a mixed culture of Gram-positive and Gram-negative cells, you realize that you omitted the iodine step. What would you expect to see if you observed the slide under the microscope?

a. pink Gram-positive and purple Gram-negative cells
b. purple Gram-positive and pink Gram-negative cells
c. pink Gram-positive and pink Gram-negative cells
d. colorless Gram-positive and pink Gram-negative cells
e. colorless Gram-positive and colorless Gram-negative cells

Answer: c
Difficulty: Difficult
Bloom's Taxonomy: Comprehension, Application

11. A Gram-positive bacterium is characterized by the

a. presence of a thin cell wall.
b. presence of a thick cell wall.
c. absence of a cell wall.
d. presence of an outer membrane.
e. absence of ribosomes.

Answer: b
Difficulty: Easy
Bloom's Taxonomy: Knowledge

12. The lipopolysaccharides are associated with the

a. outer membrane of Gram-positive bacteria.
b. outer membrane of Gram-negative bacteria.
c. plasma membrane of Gram-positive bacteria.
d. plasma membrane of Gram-negative bacteria.
e. plasma membrane of both Gram-positive and Gram-negative bacteria.

Answer: b
Difficulty: Easy
Bloom's Taxonomy: Knowledge

13. Capsules may protect the bacterium against all of the following *except*

a. desiccation.
b. antibiotics.
c. antibodies.
d. osmotic lysis.
e. viruses.

Answer: d
Difficulty: Moderate
Bloom's Taxonomy: Knowledge

14. Ribosomes are composed of

a. proteins only.
b. proteins and DNA.
c. proteins and RNA.
d. glycolipids and DNA.
e. glycolipids and RNA.

Answer: c
Difficulty: Easy
Bloom's Taxonomy: Knowledge

15. Which of the following does *not* describe bacterial flagella?

a. smaller and simpler than eukaryotic flagella
b. consist of a helical fiber of protein
c. found in all motile bacteria
d. powered by a gradient of hydrogen ions
e. lacks microtubules.

Answer: c
Difficulty: Moderate
Bloom's Taxonomy: Knowledge

16.   Arrange the following layers from most external to most internal.

    1 = Cell membrane
    2 = Capsule
    3 = Cell wall
a.   1, 2, 3
b.   2, 1, 3
c.   2, 3, 1
d.   3, 1, 2
e.   3, 2, 1

Answer: c
Difficulty: Easy
Bloom's Taxonomy: Knowledge

Use the following key for questions 17–19.

a.   autotroph
b.   auxotroph
c.   heterotroph
d.   chemotroph
e.   phototroph

17.   An organism that obtains its energy by oxidizing organic or inorganic substances.

Answer: d
Difficulty: Easy
Bloom's Taxonomy: Knowledge

18.   An organism that obtains its carbon from CO_2.

Answer: a
Difficulty: Easy
Bloom's Taxonomy: Knowledge

19.   An organism that obtains its energy from sunlight.

Answer: e
Difficulty: Easy
Bloom's Taxonomy: Knowledge

Use the following key for questions 20–23.

a.   chemoautotroph
b.   chemoheterotroph
c.   photoautotroph
d.   photoheterotroph
e.   chemoautotroph and photoheterotroph

20.   Includes most of the bacteria that cause human disease.

Answer: b
Difficulty: Easy
Bloom's Taxonomy: Knowledge

21.   Includes the cyanobacteria.

Answer: c
Difficulty: Easy
Bloom's Taxonomy: Knowledge

22.   Includes the purple nonsulfur bacteria.

Answer: d
Difficulty: Moderate
Bloom's Taxonomy: Knowledge

23.   Unique to prokaryotes.

Answer: e
Difficulty: Moderate
Bloom's Taxonomy: Knowledge

Use the figure shown below for questions 24–30.

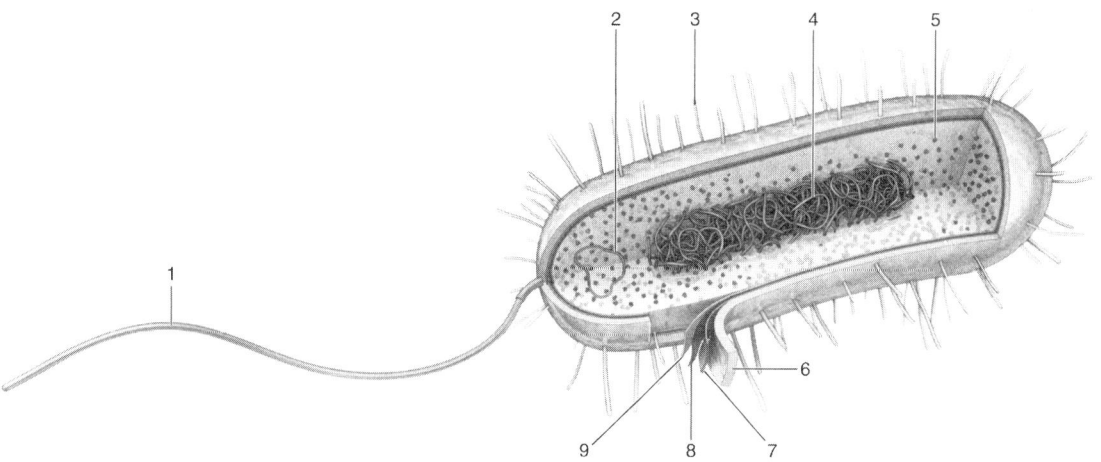

24. Which type of cell is represented in this figure?
a. a Gram-positive bacterium
b. a Gram-negative bacterium
c. either a Gram-positive or a Gram-negative bacterium
d. an archaeal cell
e. any prokaryotic cell

Answer: b
Difficulty: Difficult
Bloom's Taxonomy: Comprehension, Application

25. Which number identifies the structure that carries supplemental genetic information?
a. 1
b. 2
c. 3
d. 4
e. 5

Answer: b
Difficulty: Moderate
Bloom's Taxonomy: Knowledge

26. Which number identifies the structure that carries out protein synthesis?
a. 2
b. 4
c. 5
d. 6
e. 9

Answer: c
Difficulty: Easy
Bloom's Taxonomy: Knowledge

27. Which number identifies the structure that facilitates movement?
a. 1
b. 2
c. 3
d. 4
e. 5

Answer: a
Difficulty: Easy
Bloom's Taxonomy: Knowledge

28. Which numbers identifies a structure that aids in attachment?
a. 1
b. 2
c. 3
d. 7
e. 8

Answer: c
Difficulty: Easy
Bloom's Taxonomy: Knowledge

29. Which number identifies a structure that is composed of polysaccharides?
a. 1
b. 3
c. 6
d. 7
e. 9

Answer: c
Difficulty: Moderate
Bloom's Taxonomy: Knowledge

30. Which numbers identify structures that are found in *all* prokaryotes?

a. 2 and 4
b. 3 and 4
c. 1 and 5
d. 5 and 9
e. 7 and 8

Answer: d
Difficulty: Difficult
Bloom's Taxonomy: Comprehension

31. Which type of bacteria grows in the presence or absence of oxygen?

a. obligate aerobes
b. obligate anaerobes
c. facultative anaerobes
d. obligate aerobes and obligate anaerobes
e. obligate anaerobes and facultative anaerobes

Answer: c
Difficulty: Easy
Bloom's Taxonomy: Knowledge

32. Why is the process of nitrogen fixation by prokaryotes essential?

a. It removes nitrogen from the soil.
b. It converts atmospheric nitrogen to a non-toxic form.
c. It provides nitrogen sources for plants and animals.
d. It provides an energy source for photosynthesis.
e. It allows for the breakdown of complex macromolecules.

Answer: c
Difficulty: Moderate
Bloom's Taxonomy: Knowledge

33. Which of the following *best* describes the process of nitrification?

a. $N_2 \rightarrow NH_4^+$
b. $NH_4^+ \rightarrow N_2$
c. $N_2 \rightarrow NO_3^-$
d. $NO_3^- \rightarrow N_2$
e. $NH_4^+ \rightarrow NO_3^-$

Answer: e
Difficulty: Moderate
Bloom's Taxonomy: Comprehension

34. Prokaryotes typically undergo _____ via the process of _____.

a. asexual reproduction; mitosis
b. sexual reproduction; meiosis
c. asexual reproduction; binary fission
d. sexual reproduction; binary fission
e. asexual reproduction; budding

Answer: c
Difficulty: Easy
Bloom's Taxonomy: Knowledge

35. _____ are dormant structures formed by certain bacteria in response to unfavorable environmental conditions.

a. capsids
b. endospores
c. endotoxins
d. exotoxins
e. heterocysts

Answer: b
Difficulty: Easy
Bloom's Taxonomy: Knowledge

36. Which of the following statements about biofilms is *true*?

a. They are formed by prokaryotes living in a watery environment.
b. They may develop on surgical implants.
c. An example is dental plaque.
d. a and b only.
e. a, b, and c.

Answer: e
Difficulty: Easy
Bloom's Taxonomy: Knowledge

37. How long does it take for a biofilm to form?

a. seconds to minutes
b. minutes to hours
c. hours to days
d. days to months
e. months to years

Answer: d
Difficulty: Easy
Bloom's Taxonomy: Knowledge

25.2 THE DOMAIN BACTERIA

Use figure shown below for questions 38–42.

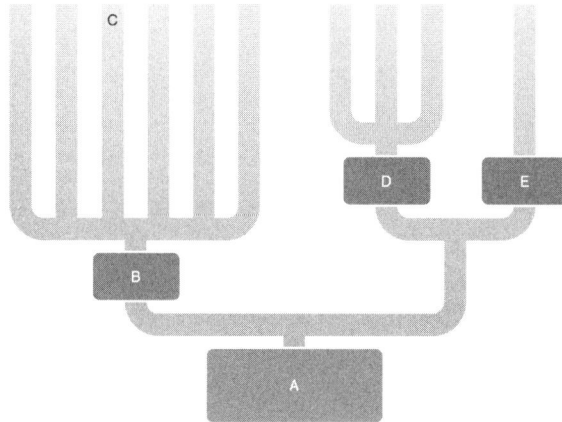

38. Which branch of this phylogenetic tree identifies the common ancestor of all modern organisms?

a.   A
b.   B
c.   C
d.   D
e.   E

Answer: a
Difficulty: Easy
Bloom's Taxonomy: Knowledge

39. Which branch of this phylogenetic tree identifies a kingdom?

a.   C
b.   D
c.   E
d.   D and E
e.   C, D, and E

Answer: a
Difficulty: Moderate
Bloom's Taxonomy: Knowledge

40. Which branch of this phylogenetic tree identifies Bacteria?

a.   A
b.   B
c.   C
d.   D
e.   E

Answer: b
Difficulty: Easy
Bloom's Taxonomy: Knowledge

41. Which branch of this phylogenetic tree identifies Archaea?

a.   A
b.   B
c.   C
d.   D
e.   E

Answer: d
Difficulty: Easy
Bloom's Taxonomy: Knowledge

42. Which branch of this phylogenetic tree identifies Eukarya?

a.   A
b.   B
c.   C
d.   D
e.   E

Answer: e
Difficulty: Easy
Bloom's Taxonomy: Knowledge

43. Which of the following best describes the number of distinct evolutionary branches in the domain Bacteria, as revealed by genome sequencing studies?

a.   3–4
b.   6–7
c.   9–10
d.   at least 12
e.   at least 20

Answer: d
Difficulty: Easy
Bloom's Taxonomy: Knowledge

44. The evolutionary ancestors of mitochondria are believed to have been ancient

a.   Archaea.
b.   cyanobacteria.
c.   proteobacteria.
d.   green bacteria.
e.   Gram-positive bacteria.

Answer: c
Difficulty: Moderate
Bloom's Taxonomy: Knowledge

45. In which of the following ways is proteobacterial photosynthesis *different* from plant photosynthesis?

a. Proteobacterial photosynthesis uses a different type of chlorophyll.
b. Proteobacterial photosynthesis does not produce oxygen.
c. Proteobacterial photosynthesis uses water as an electron donor.
d. a and b only.
e. a, b, and c.

Answer: d
Difficulty: Moderate
Bloom's Taxonomy: Comprehension

46. How is cyanobacterial photosynthesis *similar* to plant photosynthesis?

a. in type of chlorophyll
b. in splitting of water
c. in production of oxygen
d. a and b only
e. a, b, and c

Answer: e
Difficulty: Moderate
Bloom's Taxonomy: Comprehension

47. Which of the following statements describes proteobacteria?

a. Gram-negative and includes the purple bacteria
b. Gram-negative and includes the green bacteria
c. Gram-positive and includes the purple bacteria
d. Gram-positive and includes the green bacteria
e. none of the above

Answer: a
Difficulty: Moderate
Bloom's Taxonomy: Knowledge

25.3 THE DOMAIN ARCHAEA

48. Archaea were first discovered in

a. 1901.
b. 1933.
c. 1952.
d. 1977.
e. 1996.

Answer: d
Difficulty: Moderate
Bloom's Taxonomy: Knowledge

49. Prokaryotes were split into two domains based on differences in

a. staining characteristics.
b. metabolic capabilities.
c. rRNA sequences.
d. cell membrane properties.
e. cell wall composition.

Answer: c
Difficulty: Moderate
Bloom's Taxonomy: Knowledge

50. Which of the following best describes the number of distinct evolutionary branches in the domain Archaea, as revealed by genome sequencing studies?

a. 3–4
b. 6–7
c. 9–10
d. at least 12
e. at least 20

Answer: a
Difficulty: Easy
Bloom's Taxonomy: Knowledge

51. Where are the methanogens found?

a. Crenarchaeota
b. Euryarchaeota
c. Korarchaeota
d. Nanoarchaeota
e. both Crenarchaeota and Euryarchaeota

Answer: b
Difficulty: Moderate
Bloom's Taxonomy: Knowledge

52. Archaea that have been cultured in the laboratory include

a. Crenarchaeota and Euryarchaeota.
b. Crenarchaeota and Korarchaeota.
c. Euryarchaeota and Korarchaeota.
d. Euryarchaeota and Nanorarchaeota.
e. Nanoarchaeota and Korarchaeota.

Answer: a
Difficulty: Moderate
Bloom's Taxonomy: Knowledge

53. Which of the following statements about Korarchaeota is *true*?

a. includes halophiles
b. includes thermophiles
c. oldest known lineage in the Archaea
d. a and b only
e. a, b and c

Answer: c
Difficulty: Easy
Bloom's Taxonomy: Knowledge

54.   Where was *Methanococcus jannaschii* discovered?

a.   in polar ice
b.   in a volcano crater
c.   in the intestines of a cow
d.   in a surface hot spring
e.   in a deep hot-water vent

Answer: e
Difficulty: Easy
Bloom's Taxonomy: Knowledge

55.   Only about _____ of the *Methanococcus jannaschii* protein-encoding sequences show similarity to known genes from other organisms.

a.   7 percent
b.   13 percent
c.   22 percent
d.   38 percent
e.   50 percent

Answer: d
Difficulty: Moderate
Bloom's Taxonomy: Knowledge

25.4 VIRUSES, VIROIDS, AND PRIONS

56.   Viruses are considered nonliving because they

a.   lack a nucleus.
b.   lack a cell wall.
c.   cannot undergo mutations.
d.   cannot reproduce outside a host cell.
e.   contain RNA as their genetic material.

Answer: d
Difficulty: Easy
Bloom's Taxonomy: Comprehension

57.   A complete virus particle is called a

a.   viroid.
b.   virusoid.
c.   virulen.
d.   virion.
e.   virogen.

Answer: d
Difficulty: Easy
Bloom's Taxonomy: Knowledge

58.   The protein layer surrounding the viral genome is called

a.   a cell wall.
b.   a cell membrane.
c.   an envelope.
d.   a capsule.
e.   a capsid.

Answer: e
Difficulty: Easy
Bloom's Taxonomy: Knowledge

59.   Envelopes are found most commonly in viruses of

a.   bacteria.
b.   archaea.
c.   fungi.
d.   animals.
e.   plants.

Answer: d
Difficulty: Difficult
Bloom's Taxonomy: Comprehension

60.   In a(n) _____ cycle, the virus kills the infected host cell.

a.   phagocytic
b.   lysogenic
c.   lytic
d.   exotoxic
e.   endocytic

Answer: c
Difficulty: Easy
Bloom's Taxonomy: Knowledge

61.   Viral DNA that is integrated into a bacterial host chromosome is called a

a.   prophage.
b.   microphage.
c.   macrophage.
d.   spike.
e.   endospore.

Answer: a
Difficulty: Easy
Bloom's Taxonomy: Knowledge

62. The latent phase of an animal virus is

a. the time it takes to replicate the viral genome.
b. the time it takes to assemble one viral particle.
c. the period from infection to incorporation of the viral genome into the host chromosome.
d. the period from infection to host cell lysis.
e. the equivalent of the lysogenic cycle of some bacteriophages.

Answer: e
Difficulty: Moderate
Bloom's Taxonomy: Comprehension

Use the figure shown below for questions 63–70.

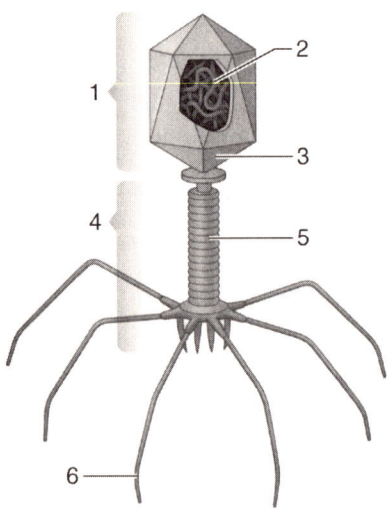

63. Which virus does this figure illustrate?

a. adenovirus
b. bacteriophage
c. herpesvirus
d. tobacco mosaic virus
e. human immunodeficiency virus

Answer: b
Difficulty: Easy
Bloom's Taxonomy: Knowledge

64. The shape of this virus can best be described as

a. helical.
b. polyhedral.
c. enveloped.
d. complex.
e. icosahedral.

Answer: d
Difficulty: Moderate
Bloom's Taxonomy: Knowledge

65. The genome of this virus is _____ and is identified by the number _____.

a. DNA; 2
b. DNA; 3
c. RNA; 2
d. RNA; 3
e. none of the above

Answer: a
Difficulty: Moderate
Bloom's Taxonomy: Knowledge

66. Which number identifies the head?

a. 1
b. 2
c. 3
d. 4
e. 5

Answer: a
Difficulty: Easy
Bloom's Taxonomy: Knowledge

67. Which number identifies the sheath?

a. 2
b. 3
c. 4
d. 5
e. 6

Answer: d
Difficulty: Easy
Bloom's Taxonomy: Knowledge

68. Which number identifies the coat?

a. 1
b. 2
c. 3
d. 4
e. none of the above

Answer: c
Difficulty: Moderate
Bloom's Taxonomy: Knowledge

69. Which number identifies the structure that binds to the host cell?

a. 1
b. 3
c. 4
d. 5
e. 6

Answer: e
Difficulty: Easy
Bloom's Taxonomy: Knowledge

70. The structure that enters the host cell is identified by number(s)

a. 1.
b. 2.
c. 3.
d. 1 and 4.
e. 2 and 5.

Answer: b
Difficulty: Moderate
Bloom's Taxonomy: Knowledge

71. Viroids are infectious _____ particles that cause diseases in _____.

a. RNA; animals
b. RNA; plants
c. protein; animals
d. protein; plants
e. none of the above

Answer: b
Difficulty: Easy
Bloom's Taxonomy: Knowledge

72. Who is credited with the discovery of prions as a new biological principle of infection?

a. Stanley Prusiner
b. Carl Louis
c. Stanley Brenner
d. Carl Woese
e. none of the above

Answer: a
Difficulty: Easy
Bloom's Taxonomy: Knowledge

73. Which of the following prion diseases is accurately described?

a. bovine spongiform encephalopathy – mad cow disease
b. scrapie – sheep disease
c. kuru – human disease
d. a and b only
e. a, b, and c

Answer: e
Difficulty: Moderate
Bloom's Taxonomy: Knowledge

UNANSWERED QUESTIONS

74. Which of the following statements about West Nile virus is *true*?

a. It first entered the United States in 1999.
b. It is spread by flies.
c. It is not fatal.
d. a and b only.
e. a, b, and c.

Answer: a
Difficulty: Moderate
Bloom's Taxonomy: Knowledge

Integrative Multiple-Choice

75. How are viruses different from bacteria?

a. They lack a nucleus.
b. They have a single type of nucleic acid.
c. They do not contain protein.
d. They are too small to be seen with the unaided eye.
e. They may be pathogenic to humans.

Answer: b
Difficulty: Moderate
Bloom's Taxonomy: Knowledge
Source: Sections 25.1 and 25.4

Use the following key for questions 76–83.

a. characteristic of Archaea only
b. characteristic of Bacteria only
c. characteristic of both Archaea and Bacteria
d. characteristic of neither Archaea nor Bacteria

76. Presence of a nuclear membrane.

Answer: d
Difficulty: Easy
Bloom's Taxonomy: Knowledge
Source: Sections 25.1, 25.2, and 25.3

77. Pathogenic to humans.

Answer: b
Difficulty: Moderate
Bloom's Taxonomy: Knowledge
Source: Sections 25.1, 25.2, and 25.3

78. First amino acid in proteins is formylmethionine.

Answer: b
Difficulty: Moderate
Bloom's Taxonomy: Knowledge
Source: Sections 25.1, 25.2, and 25.3

79. Presence of cytoplasmic organelles.

Answer: d
Difficulty: Easy
Bloom's Taxonomy: Knowledge
Source: Sections 25.1, 25.2, and 25.3

80. Stain as either Gram-positive or Gram-negative.

Answer: c
Difficulty: Moderate
Bloom's Taxonomy: Knowledge
Source: Sections 25.1, 25.2, and 25.3

81. Presence of flagella.

Answer: c
Difficulty: Easy
Bloom's Taxonomy: Knowledge
Source: Sections 25.1, 25.2, and 25.3

82. Presence of branched membrane lipids, with ether linkages.

Answer: a
Difficulty: Moderate
Bloom's Taxonomy: Knowledge
Source: Sections 25.1, 25.2, and 25.3

83. Presence of a single, circular chromosome.

Answer: c
Difficulty: Easy
Bloom's Taxonomy: Knowledge
Source: Sections 25.1, 25.2, and 25.3

Matching

Match each prokaryotic genus with its correct description.

84. _____ Azotobacter

85. _____ Bacillus

86. _____ Chlamydia

87. _____ Halobacterium

88. _____ Lactobacillus

89. _____ Methanococcus

90. _____ Neisseria

91. _____ Pyrobolus

92. _____ Salmonella

93. _____ Streptococcus

94. _____ Treponema

A. Causes anthrax

B. Fixes nitrogen

C. Causes syphilis

D. Loves very high temperatures

E. Causes typhoid fever

F. Helps in the production of yoghurt

G. Causes necrotizing fasciitis (flesh-eating disease)

H. Loves very high salt concentrations

I. Causes gonorrhea

J. Produces CH_4

K. Causes trachoma (a preventable form of blindness)

Answers: **84. B   85. A   86. K   87. H   88. F   89. J   90. I   91. D   92. E   93. G   94. C**

Difficulty: Difficult
Bloom's Taxonomy: Knowledge
Source: Sections 25.1, 25.2, 25.3

Select the Exception

95. Viruses may have all of the following *except*

a. protein.
b. DNA.
c. RNA.
d. lipids.
e. cytoplasm.

Answer: e
Difficulty: Easy
Bloom's Taxonomy: Knowledge

96. Which of the following is *not* a Gram-positive bacterium?

a. *Clostridium*
b. *Staphylococcus*
c. *Streptococcus*
d. *Salmonella*
e. *Lactobacillus*

Answer: d
Difficulty: Moderate
Bloom's Taxonomy: Knowledge

97. Which of the following statements about the botulism toxin is *not* true?

a. It is an exotoxin.
b. It is produced by the bacterium *Clostridium.*
c. It may be found as a contaminant in poorly prepared food.
d. Its brand name is Botox.
e. It causes muscles to contract.

Answer: e
Difficulty: Moderate
Bloom's Taxonomy: Knowledge

98. All of the following statements about mycoplasmas are true *except*

a. they possess a nucleoid.
b. they lack a cell wall.
c. they are the smallest known cells.
d. they are Archaea.
e. they are resistant to penicillin.

Answer: d
Difficulty: Moderate
Bloom's Taxonomy: Knowledge

99. All of the following are viral diseases *except*

a. polio.
b. diphtheria.
c. influenza.
d. herpes.
e. hepatitis.

Answer: b
Difficulty: Moderate
Bloom's Taxonomy: Knowledge

True/False

If the statement is true, write a "T" in the blank. If the statement is false, make it correct by changing the underlined word(s) and writing the correct word(s) in the answer blank.

100. _____ The first virus to be isolated and crystallized was the <u>polio</u> virus.

101. _____ Chlamydias are bacteria that possess cell walls, but lack <u>peptidoglycan</u>.

102. _____ Influenza is caused by a <u>DNA</u> virus.

103. _____ A lichen is a symbiotic association between a <u>proteobacterium</u> and a fungus.

104. _____ Animal viruses without envelopes enter the host cell via <u>fusion with the cell membrane</u>.

105. _____ Archaea have <u>multiple types</u> of RNA polymerase.

106. _____ Organisms that live optimally in cold temperatures are called <u>barophiles</u>.

107. _____ <u>Spirilli</u> are Gram-negative bacteria that possess internal flagella, which cause the entire cell to twist in a corkscrew pattern.

108. _____ Viral infections in plants are spread from cell to cell via <u>plasmodesmata</u>.

109. _____ Some <u>viruses</u> can cause cancer.

110. _____ Genome sequence comparisons have now shown that the <u>Korarchaeota</u> are most likely a subgroup of the Euryarchaeota.

111. _____ Extracellular polymer substance (EPS) is a slimy, gluelike substance produced by bacteria to facilitate the formation of <u>endospores</u>.

112. _____ Transformation involves contact between two bacterial cells, followed by the unidirectional transfer of plasmid genes.

113. _____ Creutzfeldt-Jacob disease (CJD) is a fatal prion infection in humans.

Answers: *100. F, tobacco mosaic virus*
101. T   102. F, RNA   103. F, cyanobacterium 104. F, endocytosis   105. T
106. F, psychrophiles   107. F, spirochetes
108. T   109. T   110. F, nanoarchaeota
111. F, biofilms   112. F, conjugation   113. T

Difficulty: Difficult
Bloom's Taxonomy: Knowledge, Comprehension
Source: Sections 25.1, 25.2, 25.3, 25.4

Essay

114.   Discuss why Gram-positive and Gram-negative bacteria stain differently with the Gram stain? How is this difference reflected in antibiotic sensitivity/resistance?

Answer: Gram-positive bacteria are those that retain crystal violet, the primary dye of the Gram stain, and thus appear purple. Gram-negative bacteria do not retain the crystal violet dye, and stain pink because of safranin, the second dye. The staining difference reflects differences in the cell wall structure. Gram-positive bacteria have a thick peptidoglycan layer, which absorbs and retains the crystal violet pigment. On the other hand, Gram-negative bacteria have a thin peptidoglycan layer, and therefore lose the crystal violet pigment during the alcohol wash. This difference in cell walls has a bearing on how certain antibiotics affect these two bacterial types. Gram-positive bacteria are very susceptible to penicillin, which blocks new cell wall formation by inhibiting peptidoglycan cross-linking. The weakened cell wall soon leads to the death of the bacterium. Penicillin is less effective against Gram-negative bacteria because their outer membrane inhibits entry of the antibiotic. Interestingly, this outer membrane makes Gram-negative bacteria much more sensitive than Gram-positive bacteria to antibiotics that disrupt the cell membrane (e.g., polymyxin)
Difficulty: Moderate
Bloom's Taxonomy: Knowledge, Comprehension
Source: Section 25.1

115.   List and discuss five structures in bacteria that allow them to cause disease.

Answer:

1. Capsules: Many bacteria are surrounded by a capsule, which is typically composed of polysaccharides. The capsule protects the bacterium from the host animal's immune system. For example, Streptococcus pneumoniae strains that are encapsulated can cause pneumonia, while those without capsules are nonvirulent.

2. Pili: Pili are short protein appendages found in a number of bacteria. These help bacteria to bind to animal cells. Neisseria gonorrhoeae, which causes gonorrhea, has pili that allow it to attach to cells of the throat, eye, urogenital tract, or rectum in humans.

3. Exotoxins: A number of bacterial lineages produce exotoxins. These are toxic proteins that are secreted from the bacterium and which interfere with the biochemical processes of body cells in various ways. For example, the botulinum exotoxin causes muscle paralysis by interfering with the transmission of nerve impulses.

4. Endotoxins: Gram-negative bacteria can cause disease through endotoxins, which are lipopolysaccharides released from the outer membrane when bacteria die and lyse. Endotoxins cause disease by overstimulating the host's immune response, often triggering inflammation.

5. Exoenzymes: Some bacteria release exoenzymes, which are catalytic proteins. These may breakdown cell membranes (causing cells to rupture), collagen (causing connective tissue disorders), red and white blood cells (leading to anemias and weakening of the immune system). Bacteria that release exoenzymes include Streptococcus and Clostridium.
Difficulty: Moderate
Bloom's Taxonomy: Knowledge, Comprehension
Source: Sections 25.1, 25.2

26
PROTISTS

Multiple-Choice

WHY IT MATTERS

1. Which of the following statements about protists is *true*?

a. They are the simplest of eukaryotes.
b. They are generally aquatic.
c. They are members of the Kingdom Protoctista.
d. a and b only.
e. a, b, and c.

Answer: e
Difficulty: Moderate
Bloom's Taxonomy: Knowledge

2. The cause of the Irish potato famine of the 1840s was the protist

a. *Pseudomonas aeruginosa.*
b. *Phytophthora infestans.*
c. *Plasmodium falciparum.*
d. *Physarum polycephalum.*
e. *Proteus mirabilis.*

Answer: b
Difficulty: Easy
Bloom's Taxonomy: Knowledge

26.1 WHAT IS A PROTIST?

3. The protists are a diverse group of organisms that have traditionally been grouped together because they all

a. have the same type of nutrition.
b. live in the same environments.
c. have very similar DNA sequences.
d. have very similar shapes.
e. are not prokaryotes, fungi, plants, or animals.

Answer: e
Difficulty: Moderate
Bloom's Taxonomy: Knowledge

4. The majority of protists are

a. anaerobic.
b. aerobic.
c. autotrophic.
d. parasitic.
e. photosynthetic.

Answer: b
Difficulty: Moderate
Bloom's Taxonomy: Knowledge

5. All protists possess

a. multiple, circular chromosomes.
b. mitochondria.
c. microtubules and microfilaments.
d. chloroplasts.
e. cell walls.

Answer: c
Difficulty: Moderate
Bloom's Taxonomy: Knowledge

6. Until recently, protists have been classified according to which of the following criteria?

a. modes of nutrition
b. mechanisms of movement
c. types of life cycle
d. a and b only
e. a, b, and c

Answer: e
Difficulty: Moderate
Bloom's Taxonomy: Knowledge

7. Which of the following best describes the phylogenetic relationship between protists, fungi, plants, and animals?

a. Fungi and animals share a common protist lineage, while plants arose from a different protist ancestor.
b. Fungi and plants share a common protist lineage, while animals arose from a different protist ancestor.
c. Animals and plants share a common protist lineage, while fungi arose from a different protist ancestor.
d. Plants, animals, and fungi each arose from a different protist lineage.
e. Plants, animals, fungi, and protists each arose from the same prokaryotic ancestor.

Answer: a
Difficulty: Moderate
Bloom's Taxonomy: Comprehension

8. Small, photosynthetic protists found in aquatic habitats are collectively called

a. zooplankton.
b. phytoplankton.
c. phagoplankton.
d. bryophytes.
e. chrysophytes.

Answer: b
Difficulty: Easy
Bloom's Taxonomy: Knowledge

9. Some protists have a pellicle, which is used for

a. defensive purposes.
b. sexual reproduction.
c. structural support.
d. water absorption.
e. food storage.

Answer: c
Difficulty: Easy
Bloom's Taxonomy: Knowledge

10. A _____ is an extension of a lobe of the cytoplasm, which is used by some protists for movement.

a. gullet
b. pseudopod
c. cilium
d. pilus
e. flagellum

Answer: b
Difficulty: Easy
Bloom's Taxonomy: Knowledge

Use figure below for questions 11–15.

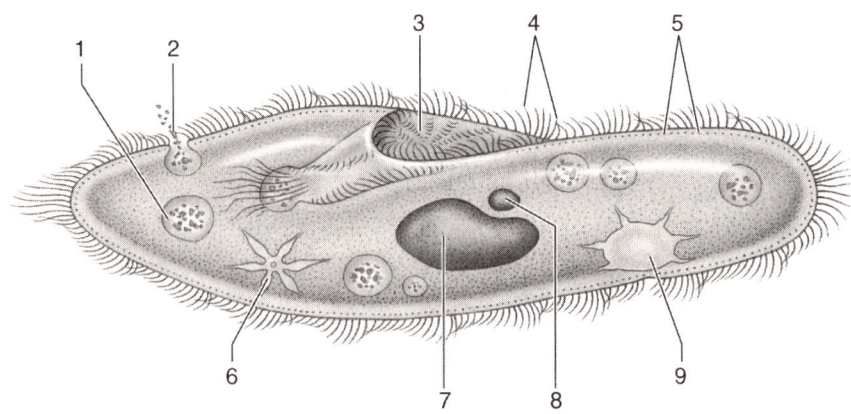

11. The function of the structure labeled 2 is

a. food intake.
b. digestion.
c. waste elimination.
d. water discharge.
e. asexual reproduction.

Answer: c
Difficulty: Moderate
Bloom's Taxonomy: Knowledge

12. The structures labeled 4 are

a. trichocysts.
b. cilia.
c. flagella.
d. pili.
e. pellicles.

Answer: b
Difficulty: Easy
Bloom's Taxonomy: Knowledge

13. The structures labeled 5 are

a. trichocysts.
b. cilia.
c. flagella.
d. pili.
e. pellicles.

Answer: a
Difficulty: Moderate
Bloom's Taxonomy: Knowledge

14. The structure labeled 7 is the

a. food vacuole.
b. gullet.
c. macronucleus.
d. micronucleus.
e. mitochondrion.

Answer: c
Difficulty: Easy
Bloom's Taxonomy: Knowledge

15. Water balance is maintained by the structures labeled

a. 1 and 3.
b. 5 and 9.
c. 7 and 8.
d. 6 and 9.
e. 1 and 6.

Answer: d
Difficulty: Moderate
Bloom's Taxonomy: Comprehension

26.2 THE PROTIST GROUPS

16. You discover an organism that has a nucleus but no mitochondria. This organism could be a

a. euglenoid.
b. diplomonad.
c. kinetoplastid.
d. ciliate.
e. dinoflagellate.

Answer: b
Difficulty: Moderate
Bloom's Taxonomy: Knowledge

17. The undulating membrane of Parabasala is used for

a. defensive purposes.
b. asexual reproduction.
c. structural support.
d. food absorption.
e. movement.

Answer: e
Difficulty: Moderate
Bloom's Taxonomy: Knowledge

18. *Giardia lamblia* forms resistant _____ that can survive in water contaminated with feces.

a. endospores
b. exospores
c. cysts
d. sporangia
e. trophozoites

Answer: c
Difficulty: Easy
Bloom's Taxonomy: Knowledge

19. The term *protozoa* means "first animal." It has been used to refer to protists of the

a. Alveolates.
b. Amoebozoa.
c. Heterokonts.
d. Discicristates.
e. Excavates.

Answer: d
Difficulty: Moderate
Bloom's Taxonomy: Knowledge

20. Which of the following statements about euglenoids is *false*?

a. They are much more common in fresh water than marine environments.
b. Most are photosynthetic.
c. Their cell body is surrounded by a pellicle.
d. Some absorb nutrients from their environment in a heterotrophic manner.
e. They move by pseudopodia.

Answer: e
Difficulty: Moderate
Bloom's Taxonomy: Knowledge

21. Which of the following terms describes kinetoplastida?

a. non-photosynthetic
b. parasitic
c. ciliated
d. a and b only
e. a, b, and c

Answer: d
Difficulty: Moderate
Bloom's Taxonomy: Knowledge

22. Ciliates include the genus

a. *Euglena.*
b. *Anopheles.*
c. *Trichomonas.*
d. *Paramecium.*
e. *Entamoeba.*

Answer: d
Difficulty: Moderate
Bloom's Taxonomy: Knowledge

23. When under attack, ciliates can eject dartlike protein threads from surface organelles called

a. vacuoles.
b. gullets.
c. trichocysts.
d. phagosomes.
e. alveoli.

Answer: c
Difficulty: Easy
Bloom's Taxonomy: Knowledge

Use figure shown below for questions 24–30.

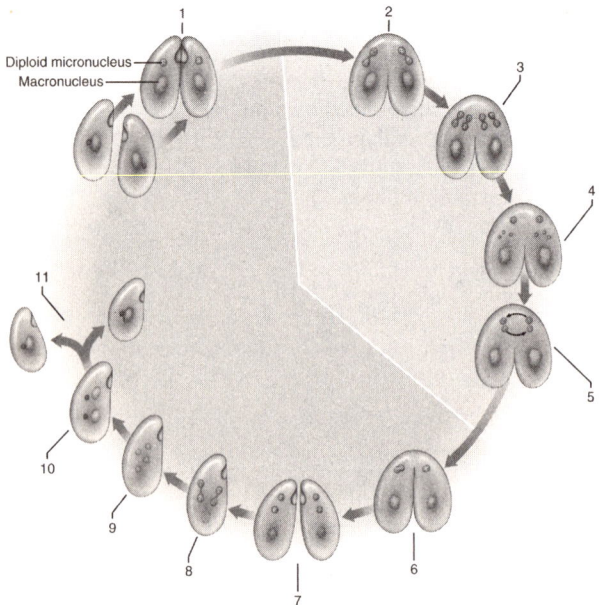

24. This figure illustrates the process of

a. conjugation.
b. transformation.
c. transduction.
d. translocation.
e. excystation.

Answer: a
Difficulty: Easy
Bloom's Taxonomy: Knowledge

25. Which of the following does the macronucleus control?

a. asexual reproduction
b. sexual reproduction
c. growth and metabolism
d. a and b
e. a, b, and c

Answer: c
Difficulty: Moderate
Bloom's Taxonomy: Knowledge

26. What is occurring in step 2?

a. mitosis
b. the first meiotic division
c. the second meiotic division
d. fertilization
e. fusion

Answer: b
Difficulty: Easy
Bloom's Taxonomy: Knowledge

27. The degeneration of nuclei is occurring in step(s)

a. 4.
b. 5.
c. 8.
d. 4 and 5.
e. 4 and 8.

Answer: e
Difficulty: Moderate
Bloom's Taxonomy: Knowledge, Comprehension

28. What is occurring in step 6?

a. fission
b. fusion
c. fertilization
d. the first meiotic division
e. the second meiotic division

Answer: b
Difficulty: Moderate
Bloom's Taxonomy: Knowledge

29. Mitosis is occurring in step(s)

a. 7.
b. 8.
c. 10.
d. 7 and 8.
e. 7, 8, and 10.

Answer: d
Difficulty: Moderate
Bloom's Taxonomy: Knowledge, Comprehension

30. What is occurring in step 11?

a. cytoplasmic division
b. fusion
c. fertilization
d. the first meiotic division
e. the second meiotic division

Answer: a
Difficulty: Moderate
Bloom's Taxonomy: Knowledge

31. The least mobile protists are the

a. apicomplexans.
b. dinoflagellates.
c. diatoms.
d. heterokonts.
e. euglenoids.

Answer: a
Difficulty: Moderate
Bloom's Taxonomy: Knowledge

32. Red tide and extensive fish kills are caused by population blooms of

a. euglenoids.
b. dinoflagellates.
c. diatoms.
d. ciliates.
e. brown algae.

Answer: b
Difficulty: Easy
Bloom's Taxonomy: Knowledge

33. The apical complex of apicomplexa is used for

a. defensive purposes.
b. asexual reproduction.
c. structural support.
d. food absorption.
e. attachment.

Answer: e
Difficulty: Easy
Bloom's Taxonomy: Knowledge

34. Heterokonts are characterized by having

a. two identical flagella.
b. two different flagella.
c. two types of nuclei.
d. cells walls of chitin.
e. cell walls of silica.

Answer: b
Difficulty: Easy
Bloom's Taxonomy: Knowledge

35. Which of the following are members of the Oomycota?

a. water molds
b. white rusts
c. mildews
d. a and b only
e. a, b, and c

Answer: e
Difficulty: Moderate
Bloom's Taxonomy: Knowledge

36. How do water molds differ from true fungi?

a. Water molds have cell walls made up of chitin; fungi have cell walls made up of cellulose.
b. Water molds form hyphae; fungi form mycelia.
c. Water molds are haploid; fungi are diploid.
d. Water molds have a flagellated stage in their life cycle; fungi do not.
e. Water molds do not cause plant diseases; fungi do.

Answer: d
Difficulty: Moderate
Bloom's Taxonomy: Knowledge

37. Diatoms belong to the group

a. Bacillariophyta.
b. Phaeophyta.
c. Chrysophyta.
d. Rhodophyta.
e. Chlorophyta.

Answer: a
Difficulty: Moderate
Bloom's Taxonomy: Knowledge

38. Which of the following statement(s) about the shells of diatoms is(are) *true*?

a. They are symmetrical.
b. They are made up of silica.
c. They have perforations.
d. a and b only.
e. a, b, and c.

Answer: e
Difficulty: Moderate
Bloom's Taxonomy: Knowledge

39. Diatoms can best be described as

a. unicellular and autotrophic.
b. unicellular and heterotrophic.
c. multicellular and autotrophic.
d. multicellular and heterotrophic.
e. multicellular and parasitic.

Answer: a
Difficulty: Moderate
Bloom's Taxonomy: Knowledge

40. Algin is a substance used to thicken products as diverse as ice cream, cosmetics, and floor polish. It is extracted from the cell walls of _____ algae.

a. green
b. golden
c. blue-green
d. brown
e. red

Answer: d
Difficulty: Easy
Bloom's Taxonomy: Knowledge

41. Agar is a substance used as a culture medium in the laboratory. It is extracted from the cell walls of _____ algae.

a. green
b. golden
c. blue-green
d. brown
e. red

Answer: e
Difficulty: Easy
Bloom's Taxonomy: Knowledge

42. Most golden algae are

a. unicellular and autotrophic.
b. unicellular and heterotrophic.
c. colonial and autotrophic.
d. colonial and heterotrophic.
e. multicellular and heterotrophic.

Answer: c
Difficulty: Moderate
Bloom's Taxonomy: Knowledge

43. Fucoxanthin is the pigment responsible for the color of

a. brown algae.
b. golden algae.
c. red algae.
d. red algae and golden algae.
e. brown algae and golden algae.

Answer: e
Difficulty: Moderate
Bloom's Taxonomy: Knowledge

44. Brown algae lack true roots but have similar structures called _____ that anchor them to their substrate.

a. stipes
b. blades
c. holdfasts
d. thalli
e. rhizoids

Answer: c
Difficulty: Moderate
Bloom's Taxonomy: Knowledge

45. Many algal species have life cycles consisting of alternating haploid and diploid generations. The most common life cycle in brown algae consists of organisms that are

a. unicellular in the haploid stage and multicellular in the diploid stage.
b. multicellular in the haploid stage and unicellular in the diploid stage.
c. unicellular in both the haploid and diploid stages.
d. multicellular in both the haploid and diploid stages.
e. colonial in both the haploid and diploid stages.

Answer: d
Difficulty: Moderate
Bloom's Taxonomy: Comprehension

46. Which of the following statements about the sporophyte in brown algae is *true*?

a. It is diploid and gives rise to spores by meiosis.
b. It is diploid and gives rise to spores by mitosis.
c. It is haploid and gives rise to spores by meiosis.
d. It is haploid and gives rise to spores by mitosis.
e. None of the above.

Answer: a
Difficulty: Moderate
Bloom's Taxonomy: Comprehension

47. Radiolarians have characteristic _____, which are thin, raylike strands of cytoplasm supported internally by bundles of _____.

a. axopods ; microfilaments
b. axopods ; microtubules
c. spicules ; microfilaments
d. spicules ; microtubules
e. none of the above is true

Answer: b
Difficulty: Moderate
Bloom's Taxonomy: Knowledge

48. Shells of forams consist of organic matter reinforced by

a. silica.
b. cellulose.
c. calcium carbonate.
d. copper sulfate.
e. hydrogen sulfide.

Answer: c
Difficulty: Easy
Bloom's Taxonomy: Knowledge

49. The majority of amoebas are found in the Amoebozoa. The rest are found in the

a. Cercozoa.
b. Opisthokonts.
c. Heterokonts.
d. Alveolata.
e. Archaeplastida.

Answer: a
Difficulty: Moderate
Bloom's Taxonomy: Knowledge

50. Amoebas use pseudopodia for movement and

a. sexual reproduction.
b. excretion.
c. feeding.
d. avoiding predation.
e. asexual reproduction.

Answer: c
Difficulty: Easy
Bloom's Taxonomy: Knowledge

51. Cellular slime molds can be distinguished from plasmodial slime molds on the basis of

a. reproductive structures.
b. spore formation.
c. nuclei per cell.
d. slime trails.
e. food requirements.

Answer: c
Difficulty: Moderate
Bloom's Taxonomy: Comprehension

52. Under favorable conditions, plasmodial slime molds exist primarily as

a. a coordinated, multicellular blob.
b. individual amoeba-like cells.
c. individual flagellated cells.
d. a single, multinucleated cellular blob.
e. spore-forming fruiting bodies.

Answer: d
Difficulty: Moderate
Bloom's Taxonomy: Comprehension

Use the figure shown below for questions 53–57.

53. This figure illustrates the life cycle of a

a. brown algae.
b. green algae.
c. water mold.
d. plasmodial slime mold.
e. cellular slime mold.

Answer: e
Difficulty: Easy
Bloom's Taxonomy: Knowledge

54. The fruiting body is the structure with the label

a. A.
b. B.
c. C.
d. D.
e. E.

Answer: a
Difficulty: Easy
Bloom's Taxonomy: Knowledge

55. The slug is the structure with the label

a. A.
b. B.
c. C.
d. D.
e. E.

Answer: e
Difficulty: Easy
Bloom's Taxonomy: Knowledge

56. The structure labeled E is formed

a. under favorable conditions.
b. under unfavorable conditions.
c. by mitosis.
d. by meiosis.
e. by fusion of individual cells.

Answer: b
Difficulty: Moderate
Bloom's Taxonomy: Comprehension

57. Which of the following statements is *true*?

a. The left and right portions of this life cycle illustrate asexual and sexual reproduction, respectively.
b. The left and right portions of this life cycle illustrate sexual and asexual reproduction, respectively.
c. Both portions of this life cycle illustrate asexual reproduction, because this organism does not reproduce sexually.
d. Both portions of this life cycle illustrate sexual reproduction, because this organism does not reproduce asexually.

Answer: c
Difficulty: Easy
Bloom's Taxonomy: Comprehension

58. Rhodophyta are _____ algae.

a. brown
b. red
c. golden
d. green
e. blue-green

Answer: b
Difficulty: Easy
Bloom's Taxonomy: Knowledge

59. The green algae include

a. unicellular species only.
b. multicellular species only.
c. unicellular and multicellular species.
d. unicellular and colonial species.
e. unicellular, multicellular, and colonial species.

Answer: e
Difficulty: Easy
Bloom's Taxonomy: Knowledge

60. Many algal species have life cycles consisting of alternating haploid and diploid generations. The most common life cycle in green algae consists of organisms that are

a. unicellular in the haploid stage and multicellular in the diploid stage.
b. multicellular in the haploid stage and unicellular in the diploid stage.
c. unicellular in both the haploid and diploid stages.
d. multicellular in both the haploid and diploid stages.
e. colonial in both the haploid and diploid stages.

Answer: b
Difficulty: Moderate
Bloom's Taxonomy: Comprehension

61. The algal ancestors of land plants are most likely similar to the

a. Rhodophyta.
b. Phaeophyta.
c. Chrysophyta.
d. Charophyta.
e. Radiolaria.

Answer: d
Difficulty: Moderate
Bloom's Taxonomy: Knowledge

62. You find a protist that has a single flagellum surrounded by a collar of microvilli. You correctly conclude that this organism is

a. a cellular slime mold.
b. a plasmodial slime mold.
c. a choanoflagellate.
d. *not* an opisthokont.
e. *not* motile.

Answer: c
Difficulty: Moderate
Bloom's Taxonomy: Application

FOCUS ON RESEARCH—APPLIED RESEARCH: MALARIA AND THE PLASMODIUM LIFE CYCLE

63. Malaria is caused by _____ species of *Plasmodium*.
   a. a single
   b. two
   c. three
   d. four
   e. five

Answer: d
Difficulty: Easy
Bloom's Taxonomy: Knowledge

64. Which of the following statements about malaria is *true*?

   a. It kills about 2 million people each year.
   b. It can be treated with drugs such as chloroquine.
   c. An effective vaccine exists to protect from it.
   d. a and b only
   e. a, b, and c

Answer: d
Difficulty: Moderate
Bloom's Taxonomy: Knowledge

Use the figure shown below for questions 65–70.

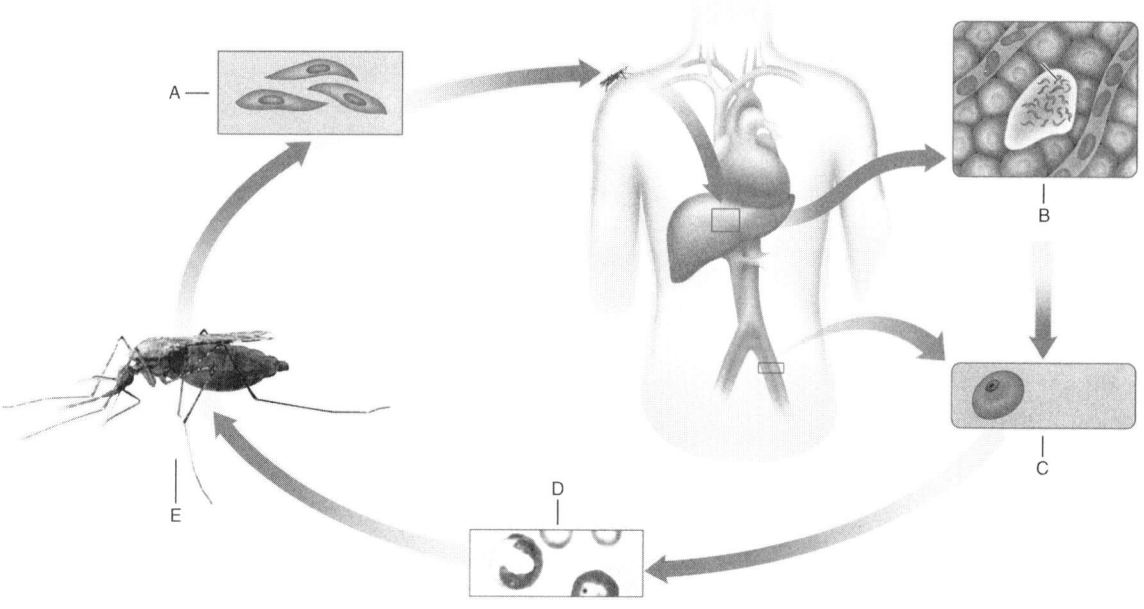

65. Which stage of the *Plasmodium* life cycle does label A represent?

   a. sporozoite
   b. gametocyte
   c. cyst
   d. zygote
   e. merozoite

Answer: a
Difficulty: Easy
Bloom's Taxonomy: Knowledge

66. Label B illustrates *Plasmodium* infection of the

   a. kidney.
   b. liver.
   c. gall bladder.
   d. stomach.
   e. lung.

Answer: b
Difficulty: Easy
Bloom's Taxonomy: Knowledge

67. Which stage(s) produce(s) the characteristic chills and fever of malaria?

a. A
b. B
c. C
d. A and B
e. B and C

Answer: c
Difficulty: Moderate
Bloom's Taxonomy: Comprehension

68. Which stage of the *Plasmodium* life cycle does label D represent?

a. sporozoite
b. gametocyte
c. cyst
d. zygote
e. merozoite

Answer: b
Difficulty: Easy
Bloom's Taxonomy: Knowledge

69. The animal labeled E is a

a. mosquito of the genus *Aedes*.
b. mosquito of the genus *Anopheles*.
c. mosquito of the genus *Culex*.
d. fly of the genus *Musca*.
e. tick of the genus *Amblyoma*.

Answer: b
Difficulty: Easy
Bloom's Taxonomy: Knowledge

70. Which of the following processes occurs in the animal labeled E?

a. zygote formation
b. meiosis
c. mitosis
d. a and b only
e. a, b, and c

Answer: d
Difficulty: Moderate
Bloom's Taxonomy: Comprehension

INSIGHTS FROM THE MOLECULAR REVOLUTION: GETTING THE SLIME MOLD ACT TOGETHER

71. The signal for aggregation and communal activity by cellular slime molds is

a. auxin.
b. glycoprotein slime.
c. GTP.
d. cyclic AMP.
e. unidentified as yet.

Answer: d
Difficulty: Easy
Bloom's Taxonomy: Knowledge

72. In the presence of the proper chemical signal, a protein _____ is activated. It then adds _____ groups to cellular target proteins, which become _____, and ultimately trigger cellular development and differentiation.

a. phosphatase; phosphate; activated
b. kinase; methyl; activated
c. phosphatase; methyl; activated
d. kinase; phosphate; activated or inhibited
e. phosphatase; methyl; activated or inhibited

Answer: d
Difficulty: Moderate
Bloom's Taxonomy: Comprehension

UNANSWERED QUESTIONS

73. According to the latest research, which of the following *best* describes the first eukaryote?

a. It lacked a nucleus, a mitochondrion, or a chloroplast.
b. It lacked a nucleus, but contained a mitochondrion and a chloroplast.
c. It contained a nucleus and a chloroplast, but not a mitochondrion.
d. It contained a nucleus and a mitochondrion, but not a chloroplast.
e. It contained a nucleus, a mitochondrion, and a chloroplast.

Answer: d
Difficulty: Moderate
Bloom's Taxonomy: Comprehension

74. Secondary endosymbiosis occurs when a

a. photosynthetic eukaryote engulfs a photosynthetic prokaryote.
b. photosynthetic eukaryote engulfs a non-photosynthetic prokaryote.
c. photosynthetic eukaryote engulfs a non-photosynthetic eukaryote.
d. nonphotosynthetic eukaryote engulfs a photosynthetic prokaryote.
e. nonphotosynthetic eukaryote engulfs a photosynthetic eukaryote.

Answer: e
Difficulty: Moderate
Bloom's Taxonomy: Comprehension

Integrative Multiple-Choice

75. All algae

a. are unicellular.
b. are multicellular.
c. have chlorophyll.
d. are motile at all stages of their life cycle.
e. have life cycles with alternating generations.

Answer: c
Difficulty: Moderate
Bloom's Taxonomy: Knowledge
Source: Section 26.2

76. You find a protist that is unicellular, heterotrophic, and non-flagellated. It has two nuclei of different sizes. You correctly conclude that this organism is a(n)

a. kinetoplastid.
b. plasmodial slime mold.
c. ciliate.
d. foraminiferan.
e. ameobozoan.

Answer: c
Difficulty: Moderate
Bloom's Taxonomy: Application
Source: Section 26.2

77. Which of the following protist groups includes human pathogens?

a. parabasalids
b. euglenoids
c. forams
d. a and b only
e. a, b, and c

Answer: a
Difficulty: Moderate
Bloom's Taxonomy: Knowledge
Source: Section 26.2

Matching

Match each protistan genus with its correct description.

78. _____ Balantidium
79. _____ Dictyostelium
80. _____ Entamoeba
81. _____ Euglena
82. _____ Giardia
83. _____ Physarum
84. _____ Toxoplasma
85. _____ Trichomonas
86. _____ Trypanosoma
87. _____ Ulva

A. Causes African sleeping sickness
B. A unicellular photosynthetic protist
C. A multicellular photosynthetic protist
D. Causes amoebic dysentery
E. Causes a sexually transmitted disease
F. A plasmodial slime mold
G. A cellular slime mold
H. A ciliate that is a human intestinal parasite
I. An excavate that is a human intestinal parasite
J. The reason why pregnant women should avoid cats

Answers: **78. H 79. G 80. D 81. B**
82. I 83. F 84. J 85. E
86. A 87. C

Difficulty: Difficult
Bloom's Taxonomy: Knowledge
Source: Section 26.2

Select the Exception

88. Which of the following protist groups does *not* include members that move by means of flagella?

a. Discicristates
b. Excavates
c. Alveolates
d. Cercozoa
e. Heterokonts

Answer: d
Difficulty: Difficult
Bloom's Taxonomy: Knowledge
Source: Section 26.2

89. Which of the following groups does *not* belong to the Alveolates?

a. apicomplexans
b. diatoms
c. dinoflagellates
d. parasitic ciliates
e. free-living ciliates

Answer: b
Difficulty: Moderate
Bloom's Taxonomy: Knowledge
Source: Section 26.2

90. Heterokonts include all of the following *except*

a. Rhodophyta.
b. Chrysophyta.
c. Phaeophyta.
d. Bacillariophyta.
e. Oomycota.

Answer: b
Difficulty: Moderate
Bloom's Taxonomy: Knowledge
Source: Section 26.2

91. Which of the following terms does *not* belong with the other four?

a. Radiolaria
b. Chlorarachniophyta
c. Foraminifera
d. Charophyta
e. Cercozoa

Answer: d
Difficulty: Difficult
Bloom's Taxonomy: Knowledge
Source: Section 26.2

92. Groups that contain planktonic species include all the following *except*

a. golden algae.
b. apicomplexans.
c. dinoflagellates.
d. diatoms.
e. red algae.

Answer: b
Difficulty: Difficult
Bloom's Taxonomy: Knowledge
Source: Section 26.2

Choice

Choose the type of algae that corresponds to each of the characteristics listed below.

a. blue-green   b. green
c. brown   d. red
e. golden

93. _____ largest and most complex

94. _____ most diverse

95. _____ grows at the deepest ocean levels

96. _____ ancestors of plants

97. _____ source of carrageenan, a thickening agent

98. _____ source of nori, the sushi "wrapper"

99. _____ Chrysophyta

100. _____ Prokaryotes

Answers: ***93. c***   ***94. b***   ***95. d***   ***96. b***
                97. d   ***98. d***   ***99. e***   ***100. a***

Difficulty: Difficult
Bloom's Taxonomy: Knowledge, Comprehension
Source: Section 26.2

True/False

If the statement is true, write a "T" in the blank. If the statement is false, make it correct by changing the underlined words(s) and writing the correct word(s) in the answer blank.

101. _____ The Discicristates are named after the disc-shaped cristae of their chloroplasts.

102. _____ The eyespot is a light-sensitive structure found in photosynthetic kinetoplastids.

103. _____ Diatomaceous earth is the fine powder produced from grinding the fossilized shells of diatoms.

104. _____ Some dinoflagellates exhibit bioluminescence with the help of the enzyme photolyase.

105. _____ Colonies are aggregates of potentially independent protist cells that show little or no differentiation.

106. _____ Amoeba is a term used to describe unicellular protists that move by means of flagella.

107. _____ Radiolarians have vacuoles, which provide them with buoyancy in marine environments.

108. _____ The White Cliffs of Dover, England, are composed of shells of ancient diatoms.

109. _____ Slime molds are classified as Amoebozoa.

110. _____ The land plants are termed viridaeplantae.

111. _____ The pigments responsible for the color of red algae are the prodigiosins.

112. _____ The Chlorophyta are believed to be the ancestors of fungi and animals.

113. _____ The parasite that causes African sleeping sickness is transmitted through the bite of a mosquito.

Answers: *101. F, mitochondria   102. F, euglenoids   103. T   104. F, luciferase   105. T   106. F, pseudopodia   107. T   108. F, foraminiferans   109. T   110. T   111. F, phycobilins   112. F, Choanoflagellata   113. F, tsetse fly*

Difficulty: Difficult
Bloom's Taxonomy: Knowledge, Comprehension
Source: Sections 26.1, 26.2

Essay

114.   Discuss how protists can be distinguished from each of the following: viruses, bacteria, fungi, plants, and animals.

Answer: First, protists can be distinguished from viruses because they have a cellular structure (cell membrane, cytoplasm, ribosomes, etc.) and viruses don't. They also differ from bacteria simply because they are eukaryotes, and bacteria are prokaryotes. Protists possess a nucleus and various organelles, all of which are lacking in bacteria. Protists do have features in common with fungi, animals, and plants. Indeed, these three eukaryotic lineages arose from protist ancestors. However, protists have several distinctive characteristics. For example, many protists lack cell walls, while true fungi possess them. The cell walls that are found in protists typically contain different components than those found in fungi (cellulose versus chitin). In contrast to land plants, protists lack highly differentiated structures equivalent to true roots, stems, and leaves. They also lack the protective structures that encase developing embryos in plants. Finally, protists are distinguished from animals by their lack of highly differentiated structures such as limbs and a heart, and by the absence of features such as nerve cells, complex developmental stages, and an internal digestive tract. Protists also lack collagen, the characteristic extracellular support protein of animals.
Difficulty: Moderate
Bloom's Taxonomy: Knowledge, Comprehension
Source: Section 26.1

115. Discuss the ecological significance of the dinoflagellates.

Answer: More than 4000 species of dinoflagellates have been discovered so far. Most are unicellular organisms that are autotrophic or heterotrophic (Some species can actually carry out both types of nutrition). Dinoflagellates are abundant in marine phytoplankton, and as such are a major primary producer in ocean ecosystems. Some species live as symbionts in the tissues of other marine organisms, most notably corals. For example, dinoflagellates in corals use the coral's carbon dioxide and nitrogenous waste, while supplying 90 percent of the coral's nutrition. This symbiotic relationship allows tropical coral reefs to reach massive size. Indeed, without the dinoflagellates many coral species would die. At times, population explosions (blooms) of dinoflagellates may color the sea red. Some red-tide dinoflagellates produce a toxin that interferes with nerve function in animals that ingest these protists. Fish that feed on plankton may be killed in large numbers by the toxin. Dinoflagellate toxin does not noticeably affect clams, oysters, and other mollusks, although it becomes concentrated in their tissues. Eating the tainted mollusks can cause respiratory failure and death for humans and other animals.
Difficulty: Moderate
Bloom's Taxonomy: Knowledge, Comprehension
Source: Section 26.2

27
PLANTS

Multiple-Choice

WHY IT MATTERS

1.   The first organisms to adapt to life in intertidal zones were probably _____.

a.   plants
b.   cyanobacteria
c.   green algae
d.   fungi
e.   animals

Answer: b
Difficulty: Moderate
Bloom's Taxonomy: Knowledge

2.   Which of the following is NOT typically used by plants to produce their own food?

a.   carbon dioxide
b.   dissolved minerals
c.   water
d.   oxygen
e.   sunlight energy

Answer: d
Difficulty: Easy
Bloom's Taxonomy: Comprehension

27.1 THE TRANSITION TO LIFE ON LAND

3.   The most direct ancestors of modern plants were

a.   sponges.
b.   animals.
c.   green algae.
d.   cyanobacteria.
e.   fungi.

Answer: c
Difficulty: Easy
Bloom's Taxonomy: Knowledge

4.   Which of the following is NOT a trait shared between green algae and land plants?

a.   stomata
b.   chloroplasts with chlorophyll *a*
c.   cellulose in their cell walls
d.   store energy captured during photosynthesis as starch
e.   chloroplasts with chlorophyll *b*

Answer: a
Difficulty: Moderate
Bloom's Taxonomy: Application

5.   The zygotes of modern charophytes and reproductive spores in land plants are protect from drying out and other damage in part by _____.

a.   stomata
b.   chlorophyll *a*
c.   a cuticle
d.   starch
e.   sporopollenin

Answer: e
Difficulty: Moderate
Bloom's Taxonomy: Knowledge

6.   These are small passages through the cuticle in plants that can open and close, allowing for some control of water loss by evaporation. In plants that have them, they are the main route for uptake of carbon dioxide.

a.   gemmae
b.   tracheophytes
c.   sporopollenins
d.   stomata
e.   thalli

Answer: d
Difficulty: Easy
Bloom's Taxonomy: Knowledge

7.   Which of these features is found in both nonvascular and vascular plants?

a.   lignin
b.   apical meristem
c.   cuticle
d.   xylem
e.   phloem

Answer: c
Difficulty: Moderate
Bloom's Taxonomy: Application

8.   Fossils of *Cooksonia* represent the oldest known fossils of vascular plants. Such fossils date to about

a.   1.5 billion years ago.
b.   420 million years ago.
c.   250 million years ago.
d.   125 million years ago.
e.   17 million years ago.

Answer: b
Difficulty: Moderate
Bloom's Taxonomy: Knowledge

9.   A structure with stems and leaves that arise from apical meristems is a _____.

a.   vascular system
b.   protonema
c.   sporangium
d.   rhizome
e.   shoot system

Answer: e
Difficulty: Easy
Bloom's Taxonomy: Comprehension

10.   A horizontal stem modified to penetrate the substrate and anchor a plant is a _____.

a.   vascular system
b.   protonema
c.   sporangium
d.   rhizome
e.   shoot system

Answer: d
Difficulty: Easy
Bloom's Taxonomy: Knowledge

11.   A special capsule or chamber where a plant will make the first cells of the gametophyte generation is called a _____.

a.   vascular system
b.   protonema
c.   sporangium
d.   rhizome
e.   shoot system

Answer: c
Difficulty: Moderate
Bloom's Taxonomy: Comprehension

12.   Lignified, tubelike structures that branch throughout the body of some plants, conducting water and solutes, are called _____.

a.   vascular tissues
b.   apical meristems
c.   root systems
d.   rhizomes
e.   shoot systems

Answer: a
Difficulty: Easy
Bloom's Taxonomy: Comprehension

13.   Heterospory refers to having

a.   male and female gametes.
b.   two spore types.
c.   both male and female parts on the same plant.
d.   separate male and female plants.
e.   both sporophyte and gametophyte generations.

Answer: b
Difficulty: Moderate
Bloom's Taxonomy: Comprehension

14.   Consider a plant where a multicellular generation is able to make both sperm and eggs. Which of the following terms would describe such a plant?

a.   homosporous
b.   bryophyte
c.   heterosporous
d.   lycophyte
e.   none of these

Answer: a
Difficulty: Moderate
Bloom's Taxonomy: Analysis

15.   A pollen grain represents the male _____ generation and develops from the _____ of two different spore types.

a.   sporophyte; larger
b.   gametophyte; larger
c.   gametophyte; smaller
d.   sporophyte; smaller
e.   none of these

Answer: c
Difficulty: Moderate
Bloom's Taxonomy: Analysis

16.   In _____ the sporophyte generation is clearly larger, more complex, and longer living than the gametophyte generation.

a.   bryophytes
b.   seedless vascular plants
c.   seed plants
d.   seedless vascular plants and seed plants
e.   all plants

Answer: d
Difficulty: Moderate
Bloom's Taxonomy: Synthesis

27.2 BRYOPHYTES: NONVASCULAR LAND PLANTS

17.   The _____ is a protective structure where egg cells are formed in bryophytes and some other plants.

a.   antheridium
b.   stoma
c.   strobilus
d.   sporangium
e.   archegonium

Answer: e
Difficulty: Moderate
Bloom's Taxonomy: Knowledge

18. The _____ is a protective structure where sperm cells are formed in bryophytes and some other plants.

a. antheridium
b. stoma
c. strobilus
d. sporangium
e. archegonium

Answer: a
Difficulty: Moderate
Bloom's Taxonomy: Knowledge

19. Which of the following phyla is NOT made up of bryophytes?

a. Pterophyta
b. Hepatophyta
c. Bryophyta
d. Anthocerophyta
e. all of these are bryophytes

Answer: a
Difficulty: Moderate
Bloom's Taxonomy: Synthesis

20. This phylum of nonvascular plants has about 6000 known living species. It includes the genus *Marchantia*, whose members can reproduce asexually via cups called gemmae.

a. Hepatophyta
b. Pterophyta
c. Bryophyta
d. Lycophyta
e. Anthocerophyta

Answer: a
Difficulty: Moderate
Bloom's Taxonomy: Knowledge

21. This simple body structure of a flat, branching, ribbonlike plate of tissue closely pressed against damp soil is commonly found as the gametophyte generation for many liverworts.

a. rhizoid
b. gemma
c. antheridium
d. archegonium
e. thallus

Answer: e
Difficulty: Easy
Bloom's Taxonomy: Knowledge

22. This phylum of nonvascular plants has about 100 known living species. They produce hornlike sporophytes, and they are the only land plants whose chloroplasts contain alga-like protein bodies called pyrenoids.

a. Hepatophyta
b. Pterophyta
c. Bryophyta
d. Lycophyta
e. Anthocerophyta

Answer: e
Difficulty: Moderate
Bloom's Taxonomy: Knowledge

23. This phylum of nonvascular plants has about 10,000 known living species, including those in the genus *Sphagnum*, a peat moss used by humans for its absorbent property and also as a fuel.

a. Hepatophyta
b. Pterophyta
c. Bryophyta
d. Lycophyta
e. Anthocerophyta

Answer: c
Difficulty: Easy
Bloom's Taxonomy: Knowledge

24. Members of the phylum _____ are considered by many researchers to be the living plants most closely related to the first land plants, based on their morphology and mitochondrial gene sequence data.

a. Hepatophyta
b. Pterophyta
c. Bryophyta
d. Lycophyta
e. Anthocerophyta

Answer: a
Difficulty: Difficult
Bloom's Taxonomy: Knowledge

25. Bryophytes have which of the following?

a. true leaves
b. lignified tissue
c. true roots
d. true stems
e. none of these

Answer: e
Difficulty: Moderate
Bloom's Taxonomy: Application

26. Which of the following is true in the moss life cycle?

a. Gametophytes originate within and are dependent upon sporophytes.
b. Sporophytes originate within and are dependent upon gametophytes.
c. Both gametophytes and sporophytes are totally independent from each other and are equally dominant.
d. Gametophytes are free-living and photosynthetic, but are replaced by a dominant sporophyte generation.
e. Gametophytes are photosynthetic and partially independent from the sporophytes.

Answer: b
Difficulty: Moderate
Bloom's Taxonomy: Application

27. Which of the following best describes fertilization in the moss life cycle?

a. Pollen is blown by the wind to a female cone, where it forms a pollen tube that grows toward where the egg will form.
b. Flagellated sperm swim in a film of water on the surface of the plant to reach an egg.
c. Flagellated sperm are blown by the wind to a location near an egg, then swim through plant fluids to reach the egg.
d. Flagellated sperm swim through plant fluids to reach an egg.

Answer: b
Difficulty: Moderate
Bloom's Taxonomy: Comprehension

28. A moss spore develops into a young gametophyte called a(n) _____.

a. capsule
b. antheridium
c. protonema
d. prothallus
e. archegonium

Answer: c
Difficulty: Difficult
Bloom's Taxonomy: Knowledge

29. The slender, rootlike structures found in nonvascular plants are called _____.

a. thalli
b. rhizomes
c. gametangia
d. protonemata
e. rhizoids

Answer: e
Difficulty: Easy
Bloom's Taxonomy: Knowledge

27.3 SEEDLESS VASCULAR PLANTS

30. Although there are fewer than 14,000 living species of seedless vascular plants, they were the dominant plants on Earth for nearly 200 million years. Their dominance ended at the end of the Carboniferous period, about

a. 1.5 billion years ago.
b. 420 million years ago.
c. 250 million years ago.
d. 125 million years ago.
e. 17 million years ago.

Answer: c
Difficulty: Moderate
Bloom's Taxonomy: Knowledge

31. Which of the following best describes fertilization in the fern life cycle?

a. Pollen is blown by the wind to a female cone, where it forms a pollen tube that grows toward where the egg will form.
b. Flagellated sperm swim in a film of water on the surface of the plant to reach an egg.
c. Flagellated sperm are blown by the wind to a location near an egg, then swim through plant fluids to reach the egg.
d. Flagellated sperm swim through plant fluids to reach an egg.

Answer: b
Difficulty: Moderate
Bloom's Taxonomy: Comprehension

32. Which of the following best explains why modern seedless vascular plants are confined largely to wet or humid environments?

a. They require external water for reproduction.
b. They do not have true roots.
c. They lack stomata.
d. They have a dominate sporophyte generation.
e. They do not have flowers or fruits.

Answer: a
Difficulty: Moderate
Bloom's Taxonomy: Comprehension

33. Much of the world's coal reserves were formed from the buried remains of the dominant plants of the Carboniferous period, which were _____.

a. bryophytes
b. seedless vascular plants
c. gymnosperms
d. angiosperms
e. gymnosperms and angiosperms

Answer: b
Difficulty: Easy
Bloom's Taxonomy: Comprehension

34. Which of the following is true in the fern life cycle?

a. Gametophytes originate within and are dependent upon sporophytes
b. Sporophytes originate within and are dependent upon gametophytes
c. Both gametophytes and sporophytes are totally independent from each other and are equally dominant
d. Gametophytes are free-living and photosynthetic, but are replaced by a dominant sporophyte generation
e. Gametophytes are photosynthetic and partially independent from the sporophytes

Answer: d
Difficulty: Moderate
Bloom's Taxonomy: Application

35. Club mosses, spike mosses, and quillworts are members of this phylum.

a. Hepatophyta
b. Pterophyta
c. Bryophyta
d. Lycophyta
e. Anthocerophyta

Answer: d
Difficulty: Easy
Bloom's Taxonomy: Knowledge

36. Ferns, whisk ferns, and horsetails are members of this phylum, which has about 13,000 described living species.

a. Hepatophyta
b. Pterophyta
c. Bryophyta
d. Lycophyta
e. Anthocerophyta

Answer: b
Difficulty: Easy
Bloom's Taxonomy: Knowledge

37. Club moss sporangia form on specialized leaves called _____ that occur near stem tips.

a. cladophylls
b. archegonia
c. thalli
d. antheridia
e. sporophylls

Answer: e
Difficulty: Difficult
Bloom's Taxonomy: Knowledge

38. The mature body form of the sporophyte of a typical _____ has fronds growing from a rhizome, with sori on the lower surface or margins of some of the fronds.

a. moss
b. whisk fern
c. fern
d. club moss
e. horsetail

Answer: c
Difficulty: Easy
Bloom's Taxonomy: Application

39. A sorus is a cluster of _____.

a. sporangia
b. fronds
c. strobili
d. gametangia
e. rhizomes

Answer: a
Difficulty: Easy
Bloom's Taxonomy: Knowledge

40. This phylum of seedless vascular plants has about 1000 known living species, including those in the genera *Lycopodium* and *Selaginella*. The gametophyte of some of these species cannot photosynthesize, and instead uses mycorrhizae to obtain nutrients.

a. Hepatophyta
b. Pterophyta
c. Bryophyta
d. Lycophyta
e. Anthocerophyta

Answer: d
Difficulty: Difficult
Bloom's Taxonomy: Knowledge

41. Represented by only two living genera, *Psilotum* and *Tmesipteris*, these plants resemble the extinct *Cooksonia*. Having no true roots or leaves, these seedless vascular plants are essentially forking green stems.

a. mosses
b. whisk ferns
c. ferns
d. club mosses
e. horsetails

Answer: b
Difficulty: Difficult
Bloom's Taxonomy: Knowledge

42. Represented by only one living genus, *Equisetum*, these seedless vascular plants have leaves in whorls around a photosynthetic stem that is stiff and gritty because they accumulate silica in their tissues.

a. mosses
b. whisk ferns
c. ferns
d. club mosses
e. horsetails

Answer: e
Difficulty: Moderate
Bloom's Taxonomy: Knowledge

27.4 GYMNOSPERMS: THE FIRST SEED PLANTS

43. The word gymnosperm means _____.

a. covered seed
b. active sperm
c. naked egg
d. naked seed
e. covered sperm

Answer: d
Difficulty: Easy
Bloom's Taxonomy: Knowledge

44. Which of the following is true in the pine life cycle?

a. Gametophytes originate within and are dependent upon sporophytes.
b. Sporophytes originate within and are dependent upon gametophytes.
c. Both gametophytes and sporophytes are totally independent from each other and are equally dominant.
d. Gametophytes are free-living and photosynthetic, but are replaced by a dominant sporophyte generation.
e. Gametophytes are photosynthetic and partially independent from the sporophytes.

Answer: a
Difficulty: Moderate
Bloom's Taxonomy: Synthesis

45. The specialized male gametophyte of seed plants, such as gymnosperms, is called a(n) _____.

a. sporopollenin
b. ovule
c. seed
d. strobilus
e. pollen grain

Answer: e
Difficulty: Easy
Bloom's Taxonomy: Knowledge

46. In seed plants, the transfer of male gametophytes to female reproductive parts is a process called _____.

a. meiosis
b. pollination
c. fertilization
d. syngamy
e. sporulation

Answer: b
Difficulty: Easy
Bloom's Taxonomy: Knowledge

47. The dominant land plants during the Mesozoic era, from about 250 million years ago to 65 million years ago, were _____.

a. lycophytes
b. ferns
c. gymnosperms
d. bryophytes
e. angiosperms

Answer: c
Difficulty: Moderate
Bloom's Taxonomy: Comprehension

48. An embryo sporophyte surrounded by nutritive tissue and then a tough, protective outer coat makes up a(n) _____ that shelters the embryo from drought, cold, and other adverse conditions.

a. pollen grain
b. spore
c. ovule
d. seed
e. strobilus

Answer: d
Difficulty: Easy
Bloom's Taxonomy: Comprehension

49. Living members of this group are often the dominant land plants in cool temperate zones, especially in areas with poor soils. Among other products, they are important sources of lumber, paper pulp, resins, and turpentine.

a. lycophytes
b. ferns
c. gymnosperms
d. bryophytes
e. angiosperms

Answer: c
Difficulty: Easy
Bloom's Taxonomy: Knowledge

50. This phylum of vascular, seed-bearing plants has about 185 known living species. Many of them look like palm trees, and some produce massive strobili. Their male and female strobili are found on separate plants.

a. Cycadophyta
b. Anthophyta
c. Gnetophyta
d. Coniferophyta
e. Ginkgophyta

Answer: a
Difficulty: Moderate
Bloom's Taxonomy: Knowledge

51. This phylum of vascular, seed-bearing plants has about 550 known living species, the most among gymnosperms. Many of them have needlelike leaves with thick cuticles and sunken stomata, and are thus adapted to dry conditions.

a. Cycadophyta
b. Anthophyta
c. Gnetophyta
d. Coniferophyta
e. Ginkgophyta

Answer: d
Difficulty: Moderate
Bloom's Taxonomy: Comprehension

52. This phylum of vascular, seed-bearing plants includes pines, spruces firs, hemlocks, junipers, cypresses, and redwoods. Most of them are evergreens that form woody reproductive cones.

a. Cycadophyta
b. Anthophyta
c. Gnetophyta
d. Coniferophyta
e. Ginkgophyta

Answer: d
Difficulty: Easy
Bloom's Taxonomy: Knowledge

53. In pine trees the megaspores develop within a(n) _____ .

a. sporopollenin
b. ovule
c. seed
d. strobilus
e. pollen grain

Answer: b
Difficulty: Moderate
Bloom's Taxonomy: Comprehension

54. In the pine life cycle, fertilization often takes place _____ pollination.

a. at the same time as
b. a few days after
c. a few weeks after
d. a few days before
e. months to a year after

Answer: e
Difficulty: Moderate
Bloom's Taxonomy: Knowledge

55. This phylum of vascular, seed-bearing plants is limited to one living species. That species is a deciduous tree with fan-shaped leaves. It has separate male and female plants, and the males are often planted in cities because of their resistance to pollution, insects, and disease.

a. Cycadophyta
b. Anthophyta
c. Gnetophyta
d. Coniferophyta
e. Ginkgophyta

Answer: e
Difficulty: Moderate
Bloom's Taxonomy: Knowledge

56. This phylum of vascular, seed-bearing plants has only about 70 living species in three genera, including *Welwitschia* and *Ephedra*. Some species have a two-step fertilization process, similar to fertilization in angiosperms.

a. Cycadophyta
b. Anthophyta
c. Gnetophyta
d. Coniferophyta
e. Ginkgophyta

Answer: c
Difficulty: Difficult
Bloom's Taxonomy: Knowledge

27.5 ANGIOSPERMS: FLOWERING PLANTS

57. This phylum of vascular, seed-bearing plants has over 260,000 known living species, and is the dominant group of land plants today.

a. Cycadophyta
b. Anthophyta
c. Gnetophyta
d. Coniferophyta
e. Ginkgophyta

Answer: b
Difficulty: Easy
Bloom's Taxonomy: Comprehension

58. These land plants are the only ones to make true flowers and fruits.

a. bryophytes
b. seedless vascular plants
c. gymnosperms
d. angiosperms
e. gymnosperms and angiosperms

Answer: d
Difficulty: Easy
Bloom's Taxonomy: Comprehension

59. The oldest well-documented flowering plant fossils are *Archaefructus*, dating to about

a. 1.5 billion years ago.
b. 420 million years ago.
c. 250 million years ago.
d. 125 million years ago.
e. 17 million years ago.

Answer: d
Difficulty: Moderate
Bloom's Taxonomy: Knowledge

60. One of the two main angiosperm clades, this group has about 60,000 species that are characterized by having a single cotyledon and by pollen grains with one groove. Most have parallel-veined leaves.

a. eudicots
b. the star anise group
c. monocots
d. water lilies
e. magnoliids

Answer: c
Difficulty: Moderate
Bloom's Taxonomy: Knowledge

61. One of the two main angiosperm clades, this group has about 200,000 species. Members of the group all have pollen grains with three grooves, and most have two cotyledons. Most also have net-veined leaves.

a. eudicots
b. the star anise group
c. monocots
d. water lilies
e. magnoliids

Answer: a
Difficulty: Moderate
Bloom's Taxonomy: Knowledge

62. Based on morphology and phytochrome gene sequencing data, it appears that the closest living relative of the first flowering plants is _____.

a. *Magnolia grandiflora*
b. *Illicium floridanum*
c. *Nelumbo nucifera*
d. *Archaefructus sinensis*
e. *Amborella trichopoda*

Answer: e
Difficulty: Difficult
Bloom's Taxonomy: Knowledge

63. Which of the following is true in the angiosperm life cycle?

a. Gametophytes originate within and are dependent upon sporophytes.
b. Sporophytes originate within and are dependent upon gametophytes.
c. Both gametophytes and sporophytes are totally independent from each other and are equally dominant.
d. Gametophytes are free-living and photosynthetic, but are replaced by a dominant sporophyte generation.
e. Gametophytes are photosynthetic and partially independent from the sporophytes.

Answer: a
Difficulty: Moderate
Bloom's Taxonomy: Synthesis

64. Based on gene sequencing data, it appears that magnoliids are most closely related to _____.

a. eudicots
b. star anise group
c. monocots
d. water lilies

Answer: c
Difficulty: Difficult
Bloom's Taxonomy: Comprehension

65. Which of the following contributed to the adaptive success of angiosperms over gymnosperms?

a. vessel elements
b. more efficient sugar transport
c. ovary
d. fruit
e. all of these

Answer: e
Difficulty: Easy
Bloom's Taxonomy: Synthesis

66. One advantage that angiosperms have is a unique triploid _____, a nutritive tissue produced during fertilization.

a. seed coat
b. endosperm
c. ovule
d. pollen sac
e. ovary

Answer: b
Difficulty: Moderate
Bloom's Taxonomy: Comprehension

67. Many angiosperm species have _____ animal pollinators.

a. competed with
b. parasitized
c. been driven to extinction by
d. coevolved with
e. antagonistic relationships with

Answer: d
Difficulty: Easy
Bloom's Taxonomy: Comprehension

68. If you see a yellow flower with "nectar guide" stripes best seen in ultraviolet light and with a strong, sweet odor, it is likely pollinated by _____.

a. flies
b. birds
c. bats or moths
d. butterflies
e. bees

Answer: e
Difficulty: Moderate
Bloom's Taxonomy: Application

69. If you see a red flower with very little odor, it is likely pollinated by _____.

a. flies
b. birds
c. bats or moths
d. butterflies
e. bees

Answer: b
Difficulty: Difficult
Bloom's Taxonomy: Application

INSIGHTS FROM THE MOLECULAR REVOLUTION: THE POWERFUL GENETIC TOOLKIT FOR STUDYING PLANT EVOLUTION

70. Plants genes are found in _____.

a. nuclear DNA
b. mitochondrial and nuclear DNA
c. mitochondrial, chloroplast, and nuclear DNA
d. chloroplast and nuclear DNA
e. mitochondrial and chloroplast DNA

Answer: c
Difficulty: Easy
Bloom's Taxonomy: Application

71. The *rbcL* gene, which has been useful in evolutionary studies in plants, is located in _____.

a. mitochondrial DNA
b. chloroplast DNA
c. nuclear DNA
d. mitochondrial and chloroplast DNA

Answer: b
Difficulty: Easy
Bloom's Taxonomy: Comprehension

72. Compared to other genes, the *rbcL* gene in plants has a mutation rate that is _____ most other plant genes.

a. slower than
b. similar to
c. faster than
d. more random than

Answer: a
Difficulty: Moderate
Bloom's Taxonomy: Comprehension

UNANSWERED QUESTIONS

73. What did Darwin famously refer to as an "abominable mystery"?

a. the lack of transitional fossils between algae and plants
b. alternation of generations in plants
c. associations between specific types of flowers and specific animal pollinators
d. the sudden appearance and immediate diversification of flowering plants in the fossil record
e. the development of vascular plants

Answer: d
Difficulty: Difficult
Bloom's Taxonomy: Comprehension

74. Although some gnetophytes have features similar to angiosperms, data from _____ indicates that gnetophytes and angiosperms are not closely related.

a. DNA sequencing
b. electron microscopy
c. the fossil record
d. studies of fertilization

Answer: a
Difficulty: Moderate
Bloom's Taxonomy: Comprehension

Integrative Multiple-Choice

75. You are a botanist surveying a previously unstudied island. You discover a plant that has a waxy cuticle and a dominant gametophyte, but no xylem or phloem. Into which group should you classify this plant?

a. bryophytes
b. seedless vascular plants
c. gymnosperms
d. angiosperms

Answer: a
Difficulty: Moderate
Bloom's Taxonomy: Application
Source: Sections 27.2–27.5

76. You are a botanist surveying a previously unstudied island. You discover a plant that has large blue flowers and a woody fruit. Into which group should you classify this plant?

a. bryophytes
b. seedless vascular plants
c. gymnosperms
d. angiosperms

Answer: d
Difficulty: Easy
Bloom's Taxonomy: Application
Source: Sections 27.2–27.5

77. You are a botanist surveying a previously unstudied island. You discover a plant that has xylem with tracheids but no vessel elements. The plant does not require any external water for fertilization. Into which group should you classify this plant?

a. bryophytes
b. seedless vascular plants
c. gymnosperms
d. angiosperms

Answer: c
Difficulty: Difficult
Bloom's Taxonomy: Application
Source: Sections 27.2–27.5

78. You are a botanist surveying a previously unstudied island. You discover a plant that has a small free-living gametophyte that requires external water for fertilization. After fertilization, a large sporophyte with xylem and phloem develops. The gametophyte dies, and the free-living sporophyte eventually releases spores from sporangia on the margins of its leaves. Into which group should you classify this plant?

a. bryophytes
b. seedless vascular plants
c. gymnosperms
d. angiosperms

Answer: b
Difficulty: Moderate
Bloom's Taxonomy: Application
Source: Sections 27.2–27.5

Choice

Choose the term that best describes cells in the following stages of the plant life cycle or structures in a plant. Each term may be used once, more than once, or not at all.

a. haploid (*n*)
b. diploid (*2n*)
c. some haploid and some diploid cells
d. triploid (*3n*)

79. _____ spore

80. _____ pine seed

81. _____ sporophyte

82. _____ gamete

83. _____ pollen grain

84. _____ angiosperm endosperm

85. _____ gametophyte

86. _____ zygote

87. _____ true leaf

88. _____ protonema

Answers: 79. *a* 80. *c* 81. *b* 82. *a* 83. *a*
84. *d* 85. *a* 86. *b* 87. *b* 88. *a*

Difficulty: Moderate
Bloom's Taxonomy: Knowledge
Source: Sections 27.1–27.5

Short Answer

89. A seed is often described as "a baby plant in a box with its lunch." Explain this statement using more proper terms for *baby plant*, *box*, and *lunch*.

Answer: A seed has a "baby plant" called an embryo, which is the developing sporophyte. It is surrounded by its "lunch," food supplies in the form of nutritive tissue such as endosperm. The embryo and the nutritive tissue are in a "box," hardened tissue called a seed coat that protects the embryo from drying out.
Difficulty: Moderate
Bloom's Taxonomy: Synthesis
Source: Section 27.4, 27.5

Essay

90. Describe the major differences between bryophytes and vascular plants.

Answer: The major difference between bryophytes and vascular plants is that vascular plants have xylem and phloem, specialized lignified tubes for conducting water and nutrients; bryophytes do not have these. Vascular plants also have apical meristems, constantly growing tips that bryophytes do not have. The presence of vascular tissue and apical meristems allows vascular plants to have root and shoot systems that develop true roots, stems, and leaves—again, something not found in bryophytes. All of this allows vascular plants to typically grow larger and live in more arid environments than bryophytes. Bryophytes also have a dominant gametophyte generation, whereas vascular plants have a dominant sporophyte generation.
Difficulty: Moderate
Bloom's Taxonomy: Synthesis
Source: Sections 27.1–27.3

91.   Describe the major differences between seedless vascular plants, gymnosperms, and angiosperms.

Answer: Gymnosperms and angiosperms (seed plants) both produce seeds, specialized structures for protecting the developing sporophyte embryo from drying out and other dangers; this is something that seedless vascular plants do not have. Seed plants also produce gametophytes that develop within protective tissues in the sporophyte, something that does not happen in seedless vascular plants; instead, seedless vascular plants release gametophytes as single-celled spores that then must develop on their own. Seed plants also differ from seedless vascular plants in making pollen grains, specialized male gametophytes that are protected from drying out and that do not need external water to reach an egg. All of these adaptations allow seed plants to generally live in drier conditions than seedless vascular plants can handle. Within the seed plants, angiosperms differ significantly from gymnosperms in many ways. Angiosperms have enhanced xylem and phloem that provide more efficient transport of water and nutrients. Angiosperms have enhanced nutrition for embryos, with double fertilization that produces a unique triploid endosperm to nourish the embryo. They have flowers that often attract animal pollinators, providing more efficient means for pollination. Also, angiosperms have ovaries, tissues that shelter the ovule from desiccation and attack. Ovaries typically develop into fruits that can help protect seeds as well as aid in seed dispersal.
Difficulty: Moderate
Bloom's Taxonomy: Synthesis
Source: Sections 27.1, 27.4, and 27.5

28
FUNGI

Multiple-Choice

WHY IT MATTERS

1.   Worldwide, decay of organic matter is important to the global carbon cycle because it _____ tens of billions of tons of carbon dioxide each year.

a.   removes from the atmosphere
b.   prevents the planet from losing to space
c.   releases to the atmosphere
d.   prevents plants from having access to
e.   releases to space

Answer: c
Difficulty: Easy
Bloom's Taxonomy: Comprehension

2.   Based on molecular evidence, fungi were present on land at least

a.   1 billion years ago.
b.   500 million years ago.
c.   250 million years ago.
d.   125 million years ago.
e.   10 million years ago.

Answer: b
Difficulty: Moderate
Bloom's Taxonomy: Knowledge

3.   The study of fungi is called _____.

a.   ecology
b.   botany
c.   mycology
d.   microbiology
e.   parasitology

Answer: c
Difficulty: Easy
Bloom's Taxonomy: Knowledge

28.1 GENERAL CHARACTERISTICS OF FUNGI

4.   Yeast are

a.   single-celled fungi.
b.   multicellular plants.
c.   single-celled plants.
d.   single-celled protists.
e.   multicellular fungi.

Answer: a
Difficulty: Easy
Bloom's Taxonomy: Knowledge

5.   The substance that typically provides rigidity to fungal cell walls is _____.

a.   lignin
b.   cellulose
c.   starch
d.   collagen
e.   chitin

Answer: e
Difficulty: Moderate
Bloom's Taxonomy: Comprehension

6.   Multicellular fungi grow as branching filaments called _____.

a.   mycorrhizae
b.   conidia
c.   asci
d.   hyphae
e.   basidia

Answer: d
Difficulty: Easy
Bloom's Taxonomy: Knowledge

7.   Imagine that you are a researcher who has discovered a mutant fungus that produces extra-thick septa that allow for only minimal transfer of nutrients through them. You should expect that, when compared to the normal fungus, this mutant will

a.   be extremely resistant to antibiotics used to treat bacterial infections.
b.   digest food more efficiently.
c.   grow more slowly.
d.   divide more rapidly.
e.   make more extensive use of cytoplasmic streaming.

Answer: c
Difficulty: Difficult
Bloom's Taxonomy: Evaluation

8.   The process in fungi that allows nutrients to flow from food-absorbing parts of the fungal body to other, nonabsorptive parts is _____.

a.   osmosis
b.   karyogamy
c.   cytoplasmic streaming
d.   plasmogamy
e.   symbiosis

Answer: c
Difficulty: Moderate
Bloom's Taxonomy: Knowledge

9. A mycelium is best defined as

a. a mushroom.
b. a mass of hyphae.
c. an association between a fungus and a plant root.
d. a reproductive structure of a fungus.
e. a fungal cell.

Answer: b
Difficulty: Moderate
Bloom's Taxonomy: Knowledge

10. Your favorite tree is dying, and you find that there is a fungal infection on its leaves. If you were to look at the cells of the infected leaves of your tree, which of the following should you expect to find?

a. mycorrhizae
b. arbuscules
c. haustoria
d. both mycorrhizae and arbuscules
e. both mycorrhizae and haustoria

Answer: c
Difficulty: Difficult
Bloom's Taxonomy: Analysis

11. Fungi generally digest large organic molecules

a. in special compartments called arbuscules.
b. within feeding hyphae.
c. in visible mushrooms.
d. in special compartments called haustoria.
e. outside of their cells.

Answer: e
Difficulty: Moderate
Bloom's Taxonomy: Application

12. The dikaryotic stage of a fungal life cycle is described as _____.

a. n
b. $n + n$
c. $2n + 2n$
d. $2n$
e. $2n + n$

Answer: b
Difficulty: Easy
Bloom's Taxonomy: Knowledge

13. Before examining a fungal cell under a microscope, you are told that it is a dikaryotic cell. Given that, you should expect the cell to have

a. two complete sets of chromosomes.
b. two nuclei.
c. two hyphae.
d. two complete sets of chromosomes and two nuclei.
e. two nuclei and two hyphae.

Answer: d
Difficulty: Easy
Bloom's Taxonomy: Knowledge

14. All fungi can reproduce via _____.

a. asci
b. fruiting bodies
c. spores
d. conidia
e. basidia

Answer: c
Difficulty: Easy
Bloom's Taxonomy: Comprehension

15. Mature fungal spores

a. are always haploid.
b. are always diploid.
c. are always dikaryotic.
d. can be either diploid or dikaryotic.
e. can be either haploid or diploid.

Answer: a
Difficulty: Difficult
Bloom's Taxonomy: Synthesis

16. Asexually produced spores will result in offspring that

a. have no mating type.
b. are genetically distinct from the parent fungus.
c. are clones of the parent fungus.
d. have a different mating type from the parent fungus.
e. are unable to reproduce.

Answer: c
Difficulty: Easy
Bloom's Taxonomy: Knowledge

17. The fusion of the cytoplasms of two genetically different cells is _____.

a. osmosis
b. karyogamy
c. cytoplasmic streaming
d. plasmogamy
e. symbiosis

Answer: d
Difficulty: Moderate
Bloom's Taxonomy: Knowledge

18. The fusion of the nuclei is _____.

a. osmosis
b. karyogamy
c. cytoplasmic streaming
d. plasmogamy
e. symbiosis

Answer: b
Difficulty: Difficult
Bloom's Taxonomy: Knowledge

19. Mating types in fungi are generally termed

a. sperm and eggs.
b. α and β.
c. male and female.
d. plus and minus.
e. Fungi do not have specific mating types.

Answer: d
Difficulty: Difficult
Bloom's Taxonomy: Knowledge

28.2 MAJOR GROUPS OF FUNGI

20. Based on the phylogeny of fungi that is currently the most widely accepted, organisms of the phylum _____ belong to the fungal group that diverged the earliest from the rest of the fungi.

a. Glomeromycota
b. Basidiomycota
c. Chytridiomycota
d. Zygomycota
e. Ascomycota

Answer: c
Difficulty: Difficult
Bloom's Taxonomy: Analysis

21. Based on gene sequencing data, it appears that the lineages leading to animals and fungi diverged from each other about

a. 1 billion years ago.
b. 500 million years ago.
c. 250 million years ago.
d. 125 million years ago.
e. 10 million years ago.

Answer: a
Difficulty: Difficult
Bloom's Taxonomy: Knowledge

22. The first fungi were most likely _____.

a. airborne
b. terrestrial
c. arboreal
d. aquatic
e. amphibious

Answer: d
Difficulty: Easy
Bloom's Taxonomy: Knowledge

23. Traditionally, classification of fungi has been based on

a. DNA sequences.
b. body type.
c. mode of feeding.
d. whether the fungus is a saprobe, parasite, or in a mutual symbiosis.
e. structures that release sexual spores.

Answer: e
Difficulty: Moderate
Bloom's Taxonomy: Comprehension

24. Members of which fungal phylum are the only ones to produce motile spores?

a. Glomeromycota
b. Basidiomycota
c. Chytridiomycota
d. Zygomycota
e. Ascomycota

Answer: c
Difficulty: Moderate
Bloom's Taxonomy: Knowledge

25. The chytrid *Batrachochytrium dendrobatidis* is responsible for a disease that

a. is sexually transmitted in humans.
b. has wiped out many species of frogs.
c. infects wheat, rye, and other grain crops.
d. is killing trees in temperate forests.
e. kills millions of humans each year.

Answer: b
Difficulty: Difficult
Bloom's Taxonomy: Knowledge

26. Most chytrids are _____.

a. saprobes in soil
b. parasites of plants
c. symbiotic partners of cattle and other herbivores
d. parasites of animals
e. aquatic

Answer: e
Difficulty: Moderate
Bloom's Taxonomy: Knowledge

27. Creation of a special, resistant zygospore during sexual reproduction is characteristic of members of which fungal phylum?

a. Glomeromycota
b. Basidiomycota
c. Chytridiomycota
d. Zygomycota
e. Ascomycota

Answer: d
Difficulty: Easy
Bloom's Taxonomy: Knowledge

28. The black bread mold *Rhizopus stolonifer* and dung-infesting fungi of the genus *Pilobolus* are members of which fungal phylum?

a. Glomeromycota
b. Basidiomycota
c. Chytridiomycota
d. Zygomycota
e. Ascomycota

Answer: d
Difficulty: Difficult
Bloom's Taxonomy: Comprehension

29. Having aseptate hyphae, members of this phylum are described as coenocytic since they have numerous nuclei in a common cytoplasm.

a. Glomeromycota
b. Basidiomycota
c. Chytridiomycota
d. Zygomycota
e. Ascomycota

Answer: d
Difficulty: Difficult
Bloom's Taxonomy: Comprehension

30. Where does karyogamy occur in zygomycetes?

a. basidia
b. conidiophores
c. sporangia
d. asci
e. zygospores

Answer: e
Difficulty: Moderate
Bloom's Taxonomy: Comprehension

31. All known members of which fungal phylum form mutualistic associations with plant roots?

a. Glomeromycota
b. Basidiomycota
c. Chytridiomycota
d. Zygomycota
e. Ascomycota

Answer: a
Difficulty: Easy
Bloom's Taxonomy: Knowledge

32. Which fungal phylum has the fewest known living species?

a. Glomeromycota
b. Basidiomycota
c. Chytridiomycota
d. Zygomycota
e. Ascomycota

Answer: a
Difficulty: Difficult
Bloom's Taxonomy: Synthesis

33. Members of which fungal phylum make up roughly half of the fungi in the soil and help 80–90 percent of land plants by forming mycorrhizae with them?

a. Glomeromycota
b. Basidiomycota
c. Chytridiomycota
d. Zygomycota
e. Ascomycota

Answer: a
Difficulty: Moderate
Bloom's Taxonomy: Comprehension

34. Reproduction in glomeromycetes is typically _____ via spores that form _____.

a. sexual; in sporangia
b. asexual; at the tips of hyphae
c. sexual; at the tips of hyphae
d. asexual; in sporangia

Answer: b
Difficulty: Difficult
Bloom's Taxonomy: Knowledge

35. Where does karyogamy occur in ascomycetes?

a. basidia
b. conidiophores
c. sporangia
d. asci
e. zygospores

Answer: d
Difficulty: Moderate
Bloom's Taxonomy: Comprehension

36. Members of which fungal phylum are also called sac fungi?

a. Glomeromycota
b. Basidiomycota
c. Chytridiomycota
d. Zygomycota
e. Ascomycota

Answer: e
Difficulty: Easy
Bloom's Taxonomy: Knowledge

37. Pathogenic members of which fungal phylum are responsible for such things as Dutch elm disease, ergotism, athlete's foot, aflatoxins, thrush, and vaginal yeast infections?

a. Glomeromycota
b. Basidiomycota
c. Chytridiomycota
d. Zygomycota
e. Ascomycota

Answer: e
Difficulty: Moderate
Bloom's Taxonomy: Knowledge

38. Organisms of significant value to humans from this fungal phylum include the yeast *Saccharomyces cerevisiae*, antibiotic-producing species of *Penicillium*, truffles, and morels.

a. Glomeromycota
b. Basidiomycota
c. Chytridiomycota
d. Zygomycota
e. Ascomycota

Answer: e
Difficulty: Moderate
Bloom's Taxonomy: Knowledge

39. Production of asexual spores in ascomycetes often occurs on _____.

a. basidia
b. conidiophores
c. sporangia
d. asci
e. zygospores

Answer: b
Difficulty: Moderate
Bloom's Taxonomy: Comprehension

40. Production of sexual spores in ascomycetes occurs in _____.

a. basidia
b. conidiophores
c. sporangia
d. asci
e. zygospores

Answer: d
Difficulty: Easy
Bloom's Taxonomy: Comprehension

41. A trapping behavior where small worms are ensnared and then digested is found in some species of which fungal phylum?

a. Glomeromycota
b. Basidiomycota
c. Chytridiomycota
d. Zygomycota
e. Ascomycota

Answer: e
Difficulty: Difficult
Bloom's Taxonomy: Knowledge

42. Which fungal phylum has the most known living species?

a. Glomeromycota
b. Basidiomycota
c. Chytridiomycota
d. Zygomycota
e. Ascomycota

Answer: e
Difficulty: Moderate
Bloom's Taxonomy: Synthesis

43. Organisms traditionally called mushrooms, as well as shelf fungi, coral fungi, bird's-nest fungi, stinkhorns, and puffballs, are all members of which fungal phylum?

a. Glomeromycota
b. Basidiomycota
c. Chytridiomycota
d. Zygomycota
e. Ascomycota

Answer: b
Difficulty: Easy
Bloom's Taxonomy: Knowledge

44. Members of which fungal phylum are also called club fungi?

a. Glomeromycota
b. Basidiomycota
c. Chytridiomycota
d. Zygomycota
e. Ascomycota

Answer: b
Difficulty: Easy
Bloom's Taxonomy: Knowledge

45. A basidiocarp is the _____ of a basidiomycete.

a. feeding stage
b. asexual sporangium
c. reproductive body
d. infecting vessel
e. haploid life cycle stage

Answer: c
Difficulty: Easy
Bloom's Taxonomy: Knowledge

46. Where does karyogamy occur in basidiomycetes?

a. basidia
b. conidiophores
c. sporangia
d. asci
e. zygospores

Answer: a
Difficulty: Moderate
Bloom's Taxonomy: Comprehension

47. Production of sexual spores in basidiomycetes occurs on _____.

a. basidia
b. conidiophores
c. sporangia
d. asci
e. zygospores

Answer: a
Difficulty: Easy
Bloom's Taxonomy: Comprehension

48. The fly agaric mushroom (*Amanita muscaria*) and the death cap mushroom (*Amanita phalloides*) are members of which fungal phylum?

a. Glomeromycota
b. Basidiomycota
c. Chytridiomycota
d. Zygomycota
e. Ascomycota

Answer: b
Difficulty: Moderate
Bloom's Taxonomy: Knowledge

49. In eastern Oregon, the mycelium of a single individual *Armillaria ostoyae*, a member of the fungal phylum _____, covers an area equivalent to 1665 football fields (nearly 6000 m across).

a. Glomeromycota
b. Basidiomycota
c. Chytridiomycota
d. Zygomycota
e. Ascomycota

Answer: b
Difficulty: Difficult
Bloom's Taxonomy: Knowledge

50. The grouping "conidial fungi," also known as "imperfect fungi" or deuteromycetes, is a convenience grouping for fungi that are not classified in other groups, because the conidial fungi

a. have no known sexual phase
b. are unicellular
c. have a dikaryotic phase in their life cycle
d. have a haploid phase in their life cycle
e. all of these

Answer: a
Difficulty: Easy
Bloom's Taxonomy: Knowledge

51. Molecular relationships and other studies have allowed for many conidial fungi to be reclassified into a fungal phylum. Most of these reclassified conidial fungi have turned out to belong to which fungal phylum?

a. Glomeromycota
b. Basidiomycota
c. Chytridiomycota
d. Zygomycota
e. Ascomycota

Answer: e
Difficulty: Difficult
Bloom's Taxonomy: Knowledge

52. Grouped by some with the fungi based on molecular studies, these single-celled parasites are known to infect insects, fish, and humans (especially those with compromised immune systems). They physically resemble spores, but they lack mitochondria.

a. amoebae
b. red algae
c. microsporidia
d. dinoflagellates
e. apicomplexans

Answer: c
Difficulty: Easy
Bloom's Taxonomy: Knowledge

53. Microsporidia appear to be most closely related to members of which fungal phylum?

a. Glomeromycota
b. Basidiomycota
c. Chytridiomycota
d. Zygomycota
e. Ascomycota

Answer: d
Difficulty: Difficult
Bloom's Taxonomy: Comprehension

INSIGHTS FROM THE MOLECULAR REVOLUTION: THERE WAS PROBABLY A FUNGUS AMONG US

54. Studies of ribosomal RNA (rRNA) sequences indicate that _____ are more closely related to each other than either one is to _____.

a. plants and fungi; animals
b. plants and animals; fungi
c. animals and fungi; plants

Answer: c
Difficulty: Easy
Bloom's Taxonomy: Synthesis

55. Which of the following is a biological similarity that fungi and animals have with each?

a. pathways for producing chitin
b. pathways for making hydroxyproline
c. pathways for making the protein ferritin
d. all of these
e. none of these

Answer: d
Difficulty: Difficult
Bloom's Taxonomy: Knowledge

56. You discover a fungus that only makes sexual spores (no asexual spores). The spores are made in groups of four on the tips of club-shaped structures. Into which group should you classify this fungus?

a. Glomeromycota
b. Basidiomycota
c. Chytridiomycota
d. Zygomycota
e. Ascomycota

Answer: b
Difficulty: Moderate
Bloom's Taxonomy: Application

57. You discover a fungus that makes both sexual and asexual spores. The sexual spores are made in groups of eight within a structure that bursts open to release the spores. The structures containing sexual spores are found within a multicellular structure that is about five inches tall. Into which group should you classify this fungus?

a. Glomeromycota
b. Basidiomycota
c. Chytridiomycota
d. Zygomycota
e. Ascomycota

Answer: e
Difficulty: Difficult
Bloom's Taxonomy: Application

58. You discover a fungus that makes mainly asexual spores, but that occasionally makes sexual spores. When produced, the sexual spores come out of a thickened, hardened structure. A close examination of hyphae of this fungus reveals no septa. Into which group should you classify this fungus?

a. Glomeromycota
b. Basidiomycota
c. Chytridiomycota
d. Zygomycota
e. Ascomycota

Answer: d
Difficulty: Moderate
Bloom's Taxonomy: Application

59. You discover an aquatic fungus that makes spores that use flagella to swim. Into which group should you classify this fungus?

a. Glomeromycota
b. Basidiomycota
c. Chytridiomycota
d. Zygomycota
e. Ascomycota

Answer: c
Difficulty: Moderate
Bloom's Taxonomy: Application

60. The toxin α-amanitin has deadly effects on humans because it inhibits _____.

a. DNA replication
b. RNA polymerase
c. peptide bond formation
d. muscle contraction
e. active transport

Answer: b
Difficulty: Difficult
Bloom's Taxonomy: Comprehension

28.3 FUNGAL ASSOCIATIONS

61. The general term for a state where two organisms live together in close association is _____.

a. commensalism
b. parasitism
c. mutualism
d. predation
e. symbiosis

Answer: e
Difficulty: Easy
Bloom's Taxonomy: Knowledge

62. A single vegetative body that contains both a fungus and a green alga would be called a(n) _____.

a. ectomycorrhiza
b. lichen
c. arbuscule
d. endomycorrhiza
e. haustorium

Answer: b
Difficulty: Easy
Bloom's Taxonomy: Knowledge

63. In a lichen that is a mutually beneficial relationship, the mycobiont gets

a. nutrients.
b. protection from herbivory.
c. aid in sexual reproduction.
d. enhanced dispersal of offspring.
e. shelter from radiation and desiccation.

Answer: a
Difficulty: Moderate
Bloom's Taxonomy: Comprehension

64. In a lichen that is a mutually beneficial relationship, the photobiont usually gets

a. nutrients.
b. protection from herbivory.
c. aid in sexual reproduction.
d. enhanced dispersal of offspring.
e. shelter from radiation and desiccation.

Answer: e
Difficulty: Moderate
Bloom's Taxonomy: Comprehension

65. Which of the following is NOT associated with lichens?

a. primary producers in the arctic tundra
b. monitoring of air pollution
c. creation of soil from bare rock
d. enhanced photosynthesis on trees with lichens
e. nest building material for many birds

Answer: d
Difficulty: Difficult
Bloom's Taxonomy: Synthesis

66. The tough, pliable body of a lichen is called a(n)

a. soredium.
b. arbuscule.
c. thallus.
d. sporangium.
e. haustorium.

Answer: c
Difficulty: Moderate
Bloom's Taxonomy: Knowledge

67. A cell cluster with both algal and hyphal cells that is used for asexual reproduction in lichens is called a(n)

a. soredium.
b. arbuscule.
c. thallus.
d. sporangium.
e. haustorium.

Answer: a
Difficulty: Difficult
Bloom's Taxonomy: Knowledge

68. A mutualistic symbiosis where fungal hyphae penetrate plant root cells is called a(n) _____.

a. ectomycorrhiza
b. lichen
c. arbuscule
d. endomycorrhiza
e. haustorium

Answer: d
Difficulty: Moderate
Bloom's Taxonomy: Knowledge

69. A mutualistic symbiosis where fungal hyphae grow between and around plant roots but do not penetrate plant root cells is called a(n) _____.

a. ectomycorrhiza
b. lichen
c. arbuscule
d. endomycorrhiza
e. haustorium

Answer: a
Difficulty: Moderate
Bloom's Taxonomy: Knowledge

70. Arbuscules are structures associated with which fungal phylum?

a. Glomeromycota
b. Basidiomycota
c. Chytridiomycota
d. Zygomycota
e. Ascomycota

Answer: a
Difficulty: Moderate
Bloom's Taxonomy: Knowledge

71. The main benefit to a plant from a mycorrhizal association is

a. enhanced water uptake.
b. removal of excess carbohydrates.
c. enhanced flow of carbohydrates to the roots.
d. enhanced mineral ion uptake.
e. none of these

Answer: d
Difficulty: Moderate
Bloom's Taxonomy: Comprehension

72. In a mycorrhizal association, the fungus benefits mainly by

a. absorbing water from the plant.
b. using the plant roots as a growth surface.
c. absorbing mineral ions from the plant.
d. providing water to the plant.
e. absorbing carbohydrates from the plant.

Answer: e
Difficulty: Easy
Bloom's Taxonomy: Knowledge

73. Acid rain _____ mycorrhizae.

a. has no noticeable effect on
b. damages
c. enhances the use of
d. stimulates

Answer: b
Difficulty: Easy
Bloom's Taxonomy: Knowledge

74. Fossils show that _____ were common among ancient land plants.

a. fungal infections
b. endomycorrhizae
c. lichens
d. haustoria
e. ectomycorrhizae

Answer: b
Difficulty: Moderate
Bloom's Taxonomy: Knowledge

75. You discover a fungus that is associated with plant roots, with hyphae that actually penetrate the root cells and produces tree-like structures within the root cells. Into which group should you classify this fungus?

a. Glomeromycota
b. Basidiomycota
c. Chytridiomycota
d. Zygomycota
e. Ascomycota

Answer: a
Difficulty: Moderate
Bloom's Taxonomy: Application

FOCUS ON RESEARCH: APPLIED RESEARCH: LICHENS AS MONITORS OF AIR POLLUTION'S BIOLOGICAL DAMAGE

76. Worldwide, _____ are often used as pollution-monitoring devices.

a. fungal infections
b. mycorrhizae
c. lichens
d. yeast
e. mushrooms

Answer: c
Difficulty: Easy
Bloom's Taxonomy: Knowledge

77. Old man's beard (*Usnea trichodea*) and yellow *Evernia* lichens are sensitive to elevated levels of

a. sulfur dioxide (SO_2).
b. nitrogen oxides.
c. nitrate.
d. sunlight.
e. fluoride salts.

Answer: a
Difficulty: Moderate
Bloom's Taxonomy: Comprehension

UNANSWERED QUESTIONS

78. *Cochliobolus carbonum* is an ascomycete that secretes a toxin called _____ to infect maize hosts, causing leaf blight and ear rot disease.

a. aflatoxin
b. ergot
c. HC-toxin
d. ricin
e. α-amanitin

Answer: c
Difficulty: Difficult
Bloom's Taxonomy: Knowledge

79. Studies by Dan Ebbole and his group of how *Magnaporthe grisea* causes rice blast focus on 300 _____ predicted to be secreted by the fungus.

a. proteins
b. steroids
c. ions
d. carbohydrates
e. toxins

Answer: a
Difficulty: Moderate
Bloom's Taxonomy: Knowledge

Choice

Choose the fungal phylum most closely associated with each of the following terms. Each phylum may be used once, more than once, or not at all.

a. Glomeromycota
b. Basidiomycota
c. Chytridiomycota
d. Zygomycota
e. Ascomycota

80. _____ *Saccharomyces cerevisiae*

81. _____ motile spores

82. _____ microsporidia

83. _____ ectomycorrhizae

84. _____ basidiocarp

85. _____ *Rhizopus stolonifer*

86. _____ ascocarp

87. _____ α-amanitin

88. _____ aseptate hyphae

89. _____ endomycorrhizae

Answers: 80. e   81. c   82. d   83. b   84. b
85. d   86. e   87. b   88. d   89. a

Difficulty: Moderate
Bloom's Taxonomy: Knowledge
Source: Sections 28.2, 28.3

Short Answer

90. Explain the difference between plasmogamy and karyogamy.

Answer: Plasmogamy is the fusion of two cells so they share a common cytoplasm but keep the two original nuclei separate. It results in a cell that has two distinct nuclei, also called a dikaryon, with ploidy of n + n. Karyogamy is the fusion of the two nuclei; when this occurs, the cell ceases being an n + n dikaryon and instead becomes a true 2n diploid cell.
Difficulty: Moderate
Bloom's Taxonomy: Synthesis
Source: Section 28.1

Essay

91.   Describe the major roles of fungi that affect humans and their environment.

Answer: Fungi have many effects on humans and their environment. First, along with bacteria, the primary decomposers on Earth are fungi. Decomposers provide vital recycling of carbon and other nutrients; without such recycling, the planet would quickly face a nutrient crisis as carbon and other nutrients became locked up in leaf litter and other cast-off organic matter. Second, fungi are the cause of many diseases. They are the greatest cause of plant diseases, and many diseases in animals (including humans) are caused by fungi. Third, fungi provide extremely important direct benefits to about 90 percent of plants via mycorrhizae. Fourth, fungi are part of lichens, which begin the process of making soils from bare rocks and which are the primary producers in some harsh environments like the arctic tundra. Finally, fungi are used by humans for such things as baking, making cheeses, making alcoholic beverages, making antibiotics, and as model organisms for studies of molecular biology and genetic engineering in eukaryotes.
Difficulty: Moderate
Bloom's Taxonomy: Synthesis
Source: Why It Matters; Sections 28.1–28.3

29

ANIMAL PHYLOGENY, ACOELOMATES, AND PROTOSTOMES

Multiple-Choice

WHY IT MATTERS

1. The Burgess Shale formation in western Canada, with fossils dating to about _____ years ago, includes a wide variety of animal fossils such as *Hallucingenia* and *Opabinia*.

a. 4 billion
b. 8,000
c. 530 million
d. 1.2 billion
e. 78 million

Answer: c
Difficulty: Easy
Bloom's Taxonomy: Knowledge

2. Zoologists have described nearly _____ living species of animals.

a. 20,000
b. 2 million
c. 47,000
d. 100,000
e. 1 million

Answer: b
Difficulty: Moderate
Bloom's Taxonomy: Knowledge

29.1 WHAT IS AN ANIMAL?

3. Which of the following statements about kingdom Animalia is FALSE?

a. All adult animals are motile.
b. All animals are eukaryotes.
c. All adult animals are multicellular.
d. Animal cells lack cell walls.
e. All animals are heterotrophs.

Answer: a
Difficulty: Moderate
Bloom's Taxonomy: Synthesis

4. Which of the following is NOT a line of evidence that Kingdom Animalia is a monophyletic group?

a. The ribosomal RNA of animals is more similar to that of other animals than it is to any other form of life.
b. Animals share unique similarities in their cell-to-cell junctions.
c. All animals use the same genetic code.
d. Animals share unique proteins in their extracellular matrices.
e. All of these are lines of evidence that kingdom Animalia is a monophyletic group.

Answer: c
Difficulty: Moderate
Bloom's Taxonomy: Synthesis

5. Most biologists agree that the most recent common ancestor of all animals was probably a _____, an idea first proposed by Ernst Haeckel in 1874.

a. jellyfish
b. ciliated protozoan
c. sessile sponge
d. chytrid
e. colonial, flagellated protist

Answer: e
Difficulty: Easy
Bloom's Taxonomy: Comprehension

6. An organism that is unable to move on its own power is said to be

a. sessile.
b. a zygote.
c. sedentary.
d. a larva.
e. a gamete.

Answer: a
Difficulty: Easy
Bloom's Taxonomy: Comprehension

29.2 KEY INNOVATIONS IN ANIMAL EVOLUTION

7. Groups of cells that share a common structure and function are considered to be

a. organs.
b. mesenteries.
c. colonies.
d. tissues.
e. organ systems.

Answer: d
Difficulty: Moderate
Bloom's Taxonomy: Knowledge

8. In most eumetazoans the outermost tissue layer during development is the

a. mesoderm.
b. epiderm.
c. mesoglea.
d. endoderm.
e. ectoderm.

Answer: e
Difficulty: Moderate
Bloom's Taxonomy: Comprehension

9. In most eumetazoans the innermost tissue layer during development is the

a. mesoderm.
b. epiderm.
c. mesoglea.
d. endoderm.
e. ectoderm.

Answer: d
Difficulty: Easy
Bloom's Taxonomy: Comprehension

10. In most eumetazoans the middle tissue layer during development is the

a. mesoderm.
b. epiderm.
c. mesoglea.
d. endoderm.
e. ectoderm.

Answer: a
Difficulty: Moderate
Bloom's Taxonomy: Comprehension

11. Which embryonic cell layer develops into the lining of the gut in most eumetazoans?

a. mesoderm
b. epiderm
c. mesoglea
d. endoderm
e. ectoderm

Answer: d
Difficulty: Easy
Bloom's Taxonomy: Comprehension

12. The nervous system is derived mainly from which embryonic cell layer in most eumetazoans?

a. mesoderm
b. epiderm
c. mesoglea
d. endoderm
e. ectoderm

Answer: e
Difficulty: Moderate
Bloom's Taxonomy: Comprehension

13. The muscles of the body wall are derived mainly from which embryonic cell layer in most eumetazoans?

a. mesoderm
b. epiderm
c. mesoglea
d. endoderm
e. ectoderm

Answer: a
Difficulty: Moderate
Bloom's Taxonomy: Comprehension

14. Members of the phylum _____ exhibit radial symmetry, with their body parts arranged regularly around a central axis.

a. Porifera
b. Mollusca
c. Cnidaria
d. Platyhelminthes
e. Chordata

Answer: c
Difficulty: Moderate
Bloom's Taxonomy: Comprehension

15. Animals that do not have a body cavity that separates the gut from the muscles of the body wall are said to be

a. pseudocoelomate.
b. mesenteries.
c. acoelomate.
d. indeterminate.
e. coelomate.

Answer: c
Difficulty: Easy
Bloom's Taxonomy: Comprehension

16. Animals that have a fluid-filled cavity that separates the gut from the muscles of the body wall and have that cavity completely lined by the peritoneum are said to be

a. pseudocoelomate.
b. mesenteries.
c. acoelomate.
d. indeterminate.
e. coelomate.

Answer: e
Difficulty: Moderate
Bloom's Taxonomy: Comprehension

17. Which of the following is found most often for protostomes?

a. spiral, indeterminate cleavage
b. radial, indeterminate cleavage
c. spiral, determinate cleavage
d. radial, determinate cleavage
e. none of these

Answer: c
Difficulty: Moderate
Bloom's Taxonomy: Synthesis

18. Which of the following is found most often for deuterostomes?

a. spiral, indeterminate cleavage
b. radial, indeterminate cleavage
c. spiral, determinate cleavage
d. radial, determinate cleavage
e. none of these

Answer: b
Difficulty: Moderate
Bloom's Taxonomy: Synthesis

19. The _____ is the developing gut for most animal embryos.

a. schizocoelom
b. mesoglea
c. archenteron
d. enterocoelom
e. blastopore

Answer: c
Difficulty: Moderate
Bloom's Taxonomy: Knowledge

20. The _____ is the initial opening between the developing gut and the outside environment for most animal embryos.

a. schizocoelom
b. mesoglea
c. archenteron
d. enterocoelom
e. blastopore

Answer: e
Difficulty: Easy
Bloom's Taxonomy: Knowledge

21. The _____ is the developmental origin of the body cavity between the gut and muscles of the body wall for most protostomes.

a. schizocoelom
b. mesoglea
c. archenteron
d. enterocoelom
e. blastopore

Answer: a
Difficulty: Difficult
Bloom's Taxonomy: Comprehension

22. Ringlike patterns on an earthworm and "six-pack abs" in humans are evidence of _____ in an animal's body plan.

a. cephalization
b. segmentation
c. radial symmetry
d. mesenteries
e. cleavage

Answer: b
Difficulty: Moderate
Bloom's Taxonomy: Comprehension

29.3 AN OVERVIEW OF ANIMAL PHYLOGENY AND CLASSIFICATION

23. Based on molecular analysis, which of the following appears to be least meaningful for inferring animal phylogenic relationships?

a. presence or absence of tissues
b. mitochondrial DNA sequences
c. radial symmetry in the Radiata
d. body cavity differences
e. *Hox* gene sequences

Answer: d
Difficulty: Moderate
Bloom's Taxonomy: Evaluation

24. Which animal taxon is named for the cuticle or external skeleton that its members secrete and periodically molt?

a. Deuterostomia
b. Ecdysozoa
c. Reptilia
d. Lophotrochozoa
e. Protostomia

Answer: b
Difficulty: Moderate
Bloom's Taxonomy: Comprehension

25. Based on modern molecular phylogeny, which of the following is a derived condition within Lophotrochozoa?

a. acoelomate species
b. bilateral symmetry
c. tissues
d. multicellularity
e. schizocoelom

Answer: a
Difficulty: Difficult
Bloom's Taxonomy: Synthesis

26. Based on modern molecular phylogeny, which of the following animal taxons does NOT have any members with segmentation as a derived condition?

a. Deuterostomia
b. Ecdysozoa
c. Lophotrochozoa
d. Protostomia
e. All of these have some members with segmentation as a derived condition.

Answer: e
Difficulty: Moderate
Bloom's Taxonomy: Synthesis

29.4 ANIMALS WITHOUT TISSUES: PARAZOA

27.   Which of the following phyla is a member of the Parazoa?

a.   Nemertea
b.   Porifera
c.   Rotifera
d.   Ctenophora
e.   Echinodermata

Answer: b
Difficulty: Moderate
Bloom's Taxonomy: Comprehension

28.   Asymmetry, choanocytes, filter feeding, osculum, and spicules are all terms associated with which phylum?

a.   Nematoda
b.   Ctenophora
c.   Nemertea
d.   Porifera
e.   Cnidaria

Answer: d
Difficulty: Easy
Bloom's Taxonomy: Synthesis

29.   Spicules are secreted by

a.   amoeboid cells.
b.   choanocytes.
c.   sperm.
d.   porocytes.
e.   pinacoderm.

Answer: a
Difficulty: Moderate
Bloom's Taxonomy: Knowledge

30.   Spicules are _____ that are found in sponges.

a.   fibrous skeletal proteins
b.   specialized digestive cells
c.   complicated pore openings and passageways
d.   needlelike skeletal structures made up of silica or calcium carbonate
e.   feeding cells with a flagellum surrounded by a collar

Answer: d
Difficulty: Easy
Bloom's Taxonomy: Comprehension

31.   The beating flagellae of _____ maintain a constant flow of water through sponges.

a.   amoeboid cells
b.   choanocytes
c.   sperm
d.   porocytes
e.   pinacoderm

Answer: b
Difficulty: Moderate
Bloom's Taxonomy: Comprehension

29.5 EUMETAZOANS WITH RADIAL SYMMETRY

32.   Which of the following phyla is a member of the Radiata?

a.   Nemertea
b.   Porifera
c.   Rotifera
d.   Ctenophora
e.   Echinodermata

Answer: d
Difficulty: Moderate
Bloom's Taxonomy: Comprehension

33.   Members of this phylum have radial symmetry and nematocysts; they may have polyp or medusa forms (or both) as adults.

a.   Nematoda
b.   Cnidaria
c.   Porifera
d.   Rotifera
e.   Ctenophora

Answer: b
Difficulty: Easy
Bloom's Taxonomy: Comprehension

34.   Like other members of their phylum, jellyfishes have specialized stinging cells called _____ as part of their epidermis, which they use to paralyze small prey.

a.   choanocytes
b.   polyps
c.   cnidocytes
d.   flame cells
e.   tentacles

Answer: c
Difficulty: Easy
Bloom's Taxonomy: Comprehension

35.   The members of this lineage of Cnidarians exist only as polyps as adults. Many build calcium carbonate exoskeletons and have a mutualistic relationship with photosynthetic protists.

a.   Hydrozoa
b.   Scyphozoa
c.   Trematoda
d.   Cubozoa
e.   Anthozoa

Answer: e
Difficulty: Moderate
Bloom's Taxonomy: Comprehension

36. Adult members of this lineage of Cnidarians are predominately medusae. They are the true jellyfishes, and are not active swimmers.

a. Hydrozoa
b. Scyphozoa
c. Trematoda
d. Cubozoa
e. Anthozoa

Answer: b
Difficulty: Easy
Bloom's Taxonomy: Comprehension

37. Members of this lineage of Cnidarians exist primarily as cube-shaped medusae. They are active swimmers, and they produce one of the deadliest toxins made by animals.

a. Hydrozoa
b. Scyphozoa
c. Trematoda
d. Cubozoa
e. Anthozoa

Answer: d
Difficulty: Moderate
Bloom's Taxonomy: Comprehension

38. Most members of this lineage of Cnidarians have both polyp and medusa stages in their life cycles, with the polyp stage typically forming a colony. Species of *Obelia* are members of this lineage.

a. Hydrozoa
b. Scyphozoa
c. Trematoda
d. Cubozoa
e. Anthozoa

Answer: a
Difficulty: Difficult
Bloom's Taxonomy: Comprehension

39. Members of this phylum have radial symmetry, and use their two tentacles to capture prey. They have eight longitudinal plates of cilia that they use for movement, making them the largest animals to use cilia as their primary means of locomotion.

a. Nematoda
b. Cnidaria
c. Porifera
d. Rotifera
e. Ctenophora

Answer: e
Difficulty: Moderate
Bloom's Taxonomy: Comprehension

29.6 LOPHOTROCHOZOAN PROTOSTOMES

40. Members of the phyla Ectoprocta, Brachipoda, and Phoronida all possess a lophophore, which is

a. a cluster of pores along the aboral surface, used for expelling wastes from the digestive tract.
b. a hard shell made of calcium carbonate.
c. an outpocket of the digestive tract, where metabolic wastes are processed before emptying into the digestive tract for expulsion.
d. a rounded fold with one or two rows of hollow, ciliated tentacles.
e. an extra body cavity surrounding the digestive tract filled with muscles used for locomotion.

Answer: d
Difficulty: Moderate
Bloom's Taxonomy: Comprehension

41. Members of this Lophotrochozoan phylum are acoelomate and dorsoventrally flattened. They do not have circulatory or respiratory systems.

a. Rotifera
b. Mollusca
c. Platyhelminthes
d. Annelida
e. Nemertea

Answer: c
Difficulty: Moderate
Bloom's Taxonomy: Comprehension

42. Flatworms have specialized cells called _____ that serve as a simple excretory system.

a. choanocytes
b. polyps
c. cnidocytes
d. flame cells
e. tentacles

Answer: d
Difficulty: Moderate
Bloom's Taxonomy: Comprehension

43. Members of this lineage of flatworms are parasites of vertebrates. They have a head modified into a scolex with hooks and suckers for attaching to places like the intestinal wall of hosts. Most of their body is a series of reproductive units called proglottids.

a. Trematoda
b. Turbellaria
c. Polychaeta
d. Cestoda
e. Hirudinea

Answer: d
Difficulty: Moderate
Bloom's Taxonomy: Comprehension

44. Members of this lineage of flatworms are free-living and are found mostly in marine environments. Most acquire food with a muscular pharynx and are hermaphroditic.

a. Trematoda
b. Turbellaria
c. Polychaeta
d. Cestoda
e. Hirudinea

Answer: b
Difficulty: Difficult
Bloom's Taxonomy: Comprehension

45. Found mostly in freshwater, members of this Lophotrochozoan phylum have a wheel-like, ciliated food-gathering organ around their head called a corona. Some members of this phylum make extensive use of parthenogenesis for reproduction.

a. Rotifera
b. Mollusca
c. Platyhelminthes
d. Annelida
e. Nemertea

Answer: a
Difficulty: Easy
Bloom's Taxonomy: Comprehension

46. Found mostly in marine environments, members of this Lophotrochozoan phylum are called ribbon worms or proboscis worms. They have a muscular, mucus-covered tube that they can turn inside out to capture prey.

a. Rotifera
b. Mollusca
c. Platyhelminthes
d. Annelida
e. Nemertea

Answer: e
Difficulty: Moderate
Bloom's Taxonomy: Comprehension

47. This is the Lophotrochozoan phylum with the most species, having about 100,000 described living species.

a. Rotifera
b. Mollusca
c. Platyhelminthes
d. Annelida
e. Nemertea

Answer: b
Difficulty: Difficult
Bloom's Taxonomy: Knowledge

48. Species in this Lophotrochozoan phylum have a body divided into three regions: visceral mass, head-foot, and mantle.

a. Rotifera
b. Mollusca
c. Platyhelminthes
d. Annelida
e. Nemertea

Answer: b
Difficulty: Easy
Bloom's Taxonomy: Comprehension

49. Polyplacophora, Gastropoda, Bivalvia, and Cephalopoda are all lineages within the phylum

a. Rotifera.
b. Mollusca.
c. Platyhelminthes.
d. Annelida.
e. Nemertea.

Answer: b
Difficulty: Easy
Bloom's Taxonomy: Comprehension

50. Members of the lineage _____ are commonly called chitons. They have a dorsal shell divided into eight plates.

a. Cephalopoda
b. Polyplacophora
c. Bivalvia
d. Gastropoda
e. Scaphopoda

Answer: b
Difficulty: Moderate
Bloom's Taxonomy: Comprehension

51. Members of the lineage _____ include slugs and snails. Most have a single coiled shell, but some have no shell. Most undergo torsion during development, a twisting of their visceral mass that leads to placement of their anus practically over their heads.

a. Cephalopoda
b. Polyplacophora
c. Bivalvia
d. Gastropoda
e. Scaphopoda

Answer: d
Difficulty: Moderate
Bloom's Taxonomy: Comprehension

52. Members of the lineage _____ have a pair of shells that are hinged together. They close the shells together by contraction of one or two adductor muscles.

a. Cephalopoda
b. Polyplacophora
c. Bivalvia
d. Gastropoda
e. Scaphopoda

Answer: c
Difficulty: Easy
Bloom's Taxonomy: Comprehension

53.   Members of the lineage _____ include octopuses, squids, and nautiluses. Their fused head-foot body has the "foot" modified into a set of arms, or tentacles.

a.   Cephalopoda
b.   Polyplacophora
c.   Bivalvia
d.   Gastropoda
e.   Scaphopoda

Answer: a
Difficulty: Easy
Bloom's Taxonomy: Comprehension

54.   Due to their high activity and resulting need for lots of oxygen, members of the lineage _____ have a closed circulatory system.

a.   Cephalopoda
b.   Polyplacophora
c.   Bivalvia
d.   Gastropoda
e.   Scaphopoda

Answer: a
Difficulty: Moderate
Bloom's Taxonomy: Comprehension

55.   Members of this Lophotrochozoan phylum have highly segmented bodies, with many repeating units that are usually separated by transverse partitions called septa.

a.   Rotifera
b.   Mollusca
c.   Platyhelminthes
d.   Annelida
e.   Nemertea

Answer: d
Difficulty: Moderate
Bloom's Taxonomy: Comprehension

56.   Members of this Lophotrochozoan phylum include earthworms, bristle worms, and leeches.

a.   Rotifera
b.   Mollusca
c.   Platyhelminthes
d.   Annelida
e.   Nemertea

Answer: d
Difficulty: Easy
Bloom's Taxonomy: Comprehension

57.   Paired, fleshy extensions of the body wall in bristle worms that are used for locomotion and gas exchange are called

a.   metanephridia.
b.   parapodia.
c.   septa.
d.   clitella.
e.   setae.

Answer: b
Difficulty: Moderate
Bloom's Taxonomy: Comprehension

58.   Paired excretory organs found in most body segments of segmented worms are called

a.   metanephridia.
b.   parapodia.
c.   septa.
d.   clitella.
e.   setae.

Answer: a
Difficulty: Difficult
Bloom's Taxonomy: Comprehension

59.   Chitin-reinforced bristles that protrude outward from the body wall and are used by most segmented worms to anchor themselves against a substrate are called

a.   metanephridia.
b.   parapodia.
c.   septa.
d.   clitella.
e.   setae.

Answer: e
Difficulty: Moderate
Bloom's Taxonomy: Comprehension

60.   Members of the lineage _____ are found primarily in marine environments. Called bristle worms, they characteristically have parapodia.

a.   Hirudinea
b.   Polychaeta
c.   Monogenoidea
d.   Cestoda
e.   Oligochaeta

Answer: b
Difficulty: Moderate
Bloom's Taxonomy: Comprehension

61.   Most members of the lineage _____ are terrestrial earthworms.

a.   Hirudinea
b.   Polychaeta
c.   Monogenoidea
d.   Cestoda
e.   Oligochaeta

Answer: e
Difficulty: Moderate
Bloom's Taxonomy: Comprehension

62.   Members of the lineage _____ are mostly freshwater parasites. Called leeches, they are dorsoventrally flattened and do not have septa.

a.   Hirudinea
b.   Polychaeta
c.   Monogenoidea
d.   Cestoda
e.   Oligochaeta

Answer: a
Difficulty: Moderate
Bloom's Taxonomy: Comprehension

FOCUS ON RESEARCH: APPLIED RESEARCH: A ROGUE'S GALLERY OF PARASITIC WORMS

63. Members of this lineage of flatworms are endoparasites of vertebrates. Some, called blood flukes, cause an often deadly disease in humans called schistosomiasis.

a. Trematoda
b. Turbellaria
c. Polychaeta
d. Cestoda
e. Hirudinea

Answer: a
Difficulty: Moderate
Bloom's Taxonomy: Comprehension

64. A blood fluke infection is most likely to be found in persons who

a. eat undercooked beef.
b. walk barefoot in a rice field.
c. eat undercooked pork.
d. are bitten by mosquitoes.
e. eat undercooked fish.

Answer: b
Difficulty: Moderate
Bloom's Taxonomy: Application

65. The disease trichinosis is most commonly found in persons who

a. eat undercooked beef.
b. walk barefoot in a rice field.
c. eat undercooked pork.
d. are bitten by mosquitoes.
e. eat undercooked fish.

Answer: c
Difficulty: Moderate
Bloom's Taxonomy: Application

66. The disease elephantiasis is most commonly found in persons who

a. eat undercooked beef.
b. walk barefoot in a rice field.
c. eat undercooked pork.
d. are bitten by mosquitoes.
e. eat undercooked fish.

Answer: d
Difficulty: Moderate
Bloom's Taxonomy: Application

29.7 ECDYSOZOAN PROTOSTOMES

67. Members of this Ecdysozoan phylum are so plentiful in soil and other places that they may be the most abundant animals on Earth. Commonly called roundworms, it is estimated that more than half a million living species of them exist, although only about 80,000 have been described.

a. Arthropoda
b. Onychophora
c. Nemertea
d. Annelida
e. Nematoda

Answer: e
Difficulty: Moderate
Bloom's Taxonomy: Comprehension

68. Members of this Ecdysozoan phylum include many important decomposers, but also many damaging parasites. Many parasitize the roots of plants, while others cause disease in animals. More than 1 billion people worldwide are estimated to be infected by members of this group such as pinworms, trichinas, and filarial worms.

a. Arthropoda
b. Onychophora
c. Nemertea
d. Annelida
e. Nematoda

Answer: e
Difficulty: Difficult
Bloom's Taxonomy: Comprehension

69. Only 65 living species are known in this Ecdysozoan phylum. Called velvet worms, they have jaws, an open circulatory system, and relatively large brains.

a. Arthropoda
b. Onychophora
c. Nemertea
d. Annelida
e. Nematoda

Answer: b
Difficulty: Moderate
Bloom's Taxonomy: Comprehension

70. This Ecdysozoan phylum contains more than half of the known animal species.

a. Arthropoda
b. Onychophora
c. Nemertea
d. Annelida
e. Nematoda

Answer: a
Difficulty: Moderate
Bloom's Taxonomy: Knowledge

71. Members of this Ecdysozoan phylum have a segmented body encased in a rigid exoskeleton made of chitin plus other materials. Their phylum name means "jointed feet."

a. Arthropoda
b. Onychophora
c. Nemertea
d. Annelida
e. Nematoda

Answer: a
Difficulty: Easy
Bloom's Taxonomy: Comprehension

72. Members of this Ecdysozoan phylum include insects, spiders, crustaceans, millipedes, and centipedes.

a. Arthropoda
b. Onychophora
c. Nemertea
d. Annelida
e. Nematoda

Answer: a
Difficulty: Easy
Bloom's Taxonomy: Comprehension

73. Members of this subphylum are now all extinct, but they were extremely numerous in shallow Paleozoic seas. Most of them were dorsoventrally flattened and heavily armored, with two deep longitudinal grooves that gave their body a three-lobed appearance.

a. Myriapoda
b. Trilobita
c. Chelicerata
d. Hexapoda
e. Crustacea

Answer: b
Difficulty: Easy
Bloom's Taxonomy: Comprehension

74. Members of the subphylum include spiders, ticks, mites, scorpions, and horseshoe crabs. Their first pair of appendages are fanglike structures used for biting prey, and their second pair of appendages, called pedipalps, are used for many different activities.

a. Myriapoda
b. Trilobita
c. Chelicerata
d. Hexapoda
e. Crustacea

Answer: c
Difficulty: Moderate
Bloom's Taxonomy: Comprehension

75. Members of the subphylum include shrimp, lobsters, and crabs. They generally have five pairs of appendages on their heads: two pairs of antennae and three pairs of mouthparts, including mandibles.

a. Myriapoda
b. Trilobita
c. Chelicerata
d. Hexapoda
e. Crustacea

Answer: e
Difficulty: Easy
Bloom's Taxonomy: Comprehension

76. Members of the subphylum are the centipedes and millipedes. They have two body regions, a head with one pair of antennae and a segmented trunk with one or two pairs of legs per segment.

a. Myriapoda
b. Trilobita
c. Chelicerata
d. Hexapoda
e. Crustacea

Answer: a
Difficulty: Moderate
Bloom's Taxonomy: Comprehension

77. The members of the subphylum _____ are mostly insects.

a. Myriapoda
b. Trilobita
c. Chelicerata
d. Hexapoda
e. Crustacea

Answer: d
Difficulty: Easy
Bloom's Taxonomy: Comprehension

78. The members this subphylum have a three-part body plan: a head, thorax, and abdomen. Among other structures, their head has one pair of antennae and a pair of mandibles, and their thorax has three pairs of walking legs and often one or two pairs of wings.

a. Myriapoda
b. Trilobita
c. Chelicerata
d. Hexapoda
e. Crustacea

Answer: d
Difficulty: Easy
Bloom's Taxonomy: Comprehension

79. A Malpighian tubule is

a. a rasping tongue with chitinous teeth.
b. a muscular sac used for extending tube feet in echinoderms.
c. a filamentous projection of the mantle.
d. a specialized tube in some arthropods used for waste processing.
e. an abdominal appendage in some arthropods that secretes silk.

Answer: d
Difficulty: Moderate
Bloom's Taxonomy: Comprehension

80. Members of the order Hemiptera (true bugs) have postembryonic development called incomplete metamorphosis, where they have

a. a series of nymph instars that lack functional wings, ending with an adult instar with functional wings.
b. a series of young instars that become more and more like an adult, ending with a wingless adult instar.
c. a single, wingless nymph instar followed by several winged adult instars.
d. several instars of often worm-like larvae, then a stage where they are a sessile pupa, before ending in an adult stage that is very different from the larval stages.

Answer: a
Difficulty: Difficult
Bloom's Taxonomy: Comprehension

81. Members of the order Diptera (flies) have postembryonic development called complete metamorphosis, where they have

a. a series of nymph instars that lack functional wings, ending with an adult instar with functional wings.
b. a series of young instars that become more and more like an adult, ending with a wingless adult instar.
c. a single, wingless nymph instar followed by several winged adult instars.
d. several instars of often worm-like larvae, then a stage where they are a sessile pupa, before ending in an adult stage that is very different from the larval instars.

Answer: d
Difficulty: Moderate
Bloom's Taxonomy: Comprehension

82. Members of the order Thysanura (silverfish) have postembryonic development that is not considered a metamorphosis. Instead, they have

a. a series of nymph instars that lack functional wings, ending with an adult instar with functional wings.
b. a series of young instars that become more and more like an adult, ending with a wingless adult instar.
c. a single, wingless nymph instar followed by several winged adult instars.
d. several instars of often worm-like larvae, then a stage where they are a sessile pupa, before ending in an adult stage that is very different from the larval stages.

Answer: b
Difficulty: Moderate
Bloom's Taxonomy: Comprehension

FOCUS ON RESEARCH: MODEL RESEARCH ORGANISMS: *Caenorhabditis elegans*

83. The organism *Caenorhabditis elegans*, a member of the phylum _____, is an important model organism for studies of the genetic control of development.

a. Arthropoda
b. Onychophora
c. Nemertea
d. Annelida
e. Nematoda

Answer: e
Difficulty: Moderate
Bloom's Taxonomy: Knowledge

INSIGHTS FROM THE MOLECULAR REVOLUTION: UNSCRAMBLING THE ARTHROPODS

84. According to modern phylogeny based on such data as the arrangement of genes in mitochondrial DNA and the sequences of ribosomal RNA genes, hexapods are most closely related to which of the following groups?

a. trilobites
b. myriapods
c. crustaceans
d. annelids
e. chelicerates

Answer: c
Difficulty: Moderate
Bloom's Taxonomy: Analysis

85. Comparisons of genetic information from various invertebrate groups has

a. confirmed nearly all classifications that were based on phenotypic resemblances.
b. provided a complete and final picture of animal phylogeny.
c. led to a decision to forgo formal classification of animals, as animal relationships are too complex to determine.
d. led to a reworking of the classification of animals that is still being refined.

Answer: d
Difficulty: Moderate
Bloom's Taxonomy: Synthesis

86. Research into the genetics of invertebrates

a. is mostly complete, and is considered by most biologists to be a field of the past.
b. holds the promise to address many important basic and applied research questions that could have profound positive effects on human health.
c. should be largely abandoned in favor of research into the genetics of vertebrates, since findings in invertebrates have little chance of making a difference for humans.
d. is at an early stage, as genetic research in the past has almost exclusively been centered on vertebrate animals and bacteria.
e. has led biologists to largely abandon hope of understanding animal development due to the complexity of developmental processes.

Answer: b
Difficulty: Moderate
Bloom's Taxonomy: Evaluation

Matching

Match each of the groups of organisms listed with the correct phylum. Each phylum will be used only once.

87. _____ ribbon worms

88. _____ spiders, insects, shrimp, and crabs

89. _____ comb jellies

90. _____ velvet worms

91. _____ sponges

92. _____ chitons, snails, slugs, and scallops

93. _____ bristle worms, earthworms, and leeches

94. _____ flatworms

95. _____ jellyfishes, corals, and sea anemones

96. _____ roundworms

A. Annelida

B. Arthropoda

C. Cnidaria

D. Ctenophora

E. Mollusca

F. Nematoda

G. Nemertea

H. Onychophora

I. Platyhelminthes

J. Porifera

Answers: *87. G   88. B   89. D   90. H   91. J*
              92. E   93. A   94. I   95. C   96. F

Difficulty: Moderate
Bloom's Taxonomy: Knowledge
Source: Sections 29.4-29.7

Short Answer

97.   Contrast typical embryonic development in protostomes and deuterostomes.

Answer: Protostome embryos typically have determinate, spiral cleavage as cells divide, while deuterostome embryos have indeterminate, radial cleavage. Also, protostome mesoderm differentiates near the blastopore and forms a coelom (schizocoelom) within the mesoderm. Deuterostome mesoderm instead forms as outpocketings of the archenteron, forming the coelom (enterocoelom) from the space in the outpocketings. Finally, in protostomes the blastopore becomes the mouth and a second, later opening becomes the anus, while in deuterostomes the blastopore becomes the anus and the mouth instead forms from a later opening.
Difficulty: Moderate
Bloom's Taxonomy: Synthesis
Source: Section 29.2

30

DEUTEROSTOMES: VERTEBRATES AND THEIR CLOSEST RELATIVES

Multiple-Choice

WHY IT MATTERS

1.   Deuterostomia is a _____ lineage of animals that dates to the _____.

a.   polyphyletic; Mesozoic
b.   paraphyletic; Cenozoic
c.   monophyletic; Paleozoic
d.   paraphyletic; Paleozoic
e.   polyphyletic; Cenozoic

Answer: c
Difficulty: Moderate
Bloom's Taxonomy: Comprehension

30.1 INVERTEBRATE DEUTEROSTOMES

2.   Members of the phylum _____ have bilaterally symmetrical larvae but exhibit secondary radial symmetry as adults, usually organized around five rays, or "arms."

a.   Hemichordata
b.   Echinodermata
c.   Cnidaria
d.   Arthropoda
e.   Chordata

Answer: b
Difficulty: Easy
Bloom's Taxonomy: Comprehension

3.   Which of the following phyla has a unique water vascular system that usually includes tube feet that can be used for movement?

a.   Hemichordata
b.   Echinodermata
c.   Cnidaria
d.   Arthropoda
e.   Chordata

Answer: b
Difficulty: Moderate
Bloom's Taxonomy: Comprehension

4.   The internal skeleton of echinoderms is made of calcium-stiffened _____ that develop from mesoderm.

a.   ampullae
b.   ossicles
c.   radial canals
d.   ring canals
e.   madreporites

Answer: b
Difficulty: Moderate
Bloom's Taxonomy: Knowledge

5.   Water enters the water vascular system of a sea star through a structure called the

a.   ampulla.
b.   ossicle.
c.   radial canal.
d.   ring canal.
e.   madreporite.

Answer: e
Difficulty: Moderate
Bloom's Taxonomy: Knowledge

6.   A tube foot of a sea star moves when the small muscular bulb called the _____ contracts, forcing fluid into the tube foot.

a.   ampulla
b.   ossicle
c.   radial canal
d.   ring canal
e.   madreporite

Answer: a
Difficulty: Moderate
Bloom's Taxonomy: Knowledge

7.   This lineage of Echinoderms is made up of about 1500 species of sea stars, each with a central disk surrounded by 5 to 20 "arms."

a.   Echinoidea
b.   Ophiuroidea
c.   Crinoidea
d.   Holothuroidea
e.   Asteroidea

Answer: e
Difficulty: Easy
Bloom's Taxonomy: Comprehension

8.   This lineage of Echinoderms is made up of about 2000 species of brittle stars and basket stars, each with elongated, slender, somewhat snakelike arms surrounding a well-defined central disk.

a.   Echinoidea
b.   Ophiuroidea
c.   Crinoidea
d.   Holothuroidea
e.   Asteroidea

Answer: b
Difficulty: Moderate
Bloom's Taxonomy: Comprehension

9.   This lineage of Echinoderms is made up of about 950 species of sea urchins and sand dollars, which have their ossicles fused into solid tests that restrict flexibility but provide excellent protection.

a.   Echinoidea
b.   Ophiuroidea
c.   Crinoidea
d.   Holothuroidea
e.   Asteroidea

Answer: a
Difficulty: Moderate
Bloom's Taxonomy: Comprehension

10.   This lineage of Echinoderms is made up of about 1500 species of sea cucumbers, which are elongated along their oral-aboral axis.

a.   Echinoidea
b.   Ophiuroidea
c.   Crinoidea
d.   Holothuroidea
e.   Asteroidea

Answer: d
Difficulty: Difficult
Bloom's Taxonomy: Comprehension

11   This lineage of Echinoderms is made up of about 600 living species of sea lilies and feather stars, although the group was much more diverse and abundant 500 million years ago. Adult feather stars can swim or crawl, while adult sea lilies are sessile, with their central disk attached to a stalk.

a.   Echinoidea
b.   Ophiuroidea
c.   Crinoidea
d.   Holothuroidea
e.   Asteroidea

Answer: c
Difficulty: Moderate
Bloom's Taxonomy: Comprehension

12.   Members of the phylum _____ are called acorn worms. They use their mucus-coated proboscis to construct U-shaped burrows to trap food particles, and also trap food from water passed through gill slits in their pharynx.

a.   Hemichordata
b.   Echinodermata
c.   Cnidaria
d.   Arthropoda
e.   Chordata

Answer: a
Difficulty: Moderate
Bloom's Taxonomy: Comprehension

30.2 OVERVIEW OF THE PHYLUM CHORDATA

13.   Members of the phylum _____ all have at some point during development a hollow dorsal nerve cord, a perforated pharynx, a notochord, and segmented muscles in both the body wall and postanal tail.

a.   Hemichordata
b.   Echinodermata
c.   Cnidaria
d.   Arthropoda
e.   Chordata

Answer: e
Difficulty: Easy
Bloom's Taxonomy: Comprehension

14.   Which of the following forms the skeleton of invertebrate chordates?

a.   pharynx
b.   neural crest
c.   oral hood
d.   notochord
e.   cranium

Answer: d
Difficulty: Moderate
Bloom's Taxonomy: Comprehension

15.   Which of the following is the part of the chordate digestive system just posterior to the mouth?

a.   pharynx
b.   neural crest
c.   oral hood
d.   notochord
e.   cranium

Answer: a
Difficulty: Easy
Bloom's Taxonomy: Knowledge

16.   Members of one major class of the lineage _____ are called tunicates or sea squirts. They have tadpolelike larvae, but their sessile adult form is usually encased in a leathery "tunic."

a.   Petromyzontoidea
b.   Placodermi
c.   Cephalochordata
d.   Urochordata
e.   Myxinoidea

Answer: d
Difficulty: Moderate
Bloom's Taxonomy: Comprehension

17. Members of the lineage _____ are called lancelets. As adults they live mostly buried in the sand in warm, shallow marine habitats, but they do have well-developed segmented muscles and can move.

a. Petromyzontoidea
b. Placodermi
c. Cephalochordata
d. Urochordata
e. Myxinoidea

Answer: c
Difficulty: Moderate
Bloom's Taxonomy: Comprehension

18. Which of the following surrounds and protects the brain in all vertebrates?

a. pharynx
b. neural crest
c. oral hood
d. notochord
e. cranium

Answer: e
Difficulty: Easy
Bloom's Taxonomy: Comprehension

19. Which of the following is a cell type unique to vertebrates?

a. pharyngeal
b. neural crest
c. oral hood
d. notochordal
e. cranial

Answer: b
Difficulty: Moderate
Bloom's Taxonomy: Comprehension

30.3 THE ORIGIN AND DIVERSIFICATION OF VERTEBRATES

20. Among the following lineages, which has the largest number of *Hox* genes when all *Hox* gene clusters are taken into account?

a. cnidarians
b. cephalochordates
c. vertebrates
d. arthropods

Answer: c
Difficulty: Moderate
Bloom's Taxonomy: Analysis

21. Which of the following is NOT considered to be a key derived morphological innovation within the vertebrate lineage?

a. neural crest
b. notochord
c. bone
d. vertebrae
e. All of these are key features derived within the vertebrate lineage.

Answer: b
Difficulty: Moderate
Bloom's Taxonomy: Synthesis

22. Which of the following lineages contains all of the rest of the lineages in the list?

a. Gnathostomata
b. Vertebrata
c. Amniota
d. Tetrapoda

Answer: b
Difficulty: Moderate
Bloom's Taxonomy: Synthesis

30.4 AGNATHANS: HAGFISHES AND LAMPREYS, CONODONTS AND OSTRACODERMS

23. Living members of the lineage _____ are jawless vertebrates called hagfishes. They have only a cranium and a notochord for their axial skeleton, with no signs of vertebrae.

a. Petromyzontoidea
b. Placodermi
c. Cephalochordata
d. Urochordata
e. Myxinoidea

Answer: e
Difficulty: Moderate
Bloom's Taxonomy: Comprehension

24. Living members of the lineage _____ are jawless vertebrates called lampreys, which are mostly parasitic as adults. They have shards of cartilage that partially cover the nerve cord, which may represent an early stage of the evolution of vertebrae.

a. Petromyzontoidea
b. Placodermi
c. Cephalochordata
d. Urochordata
e. Myxinoidea

Answer: a
Difficulty: Difficult
Bloom's Taxonomy: Comprehension

25. Now extinct, tooth parts of _____ are abundant as fossils dating from the early Paleozoic to the early Mesozoic era. These jawless fishes are considered to be the earliest vertebrates with bonelike structures.

a. acanthodians
b. conodonts
c. osteolepiforms
d. placoderms
e. ostracoderms

Answer: b
Difficulty: Moderate
Bloom's Taxonomy: Comprehension

26. Now extinct, several lineages of _____ appear in the fossil record from the Ordovician through the Devonian periods. These jawless fishes had skin heavily armored with bony plates and scales, and imprints of some of their fossils indicate that their brains had three regions.

a. acanthodians
b. conodonts
c. osteolepiforms
d. placoderms
e. ostracoderms

Answer: e
Difficulty: Moderate
Bloom's Taxonomy: Comprehension

30.5 JAWED FISHES

27. A key trait that distinguishes gnathostomes from other chordate lineages is the presence of

a. bony scales.
b. claspers.
c. jaws.
d. the operculum.
e. keratin.

Answer: c
Difficulty: Moderate
Bloom's Taxonomy: Synthesis

28. Now extinct, _____ appear in the fossil record from the late Ordovician through the Permian periods. These jawed fishes, also called spiny sharks, had streamlined bodies, large jaws, and numerous teeth.

a. acanthodians
b. conodonts
c. osteolepiforms
d. placoderms
e. ostracoderms

Answer: a
Difficulty: Difficult
Bloom's Taxonomy: Comprehension

29. The fossil record shows that the jawed fishes called _____ diversified in the Devonian and Carboniferous periods, but they left no direct descendants. Their heads were covered with large, heavy plates of bone, and their jaws had sharp cutting edges.

a. acanthodians
b. conodonts
c. osteolepiforms
d. placoderms
e. ostracoderms

Answer: d
Difficulty: Difficult
Bloom's Taxonomy: Comprehension

30. Living members of the lineage _____ include the sharks, skates, and rays. They all have skeletons made entirely of cartilage.

a. Petromyzontoidea
b. Chondrichthyes
c. Actinopterygii
d. Myxinoidea
e. Sarcopterygii

Answer: b
Difficulty: Moderate
Bloom's Taxonomy: Comprehension

31. Squalene, spiral valve, electroreceptors, lateral-line system, and claspers are all adaptations associated with living members of which lineage?

a. Petromyzontoidea
b. Chondrichthyes
c. Actinopterygii
d. Myxinoidea
e. Sarcopterygii

Answer: b
Difficulty: Moderate
Bloom's Taxonomy: Comprehension

32. Living members of the lineage _____ are the ray-finned fishes, with fins supported by thin and flexible bony rays. This lineage includes over 95 percent of the living species of fishes.

a. Petromyzontoidea
b. Chondrichthyes
c. Actinopterygii
d. Myxinoidea
e. Sarcopterygii

Answer: c
Difficulty: Moderate
Bloom's Taxonomy: Comprehension

33. Which of these structures is derived from an ancestral air-breathing lung in ray-finned fishes and is used to increase buoyancy?

a. swim bladder
b. spiral valve
c. operculum
d. atrium
e. lateral-line system

Answer: a
Difficulty: Easy
Bloom's Taxonomy: Comprehension

34. Which of these structures is a covering for the gill chamber in ray-finned fishes?

a. swim bladder
b. spiral valve
c. operculum
d. atrium
e. lateral-line system

Answer: c
Difficulty: Easy
Bloom's Taxonomy: Comprehension

35. Living members of the lineage _____ are two species of coelacanths and six species of lungfish.

a. Petromyzontoidea
b. Chondrichthyes
c. Actinopterygii
d. Myxinoidea
e. Sarcopterygii

Answer: e
Difficulty: Moderate
Bloom's Taxonomy: Comprehension

36. The lineage _____ are the fleshy-finned fishes, with fins supported by muscles and an internal bony skeleton. Based on the fossil record, tetrapods apparently arose from this group.

a. Petromyzontoidea
b. Chondrichthyes
c. Actinopterygii
d. Myxinoidea
e. Sarcopterygii

Answer: e
Difficulty: Moderate
Bloom's Taxonomy: Comprehension

30.6 EARLY TETRAPODS AND MODERN AMPHIBIANS

37. The lineage _____ shared several derived traits with early tetrapods, such as the shapes and positions of bones in their appendages, and apparently gave rise to the tetrapods.

a. acanthodians
b. conodonts
c. osteolepiforms
d. placoderms
e. ostracoderms

Answer: c
Difficulty: Difficult
Bloom's Taxonomy: Comprehension

38. Nearly complete skeletal data is available from fossils of _____, an early tetrapod that differed from fishes in part by having a neck.

a. *Ichthyostega*
b. *Anolis*
c. *Dunkleosteus*
d. *Eusthenopteron*
e. *Archaeopteryx*

Answer: a
Difficulty: Moderate
Bloom's Taxonomy: Application

39. The tympanum and stapes are used in tetrapods for

a. seeing.
b. hearing.
c. heat-sensing.
d. tasting.
e. smelling.

Answer: b
Difficulty: Easy
Bloom's Taxonomy: Knowledge

40. The lineages Apoda, Urodela, and Gymnophiona are all part of which of the following groups?

a. Testudines
b. Archosauromorpha
c. Squamata
d. Sphenodontia
e. Amphibia

Answer: e
Difficulty: Moderate
Bloom's Taxonomy: Comprehension

41. Frogs and toads belong to which of the following lineages?

a. Urodela
b. Aves
c. Squamata
d. Gymnophiona
e. Apoda

Answer: e
Difficulty: Easy
Bloom's Taxonomy: Knowledge

42. Salamanders belong to which of the following lineages?

a. Urodela
b. Aves
c. Squamata
d. Gymnophiona
e. Apoda

Answer: a
Difficulty: Moderate
Bloom's Taxonomy: Knowledge

43. Legless caecilians belong to which of the following lineages?

a. Urodela
b. Aves
c. Squamata
d. Gymnophiona
e. Apoda

Answer: d
Difficulty: Difficult
Bloom's Taxonomy: Knowledge

30.7 THE ORIGIN AND MESOZOIC RADIATIONS OF AMNIOTES

44. Along with lipids, the skin of amniotes is filled with _____ to help prevent water loss.

a. squalene
b. bone
c. keratin
d. collagen
e. albumin

Answer: c
Difficulty: Moderate
Bloom's Taxonomy: Synthesis

45. Which of the following is a typical source of nutrients and water for developing amniote embryos?

a. squalene
b. bone
c. keratin
d. collagen
e. albumin

Answer: e
Difficulty: Moderate
Bloom's Taxonomy: Comprehension

46. Which of the following is a waste product of nitrogen metabolism in some amniotes that is low enough in toxicity that it can thus be excreted as a semisolid paste, reducing water loss?

a. nitric oxide
b. uric acid
c. squalene
d. urea
e. ammonium

Answer: b
Difficulty: Moderate
Bloom's Taxonomy: Knowledge

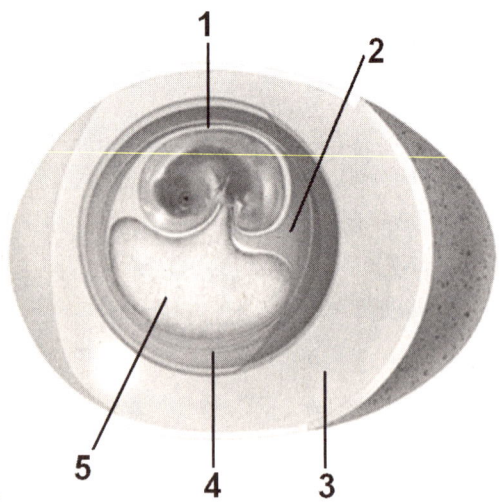

Use the figure above for questions 47–51.

47. In the amniote egg diagram shown, the item labeled "1" is the

a. yolk sac.
b. chorion.
c. allantois.
d. amnion.
e. albumin.

Answer: d
Difficulty: Moderate
Bloom's Taxonomy: Application
Source: Figure 30.22

48. In the amniote egg diagram shown, the item labeled "2" is the

a. yolk sac.
b. chorion.
c. allantois.
d. amnion.
e. albumin.

Answer: c
Difficulty: Difficult
Bloom's Taxonomy: Application
Source: Figure 30.22

49. In the amniote egg diagram shown, the item labeled "3" is the

a. yolk sac.
b. chorion.
c. allantois.
d. amnion.
e. albumin.

Answer: e
Difficulty: Moderate
Bloom's Taxonomy: Application
Source: Figure 30.22

50. In the amniote egg diagram shown, the item labeled "4" is the

a. yolk sac.
b. chorion.
c. allantois.
d. amnion.
e. albumin.

Answer: b
Difficulty: Difficult
Bloom's Taxonomy: Application
Source: Figure 30.22

51. In the amniote egg diagram shown, the item labeled "5" is the

a. yolk sac.
b. chorion.
c. allantois.
d. amnion.
e. albumin.

Answer: a
Difficulty: Easy
Bloom's Taxonomy: Application
Source: Figure 30.22

52. Turtles are thought to be living representatives of which of the following lineages?

a. anapsids
b. archosaurs
c. lepidosaurs
d. synapsids
e. none of these

Answer: a
Difficulty: Moderate
Bloom's Taxonomy: Comprehension

53. Mammals are thought to be living descendants of which of the following lineages?

a. anapsids
b. archosaurs
c. lepidosaurs
d. synapsids
e. none of these

Answer: d
Difficulty: Moderate
Bloom's Taxonomy: Comprehension

54. Crocodiles are living representatives of which of the following lineages?

a. anapsids
b. archosaurs
c. lepidosaurs
d. synapsids
e. none of these

Answer: b
Difficulty: Moderate
Bloom's Taxonomy: Comprehension

55. Lizards and snakes are living representatives of which of the following lineages?

a. anapsids
b. archosaurs
c. lepidosaurs
d. synapsids
e. none of these

Answer: c
Difficulty: Difficult
Bloom's Taxonomy: Comprehension

56. Based on modern understanding of physiology, DNA sequences, and the fossil record, which of the following types of animals is most closely related to crocodiles?

a. snakes
b. birds
c. fleshy-finned fishes
d. lizards
e. mammals

Answer: b
Difficulty: Difficult
Bloom's Taxonomy: Comprehension

30.8 TESTUDINES: TURTLES

57. Turtles are members of which of the following lineages?

a. Testudines
b. Archosauromorpha
c. Squamata
d. Sphenodontia
e. Amphibia

Answer: a
Difficulty: Easy
Bloom's Taxonomy: Comprehension

58. Turtles have all of the following except for

a. a carapace.
b. jaws.
c. teeth.
d. a plastron.
e. a keratinized beak.

Answer: c
Difficulty: Moderate
Bloom's Taxonomy: Knowledge

59. Which of the following is a reason why many species of turtles are now highly endangered?

a. Collection of young turtles for the pet trade.
b. Hunting of turtles for their meat.
c. Consumption of their eggs by humans.
d. All of these are reasons why many turtle species are endangered.

Answer: d
Difficulty: Easy
Bloom's Taxonomy: Comprehension

30.9 LIVING NONFEATHERED DIAPSIDS: SPHENODONTIDS, SQUAMATES, AND CROCODILIANS

60. While it was a diverse group in the Mesozoic, the tuatara is one of only two living members of this lineage.

a. Testudines
b. Archosauromorpha
c. Squamata
d. Sphenodontia
e. Amphibia

Answer: d
Difficulty: Difficult
Bloom's Taxonomy: Comprehension

61. Lizards and snakes are living members of this lineage.

a. Testudines
b. Archosauromorpha
c. Squamata
d. Sphenodontia
e. Amphibia

Answer: c
Difficulty: Easy
Bloom's Taxonomy: Comprehension

62. American alligators are living members of this lineage.

a. Testudines
b. Archosauromorpha
c. Squamata
d. Sphenodontia
e. Amphibia

Answer: b
Difficulty: Easy
Bloom's Taxonomy: Comprehension

FOCUS ON RESEARCH: MODEL RESEARCH ORGANISMS: *ANOLIS* LIZARDS OF THE CARIBBEAN

63. Members of this genus are commonly used as model systems for studies of ecology and evolutionary biology.

a. *Ichthyostega*
b. *Anolis*
c. *Dunkleosteus*
d. *Eusthenopteron*
e. *Archaeopteryx*

Answer: b
Difficulty: Easy
Bloom's Taxonomy: Knowledge

64. Ecomorphs are

a. not very closely related members of the same genus that appear very different from each other and that live in very different habitats.
b. closely related members of the same genus that appear very similar to each other although they live in very different habitats.
c. members of the same species that appear very similar to each other although they live in very different habitats.
d. not very closely related members of the same genus that nevertheless appear very similar to each other and that live in very similar habitats.
e. closely related members of the same genus that appear very similar to each other and that live in very similar habitats.

Answer: d
Difficulty: Moderate
Bloom's Taxonomy: Synthesis

30.10 AVES: BIRDS

65. Birds are living representatives of which of the following lineages?

a. anapsids
b. archosaurs
c. lepidosaurs
d. synapsids
e. none of these

Answer: b
Difficulty: Difficult
Bloom's Taxonomy: Comprehension

66. A four-chambered heart, hollow limb bones with supporting struts, and a keeled sternum are features associated with all or most modern

a. mammals.
b. snakes.
c. turtles.
d. lizards.
e. birds.

Answer: e
Difficulty: Moderate
Bloom's Taxonomy: Knowledge

67. Which of the following is a unique, derived trait that is present in all birds?

a. ability to fly
b. four-chambered heart
c. feathers
d. lack of a urinary bladder
e. migration

Answer: c
Difficulty: Easy
Bloom's Taxonomy: Knowledge

30.11 MAMMALIA: MONOTREMES, MARSUPIALS, AND PLACENTALS

68. Which of the following mammalian lineages contains organisms that lay eggs?

a. Strepsirhini
b. monotremes
c. Rodentia
d. Haplorhini
e. marsupials

Answer: b
Difficulty: Moderate
Bloom's Taxonomy: Comprehension

69. Two species of spiny anteaters (echidnas) and the duck-billed platypus are the only known living members of which of the following mammalian lineages?

a. Strepsirhini
b. monotremes
c. Rodentia
d. Haplorhini
e. marsupials

Answer: b
Difficulty: Moderate
Bloom's Taxonomy: Comprehension

70. The opossum and kangaroos are examples of living members of which of the following mammalian lineages?

a. Strepsirhini
b. monotremes
c. Rodentia
d. Haplorhini
e. marsupials

Answer: e
Difficulty: Easy
Bloom's Taxonomy: Comprehension

71. The majority of mammals species living today are _____, or placental mammals, including such diverse groups as rodents, bats, whales, and primates.

a. protherians
b. therapsids
c. anthropoids
d. eutherians
e. metatherians

Answer: d
Difficulty: Moderate
Bloom's Taxonomy: Synthesis

INSIGHTS FROM THE MOLECULAR REVOLUTION: THE GUINEA PIG IS NOT A RAT

72. Based on mitochondrial DNA sequence data, the guinea pig is

a. more closely related to mice than to rats.
b. more closely related to mice and rats than to any other mammal.
c. more closely related to hedgehogs than to rabbits.
d. more closely related to humans than to mice and rats.
e. more closely related to horses than to whales.

Answer: d
Difficulty: Difficult
Bloom's Taxonomy: Evaluation

30.12 NONHUMAN PRIMATES

73. Galagos, lemurs, and lorises are the living members of which of the following mammalian lineages?

a. Strepsirhini
b. monotremes
c. Rodentia
d. Haplorhini
e. marsupials

Answer: a
Difficulty: Difficult
Bloom's Taxonomy: Comprehension

74. Tarsiers, new world monkeys, old world monkeys, and apes are all living members of which of the following mammalian lineages?

a. Strepsirhini
b. monotremes
c. Rodentia
d. Haplorhini
e. marsupials

Answer: d
Difficulty: Difficult
Bloom's Taxonomy: Comprehension

75. Which of these groups is most closely related to gorillas?

a. old world monkeys
b. lemurs
c. lorises
d. new world monkeys
e. tarsiers

Answer: a
Difficulty: Moderate
Bloom's Taxonomy: Synthesis

76. Which of these groups is most closely related to bonobos?

a. tarsiers
b. humans
c. lorises
d. gorillas
e. new world monkeys

Answer: b
Difficulty: Difficult
Bloom's Taxonomy: Synthesis

77. Based on the fossil record, which of these groups appeared first?

a. chimpanzees
b. humans
c. new world monkeys
d. gorillas
e. old world monkeys

Answer: c
Difficulty: Difficult
Bloom's Taxonomy: Synthesis

78. The pattern of locomotion used by gibbons and siamangs, where they use their arms to hang below branches and swing forward, is called

a. bipedal movement.
b. peristalsis.
c. gliding.
d. brachiation.
e. orthokinesis.

Answer: d
Difficulty: Moderate
Bloom's Taxonomy: Knowledge

79. Which of these groups is most closely related to humans?

a. gorillas
b. new world monkeys
c. orangutans
d. old world monkeys
e. chimpanzees

Answer: e
Difficulty: Easy
Bloom's Taxonomy: Synthesis

30.13 THE EVOLUTION OF HUMANS

80. Which of these groups includes modern humans and their bipedal relatives but not gorillas?

a. hominids
b. primates
c. Haplorhini
d. hominoids
e. arthropoids

Answer: a
Difficulty: Moderate
Bloom's Taxonomy: Analysis

81. Dating to about 3.5 million to 3 million years ago, the famous fossil "Lucy," found in northern Ethiopia, represents a member of which species?

a. *Homo habilis*
b. *Homo erectus*
c. *Australopithecus africanus*
d. *Australopithecus afarensis*
e. *Homo neanderthalensis*

Answer: d
Difficulty: Moderate
Bloom's Taxonomy: Comprehension

82. Based on the fossil record, which of the following groups evolved about 1.8 million years ago and made fairly sophisticated tools, including the hand axe?

a. *Homo habilis*
b. *Homo erectus*
c. *Australopithecus africanus*
d. *Australopithecus afarensis*
e. *Homo neanderthalensis*

Answer: b
Difficulty: Moderate
Bloom's Taxonomy: Comprehension

83. This group occupied Europe and western Asia from about 150,000 to 28,000 years ago. They had a heavier build, more-pronounced brow ridges, and slightly larger brains than modern humans, and were culturally and technologically sophisticated.

a. *Homo habilis*
b. *Homo erectus*
c. *Australopithecus africanus*
d. *Australopithecus afarensis*
e. *Homo neanderthalensis*

Answer: e
Difficulty: Easy
Bloom's Taxonomy: Comprehension

84. Based on fossil evidence, which of the following existed the longest ago?

a. *Homo habilis*
b. *Homo erectus*
c. *Australopithecus africanus*
d. *Australopithecus afarensis*
e. *Homo neanderthalensis*

Answer: d
Difficulty: Moderate
Bloom's Taxonomy: Synthesis

UNANSWERED QUESTIONS

85. Live-bearing reproduction, or viviparity, has apparently evolved

a. just three separate times.
b. about 200 separate times in vertebrates alone.
c. only in mammals.
d. only once, although many groups lost it secondarily.
e. about 15 separate times.

Answer: b
Difficulty: Moderate
Bloom's Taxonomy: Knowledge

Matching

Match each of the groups of organisms listed with the correct lineage. Each lineage will be used only once.

86. _____ tunicates

87. _____ sea stars

88. _____ sea lilies and feather stars

89. _____ hagfishes

90. _____ frogs and salamanders

91. _____ humans

92. _____ lancelets

93. _____ sharks, skates, and rays

94. _____ duck-billed platypuses

95. _____ lampreys

96. _____ sea urchins and sand dollars

97. _____ gars, bowfins, sturgeons, and paddlefish

98. _____ brittle stars and basket stars

99. _____ turtles

100. _____ sea cucumbers

101. _____ lizards and snakes

102. _____ acorn worms

103. _____ kangaroos

104. _____ birds

A. Petromyzontoidea

B. Amphibia

C. Crinoidea

D. Metatheria

E. Chondrichthyes

F. Actinopterygii

G. Urochordata

H. Aves

I. Myxinoidea

J. Holothuroidea

K. Cephalochordata

L. Eutheria

M. Ophiuroidea

N. Testudines

O. Squamata

P. Asteroidea

Q. Protheria

R. Echinoidea

S. Hemichordata

Answers:

86. G	*87. P*	*88. C*	*89. I*	*90. B*
91. L	*92. K*	*93. E*	*94. Q*	*95. A*
96. R	*97. F*	*98. M*	*99. N*	*100. J*
101. O	*102. S*	*103. D*	*104. H*	

Difficulty: Moderate
Bloom's Taxonomy: Knowledge
Source: Sections 30.1–30.11

31

THE PLANT BODY

Multiple-Choice

WHY IT MATTERS

1.   The structure and arrangement of internal parts of a plant is the _____ of the plant.

a.   morphology
b.   ecology
c.   anatomy
d.   distribution
e.   physiology

Answer: c
Difficulty: Moderate
Bloom's Taxonomy: Comprehension

2.   The external form of a plant is the _____ of the plant.

a.   morphology
b.   ecology
c.   anatomy
d.   distribution
e.   physiology

Answer: a
Difficulty: Moderate
Bloom's Taxonomy: Comprehension

3.   The mechanisms by which the body of a plant functions in its environment is the _____ of the plant.

a.   morphology
b.   ecology
c.   anatomy
d.   distribution
e.   physiology

Answer: e
Difficulty: Easy
Bloom's Taxonomy: Comprehension

31.1 PLANT STRUCTURE AND GROWTH: AN OVERVIEW

4.   Vascular plant bodies typically consist of a _____ shoot system and a _____ root system.

a.   photosynthetic; photosynthetic
b.   nonphotosynthetic; photosynthetic
c.   photosynthetic; nonphotosynthetic
d.   nonphotosynthetic; nonphotosynthetic

Answer: c
Difficulty: Easy
Bloom's Taxonomy: Knowledge

5.   A plant stem would be considered to be

a.   a shoot system.
b.   a tissue.
c.   a root system.
d.   a meristem.
e.   an organ.

Answer: e
Difficulty: Moderate
Bloom's Taxonomy: Application

6.   Which of the following is NOT typically found in primary plant cell walls?

a.   lignin
b.   hemicelluloses
c.   pectin
d.   cellulose

Answer: a
Difficulty: Moderate
Bloom's Taxonomy: Knowledge

7.   Which of the following is often used to congeal jams and jellies?

a.   lignin
b.   hemicelluloses
c.   pectin
d.   cellulose

Answer: c
Difficulty: Difficult
Bloom's Taxonomy: Knowledge

8.   Which of the following can waterproof cell walls, allowing for structures such as watertight conduction channels?

a.   lignin
b.   hemicelluloses
c.   pectin
d.   cellulose

Answer: a
Difficulty: Moderate
Bloom's Taxonomy: Comprehension

9.   The _____ of a plant is self-perpetuating embryonic tissue typically found at the tips of shoots and roots.

a.   ground tissue
b.   protoderm
c.   vascular tissue
d.   meristem
e.   dermal tissue

Answer: d
Difficulty: Moderate
Bloom's Taxonomy: Knowledge

10. Growth from apical meristems, generally resulting in an increase in the length of a plant, is referred to as

a. horizontal growth.
b. typical growth.
c. secondary growth.
d. longitudinal growth.
e. primary growth.

Answer: e
Difficulty: Moderate
Bloom's Taxonomy: Knowledge

11. Which of the following is true of typical plant development?

a. Growing tips and zones are present throughout a plant's life, and final plant form is not influenced by the environment.
b. Plant bodies have a fixed final size, and final plant form is influenced by the environment.
c. Final plant form is not influenced by the environment, and growing tips and zones are present throughout a plant's life.
d. Growing tips and zones are present throughout a plant's life, and plant bodies do not have a fixed final size.
e. Plant bodies do not have a fixed final size, and final plant form is not influenced by the environment.

Answer: d
Difficulty: Difficult
Bloom's Taxonomy: Synthesis

12. Which term refers to plants that complete their life cycle in one growing season?

a. perennials
b. annuals
c. monocots
d. biennials
e. eudicots

Answer: b
Difficulty: Easy
Bloom's Taxonomy: Knowledge

13. Which term refers to plants that complete their life cycle in two growing seasons?

a. perennials
b. annuals
c. monocots
d. biennials
e. eudicots

Answer: d
Difficulty: Easy
Bloom's Taxonomy: Knowledge

14. Which term refers to plants that typically grow for many years?

a. perennials
b. annuals
c. monocots
d. biennials
e. eudicots

Answer: a
Difficulty: Easy
Bloom's Taxonomy: Knowledge

31.2 THE THREE PLANT TISSUE SYSTEMS

15. Among other functions, the _____ of a vascular plant performs most of the photosynthesis that is conducted by the plant.

a. ground tissue
b. protoderm
c. vascular tissue
d. meristem
e. dermal tissue

Answer: a
Difficulty: Moderate
Bloom's Taxonomy: Application

16. Cells called _____ make up the bulk of the soft primary growth of roots, stems, leaves, flowers, and fruits.

a. sclerenchyma
b. tracheids
c. collenchyma
d. parenchyma
e. vessel members

Answer: d
Difficulty: Moderate
Bloom's Taxonomy: Knowledge

17. Cells called _____ form flexible support strands such as the "strings" in celery.

a. sclerenchyma
b. tracheids
c. collenchyma
d. parenchyma
e. vessel members

Answer: c
Difficulty: Moderate
Bloom's Taxonomy: Knowledge

18. Plant cells called _____ are a type of ground tissue that provide rigid support via the thick, typically lignified cell walls that remain after they die.

a. sclerenchyma
b. tracheids
c. collenchyma
d. parenchyma
e. vessel members

Answer: a
Difficulty: Moderate
Bloom's Taxonomy: Knowledge

19. Cells called _____ form the protective coat around seeds.

a. sclerenchyma
b. tracheids
c. collenchyma
d. parenchyma
e. vessel members

Answer: a
Difficulty: Difficult
Bloom's Taxonomy: Knowledge

20. Cells called _____ are a type of vascular tissue with long, tapered, overlapping ends. They develop thick, lignified cell walls and die at maturity, leaving a water conducting tube with pits at the tapered ends that allow for lateral water transport between cells.

a. sclerenchyma
b. tracheids
c. collenchyma
d. parenchyma
e. vessel members

Answer: b
Difficulty: Moderate
Bloom's Taxonomy: Knowledge

21. Cells called _____ are a type of vascular tissue that join end to end in tubelike columns. They develop thick, lignified cell walls and die at maturity, leaving a water-conducting tube. At maturity the ends of the cells are open or have large openings, allowing for enhanced water flow compared to other types of vascular tissue.

a. sclerenchyma
b. tracheids
c. collenchyma
d. parenchyma
e. vessel members

Answer: e
Difficulty: Moderate
Bloom's Taxonomy: Knowledge

22. Plant tissue specialized for conducting fluids is

a. ground tissue.
b. protoderm.
c. vascular tissue.
d. meristem.
e. dermal tissue.

Answer: c
Difficulty: Easy
Bloom's Taxonomy: Knowledge

23. Xylem and phloem are types of

a. ground tissue.
b. protoderm.
c. vascular tissue.
d. meristem.
e. dermal tissue.

Answer: c
Difficulty: Easy
Bloom's Taxonomy: Knowledge

24. What tissue in vascular plants would be used to move water and dissolved minerals from roots to the stems and leaves?

a. sclerenchyma
b. phloem
c. collenchyma
d. parenchyma
e. xylem

Answer: e
Difficulty: Moderate
Bloom's Taxonomy: Application

25. What tissue in vascular plants would be used to move sugars from roots to the stems and leaves?

a. sclerenchyma
b. phloem
c. collenchyma
d. parenchyma
e. xylem

Answer: b
Difficulty: Moderate
Bloom's Taxonomy: Application

26. Sieve tube members are part of the

a. ground tissue.
b. phloem.
c. dermal tissue.
d. meristem.
e. xylem.

Answer: b
Difficulty: Moderate
Bloom's Taxonomy: Knowledge

27. Companion cells are connected via plasmodesmata to

a. trichomes.
b. tracheids.
c. sieve tube members.
d. guard cells.
e. vessel members.

Answer: c
Difficulty: Moderate
Bloom's Taxonomy: Knowledge

28. Plant surfaces are covered and protected by

a. ground tissue.
b. protoderm.
c. vascular tissue.
d. meristem.
e. dermal tissue.

Answer: e
Difficulty: Easy
Bloom's Taxonomy: Knowledge

29. Cuticle, guard cells, stomata, and trichomes are all terms associated with

a. ground tissue.
b. vascular tissue.
c. meristem.
d. dermal tissue.

Answer: d
Difficulty: Easy
Bloom's Taxonomy: Knowledge

30. Epidermal cells secrete a coating called the _____ that protects the plant from water loss and attacks by microbes.

a. cuticle
b. protoplast
c. stoma
d. trichome
e. bark

Answer: a
Difficulty: Easy
Bloom's Taxonomy: Knowledge

31. Specialized outgrowths of the plant epidermis called _____ are hairlike projections, such as root hairs.

a. cuticles
b. protoplasts
c. stomata
d. trichomes
e. bark

Answer: d
Difficulty: Moderate
Bloom's Taxonomy: Knowledge

INSIGHTS FROM THE MOLECULAR REVOLUTION: SHAPING UP FLOWER COLOR

32. The *mixta* gene in snapdragons helps determine flower color by regulating

a. anthocyanin production.
b. production of pigments other than anthocyanin.
c. cell shape.
d. production of anthocyanin and other pigments.
e. anthocyanin production and cell shape.

Answer: c
Difficulty: Moderate
Bloom's Taxonomy: Comprehension

31.3 PRIMARY SHOOT SYSTEMS

33. Which of the following is NOT typically a main function of stems?

a. mechanical support
b. storage of water and food
c. energy capture
d. housing vascular tissues
e. new growth

Answer: c
Difficulty: Moderate
Bloom's Taxonomy: Knowledge

34. The place on a stem where one or more leaves are attached is called the

a. terminal bud.
b. internode.
c. axil.
d. lateral bud.
e. node.

Answer: e
Difficulty: Easy
Bloom's Taxonomy: Knowledge

35. The upper angle between a stem and attached leaf is called a(n)

a. terminal bud.
b. internode.
c. axil.
d. lateral bud.
e. node.

Answer: c
Difficulty: Difficult
Bloom's Taxonomy: Knowledge

36. The site of new primary growth at the apex of a shoot is the

a. terminal bud.
b. internode.
c. axil.
d. lateral bud.
e. node.

Answer: a
Difficulty: Moderate
Bloom's Taxonomy: Knowledge

37. Apical dominance is maintained by

a. hormones released by the terminal bud.
b. sugars produced in the leaves.
c. water transported from the roots.
d. hormones released by the lateral buds.
e. sugars produced in the lateral buds.

Answer: a
Difficulty: Moderate
Bloom's Taxonomy: Comprehension

38. When an apical meristem cell divides, one daughter cell is called the initial and the other is called the derivative. What are the fates of these cells?

a. The initial is used to form primary meristems, and the derivative remains part of the apical meristem.
b. Both the initial and the derivative remain part of the apical meristem.
c. The initial remains part of the apical meristem, and the derivative is used to form primary meristems.
d. Both the initial and the derivative are used to form primary meristems.

Answer: c
Difficulty: Moderate
Bloom's Taxonomy: Knowledge

39. Which of the following most directly gives rise to primary vascular tissues?

a. protoderm
b. apical meristem
c. ground meristem
d. vascular cambium
e. procambium

Answer: e
Difficulty: Moderate
Bloom's Taxonomy: Knowledge

40. Which of the following represents the correct order of structures from outside to inside for a stem that has a vascular cylinder?

a. stele, cortex, epidermis, pith
b. epidermis, pith, cortex, stele
c. cortex, stele, pith, epidermis
d. epidermis, cortex, stele, pith
e. stele, epidermis, pith, cortex

Answer: d
Difficulty: Moderate
Bloom's Taxonomy: Synthesis

41. Rhizomes, tubers, corms, and stolons are examples of modified

a. leaves.
b. bulbs.
c. roots.
d. flowers.
e. stems.

Answer: e
Difficulty: Moderate
Bloom's Taxonomy: Knowledge

42. Often connected to a petiole, these plant organs may be either simple or compound and are typically the main organs of photosynthesis and gas exchange.

a. roots
b. leaves
c. flowers
d. stems
e. trichomes

Answer: b
Difficulty: Easy
Bloom's Taxonomy: Knowledge

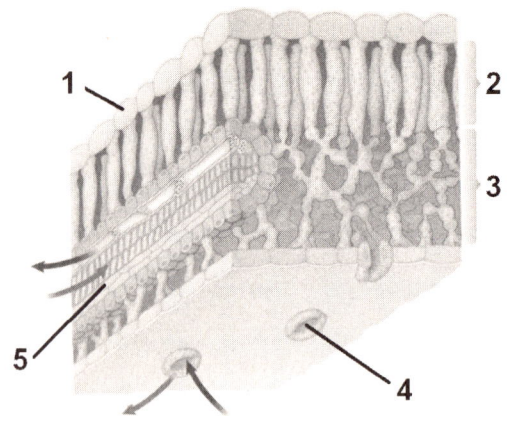

Use the figure above for questions 43–47.

43. In the internal leaf structure shown above, the item labeled "1" is

a. a vascular bundle.
b. spongy mesophyll.
c. epidermis.
d. a stoma.
e. palisade mesophyll.

Answer: c
Difficulty: Easy
Bloom's Taxonomy: Application
Source: Figure 31.17

44. In the internal leaf structure shown above, the item labeled "2" is

a. a vascular bundle.
b. spongy mesophyll.
c. epidermis.
d. a stoma.
e. palisade mesophyll.

Answer: e
Difficulty: Moderate
Bloom's Taxonomy: Application
Source: Figure 31.17

45. In the internal leaf structure shown above, the item labeled "3" is

a. a vascular bundle.
b. spongy mesophyll.
c. epidermis.
d. a stoma.
e. palisade mesophyll.

Answer: b
Difficulty: Moderate
Bloom's Taxonomy: Application
Source: Figure 31.17

46. In the internal leaf structure shown above, the item labeled "4" is

a. a vascular bundle.
b. spongy mesophyll.
c. epidermis.
d. a stoma.
e. palisade mesophyll.

Answer: d
Difficulty: Moderate
Bloom's Taxonomy: Application
Source: Figure 31.17

47. In the internal leaf structure shown above, the item labeled "5" is

a. a vascular bundle.
b. spongy mesophyll.
c. epidermis.
d. a stoma.
e. palisade mesophyll.

Answer: a
Difficulty: Easy
Bloom's Taxonomy: Application
Source: Figure 31.17

FOCUS ON RESEARCH: BASIC RESEARCH: HOMEOBOX GENES: HOW THE MERISTEM GIVES ITS MARCHING ORDERS

48. The *KN-1* gene in maize is normally expressed

a. in primary meristems, where it initiates the process of differentiation.
b. in apical meristems, where it maintains the undifferentiated state.
c. in apical meristems, where it initiates the process of differentiation.
d. in primary meristems, where it maintains the undifferentiated state.

Answer: b
Difficulty: Difficult
Bloom's Taxonomy: Knowledge

31.4 ROOT SYSTEMS

49. A single main root that is adapted for storage and that typically grows downward and fairly deep is called

a. an adventitious root.
b. a fibrous root.
c. a rhizome.
d. a taproot.
e. a tuber.

Answer: d
Difficulty: Easy
Bloom's Taxonomy: Knowledge

50. The root apical meristem is surrounded and protected by a structure called the

a. endodermis.
b. root cap.
c. quiescent center.
d. pericycle.
e. root hairs.

Answer: b
Difficulty: Easy
Bloom's Taxonomy: Knowledge

51. Lateral roots arise from the

a. endodermis.
b. root cap.
c. quiescent center.
d. pericycle.
e. root hairs.

Answer: d
Difficulty: Moderate
Bloom's Taxonomy: Knowledge

52. The innermost later of the root cortex is a selectively permeable barrier called the

a. endodermis.
b. root cap.
c. quiescent center.
d. pericycle.
e. root hairs.

Answer: a
Difficulty: Moderate
Bloom's Taxonomy: Knowledge

53. The uptake of water and mineral ions from the soil occurs primarily at the

a. endodermis.
b. root cap.
c. quiescent center.
d. pericycle.
e. root hairs.

Answer: e
Difficulty: Easy
Bloom's Taxonomy: Knowledge

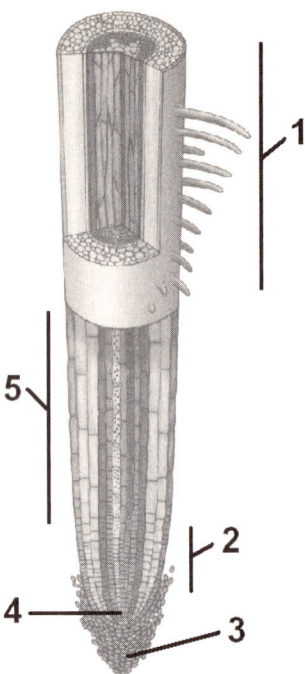

Use the figure above for questions 54–58.

54. In the diagram of a root tip shown above, the item labeled "1" is the

a. quiescent center.
b. zone of cell division.
c. zone of elongation.
d. root cap.
e. zone of maturation.

Answer: e
Difficulty: Moderate
Bloom's Taxonomy: Application
Source: Figure 31.20

55. In the diagram of a root tip shown above, the item labeled "2" is the

a. quiescent center.
b. zone of cell division.
c. zone of elongation.
d. root cap.
e. zone of maturation.

Answer: b
Difficulty: Moderate
Bloom's Taxonomy: Application
Source: Figure 31.20

56. In the diagram of a root tip shown above, the item labeled "3" is the

a. quiescent center.
b. zone of cell division.
c. zone of elongation.
d. root cap.
e. zone of maturation.

Answer: d
Difficulty: Easy
Bloom's Taxonomy: Application
Source: Figure 31.20

57. In the diagram of a root tip shown above, the item labeled "4" is the

a. quiescent center.
b. zone of cell division.
c. zone of elongation.
d. root cap.
e. zone of maturation.

Answer: a
Difficulty: Moderate
Bloom's Taxonomy: Application
Source: Figure 31.20

58. In the diagram of a root tip shown above, the item labeled "5" is the

a. quiescent center.
b. zone of cell division.
c. zone of elongation.
d. root cap.
e. zone of maturation.

Answer: c
Difficulty: Moderate
Bloom's Taxonomy: Application
Source: Figure 31.20

31.5 SECONDARY GROWTH

59. Which of the following most directly gives rise to secondary xylem and phloem?

a. protoderm
b. apical meristem
c. ground meristem
d. vascular cambium
e. procambium

Answer: d
Difficulty: Moderate
Bloom's Taxonomy: Knowledge

60. Cork cambium produces

a. vascular tissue.
b. heartwood.
c. vascular cambium.
d. sapwood.
e. secondary epidermis.

Answer: e
Difficulty: Moderate
Bloom's Taxonomy: Knowledge

61. Primary and secondary xylem that can no longer transport water and solutes, and that instead is strengthened, dry tissue where some defensive compounds are stored, is called

a. bark.
b. heartwood.
c. vascular cambium.
d. sapwood.
e. secondary epidermis.

Answer: b
Difficulty: Easy
Bloom's Taxonomy: Comprehension

62. Image that you have found a mutant oak plant that cannot produce fusiform initials. Which of the following would be the most direct result of this mutation?

a. The plant would not be able to perform photosynthesis.
b. The plant would not be able to make secondary xylem and phloem.
c. The plant would not be able to make roots.
d. The plant would not be able to transport sugar to the roots.
e. The plant would not be able to make leaves.

Answer: b
Difficulty: Difficult
Bloom's Taxonomy: Application

63. Starting from the outside and moving in toward the center, which gives the correct order of tissues in the stem of a young tree?

a. secondary phloem, primary phloem, primary xylem, secondary xylem
b. primary xylem, secondary xylem, secondary phloem, primary phloem
c. primary phloem, secondary xylem, secondary phloem, primary xylem
d. secondary phloem, primary xylem, secondary xylem, primary phloem
e. primary phloem, secondary phloem, secondary xylem, primary xylem

Answer: e
Difficulty: Difficult
Bloom's Taxonomy: Application

UNANSWERED QUESTIONS

64. Most plant cells are pluripotent, meaning that they

a. constantly switch tissue types throughout their lives.
b. cannot change after differentiating into a specific cell type.
c. end up as a hybrid of tissue types.
d. can function in any environmental condition.
e. retain the ability to change cell types when necessary.

Answer: e
Difficulty: Moderate
Bloom's Taxonomy: Comprehension

Choice

For each of the following descriptions of a plant characteristic, provide the letter that represents the plant group or groups (or no plant groups) that typically have that characteristic. Each letter may be used once, more than once, or not at all.

a. eudicots
b. monocots
c. both eudicots and monocots
d. neither eudicots nor monocots

65. _____ one pore or furrow in pollen grains

66. _____ reproduction using seeds

67. _____ vascular bundles organized in a ring in ground tissue

68. _____ three pores or furrows in pollen grains

69. _____ floral parts in threes (or multiples of three)

70. _____ parallel leaf veins

71. _____ two cotyledons

72. _____ branching fibrous root system

73. _____ vascular tissues

74. _____ one cotyledon

75. _____ lack of meristems

76. _____ leaf veins in a netlike array

77. _____ floral parts in fours or fives (or multiples of four or five)

Answers: **65.** *b* **66.** *c* **67.** *a* **68.** *a* **69.** *b*
 70. *b* **71.** *a* **72.** *b* **73.** *c* **74.** *b*
 75. *d* **76.** *a* **77.** *a*

Difficulty: Moderate
Bloom's Taxonomy: Knowledge
Source: Section 31.1

Short Answer

78. Explain the difference between primary and secondary growth.

Answer: Primary growth, which is found in all vascular plants, comes from apical meristems and results mainly in elongation of the plant in both the root and shoot systems. Secondary growth occurs in some plants, and when it does occur it primarily increases the girth of the plant through growth that comes from the vascular cambium and the cork cambium. Woody tissues are a hallmark of secondary growth.
Difficulty: Moderate
Bloom's Taxonomy: Synthesis
Source: Sections 31.1 and 31.5

32
TRANSPORT IN PLANTS

Multiple-Choice

WHY IT MATTERS

1.   The process that a tall tree uses to move water to the leaves at the top of the tree involves primarily

a.   cohesion and evaporation.
b.   evaporation and pumps.
c.   capillary action and pumps.
d.   positive pressure from roots and cohesion.
e.   capillary action and positive pressure from roots.

Answer: a
Difficulty: Easy
Bloom's Taxonomy: Comprehension

32.1 PRINCIPLES OF WATER AND SOLUTE MOVEMENT IN PLANTS

2.   Facilitated diffusion requires

a.   a channel protein.
b.   both a channel protein and a pore.
c.   a hole in the plasma membrane.
d.   either a channel protein or a carrier protein.
e.   a carrier protein.

Answer: d
Difficulty: Easy
Bloom's Taxonomy: Comprehension

3.   The membrane potential refers to the _____ across a cell membrane.

a.   difference in water concentration
b.   charge difference
c.   rate of movement
d.   distance
e.   difference in salt concentration

Answer: b
Difficulty: Moderate
Bloom's Taxonomy: Knowledge

4.   Which of the following should you expect to find for a living plant cell in a living plant?

a.   a cytoplasm that is much more negatively charged than the fluid outside the cell
b.   a cytoplasm that has essentially the same charge as the fluid outside the cell
c.   a cytoplasm that is slightly more positively charged than the fluid outside the cell
d.   a cytoplasm that is much more positively charged than the fluid outside the cell
e.   a cytoplasm that is slightly more negatively charged than the fluid outside the cell

Answer: e
Difficulty: Moderate
Bloom's Taxonomy: Application

5.   Generally, the energy for pumping protons outside of a plant cell comes most directly from

a.   sunlight.
b.   starch.
c.   ATP.
d.   diffusion.
e.   glucose.

Answer: c
Difficulty: Moderate
Bloom's Taxonomy: Comprehension

6.   Which of the following should you expect to be occurring in a living plant cell in a living plant?

a.   protons moving out of the cell via passive diffusion
b.   protons being actively pumped into the cell
c.   protons moving out of the cell via facilitated diffusion
d.   protons being actively pumped out of the cell
e.   protons moving into the cell via passive diffusion

Answer: d
Difficulty: Moderate
Bloom's Taxonomy: Application

7.   The simultaneous movement of sucrose into a cell against its concentration gradient and protons into a cell with their concentration gradient is best described as an example of

a.   symport.
b.   passive diffusion.
c.   antiport.
d.   facilitated diffusion.
e.   active transport.

Answer: a
Difficulty: Moderate
Bloom's Taxonomy: Application

8.   The simultaneous movement of Na^+ out of a cell against its concentration gradient and protons into a cell with their concentration gradient is best described as an example of

a.   symport.
b.   passive diffusion.
c.   antiport.
d.   facilitated diffusion.
e.   active transport.

Answer: c
Difficulty: Moderate
Bloom's Taxonomy: Application

9. The movement of NH_4^+ ions into a cell down an electrochemical gradient (where, when compared to the outside of the cell, the inside of the cell has less NH_4^+ ions) is best described as an example of

a. symport.
b. passive diffusion.
c. antiport.
d. facilitated diffusion.
e. active transport.

Answer: d
Difficulty: Difficult
Bloom's Taxonomy: Application

10. The typical movement of xylem sap from roots to shoot parts is best described as

a. facilitated diffusion.
b. symport.
c. guttation.
d. osmosis.
e. bulk flow.

Answer: e
Difficulty: Moderate
Bloom's Taxonomy: Knowledge

11. Individual plant cells typically gain or lose water mainly via

a. facilitated diffusion.
b. symport.
c. guttation.
d. osmosis.
e. bulk flow.

Answer: d
Difficulty: Easy
Bloom's Taxonomy: Knowledge

12. Water potential is typically represented by the Greek letter _____.

a. α
b. γ
c. ψ
d. π
e. θ

Answer: c
Difficulty: Easy
Bloom's Taxonomy: Knowledge

13. Consider a living plant cell in a living leaf that is not wilted. The solute potential (ψ_S) of such a cell _____ than that in the fluids surrounding the cell.

a. is typically lower
b. varies widely from higher to lower
c. is typically higher
d. is typically about the same as

Answer: a
Difficulty: Moderate
Bloom's Taxonomy: Application

14. Consider a living plant cell in a living leaf that is not wilted. The pressure potential (ψ_P) of such a cell _____ than that in the fluids surrounding the cell.

a. is typically lower
b. varies widely from higher to lower
c. is typically higher
d. is typically about the same as

Answer: c
Difficulty: Difficult
Bloom's Taxonomy: Synthesis

15. Consider a living plant cell in a living leaf that has a higher water potential than the fluids surrounding the cell. In such a situation there should be

a. a net flow of water out of the cell.
b. gain of water by the cell only if the cell's pressure potential is high enough.
c. no net flow of water into or out of the cell.
d. a net flow of water into the cell.
e. gain of water by the cell only if the cell's pressure potential is low enough.

Answer: a
Difficulty: Difficult
Bloom's Taxonomy: Application

16. Which plant organelle stores solutes and plays a major role in maintaining turgor pressure?

a. nucleus
b. cell wall
c. chloroplast
d. vacuole
e. mitochondrion

Answer: d
Difficulty: Moderate
Bloom's Taxonomy: Application

17. The tonoplast is another name for

a. the plant nuclear envelope.
b. a chloroplast.
c. the cell wall.
d. an undifferentiated plastid.
e. the vacuolar membrane.

Answer: e
Difficulty: Moderate
Bloom's Taxonomy: Knowledge

18. Suppose you discover a mutant plant that cannot actively transport solutes across the tonoplast. Which of the following should you expect?

a. The plant will wilt easily.
b. The plant will generally be like a normal plant.
c. The plant will not have leaves.
d. The plant will not have roots.
e. The plant cells will not be able to divide.

Answer: a
Difficulty: Moderate
Bloom's Taxonomy: Application

19. Aquaporins are

a. channel proteins for water.
b. carrier proteins for solutes.
c. channel proteins for solutes.
d. carrier proteins for water.
e. carrier proteins for water and solutes.

Answer: a
Difficulty: Moderate
Bloom's Taxonomy: Knowledge

20. If the ψ of surrounding soil is higher than that in living root epidermal cells, then water should

a. leave the root cells, making them flaccid.
b. enter the root cells, making them turgid.
c. leave the root cells, making them turgid.
d. enter the root cells, making them flaccid.

Answer: b
Difficulty: Easy
Bloom's Taxonomy: Comprehension

21. If a living plant cell is placed in a beaker with pure water, it will take up water until

a. within the cell $\psi_P = -\psi_S$.
b. it bursts.
c. its $\psi_P = 0$.
d. its $\psi_S = \psi_S$ of the pure water.
e. its $\psi_S = 0$.

Answer: a
Difficulty: Moderate
Bloom's Taxonomy: Application

22. If a living plant cell is placed in a beaker with a solution that has a ψ value lower than the ψ value of the plant cell, the plant cell will

a. take up water until it bursts.
b. lose water until its $\psi = \psi$ of the solution.
c. take up water until within the cell $\psi_P = -\psi_S$.
d. take up water until its $\psi = \psi_S$ of the solution.
e. lose water until its $\psi_S = $ its ψ_P.

Answer: b
Difficulty: Difficult
Bloom's Taxonomy: Application

23. Suppose a living plant cell has $\psi = \psi_S = -0.5$ MPa. If the plant cell is placed into a beaker filled with a solution with $\psi = 0$, then the cell should

a. take up water until it bursts.
b. lose water until its $\psi = \psi$ of the solution.
c. take up water until within the cell $\psi_P = -\psi_S$.
d. take up water until its $\psi_S = \psi_S$ of the solution.
e. lose water until its $\psi_S = $ its ψ_P.

Answer: c
Difficulty: Difficult
Bloom's Taxonomy: Application

24. Suppose a living plant cell has $\psi_P = 0.4$ MPa and $\psi_S = -0.5$ MPa. If the plant cell is placed into a beaker filled with a solution with $\psi = \psi_S = -0.9$ MPa, then the cell should

a. take up water until it bursts.
b. lose water until its $\psi = \psi$ of the solution.
c. take up water until within the cell $\psi_P = -\psi_S$.
d. take up water until its $\psi_S = \psi_S$ of the solution.
e. lose water until its $\psi_S = $ its ψ_P.

Answer: b
Difficulty: Difficult
Bloom's Taxonomy: Application

25. Suppose two living plant cells are in contact with each other so that water, but not solutes, can pass between them. The cells have the same water potential. One cell has $\psi_P = 0.3$ MPa and $\psi_S = -0.5$ MPa, while the other cell has $\psi_S = -0.3$ MPa. What is the ψ_P of the second cell?

a. $\psi_P = -0.8$ MPa
b. $\psi_P = 0$ MPa
c. $\psi_P = 0.5$ MPa
d. $\psi_P = 0.8$ MPa
e. $\psi_P = 0.1$ MPa

Answer: e
Difficulty: Moderate
Bloom's Taxonomy: Application

26. What is the ψ_P of a living plant cell that has $\psi = -0.2$ MPa and $\psi_S = -0.4$ MPa?

a. $\psi_P = -0.6$ MPa
b. $\psi_P = 0.08$ MPa
c. $\psi_P = -0.2$ MPa
d. $\psi_P = 0.2$ MPa
e. $\psi_P = 0.5$ MPa

Answer: d
Difficulty: Moderate
Bloom's Taxonomy: Application

INSIGHTS FROM THE MOLECULAR REVOLUTION: A PLANT WATER CHANNEL GIVES OOCYTES A DRINK

27. Studies where mRNA for tonoplastintrinsic protein from a plant were injected into frog eggs resulted in _____ the egg cells when they were placed in a hypotonic medium.

a. rapid water loss by
b. no effect on
c. slight swelling of
d. slow water loss by
e. swelling and bursting of

Answer: e
Difficulty: Moderate
Bloom's Taxonomy: Knowledge

28. In its normal location in plant cells, tonoplastintrinsic protein apparently directly allows

a. water flow only into the chloroplast.
b. water flow into the cell across the cell wall.
c. water flow out of the cell across the cell wall.
d. water flow both into and out of the vacuole.
e. water flow only out of the chloroplast.

Answer: d
Difficulty: Moderate
Bloom's Taxonomy: Knowledge

32.2 TRANSPORT IN ROOTS

29. Water can move inside a root via the

a. symplastic pathway.
b. transmembrane and symplastic pathways.
c. transmembrane pathway.
d. apoplastic, transmembrane, and symplastic pathways.
e. apoplastic pathway.

Answer: d
Difficulty: Easy
Bloom's Taxonomy: Knowledge

30. Water that moves through living cells in a root is following the

a. symplastic pathway.
b. transmembrane and symplastic pathways.
c. transmembrane pathway.
d. apoplastic, transmembrane, and symplastic pathways.
e. apoplastic pathway.

Answer: b
Difficulty: Moderate
Bloom's Taxonomy: Application

31. Water that moves through nonliving regions of root, such as air spaces in root tissue, is following the

a. symplastic pathway.
b. transmembrane and symplastic pathways.
c. transmembrane pathway.
d. apoplastic, transmembrane, and symplastic pathways.
e. apoplastic pathway.

Answer: e
Difficulty: Moderate
Bloom's Taxonomy: Application

32. The Casparian strip of the endodermis stops water from the _____ before it reaches the stele.

a. symplastic pathway
b. transmembrane and symplastic pathways
c. transmembrane pathway
d. apoplastic, transmembrane, and symplastic pathways
e. apoplastic pathway

Answer: e
Difficulty: Moderate
Bloom's Taxonomy: Knowledge

33. Suberin is a waxy substance that is typically associated with the

a. primary xylem.
b. root hairs.
c. endodermis.
d. tonoplasts.
e. pericycle.

Answer: c
Difficulty: Moderate
Bloom's Taxonomy: Knowledge

34. Most mineral ions that plants need are

a. moved into the symplast via active transport.
b. moved into the transmembrane pathway via bulk flow.
c. moved into the apoplast via facilitated diffusion.
d. moved into the symplast via facilitated diffusion.
e. moved into the apoplast via bulk flow.

Answer: a
Difficulty: Difficult
Bloom's Taxonomy: Knowledge

35. The exodermis is

a. the substitute for xylem in nonvascular plants.
b. the above-ground continuation of endodermis.
c. the outer layer of roots, including root hairs.
d. in the roots of most flowering plants.
e. the substitute for endodermis in gymnosperms.

Answer: d
Difficulty: Difficult
Bloom's Taxonomy: Knowledge

36. Active transport in plant root cells requires that those cells have access to O_2. Normally there is enough O_2 available in air pockets in the soil, but flooded soil has very little O_2. Thus, unless they have special adaptations, plants in flooded soil effectively have no active transport in their roots. Which of the following should you expect to occur for trees without special adaptations to flooding after several days in flooded soil?

a. excess uptake of water and minerals in the xylem
b. wilting of their leaves
c. bursting of leaf cells due to excess water flow
d. excess water pushed out at the margins of their leaves
e. excess uptake of water and minerals in the xylem, and excess water pushed out at the margins of their leaves

Answer: b
Difficulty: Difficult
Bloom's Taxonomy: Synthesis

37. Many plants wind up with a Na$^+$ concentration that is considerably lower than that of the surrounding soil. Which of these plays a key role in allowing for such a difference to exist?

a. water potential
b. aquaporins
c. root cap
d. endodermis
e. tonoplast

Answer: d
Difficulty: Difficult
Bloom's Taxonomy: Application

32.3 TRANSPORT OF WATER AND MINERALS IN THE XYLEM

38. The principal driving force for movement of water into and through the plant shoot is

a. passive transport.
b. sunlight.
c. root pressure.
d. upward pressure as sugar is forced into the roots.
e. pumping by the xylem cells.

Answer: b
Difficulty: Moderate
Bloom's Taxonomy: Knowledge

39. The process of transpiration is driven by

a. passive transport.
b. sunlight.
c. root pressure.
d. upward pressure as sugar is forced into the roots.
e. pumping by the xylem cells.

Answer: b
Difficulty: Easy
Bloom's Taxonomy: Knowledge

40. The majority of the water in xylem sap typically

a. is used in capturing light energy.
b. becomes part of new plant cells.
c. is used to make sugars.
d. is stored in older plant cells.
e. evaporates into the air.

Answer: e
Difficulty: Easy
Bloom's Taxonomy: Knowledge

41. The evaporation of water into the air from plant tissues is called

a. cohesion-tension.
b. photorespiration.
c. transpiration.
d. root pressure.
e. guttation.

Answer: c
Difficulty: Easy
Bloom's Taxonomy: Knowledge

42. According to the cohesion-tension mechanism of water transport, cohesion occurs because

a. waxy coatings on insides of xylem keep water molecules together in the xylem.
b. evaporation removes water from the leaves.
c. water molecules tend to form hydrogen bonds with each other.
d. water is pushed into the xylem by bulk flow.
e. all of these

Answer: c
Difficulty: Moderate
Bloom's Taxonomy: Knowledge

43. According to the cohesion-tension mechanism of water transport, which of the following contributes to the flow of the xylem sap?

a. hydrogen bonds between water molecules
b. adhesion of water molecules to the xylem vessel walls
c. evaporation of water from the leaves
d. water potential in leaf cells below that of the leaf xylem
e. all of these

Answer: e
Difficulty: Easy
Bloom's Taxonomy: Synthesis

44. Most leaf cells are no more than _____ from a xylem vein.

a. one inch
b. one centimeter
c. half a millimeter
d. 50 micrometers
e. two inches

Answer: c
Difficulty: Moderate
Bloom's Taxonomy: Knowledge

45. Theoretically, based on the cohesion-tension mechanism, the maximum height for the tallest trees should be about

a. 50 ft.
b. 130 m.
c. 6 ft.
d. 3 m.
e. 75 m.

Answer: b
Difficulty: Difficult
Bloom's Taxonomy: Knowledge

46. The principal driving force for guttation is

a. passive transport.
b. sunlight.
c. root pressure.
d. upward pressure as sugar is forced into the roots.
e. pumping by the xylem cells.

Answer: c
Difficulty: Moderate
Bloom's Taxonomy: Knowledge

47. An air temperature rise of 10°C will typically _____ evaporation from leaves.

a. completely eliminate
b. double the rate of
c. mostly eliminate
d. triple the rate of
e. quadruple the rate of

Answer: b
Difficulty: Moderate
Bloom's Taxonomy: Knowledge

48. Which of the following has a major effect on the rate of transpiration?

a. air temperature
b. relative humidity
c. air movement
d. all of these

Answer: d
Difficulty: Moderate
Bloom's Taxonomy: Knowledge

49. Under which condition would you expect essentially no transpiration to occur?

a. brisk winds
b. 100 percent relative humidity
c. no winds
d. near freezing air temperature
e. hot air temperature and no winds

Answer: b
Difficulty: Moderate
Bloom's Taxonomy: Application

50. The major cost to plants for having cuticle-covered epidermis in the shoot system to reduce water loss is reduced _____ uptake.

a. carbon dioxide
b. potassium
c. nitrogen
d. calcium
e. oxygen

Answer: a
Difficulty: Moderate
Bloom's Taxonomy: Comprehension

51. A plant stoma is found between two

a. leaves.
b. root hairs.
c. xylem veins.
d. trichomes.
e. guard cells.

Answer: e
Difficulty: Easy
Bloom's Taxonomy: Knowledge

52. Stomata are opened when _____ guard cells and they become _____.

a. active transport pumps H^+ into; turgid
b. active transport of H^+ stops in; flaccid
c. active transport pumps H^+ out of; turgid
d. active transport pumps H^+ into; flaccid
e. active transport of H^+ stops in; turgid

Answer: c
Difficulty: Moderate
Bloom's Taxonomy: Comprehension

53. Stomata close when _____ guard cells and they become _____.

a. active transport pumps H^+ into; turgid
b. active transport of H^+ stops in; flaccid
c. active transport pumps H^+ out of; turgid
d. active transport pumps H^+ into; flaccid
e. active transport of H^+ stops in; turgid

Answer: b
Difficulty: Moderate
Bloom's Taxonomy: Comprehension

54. In most plants stomata are

a. open during the night and closed during the day.
b. nearly always open.
c. closed only when guard cells are not exposed to sunlight.
d. almost never open.
e. open during the day and closed during the night.

Answer: e
Difficulty: Easy
Bloom's Taxonomy: Knowledge

55. Which of the following generally causes stomata to open?

a. release of abscisic acid by the roots
b. a drop in CO_2 concentration in leaf air spaces
c. exposure to red light
d. a drop in O_2 concentration in leaf air spaces
e. all of these

Answer: b
Difficulty: Moderate
Bloom's Taxonomy: Comprehension

56. Which of the following generally causes stomata to close?

a. release of abscisic acid by the roots
b. a drop in CO_2 concentration in leaf air spaces
c. exposure to red light
d. a drop in O_2 concentration in leaf air spaces
e. all of these

Answer: a
Difficulty: Moderate
Bloom's Taxonomy: Comprehension

57. Which has the strongest effect on whether stomata are open or closed?

a. sunlight
b. ion concentrations in the xylem
c. CO_2 concentration
d. O_2 concentration
e. abscisic acid

Answer: e
Difficulty: Difficult
Bloom's Taxonomy: Evaluation

58. Stomata open when K^+ concentration in guard cells _____, followed by water _____ the guard cells by osmosis.

a. increases, entering
b. decreases, entering
c. decreases, leaving
d. increases, leaving

Answer: a
Difficulty: Moderate
Bloom's Taxonomy: Comprehension

59. Which of the following ions plays the most prominent known role in the opening and closing of stomata?

a. Ca^{2+}
b. Cl^-
c. K^+
d. NH_4^+
e. Na^+

Answer: c
Difficulty: Moderate
Bloom's Taxonomy: Knowledge

60. In CAM plants, such as cacti, stomata are

a. open during the night and closed during the day.
b. nearly always open.
c. closed only when guard cells are not exposed to sunlight.
d. almost never open.
e. open during the day and closed during the night.

Answer: a
Difficulty: Difficult
Bloom's Taxonomy: Knowledge

32.4 TRANSPORT OF ORGANIC SUBSTANCES IN THE PHLOEM

61. Which of the following would you not expect to find in phloem sap?

a. Carbohydrates.
b. Hormones.
c. Water.
d. Amino acids.
e. All of these are found in phloem sap.

Answer: e
Difficulty: Easy
Bloom's Taxonomy: Knowledge

62. The main form in which sugars are transported in the phloem sap is

a. glucose.
b. lactose.
c. starch.
d. sucrose.
e. fructose.

Answer: d
Difficulty: Easy
Bloom's Taxonomy: Knowledge

63. The general term for long-distance transport of substances in plants is called

a. cohesion-tension.
b. translocation.
c. pressure flow.
d. osmosis.
e. transpiration.

Answer: b
Difficulty: Moderate
Bloom's Taxonomy: Knowledge

64. The hypothesis that high pressure forces phloem sap to flow was tested and supported by studies using

a. caterpillars.
b. radiolabeled hormones.
c. vacuum chambers.
d. aphids.
e. radiolabeled sugars.

Answer: d
Difficulty: Easy
Bloom's Taxonomy: Knowledge

65. Honeydew is essentially

a. xylem sap forced out of the margins of leaves.
b. xylem sap harvested by honeybees.
c. phloem sap harvested as syrup.
d. xylem sap separated from syrup.
e. phloem sap leaving the anus of an aphid.

Answer: e
Difficulty: Moderate
Bloom's Taxonomy: Knowledge

66. Loading of most carbohydrates into companion cells at a source occurs by

a. osmosis.
b. proton pumping.
c. suction from unloading at a sink.
d. active transport.
e. diffusion of water from the xylem.

Answer: d
Difficulty: Moderate
Bloom's Taxonomy: Comprehension

67. Companion cells load most of the carbohydrates into sieve tube members

a. through plasmodesmata.
b. by osmosis.
c. by antiport.
d. through carrier proteins.
e. by symport.

Answer: a
Difficulty: Difficult
Bloom's Taxonomy: Knowledge

68. In phloem movement, which of the following can serve as a source?

a. photosynthesizing cells in a leaf
b. roots
c. photosynthesizing cells in a stem
d. food storage cells in a stem
e. all of these

Answer: e
Difficulty: Moderate
Bloom's Taxonomy: Application

69. At a source the phloem typically has a water potential that is _____ that in surrounding xylem.

a. lower than
b. about the same as
c. higher than
d. exactly the same as

Answer: a
Difficulty: Difficult
Bloom's Taxonomy: Comprehension

70. When solutes are unloaded from phloem, water

a. is pumped into the phloem by active transport.
b. leaves the xylem by osmosis.
c. moves into the phloem by facilitated diffusion.
d. is pumped out of the phloem by active transport.
e. leaves the phloem by osmosis.

Answer: e
Difficulty: Moderate
Bloom's Taxonomy: Knowledge

UNANSWERED QUESTIONS

71. Plasmodesmata are considered part of the

a. symplastic pathway.
b. transmembrane and symplastic pathways.
c. transmembrane pathway.
d. apoplastic, transmembrane, and symplastic pathways.
e. apoplastic pathway.

Answer: a
Difficulty: Difficult
Bloom's Taxonomy: Knowledge

72. Plasmodesmata are

a. only found in root cells.
b. connections between xylem and phloem.
c. incapable of structural change.
d. able to be opened and closed.
e. only found in leaf cells.

Answer: d
Difficulty: Difficult
Bloom's Taxonomy: Comprehension

Integrative Multiple-Choice

73. The most widely accepted and supported model that explains the movement of xylem sap is the _____ mechanism.

a. cohesion-tension
b. translocation
c. pressure flow
d. osmosis
e. transpiration

Answer: a
Difficulty: Easy
Bloom's Taxonomy: Comprehension
Source: Sections 32.3 and 32.4

74. The most widely accepted and supported model that explains the movement of phloem sap in flowering plants is the _____ mechanism.

a. cohesion-tension
b. translocation
c. pressure flow
d. osmosis
e. transpiration

Answer: c
Difficulty: Moderate
Bloom's Taxonomy: Comprehension
Source: Sections 32.3 and 32.4

75. The movement of phloem sap is best described from _____ to _____.

a. the shoot system; the root system
b. sinks; sources
c. the root system; the shoot system
d. sources; sinks
e. none of these

Answer: d
Difficulty: Moderate
Bloom's Taxonomy: Comprehension
Source: Sections 32.3 and 32.4

76. The movement of xylem sap is best described from _____ to _____.
a. the shoot system; the root system
b. sinks; sources
c. the root system; the shoot system
d. sources; sinks
e. none of these

Answer: c
Difficulty: Easy
Bloom's Taxonomy: Comprehension
Source: Sections 32.3 and 32.4

Choice

Provide the letter that represents the plant tissue or tissues that are directly associated with each of the items given. Each letter may be used once, more than once, or not at all.

a. xylem
b. phloem
c. both xylem and phloem
d. neither xylem nor phloem

77. _____ bulk flow

78. _____ transport of hormones from growing shoot tips

79. _____ primary transport of ions from leaves

80. _____ primary transport of water from roots

81. _____ dead cells throughout conducting portions

82. _____ companion cells

83. _____ stomata

84. _____ cohesion-tension mechanism

85. _____ transport of sugars

86. _____ trichomes

Answers: 77. c   78. b   79. b   80. a   81. a
82. b   83. d   84. a   85. b   86. d

Difficulty: Moderate
Bloom's Taxonomy: Knowledge
Source: Sections 32.1–32.4

Essay

87. Explain the process that occurs when a stoma goes from closed to open.

Answer: When triggered by items such as blue light or a decrease in carbon dioxide concentration, guard cells will undergo activity changes that will change them from flaccid to turgid. Turgid guard cells bend away from each other, opening the space between them (the stoma). The process that leads guard cells to become turgid is as follows. First, protons are pumped out of the guard cells by active transport. Then, protons flow back into the cells, providing energy for active transport of K^+ into the cells. This increases the solute concentration inside the cells, so water then follows into the cells by osmosis. The movement of water into the cells causes turgor pressure to build until the guard cells become turgid, thus opening the stoma.
Difficulty: Moderate
Bloom's Taxonomy: Synthesis
Source: Sections 32.3

88. Explain the processes of phloem loading and unloading in a flowering plant.

Answer: At a source, such as a photosynthesizing leaf, an excess of solutes such as sucrose will generally be available. The phloem is loaded with solutes by active transport mechanisms. Mostly this occurs through companion cells, which take up solutes by active transport. Solutes then pass into sieve tube members mainly through plasmodesmata. As the sieve tube members fill with solute, their water potential is lowered compared to surrounding tissues; thus, water then enters them via osmosis. The entry of water creates pressure potential, which then pushes solutes and water away from the source by bulk flow. Pressure gradually decreases away from the source at a sink or sinks, so the bulk flow is toward the sink or sinks. Sinks include locations such as growing shoot tips or roots. At a sink solutes are unloaded into nearby cells. This unloading of solutes causes water to flow out by osmosis, further lowering the pressure potential at the sink.
Difficulty: Moderate
Bloom's Taxonomy: Synthesis
Source: Sections 32.4

33
PLANT NUTRITION

Multiple-Choice

WHY IT MATTERS

1. Which of the following is typically FALSE for tropical rainforests?

a. Tropical rainforests are biologically diverse ecosystems.
b. Tropical rainforest soils are highly acidic.
c. Minerals such as calcium and phosphorus are subject to leaching from the upper tropical rainforest soil level.
d. Tropical rainforest soils are extremely rich in nutrients.
e. Organic remains are decomposed rapidly in tropical rainforests.

Answer: d
Difficulty: Moderate
Bloom's Taxonomy: Comprehension

33.1 PLANT NUTRITIONAL REQUIREMENTS

2. Plant tissues are more than 90 percent _____ by weight.

a. carbohydrates
b. water
c. carbon
d. nitrogen
e. cellulose

Answer: b
Difficulty: Moderate
Bloom's Taxonomy: Knowledge

3. Hydroponic culture is the process of

a. growing plants in pure water mixed with carefully measured amounts of specific minerals.
b. irrigation of dry areas to improve mineral absorption by plants.
c. growing plants in ponds and lakes to improve their access to water, allowing the plants to place more resources into the shoot system.
d. irrigation of dry areas to prevent crop plants from wilting.
e. growing plants with specific hydrocarbons added to the soil to test the effect on plant growth.

Answer: a
Difficulty: Moderate
Bloom's Taxonomy: Comprehension

4. Which of the following elements is considered an essential macronutrient for plants?

a. gold
b. manganese
c. calcium
d. nitrogen
e. copper

Answer: d
Difficulty: Moderate
Bloom's Taxonomy: Knowledge

5. Which of the following elements is considered an essential macronutrient for plants?

a. zinc
b. uranium
c. iron
d. nickel
e. carbon

Answer: e
Difficulty: Easy
Bloom's Taxonomy: Knowledge

6. Which of the following elements is considered an essential macronutrient for plants?

a. oxygen
b. boron
c. lithium
d. silicon
e. molybdenum

Answer: a
Difficulty: Easy
Bloom's Taxonomy: Knowledge

7. Which of the following elements is considered an essential micronutrient for plants?

a. gold
b. carbon
c. uranium
d. nitrogen
e. copper

Answer: e
Difficulty: Moderate
Bloom's Taxonomy: Knowledge

8. Which of the following elements is considered an essential micronutrient for plants?

a. phosphorus
b. silver
c. molybdenum
d. carbon
e. lead

Answer: c
Difficulty: Moderate
Bloom's Taxonomy: Knowledge

9.   Which of the following elements is considered an essential micronutrient for plants?

a.   zinc
b.   potassium
c.   hydrogen
d.   arsenic
e.   platinum

Answer: a
Difficulty: Moderate
Bloom's Taxonomy: Knowledge

10.   Which of the following elements is considered an essential micronutrient for plants?

a.   calcium
b.   lithium
c.   sulfur
d.   nickel
e.   oxygen

Answer: d
Difficulty: Moderate
Bloom's Taxonomy: Knowledge

11.   Experiments have shown that even if it is supplied by no other means, plants near the ocean can get enough of the essential micronutrient _____ from the air and from sweat from researchers' hands.

a.   sulfur
b.   manganese
c.   chlorine
d.   magnesium
e.   sodium

Answer: c
Difficulty: Difficult
Bloom's Taxonomy: Comprehension

12.   Which of the following statements about essential elements is FALSE?

a.   An element with only one role in plant metabolism may still be essential.
b.   An essential element is necessary for normal growth and reproduction.
c.   Plant seeds in some cases contain enough of an essential element to sustain the adult plant.
d.   An essential element may occasionally be functionally replaced by another element.
e.   Some essential elements are required in only trace amounts.

Answer: d
Difficulty: Difficult
Bloom's Taxonomy: Synthesis

13.   Which of the following elements is not considered an essential micronutrient for plants in general, but is required by horsetails and perhaps some grasses such as wheat?

a.   gold
b.   boron
c.   silicon
d.   zinc
e.   magnesium

Answer: c
Difficulty: Difficult
Bloom's Taxonomy: Knowledge

14.   A common consequence of nutrient deficiencies in plants is chlorosis, which is

a.   wilting due to a lack of chlorine in leaves.
b.   death of the growing tips.
c.   premature loss of leaves.
d.   bursting of cells from excess water uptake due to an inability to clear chlorine from them.
e.   yellowing of plant tissues due to a lack of chlorophyll.

Answer: e
Difficulty: Moderate
Bloom's Taxonomy: Knowledge

33.2 SOIL

15.   Soil particles that are decomposing organic matter are called

a.   sand.
b.   humus.
c.   silt.
d.   clay.

Answer: b
Difficulty: Easy
Bloom's Taxonomy: Knowledge

16.   Soil mineral particles that range from 2.0–0.02 mm in diameter are called

a.   sand.
b.   humus.
c.   clay.
d.   silt.

Answer: a
Difficulty: Moderate
Bloom's Taxonomy: Knowledge

17.   Soil mineral particles that range from 0.02–0.002 mm in diameter are called

a.   sand.
b.   humus.
c.   clay.
d.   silt.

Answer: d
Difficulty: Moderate
Bloom's Taxonomy: Knowledge

18. Soil mineral particles that are less than 0.002 mm in diameter are called

a. sand.
b. humus.
c. clay.
d. silt.

Answer: c
Difficulty: Moderate
Bloom's Taxonomy: Knowledge

19. A soil that is mostly _____ will tend to dry quickly compared to other soil types.

a. sand
b. humus
c. clay
d. silt

Answer: a
Difficulty: Easy
Bloom's Taxonomy: Comprehension

20. The soils in which most plants do best are

a. mixtures of mainly sand and silt.
b. sandy soils.
c. soils made mostly of humus.
d. clay soils.
e. loams.

Answer: e
Difficulty: Moderate
Bloom's Taxonomy: Knowledge

21. A soil with roughly equal amounts of humus, silt, clay, and sand is called a

a. topsoil.
b. loam.
c. subsoil.
d. compost.

Answer: b
Difficulty: Easy
Bloom's Taxonomy: Knowledge

22. Which soil region extends to the underlying bedrock, and consists of mineral particles and rock fragments but generally no organic material?

a. A horizon
b. B horizon
c. C horizon
d. D horizon
e. O horizon

Answer: c
Difficulty: Moderate
Bloom's Taxonomy: Knowledge

23. Which soil region is the most fertile soil layer and where the roots of most herbaceous plants are located?

a. A horizon
b. B horizon
c. C horizon
d. D horizon
e. O horizon

Answer: a
Difficulty: Moderate
Bloom's Taxonomy: Knowledge

24. Which soil region is the top layer of surface litter, such as twigs and leaves?

a. A horizon
b. B horizon
c. C horizon
d. D horizon
e. O horizon

Answer: e
Difficulty: Moderate
Bloom's Taxonomy: Knowledge

25. Which soil region tends to accumulate mineral ions but relatively little organic matter, and is generally penetrated by mature tree roots?

a. A horizon
b. B horizon
c. C horizon
d. D horizon
e. O horizon

Answer: b
Difficulty: Moderate
Bloom's Taxonomy: Knowledge

26. Which of the following is most important for retaining water in soil for use by plants?

a. sand
b. humus
c. clay
d. silt

Answer: b
Difficulty: Moderate
Bloom's Taxonomy: Comprehension

27. The surfaces of clay particles in soil

a. often bear positively charged ions.
b. release water to plants easily.
c. often bear negatively charged ions.
d. are generally hydrophobic.
e. acidify the soil.

Answer: c
Difficulty: Moderate
Bloom's Taxonomy: Knowledge

28. Most plants would grow best in which of the following situations?

a. a soil with very small air spaces
b. a soil with relatively large air spaces
c. a soil with air spaces filled almost entirely with water
d. a soil with various sizes of air spaces
e. a soil with no air spaces

Answer: d
Difficulty: Moderate
Bloom's Taxonomy: Application

29. Roots obtain cations through cation exchange, where soil particles adsorb _____ provided directly or indirectly by the root, and thus release cations.

a. Cl^-
b. K^+
c. H^+
d. water
e. CO_2

Answer: c
Difficulty: Moderate
Bloom's Taxonomy: Comprehension

30. Areas that receive heavy rainfall tend to have _____ soils, and arid regions tend to have _____ soils.

a. acidic; alkaline
b. pH-neutral; acidic
c. alkaline; acid
d. acidic; pH-neutral
e. pH-neutral; alkaline

Answer: a
Difficulty: Moderate
Bloom's Taxonomy: Knowledge

FOCUS ON RESEARCH: APPLIED RESEARCH: PLANTS POISED FOR ENVIRONMENTAL CLEANUP

31. Phytoremediation is

a. rotating crops so that soil minerals are replenished.
b. replacing eroded topsoil so that plants can grow in the soil again.
c. plowing under all or parts of crops to improve the soil.
d. the use of plants to remove pollutants from the environment.
e. replacing unwanted plants with plants useful to humans.

Answer: d
Difficulty: Moderate
Bloom's Taxonomy: Knowledge

32. Work by Scott Bizily, Richard Meagher, and colleagues in the 1990s produced *Arabidopsis thaliana* plants that are able to take up _____ from the soil and convert it into a much less dangerous substance.

a. arsenic
b. methylmercury
c. chelated iron
d. lead
e. uranium hexafluoride

Answer: b
Difficulty: Difficult
Bloom's Taxonomy: Knowledge

33.3 OBTAINING AND ABSORBING NUTRIENTS

33. Mycorrhizae, present for most plant species, generally help roots with the uptake of

a. water.
b. nitrogen.
c. phosphate.
d. all of these

Answer: d
Difficulty: Easy
Bloom's Taxonomy: Comprehension

34. Mycorrhizae are

a. intertwined roots of different species of plants.
b. connecting points between branching roots.
c. symbiotic associations between a fungus and plant roots.
d. specialized root hairs.
e. localized swellings in plant roots filled with nitrogen-fixing bacteria.

Answer: c
Difficulty: Moderate
Bloom's Taxonomy: Knowledge

35. The most common limit to plant growth is a lack of

a. phosphorous.
b. carbon.
c. hydrogen.
d. oxygen.
e. nitrogen.

Answer: e
Difficulty: Easy
Bloom's Taxonomy: Knowledge

36. Plants can generally absorb and make use of nitrogen in the form of

a. NO_3^- only.
b. both NO_3^- and NH_4^+.
c. NH_4^+ only.
d. N_2 only.
e. both N_2 and NH_4^+.

Answer: b
Difficulty: Moderate
Bloom's Taxonomy: Synthesis

37. The process of adding hydrogen to N_2, creating NH_3 and eventually NH_4^+, is called

a. nitrification.
b. ammonification.
c. nitrogen cycling.
d. nitrogen fixation.
e. nitrogen assimilation.

Answer: d
Difficulty: Easy
Bloom's Taxonomy: Comprehension

38. The process of producing NH_4^+ from decaying organic material is called

a. nitrification.
b. ammonification.
c. nitrogen cycling.
d. nitrogen fixation.
e. nitrogen assimilation.

Answer: b
Difficulty: Moderate
Bloom's Taxonomy: Comprehension

39. The process in which NH_4^+ is oxidized to NO_3^- is called

a. nitrification.
b. ammonification.
c. nitrogen cycling.
d. nitrogen fixation.
e. nitrogen assimilation.

Answer: a
Difficulty: Difficult
Bloom's Taxonomy: Comprehension

40. Which of the following is mostly performed by bacteria living within the roots of plants in the legume family?

a. nitrification
b. ammonification
c. nitrogen cycling
d. nitrogen fixation
e. nitrogen assimilation

Answer: d
Difficulty: Moderate
Bloom's Taxonomy: Comprehension

41. Which of the following would you expect to happen if nitrifying bacteria were not present in a soil?

a. Plants would take up nitrogen for their use mainly as NH_4^+
b. Plants would not survive because they could not get useful nitrogen
c. Plants would take up nitrogen for their use mainly as N_2
d. Plants would take up nitrogen for their use mainly as NH_3
e. Plants would take up nitrogen for their use mainly as NO_3^-

Answer: a
Difficulty: Moderate
Bloom's Taxonomy: Application

42. Plants generally take up nitrogen for their use mainly as

a. NH_4^+.
b. CN.
c. N_2.
d. NH_3.
e. NO_3^-.

Answer: e
Difficulty: Difficult
Bloom's Taxonomy: Application

43. Inside root cells, _____ is converted to _____, which is then rapidly used to synthesize organic molecules.

a. NH_4^+; NO_3^-
b. N_2; NO_3^-
c. NO_3^-; NH_4^+
d. NH_3; NH_4^+
e. NO_3^-; N_2

Answer: c
Difficulty: Difficult
Bloom's Taxonomy: Application

44. Species of *Rhizobium* and *Bradyrhizobium* bacteria are most directly associated with

a. nitrification.
b. ammonification.
c. nitrogen cycling.
d. nitrogen fixation.
e. nitrogen assimilation.

Answer: d
Difficulty: Moderate
Bloom's Taxonomy: Knowledge

45. Root nodules are

a. intertwined roots of different species of plants.
b. connecting points between branching roots.
c. symbiotic associations between a fungus and plant roots.
d. specialized root hairs.
e. localized swellings in plant roots filled with nitrogen-fixing bacteria.

Answer: e
Difficulty: Easy
Bloom's Taxonomy: Knowledge

46. Studies of a soybean plant (*Glycine max*) and the bacterium *Bradyrhizobium japonicum* have shown that the tip of a root hair curls toward the bacterium in response to

a. a flavonoid released by soybean roots.
b. *nod* gene products produced by soybean roots.
c. *nod* gene products produced by the bacterium.
d. nitrogen fixation by the bacterium.
e. a flavonoid produced by the bacterium.

Answer: c
Difficulty: Moderate
Bloom's Taxonomy: Knowledge

47. Studies of a soybean plant (*Glycine max*) and the bacterium *Bradyrhizobium japonicum* have shown that after the bacterium enters the root, cells of the root cortex begin to divide in response to

a. a flavonoid released by soybean roots.
b. *nod* gene products produced by soybean roots.
c. *nod* gene products produced by the bacterium.
d. nitrogen fixation by the bacterium.
e. a flavonoid produced by the bacterium.

Answer: c
Difficulty: Moderate
Bloom's Taxonomy: Knowledge

48. Bacteroids are

a. specialized nitrifying bacteria.
b. small buds from ammonifying bacteria.
c. specialized bacteria that help some plant roots absorb phosphorus.
d. enlarged and immobilized nitrogen-fixing bacteria.
e. interactions between plant root hairs and bacteria.

Answer: d
Difficulty: Moderate
Bloom's Taxonomy: Knowledge

49. Leghemoglobin is used to

a. remove O_2 from roots.
b. produce H_2O from O_2.
c. deliver O_2 to bacteroids.
d. transport O_2 in the xylem.
e. produce O_2 from H_2O.

Answer: c
Difficulty: Moderate
Bloom's Taxonomy: Comprehension

50. A mutant soybean plant that is unable to produce leghemoglobin will most likely suffer from _____ deficiency.

a. phosphorus
b. magnesium
c. potassium
d. copper
e. nitrogen

Answer: e
Difficulty: Moderate
Bloom's Taxonomy: Application

51. The enzyme nitrogenase is most directly involved in the process called

a. nitrification.
b. ammonification.
c. nitrogen cycling.
d. nitrogen fixation.
e. nitrogen assimilation.

Answer: d
Difficulty: Moderate
Bloom's Taxonomy: Synthesis

52. The enzyme nitrogenase is irreversibly inhibited by excess

a. O_2.
b. NO_3^-.
c. N_2.
d. CO_2.
e. NH_4^+.

Answer: a
Difficulty: Difficult
Bloom's Taxonomy: Knowledge

53. Animals trapped and digested by "carnivorous" plants such as the cobra lily are used primarily as

a. an energy supplement during winter months.
b. a carbon source for the plant.
c. a nutrient supplement in nutrient-deficient environments.
d. food for bacteria and fungi that grow symbiotically with the plant roots.
e. an energy supplement for small plants growing in shady areas.

Answer: c
Difficulty: Moderate
Bloom's Taxonomy: Comprehension

54. Haustorial roots of dodders and other nonphotosynthetic, parasitic plants rob the host plant of

a. sugars.
b. minerals.
c. water.
d. all of these
e. none of these

Answer: d
Difficulty: Easy
Bloom's Taxonomy: Knowledge

55. Which of the following plants both parasitizes other plants and performs photosynthesis?

a. snow plant
b. lady-of-the-night orchid
c. mistletoe
d. cobra lily
e. dodder

Answer: c
Difficulty: Difficult
Bloom's Taxonomy: Synthesis

56. Epiphytes are

a. plants that are not parasites but that do grow on other plants instead of the soil.
b. plants that trap animals and digest them.
c. parasitic plants that use mycorrhizae to obtain nutrients and food indirectly from other plants.
d. plants that deal with nutrient-deficient soil by growing a large, netlike mesh of roots through several layers of the soil.
e. parasitic plants that use haustoria to obtain nutrients and food directly from other plants.

Answer: a
Difficulty: Moderate
Bloom's Taxonomy: Knowledge

57. Farmers often grow legumes such as soybeans every few years and plow much of the plant parts into the soil. This practice serves mainly to replenish the soil with useful forms of

a. carbon.
b. nitrogen.
c. phosphorus.
d. oxygen.
e. manganese.

Answer: b
Difficulty: Moderate
Bloom's Taxonomy: Application

INSIGHTS FROM THE MOLECULAR REVOLUTION: GETTING TO THE ROOTS OF PLANT NUTRITION

58. Harrison and van Buuren identified a phosphate transport protein in *Glomus versiforme*, which is a

a. carnivorous plant.
b. bacterium that parasitizes plants.
c. mycorrhizal fungus.
d. bacteroid.
e. parasitic plant.

Answer: c
Difficulty: Moderate
Bloom's Taxonomy: Knowledge

UNANSWERED QUESTIONS

59. Research into mycorrhizae has shown that mycorrhizae

a. are never truly shared between different individual plants.
b. can be shared by plants as genetically different from each other as a gymnosperm and an angiosperm.
c. can only be shared between two plants of the same species.
d. can only be shared between two flowering plants.
e. can only be shared between two plants that are clones of each other.

Answer: b
Difficulty: Moderate
Bloom's Taxonomy: Comprehension

Integrative Multiple-Choice

60. Which of the following disease symptoms would you expect to see in legumes grown in sterile soil?

a. purplish veins
b. chlorosis and mottled or bronzed leaves
c. chlorosis in older leaves and stunted growth
d. burned leaf edges and curled, mottled, or spotted older leaves
e. pale green, rolled, or cupped leaves

Answer: c
Difficulty: Difficult
Bloom's Taxonomy: Synthesis
Source: Sections 33.1 and 33.3

61. Imagine that you are a farmer who chose not to rotate your crops with soybeans even though the local extension agent recommended that you do so. Which deficiency symptom in your crops should indicate to you that the extension agent's advice was correct?

a. chlorosis and mottled or bronzed leaves
b. burned leaf edges and curled, mottled, or spotted older leaves
c. purplish veins
d. pale green, rolled, or cupped leaves
e. chlorosis in older leaves and stunted growth

Answer: e
Difficulty: Difficult
Bloom's Taxonomy: Synthesis
Source: Sections 33.1 and 33.3

Choice

Choose the appropriate category of general plant nutritional requirement for each element given below.

a. nonmineral macronutrient
b. mineral macronutrient
c. micronutrient
d. not generally considered an essential nutrient

62. _____ manganese
63. _____ gold
64. _____ oxygen
65. _____ nickel
66. _____ boron
67. _____ sulfur
68. _____ magnesium
69. _____ carbon
70. _____ arsenic
71. _____ calcium
72. _____ iron
73. _____ chlorine
74. _____ lithium

75. _____ zinc

76. _____ potassium

77. _____ molybdenum

78. _____ silver

79. _____ hydrogen

80. _____ phosphorus

81. _____ copper

82. _____ nitrogen

Answers:　**62. c　63. d　64. a　65. c　66. c**
67. b　68. b　69. a　70. d　71. b
72. c　73. c　74. d　75. c　76. b
77. c　78. d　79. a　80. b　81. c
82. b

Difficulty: Moderate
Bloom's Taxonomy: Knowledge
Source: Section 33.1

Short Answer

83.　　Describe how losing leaves in the autumn can affect plant nutrient requirements, and how plants handle this issue.

*Answer:　**Much of the nutrients that a plant takes up from the soil winds up in the leaves. If those nutrients are lost when a plant sheds its leaves then the plant will need to replace those nutrients during the next growing season. Plants reduce this cost by moving significant amounts of nutrients such as nitrogen, phosphorus, potassium, and magnesium out of leaves and into twigs and branches before the leaves fall from the tree. This preserves those nutrients for use during the next growing season.***
Difficulty: Moderate
Bloom's Taxonomy: Synthesis
Source: Section 28.1

34

REPRODUCTION AND DEVELOPMENT IN FLOWERING PLANTS

Multiple-Choice

WHY IT MATTERS

1. Flowering plants
a. only reproduce asexually.
b. mainly reproduce asexually, although many will reproduce sexually under certain circumstances.
c. reproduce asexually about half the time and sexually about half the time.
d. only reproduce sexually.
e. mainly reproduce sexually, although many will reproduce asexually under certain circumstances.

Answer: e
Difficulty: Easy
Bloom's Taxonomy: Comprehension

34.1 OVERVIEW OF FLOWERING PLANT REPRODUCTION

2. After fertilization in flowering plants, an embryo in a seed
a. begins as a gametophyte and ends up as a sporophyte before germination.
b. is always a sporophyte.
c. is usually a gametophyte but can be a sporophyte.
d. begins as a sporophyte and ends up as a gametophyte before germination.
e. is always a gametophyte.

Answer: b
Difficulty: Moderate
Bloom's Taxonomy: Comprehension

3. Where in a flowering plant should you expect meiosis to occur?
a. leaves
b. roots and leaves
c. flowers
d. flowers, roots, and leaves
e. roots

Answer: c
Difficulty: Easy
Bloom's Taxonomy: Synthesis

4. Meiosis in flowering plants gives rise to cells that are
a. haploid gametophytes.
b. diploid sporophytes.
c. sporophytes that can be haploid or diploid.
d. diploid gametophytes.
e. haploid sporophytes.

Answer: a
Difficulty: Moderate
Bloom's Taxonomy: Knowledge

5. Which of the following are the male gametophytes in flowering plants?
a. stamens
b. sperm
c. shoot parts bearing male flowers
d. pollen grains
e. anthers

Answer: d
Difficulty: Easy
Bloom's Taxonomy: Comprehension

6. The female gametophyte in flowering plants is usually
a. the complete pistil.
b. seven cells embedded in floral tissues.
c. a single ovary.
d. a seed.
e. shoot parts bearing female flowers.

Answer: b
Difficulty: Easy
Bloom's Taxonomy: Comprehension

7. In flowering plants
a. the gametophytes are smaller than the sporophytes and do not nourish themselves.
b. the gametophytes arc smaller than the sporophytes but are free-living and nourish themselves for most of their lives.
c. the gametophytes are usually larger than the sporophytes, but both gametophytes and sporophytes nourish themselves.
d. the gametophytes are usually larger than the sporophytes, and the sporophytes grow out of the gametophytes and are nourished by them.
e. the gametophytes and sporophytes are roughly equal in size and each nourish themselves.

Answer: a
Difficulty: Moderate
Bloom's Taxonomy: Synthesis

8. Which of the following is a form of asexual reproduction?

a. production of new plants at nodes along stolons in strawberries
b. humans growing trees from cuttings
c. production of new plants at nodes along underground rhizomes in Bermuda grass
d. buds from an underground stem of an onion
e. all of these

Answer: e
Difficulty: Easy
Bloom's Taxonomy: Comprehension

34.2 THE FORMATION OF FLOWERS AND GAMETES

9. The end of a reproductive shoot where a flower develops is called the

a. ovule.
b. calyx.
c. receptacle.
d. filament.
e. carpel.

Answer: c
Difficulty: Moderate
Bloom's Taxonomy: Knowledge

10. The sepals of a flower make up the

a. receptacle.
b. corolla.
c. carpel.
d. calyx.
e. filament.

Answer: d
Difficulty: Moderate
Bloom's Taxonomy: Knowledge

11. The petals of a flower make up the

a. corolla.
b. filament.
c. ovule.
d. calyx.
e. receptacle.

Answer: a
Difficulty: Easy
Bloom's Taxonomy: Knowledge

12. The female gametophyte in flowering plants forms in a

a. calyx.
b. carpel.
c. receptacle.
d. corolla.
e. stamen.

Answer: b
Difficulty: Easy
Bloom's Taxonomy: Comprehension

13. The male gametophyte in flowering plants forms in a

a. corolla.
b. receptacle.
c. stamen.
d. calyx.
e. carpel.

Answer: c
Difficulty: Easy
Bloom's Taxonomy: Comprehension

14. In a flower, an anther is typically at the tip of a(n)

a. calyx.
b. receptacle.
c. carpel.
d. filament.
e. ovule.

Answer: d
Difficulty: Moderate
Bloom's Taxonomy: Knowledge

15. In flowering plants an ovary is part of a(n)

a. corolla.
b. ovule.
c. calyx.
d. stamen.
e. carpel.

Answer: e
Difficulty: Moderate
Bloom's Taxonomy: Knowledge

16. The landing platform for pollen in flowering plants is the

a. filament.
b. stigma.
c. anther.
d. style.
e. ovary.

Answer: b
Difficulty: Easy
Bloom's Taxonomy: Comprehension

17. A flower must have one or more _____ to be able to make seeds.

a. stamens
b. sepals
c. carpels
d. stamens and sepals
e. stamens and carpels

Answer: c
Difficulty: Moderate
Bloom's Taxonomy: Application

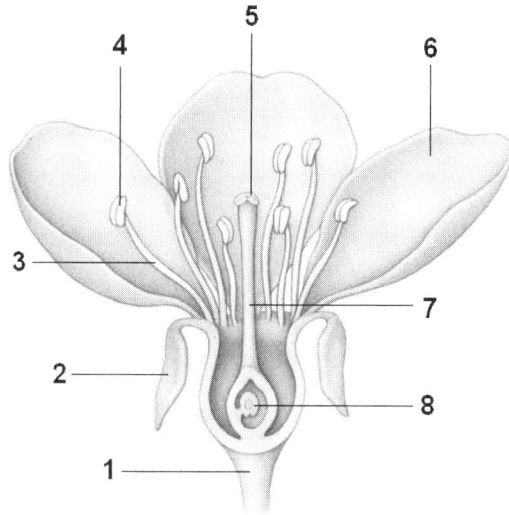

Use the figure above for questions 18–25.

18. In the flower diagram shown above, the item labeled "1" is the

a. filament.
b. sepal.
c. receptacle.
d. style.
e. ovule.

Answer: c
Difficulty: Moderate
Bloom's Taxonomy: Application
Source: Figure 34.3

19. In the flower diagram shown above, the item labeled "2" is a(n)

a. sepal.
b. anther.
c. receptacle.
d. petal.
e. ovule.

Answer: a
Difficulty: Easy
Bloom's Taxonomy: Application
Source: Figure 34.3

20. In the flower diagram shown above, the item labeled "3" is a(n)

a. anther.
b. receptacle.
c. style.
d. stigma.
e. filament.

Answer: e
Difficulty: Moderate
Bloom's Taxonomy: Application
Source: Figure 34.3

21. In the flower diagram shown above, the item labeled "4" is a(n)

a. stigma.
b. anther.
c. sepal.
d. receptacle.
e. ovule.

Answer: b
Difficulty: Moderate
Bloom's Taxonomy: Application
Source: Figure 34.3

22. In the flower diagram shown above, the item labeled "5" is the

a. receptacle.
b. ovule.
c. style.
d. stigma.
e. anther.

Answer: d
Difficulty: Moderate
Bloom's Taxonomy: Application
Source: Figure 34.3

23. In the flower diagram shown above, the item labeled "6" is a(n)

a. style.
b. sepal.
c. receptacle.
d. ovule.
e. petal.

Answer: e
Difficulty: Easy
Bloom's Taxonomy: Application
Source: Figure 34.3

24. In the flower diagram shown above, the item labeled "7" is the

a. filament.
b. stigma.
c. ovule.
d. style.
e. anther.

Answer: d
Difficulty: Moderate
Bloom's Taxonomy: Application
Source: Figure 34.3

25. In the flower diagram shown above, the item labeled "8" is a(n)

a. ovule.
b. stigma.
c. sepal.
d. anther.
e. style.

Answer: a
Difficulty: Easy
Bloom's Taxonomy: Application
Source: Figure 34.3

26. The innermost whorl of a flower typically consists of one or more

a. petals.
b. carpels.
c. stamens.
d. receptacles.
e. sepals.

Answer: b
Difficulty: Moderate
Bloom's Taxonomy: Application

27. Which part of a flower is typically the most leaflike?

a. carpels
b. receptacles
c. sepals
d. petals
e. stamens

Answer: c
Difficulty: Moderate
Bloom's Taxonomy: Application

28. The parts of a flower most likely to have distinctive colors, patterning, and shapes for attracting pollinators are the

a. petals.
b. carpels.
c. stamens.
d. receptacles.
e. sepals.

Answer: a
Difficulty: Easy
Bloom's Taxonomy: Application

29. The male reproductive whorl of a flower typically consists of one or more

a. petals.
b. carpels.
c. stamens.
d. receptacles.
e. sepals.

Answer: c
Difficulty: Easy
Bloom's Taxonomy: Application

30. The outermost whorl of a flower typically consists of one or more

a. petals.
b. carpels.
c. stamens.
d. receptacles.
e. sepals.

Answer: e
Difficulty: Moderate
Bloom's Taxonomy: Application

31. The female reproductive whorl of a flower typically consists of one or more

a. petals.
b. carpels.
c. stamens.
d. receptacles.
e. sepals.

Answer: b
Difficulty: Moderate
Bloom's Taxonomy: Application

32. A flower with carpels and stamens but no petals or sepals is

a. perfect and incomplete.
b. imperfect and complete.
c. imperfect and incomplete.
d. perfect and complete.
e. none of these

Answer: a
Difficulty: Moderate
Bloom's Taxonomy: Application

33. A flower with petals, sepals, and one carpel but no stamens is

a. perfect and incomplete.
b. imperfect and complete.
c. imperfect and incomplete.
d. perfect and complete.
e. none of these

Answer: c
Difficulty: Moderate
Bloom's Taxonomy: Application

34. A flower with petals, sepals, stamens, and one carpel is

a. perfect and incomplete.
b. imperfect and complete.
c. imperfect and incomplete.
d. perfect and complete.
e. none of these

Answer: d
Difficulty: Moderate
Bloom's Taxonomy: Application

35. A plant species where each plant can only make either male or female flowers is called a _____ species and always has _____ flowers.

a. monoecious; complete
b. dioecious; imperfect
c. dioecious; complete
d. monoecious; imperfect
e. monoecious; perfect

Answer: b
Difficulty: Difficult
Bloom's Taxonomy: Application

36.   A plant species where each plant makes some flowers that are male and some that are female is called a _____ species and always has _____ flowers.

a.   monoecious; complete
b.   dioecious; imperfect
c.   dioecious; complete
d.   monoecious; imperfect
e.   monoecious; perfect

Answer: d
Difficulty: Difficult
Bloom's Taxonomy: Application

37.   Which of the following occurs for the production of a pollen grain from a microspore mother cell?

a.   mitosis only
b.   mitosis, then meiosis
c.   mitosis, then meiosis, then mitosis
d.   meiosis only
e.   meiosis, then mitosis

Answer: e
Difficulty: Moderate
Bloom's Taxonomy: Synthesis

38.   Which of the following is the main reason why pollen withstands decay much better than typical plant parts?

a.   hemicellulose
b.   sporopollenin
c.   lignin
d.   cellulose
e.   pectin

Answer: b
Difficulty: Moderate
Bloom's Taxonomy: Evaluation

39.   The female gametophyte in flowering plants is the

a.   ovary.
b.   seed.
c.   embryo sac.
d.   megasporocyte.
e.   ovule.

Answer: c
Difficulty: Moderate
Bloom's Taxonomy: Knowledge

40.   The micropyle is

a.   the mature male gametophyte.
b.   the inner seed coat.
c.   a small opening at one end of an ovule.
d.   a passageway through the style for sperm.
e.   the eight-cell stage of a plant embryo.

Answer: c
Difficulty: Moderate
Bloom's Taxonomy: Knowledge

41.   In the part of an embryo sac farthest away from the micropyle you should find

a.   the central cell.
b.   a seed.
c.   the pollen tube.
d.   synergids.
e.   antipodal cells.

Answer: e
Difficulty: Difficult
Bloom's Taxonomy: Application

42.   Along with the egg, in the part of an embryo sac next to the micropyle you should find

a.   the central cell.
b.   a seed.
c.   the pollen tube.
d.   synergids.
e.   antipodal cells.

Answer: d
Difficulty: Difficult
Bloom's Taxonomy: Application

43.   In flowering plants how many sperm cells are typically produced from each microspore mother cell?

a.   1
b.   2
c.   3
d.   4
e.   8

Answer: e
Difficulty: Difficult
Bloom's Taxonomy: Synthesis

44.   In flowering plants how many egg cells are typically produced from each megaspore mother cell?

a.   1
b.   2
c.   3
d.   4
e.   8

Answer: a
Difficulty: Difficult
Bloom's Taxonomy: Synthesis

34.3 POLLINATION, FERTILIZATION, AND GERMINATION

45.   Self-incompatibility in flowering plants is a biochemical recognition and rejection process that

a.   prevents both self-fertilization and self-pollination.
b.   prevents self-pollination and occasionally prevents self-fertilization.
c.   prevents self-fertilization but not self-pollination.
d.   prevents self-pollination but not self-fertilization.
e.   prevents self-fertilization and occasionally prevents self-pollination.

Answer: c
Difficulty: Moderate
Bloom's Taxonomy: Comprehension

46. When pollen from one species lands on the stigma of a flower from another species

a. the pollen tube typically does not develop.
b. the pollen tube typically forms, but the sperm cannot penetrate the egg.
c. the pollen tube typically forms and grows to the ovary, but then stops.
d. the pollen tube typically forms, but the sperm are killed.
e. the pollen tube typically forms and the egg is usually fertilized, but the embryo rarely grows.

Answer: a
Difficulty: Moderate
Bloom's Taxonomy: Knowledge

47. Chemical cues that help guide a developing pollen tube toward an ovule are released by the

a. antipodal cells.
b. synergids.
c. egg.
d. central cell.
e. sperm.

Answer: b
Difficulty: Difficult
Bloom's Taxonomy: Comprehension

48. For a typical flowering plant the first cell of the triploid (3*n*) endosperm is formed from

a. one sperm fused with two antipodal cells.
b. two sperm fused with the egg.
c. one sperm fused with the central cell.
d. one sperm fused with the egg.
e. two sperm fused with a synergid.

Answer: c
Difficulty: Moderate
Bloom's Taxonomy: Application

49. Unique to flowering plants, the making of an embryo-nourishing _____ is an outcome of double fertilization.

a. hypocotyl
b. endosperm
c. seed coat
d. cotyledon
e. epicotyl

Answer: b
Difficulty: Moderate
Bloom's Taxonomy: Knowledge

50. A(n) _____ is a seed leaf.

a. suspensor
b. epicotyl
c. hypocotyl
d. endosperm
e. cotyledon

Answer: e
Difficulty: Easy
Bloom's Taxonomy: Knowledge

51. The embryonic root is called the

a. suspensor.
b. epicotyl.
c. hypocotyl.
d. radicle.
e. cotyledon.

Answer: d
Difficulty: Moderate
Bloom's Taxonomy: Knowledge

52. The fruit wall, or pericarp, develops from the

a. ovary wall.
b. petals.
c. seed coat.
d. endosperm.
e. receptacle.

Answer: a
Difficulty: Easy
Bloom's Taxonomy: Knowledge

53. A fruit that develops from a single ovary in a single flower is called a(n)

a. aggregate fruit.
b. accessory fruit.
c. multiple fruit.
d. simple fruit.
e. none of these

Answer: d
Difficulty: Easy
Bloom's Taxonomy: Knowledge

54. A fruit that develops from several ovaries in a single flower is called a(n)

a. aggregate fruit.
b. accessory fruit.
c. multiple fruit.
d. simple fruit.
e. none of these

Answer: a
Difficulty: Moderate
Bloom's Taxonomy: Knowledge

55. A fruit that develops from several ovaries in multiple flowers is called a(n)

a. aggregate fruit.
b. accessory fruit.
c. multiple fruit.
d. simple fruit.
e. none of these

Answer: c
Difficulty: Moderate
Bloom's Taxonomy: Knowledge

56. A fruit that develops from tissues of the receptacle as well as the ovary would be considered a(n)

a. aggregate fruit.
b. accessory fruit.
c. multiple fruit.
d. simple fruit.
e. none of these

Answer: b
Difficulty: Difficult
Bloom's Taxonomy: Application

57. Fleshy fruits usually aid seed dispersal by

a. catching in the hair or feathers of animals.
b. being buried by animals.
c. falling directly under the parent plant.
d. floating away from the parent plant.
e. being eaten by animals.

Answer: e
Difficulty: Moderate
Bloom's Taxonomy: Knowledge

58. Which of the following it the most typical order of events in seed germination?

a. shoot cells divide and elongate → root cells divide and elongate → seed coat splits → water imbibition
b. water imbibition → seed coat splits → root cells divide and elongate → shoot cells divide and elongate
c. water imbibition → root cells divide and elongate → seed coat splits → shoot cells divide and elongate
d. seed coat splits → water imbibition → root cells divide and elongate → shoot cells divide and elongate
e. seed coat splits → water imbibition → shoot cells divide and elongate → root cells divide and elongate

Answer: b
Difficulty: Difficult
Bloom's Taxonomy: Synthesis

34.4 ASEXUAL REPRODUCTION OF FLOWERING PLANTS

59. Many fully differentiated plant cells are totipotent, meaning that they

a. will grow roots if they touch the ground.
b. will perform photosynthesis if exposed to sunlight.
c. can fertilize an egg.
d. have the potential to form a whole, fully-functional plant.
e. can undergo meiosis.

Answer: d
Difficulty: Moderate
Bloom's Taxonomy: Knowledge

60. The process of asexual reproduction where a diploid embryo develops from an unfertilized egg or from diploid cells in ovule tissue is called

a. fragmentation.
b. protoplast fusion.
c. somaclonal section.
d. grafting.
e. apomixis.

Answer: e
Difficulty: Moderate
Bloom's Taxonomy: Knowledge

61. The pomato, a cross between a potato and a tomato, was produced by

a. fragmentation.
b. protoplast fusion.
c. somaclonal section.
d. grafting.
e. apomixis.

Answer: b
Difficulty: Difficult
Bloom's Taxonomy: Knowledge

62. Joining a scion with useful fruit traits to a stock with useful root traits is called

a. fragmentation.
b. protoplast fusion.
c. somaclonal section.
d. grafting.
e. apomixis.

Answer: d
Difficulty: Easy
Bloom's Taxonomy: Knowledge

63. Growing useful mutants that develop from callus culture is called

a. fragmentation.
b. protoplast fusion.
c. somaclonal section.
d. grafting.
e. apomixis.

Answer: c
Difficulty: Moderate
Bloom's Taxonomy: Knowledge

64. The growth of a new plant on the margin of a leaf that eventually falls to the ground and grows independently is an example of

a. fragmentation.
b. protoplast fusion.
c. somaclonal section.
d. grafting.
e. apomixis.

Answer: a
Difficulty: Easy
Bloom's Taxonomy: Knowledge

34.5 EARLY DEVELOPMENT OF PLANT FORM AND FUNCTION

65.   Studies have shown that in the flowering plant *Arabidopsis* the root-shoot axis is established

a.   by the first cell division of the embryo.
b.   within hours after the seed germinates.
c.   during the torpedo stage of embryonic development.
d.   shortly after the eight-cell embryo stage.
e.   by plant tissues breaking through the seed coat and making contact with soil.

Answer: a
Difficulty: Difficult
Bloom's Taxonomy: Comprehension

66.   Studies have shown that in the flowering plant *Arabidopsis* whether or not a root epidermal cell develops a root hair is determined

a.   by the distance of the cell from the root tip.
b.   from the timing of cell division.
c.   by positional information from cells of the root cortex.
d.   shortly after the eight-cell embryo stage.
e.   by the direction of cell division.

Answer: c
Difficulty: Difficult
Bloom's Taxonomy: Comprehension

67.   The direction of expansion for a plant cell is determined primarily by

a.   location of the cell plate.
b.   the direction of cell division.
c.   the direction to the nearest meristem.
d.   orientation of cellulose microfibrils in its primary cell wall.
e.   gravity.

Answer: d
Difficulty: Moderate
Bloom's Taxonomy: Comprehension

68.   Studies have shown that in the flowering plant *Arabidopsis* the development of parts of a flower is governed by

a.   differential uptake of hormone signals from the leaves.
b.   the oxygen gradient from outside to inside of the closed flower.
c.   differentiation of cell types as they are formed from the apical meristem.
d.   the timing of cell division.
e.   the expression patterns of several floral organ homeotic genes.

Answer: e
Difficulty: Moderate
Bloom's Taxonomy: Comprehension

69.   Almost immediately after a young leaf primordium first begins to bulge out

a.   undifferentiated vascular tissues that will later become xylem and phloem penetrate it.
b.   small xylem vessels penetrate it first, followed soon after by phloem.
c.   small phloem vessels penetrate it first, followed soon after by xylem.
d.   small xylem and phloem vessels penetrate it together.

Answer: c
Difficulty: Difficult
Bloom's Taxonomy: Comprehension

70.   The oldest cells in a mature leaf are found

a.   at the base of the leaf.
b.   near the very center of the leaf.
c.   at the stomata.
d.   at the leaf tip.
e.   as files of cells at the widest point of the leaf.

Answer: d
Difficulty: Moderate
Bloom's Taxonomy: Knowledge

FOCUS ON RESEARCH: MODEL RESEARCH ORGANISMS: *Arabidopsis thaliana*

71.   The first complete plant genome to be sequenced was that of

a.   *Phaseolus vulgaris* (kidney bean).
b.   *Arabidopsis thaliana* (thale cress).
c.   *Zea mays* (corn).
d.   *Capsella burse-pastoris* (shepherd's purse).
e.   *Oryza sativa* (rice).

Answer: b
Difficulty: Moderate
Bloom's Taxonomy: Knowledge

72.   Approximately _____ genes have been identified in the genome of *Arabidopsis thaliana*.

a.   28,000
b.   5,400
c.   3,200,000
d.   120,000
e.   67,000

Answer: a
Difficulty: Moderate
Bloom's Taxonomy: Knowledge

INSIGHTS FROM THE MOLECULAR REVOLUTION: TRICHOMES: WINDOW ON DEVELOPMENT IN A SINGLE PLANT CELL

73. Which of the following is an unusual event that occurs as trichomes differentiate?

a. DNA replication
b. endoreduplication
c. cell expansion
d. mitosis
e. regulation of the cell cycle

Answer: b
Difficulty: Difficult
Bloom's Taxonomy: Comprehension

74. In *Arabidopsis* the genes TRYPTYCHON, TUBULIN FOLDING COFACTOR C, and STICHEL are involved in determining

a. how closely trichomes will develop to each other.
b. whether or not trichomes are formed.
c. which side of a leaf will have trichomes.
d. how many branches a trichome will have.
e. cell division within trichomes.

Answer: d
Difficulty: Moderate
Bloom's Taxonomy: Comprehension

UNANSWERED QUESTIONS

75. Once a pollen tube penetrates an ovule

a. other pollen tubes are prevented from gaining access to that ovule.
b. that pollen tube continues on to another ovule.
c. other pollen tubes are attracted to that ovule.
d. other pollen tubes can penetrate that ovule more easily.

Answer: a
Difficulty: Moderate
Bloom's Taxonomy: Knowledge

76. Which statement best reflects the current understanding of pollen-tube elongation?

a. It is not even known if pollen tube growth uses chemical signals or is completely random.
b. Pollen-tube elongation will occur as long as the pistil provides moisture and energy.
c. While it is known that chemical cues from female tissues help guide pollen tube growth, the identity of these chemicals is not known.
d. Nearly all of the details of the chemical interactions between the pollen tube and pistil tissue are known.

Answer: c
Difficulty: Moderate
Bloom's Taxonomy: Knowledge

Matching

Match each of the following flower parts with its description.

77. _____ petals

78. _____ carpels

79. _____ sepals

80. _____ stamens

A. Encloses and protects all other flower parts before the flower opens

B. Male reproductive parts

C. Female reproductive parts

D. Often showy parts that function in attracting bees or other animal pollinators

Answers: 77. D   78. C   79. A   80. B

Difficulty: Easy
Bloom's Taxonomy: Knowledge
Source: Section 34.2

Short Answer

81. Describe the process of double fertilization in flowering plants.

Answer: In flowering plants fertilization is typically a double event. In one event a single sperm joins with a single egg, creating a diploid cell that will become the embryo. In the other event a single sperm joins with the central cell of the embryo sac; the central cells begins with two nuclei, so this joining produces a triploid cell. This triploid cell gives rise to endosperm, a tissue that provides nourishment for the developing embryo.
Difficulty: Moderate
Bloom's Taxonomy: Synthesis
Source: Section 34.3

82. Describe several ways that fruits can aid seed dispersal.

Answer: Fruits may aid seed dispersal in many ways. Some fruits split open and release seeds when seeds are ready to be dispersed. Some have winglike extensions that help the wind to transport seeds away from the parent plant. Others have barblike projections or sticky surfaces that easily adhere to feathers, fur, or clothing so that animals will transport them. Still other fruits rely on being eaten, with the seeds adapted for surviving in the digestive tract of the animal and then being dispersed in the animal's feces.
Difficulty: Moderate
Bloom's Taxonomy: Synthesis
Source: Section 34.3

35

CONTROL OF PLANT GROWTH AND DEVELOPMENT

Multiple-Choice

WHY IT MATTERS

1.   Which of the following is NOT known to be influenced by plant hormones?

a.   patterns of plant growth
b.   plant cell metabolism
c.   plant gene expression
d.   plant growth responses to light
e.   all of these are known to be influenced by plant hormones

Answer: e
Difficulty: Easy
Bloom's Taxonomy: Comprehension

2.   Work by Kurosawa and others showed that a fungus causes "foolish seedling" disease in rice by producing a type of plant hormone now called a(n)

a.   brassinosteroid.
b.   abscisic acid.
c.   gibberellin.
d.   jasmonate.
e.   auxin.

Answer: c
Difficulty: Moderate
Bloom's Taxonomy: Knowledge

35.1 PLANT HORMONES

3.   Plant hormones generally _____ concentrations to be effective, and usually have _____ on different plant tissues.

a.   are effective only in extremely large; the same effect
b.   require only low; different effects
c.   require only low; the same effect
d.   are effective only in extremely large; different effects

Answer: b
Difficulty: Moderate
Bloom's Taxonomy: Synthesis

4.   Work by Charles Darwin and Francis Darwin was instrumental in the discovery of which type of plant hormone?

a.   auxins
b.   brassinosteroids
c.   jasmonates
d.   cytokinins
e.   gibberellins

Answer: a
Difficulty: Moderate
Bloom's Taxonomy: Knowledge

5.   Which of the following is a natural auxin?

a.   abscisic acid
b.   gibberellic acid
c.   jasmonic acid
d.   indoleacetic acid
e.   salicylic acid

Answer: d
Difficulty: Difficult
Bloom's Taxonomy: Application

6.   Which type of plant hormone is primarily responsible for the bending of a plant shoot toward light?

a.   auxins
b.   ethylene
c.   systemin
d.   salicylic acid
e.   brassinosteroids

Answer: a
Difficulty: Easy
Bloom's Taxonomy: Comprehension

7.   The growth-promoting substance that promotes stem elongation and bending toward light is produced primarily in

a.   root tips, traveling to the stem in the xylem.
b.   the center of the stem region where elongation or bending occurs.
c.   flowers and flower buds.
d.   the shoot tip.
e.   root tips, traveling to the stem in the phloem.

Answer: d
Difficulty: Moderate
Bloom's Taxonomy: Comprehension

8.   Which of the following should prevent a grass seedling shoot from bending toward light?

a.   removing the shoot tip
b.   placing a translucent cap on the shoot tip
c.   placing an opaque cap on the shoot tip
d.   either removing the shoot tip or placing a translucent cap on the shoot tip
e.   either removing the shoot tip or placing an opaque cap on the shoot tip

Answer: e
Difficulty: Moderate
Bloom's Taxonomy: Synthesis

9. Which type of plant hormone is primarily responsible for apical dominance?

a. abscisic acid
b. cytokinins
c. auxins
d. salicylic acid
e. gibberellins

Answer: c
Difficulty: Moderate
Bloom's Taxonomy: Comprehension

10. Imagine that you want all of the orange trees in your orchard to uniformly flower and set fruit. Which type of plant hormone should you spray on your plants?

a. brassinosteroids
b. auxins
c. ethylene
d. salicylic acid
e. cytokinins

Answer: b
Difficulty: Difficult
Bloom's Taxonomy: Application

11. Which type of plant hormone is primarily responsible for inhibiting leaf abscission?

a. gibberellins
b. ethylene
c. jasmonates
d. abscisic acid
e. auxins

Answer: e
Difficulty: Difficult
Bloom's Taxonomy: Comprehension

12. Which type of plant hormone is primarily responsible for promoting the formation of lateral roots?

a. jasmonates
b. gibberellins
c. auxins
d. systemin
e. cytokinins

Answer: c
Difficulty: Moderate
Bloom's Taxonomy: Comprehension

13. The herbicide 2,4-D is a synthetic form of

a. abscisic acid.
b. ethylene.
c. salicylic acid.
d. gibberellin.
e. auxin.

Answer: e
Difficulty: Difficult
Bloom's Taxonomy: Knowledge

14. Trimming a shrub will make it become more bushy (have more lateral growth) primarily due to a reduction in the amount of _____ that inhibits growth of shoot lateral meristems.

a. auxin
b. ethylene
c. abscisic acid
d. gibberellin
e. cytokinin

Answer: a
Difficulty: Moderate
Bloom's Taxonomy: Synthesis

15. According to the most widely accepted model for polar transport of IAA, IAA enters a plant cell in the form of _____ through the _____.

a. IAA$^-$; PIN transporter
b. IAAH; AUX1 transporter
c. IAA$^-$; AUX1 transporter
d. IAAH; PIN transporter

Answer: b
Difficulty: Difficult
Bloom's Taxonomy: Synthesis

16. According to the most widely accepted model for polar transport of IAA, IAA leaves a plant cell in the form of _____ through the _____.

a. IAA$^-$; PIN transporter
b. IAAH; AUX1 transporter
c. IAA$^-$; AUX1 transporter
d. IAAH; PIN transporter

Answer: a
Difficulty: Difficult
Bloom's Taxonomy: Synthesis

17. An agar block filled with auxin is placed on top of a shoot that has had the shoot tip removed. The agar block is placed so that it covers only one side of the top of the shoot. Which of the following should occur?

a. The shoot will stop growing.
b. The shoot will grow faster on the side away from the agar block.
c. The shoot will grow straight up.
d. The shoot will grow faster on the side with the agar block.

Answer: d
Difficulty: Moderate
Bloom's Taxonomy: Application

18.   According to the acid-growth hypothesis, auxin causes increased acidity in cell walls, which then activates _____ that disrupt(s) bonds between cellulose microfibrils, allowing cell expansion.

a.   a membrane potential
b.   AUX1 transporters
c.   proteins called expansins
d.   an mRNA
e. K^+ ions

Answer: c
Difficulty: Moderate
Bloom's Taxonomy: Knowledge

19.   Which type of plant hormone is primarily responsible for helping break the dormancy of seeds and buds?

a.   gibberellins
b.   cytokinins
c.   ethylene
d.   auxins
e.   jasmonates

Answer: a
Difficulty: Moderate
Bloom's Taxonomy: Comprehension

20.   Which type of plant hormone is primarily responsible for bolting in rosette plants such as cabbages?

a.   salicylic acid
b.   cytokinins
c.   ethylene
d.   gibberellins
e.   brassinosteroids

Answer: d
Difficulty: Moderate
Bloom's Taxonomy: Comprehension

21.   Which type of plant hormone is used by commercial grape growers to get larger grapes by lengthening the stem on which fruits develop?

a.   abscisic acid
b.   ethylene
c.   auxins
d.   salicylic acid
e.   gibberellins

Answer: e
Difficulty: Difficult
Bloom's Taxonomy: Comprehension

22.   A pea plant that is not able to make GA_1 will

a.   grow away from light.
b.   not be able to make flowers.
c.   be a dwarf.
d.   not be able to make fruits.
e.   have nearly no root growth.

Answer: c
Difficulty: Moderate
Bloom's Taxonomy: Comprehension

23.   In plants that normally have separate "male" and "female" flowers both on the same plant, applications of gibberellin

a.   cause all flowers formed to be "female".
b.   encourage proportionally more "male" flowers to develop.
c.   cause all flowers formed to actually be both "male" and "female".
d.   cause all flowers formed to be "male".
e.   encourage proportionally more "female" flowers to develop.

Answer: b
Difficulty: Difficult
Bloom's Taxonomy: Comprehension

24.   Which type of plant hormone was first discovered in experiments to define the nutrient media required for plant tissue culture?

a.   gibberellins
b.   cytokinins
c.   salicylic acid
d.   auxins
e.   brassinosteroids

Answer: b
Difficulty: Easy
Bloom's Taxonomy: Knowledge

25.   Which type of plant hormone can be produced by boiling DNA?

a.   salicylic acid
b.   auxins
c.   cytokinins
d.   abscisic acid
e.   jasmonates

Answer: c
Difficulty: Moderate
Bloom's Taxonomy: Knowledge

26.   Which type of plant hormone is structurally similar to adenine?

a.   brassinosteroid
b.   abscisic acid
c.   ethylene
d.   cytokinin
e.   systemin

Answer: d
Difficulty: Difficult
Bloom's Taxonomy: Knowledge

27.   Which type of plant hormone is synthesized mainly in root tips and is apparently transported through the plant in xylem sap?

a.   gibberellins
b.   auxins
c.   oligosaccharins
d.   ethylene
e.   cytokinins

Answer: e
Difficulty: Moderate
Bloom's Taxonomy: Comprehension

28. Which type of plant hormone causes chloroplasts to mature and retards leaf aging?

a. cytokinins
b. salicylic acid
c. abscisic acid
d. ethylene
e. gibberellins

Answer: a
Difficulty: Moderate
Bloom's Taxonomy: Knowledge

29. In tobacco tissue culture, getting both roots and shoots to develop requires

a. an auxin-to-cytokinin ratio of 10:1.
b. a higher concentration of cytokinin than auxin.
c. an auxin-to-cytokinin ratio of slightly greater than 10:1.
d. an intermediate auxin-to-cytokinin ratio.
e. exactly equal amounts of auxin and cytokinin.

Answer: d
Difficulty: Difficult
Bloom's Taxonomy: Knowledge

30. Synthetic compounds similar to which type of plant hormone are used to prolong the shelf life of vegetables such as lettuces and mushrooms and to keep cut flowers fresh?

a. salicylic acid
b. jasmonates
c. abscisic acid
d. gibberellins
e. cytokinins

Answer: e
Difficulty: Difficult
Bloom's Taxonomy: Comprehension

31. Which type of plant hormone is a gas at normal temperature and pressure?

a. abscisic acid
b. ethylene
c. auxin
d. salicylic acid
e. systemin

Answer: b
Difficulty: Easy
Bloom's Taxonomy: Comprehension

32. Which type of plant hormone governs senescence in plants, such as the loss of leaves in autumn by some plants?

a. cytokinins
b. abscisic acid
c. gibberellins
d. brassinosteroids
e. ethylene

Answer: e
Difficulty: Moderate
Bloom's Taxonomy: Comprehension

33. Which type of plant hormone helps a growing seedling "find its way" through the soil to air by simultaneously slowing stem elongation and stimulating an increase in stem girth?

a. salicylic acid
b. oligosaccharins
c. gibberellins
d. ethylene
e. abscisic acid

Answer: d
Difficulty: Difficult
Bloom's Taxonomy: Comprehension

34. Which type of plant hormone promotes fruit ripening?

a. ethylene
b. jasmonates
c. gibberellins
d. brassinosteroids
e. abscisic acid

Answer: a
Difficulty: Easy
Bloom's Taxonomy: Comprehension

35. Placing a ripe apple with other apples that are not ripe will cause the other apples to ripen sooner. Which type of plant hormone is responsible for this effect?

a. brassinosteroids
b. ethylene
c. auxins
d. oligosaccharins
e. abscisic acid

Answer: b
Difficulty: Moderate
Bloom's Taxonomy: Application

36. Which type of plant hormone has been shown to both promote pollen tube elongation and inhibit root elongation?

a. jasmonates
b. ethylene
c. brassinosteroids
d. salicylic acid
e. oligosaccharins

Answer: c
Difficulty: Moderate
Bloom's Taxonomy: Comprehension

37. Which type of plant hormone was first isolated from the pollen of *Brassica napus*?

a. auxins
b. jasmonates
c. gibberellins
d. brassinosteroids
e. oligosaccharins

Answer: d
Difficulty: Easy
Bloom's Taxonomy: Knowledge

38. Which of the following types of plant hormones appears to regulate the expression of genes associated with a plant's growth responses to light?

a. salicylic acid
b. brassinosteroids
c. gibberellins
d. jasmonates
e. oligosaccharins

Answer: b
Difficulty: Difficult
Bloom's Taxonomy: Knowledge

39. Which of the following types of plant hormones is apparently synthesized from carotenoid pigments inside plastids?

a. salicylic acid
b. cytokinins
c. abscisic acid
d. jasmonates
e. brassinosteroids

Answer: c
Difficulty: Difficult
Bloom's Taxonomy: Knowledge

40. Which type of plant hormone is generally responsible for long-term inhibition of plant growth such as in buds and seeds?

a. auxins
b. gibberellins
c. oligosaccharins
d. brassinosteroids
e. abscisic acid

Answer: e
Difficulty: Easy
Bloom's Taxonomy: Knowledge

41. Imagine that you own a nursery and you want to reduce the chance that plants you ship will suffer from shipping damage. Which type of plant hormone should you apply to the plants before shipping?

a. cytokinins
b. abscisic acid
c. oligosaccharins
d. gibberellins
e. ethylene

Answer: b
Difficulty: Moderate
Bloom's Taxonomy: Application

42. Which type of plant hormone stimulates stomata to close during a drought?

a. salicylic acid
b. gibberellins
c. abscisic acid
d. auxins
e. oligosaccharins

Answer: c
Difficulty: Moderate
Bloom's Taxonomy: Comprehension

43. Which of the following types of plant hormones triggers plant responses to environmental stresses such as cold snaps, high salinity, and drought?

a. abscisic acid
b. cytokinins
c. oligosaccharins
d. jasmonates
e. ethylene

Answer: a
Difficulty: Easy
Bloom's Taxonomy: Knowledge

44. Which type of plant signaling compound is produced from structural elements of the cell wall?

a. salicylic acid
b. oligosaccharins
c. abscisic acid
d. auxins
e. brassinosteroids

Answer: b
Difficulty: Moderate
Bloom's Taxonomy: Comprehension

INSIGHTS FROM THE MOLECULAR REVOLUTION: STRESSING OUT IN PLANTS AND PEOPLE

45. Which of the following did Nam-Hai Chua and his colleagues demonstrate to be part of the abscisic acid response pathway?

a. cyclic ADP-ribose
b. Ca^{2+}
c. protein kinases
d. a and b only
e. a, b and c

Answer: e
Difficulty: Moderate
Bloom's Taxonomy: Synthesis

35.2 PLANT CHEMICAL DEFENSES

46. In some cases, when an insect begins feeding on a leaf the plant responds by activating a signaling pathway that results in the production of _____, which disrupt an insect's capacity to digest proteins.

a. gibberellins
b. lipases
c. chitinases
d. protease inhibitors
e. cytokinins

Answer: d
Difficulty: Easy
Bloom's Taxonomy: Comprehension

47. Which type of plant signaling molecule is similar in structure to aspirin?

a. abscisic acid
b. ethylene
c. salicylic acid
d. jasmonate
e. oligosaccharin

Answer: c
Difficulty: Moderate
Bloom's Taxonomy: Comprehension

48. The first peptide hormone discovered in plants was _____, which functions in wound response in the tomato.

a. cytokinin
b. systemin
c. gibberellin
d. jasmonate
e. salicylic acid

Answer: b
Difficulty: Moderate
Bloom's Taxonomy: Knowledge

49. The synthesis of _____ is triggered by systemin.

a. gibberellin
b. auxin
c. jasmonate
d. cytokinin
e. brassinosteroids

Answer: c
Difficulty: Moderate
Bloom's Taxonomy: Comprehension

50. When a plant has a hypersensitive response, production of _____ triggers effects such as the expression of PR proteins.

a. oligosaccharins
b. salicylic acid
c. ethylene
d. cytokinin
e. brassinosteroids

Answer: b
Difficulty: Moderate
Bloom's Taxonomy: Comprehension

51. Chitinases are examples of

a. systemins.
b. phytoalexins.
c. PR proteins.
d. R genes.
e. heat-shock proteins.

Answer: c
Difficulty: Moderate
Bloom's Taxonomy: Comprehension

52. Secondary metabolites that function for plants as antibiotics are

a. systemins.
b. phytoalexins.
c. PR proteins.
d. R genes.
e. heat-shock proteins.

Answer: b
Difficulty: Moderate
Bloom's Taxonomy: Comprehension

53. After a plant has survived a microbial invasion, the rest of the plant is often less vulnerable to future infections. This is called

a. systemic acquired resistance.
b. wound defense.
c. gene-for-gene recognition.
d. hypersensitive response.
e. jasmonate response.

Answer: a
Difficulty: Easy
Bloom's Taxonomy: Comprehension

54. In systemic acquired resistance, the regulatory protein NPR-1 moves from the cytoplasm to the cell nucleus, apparently in response to a buildup of _____ in the cytoplasm.

a. ethylene
b. brassinosteroids
c. oligosaccharins
d. jasmonates
e. salicylic acid

Answer: e
Difficulty: Moderate
Bloom's Taxonomy: Comprehension

55. Chaperone-type proteins that stabilize other proteins in response to environmental stresses such as salinity, drought, heat, and cold are

a. systemins.
b. phytoalexins.
c. PR proteins.
d. R genes.
e. heat-shock proteins.

Answer: e
Difficulty: Easy
Bloom's Taxonomy: Comprehension

35.3 PLANT RESPONSES TO THE ENVIRONMENT: MOVEMENTS

56. The main stimulus for phototropism is light of _____ wavelengths.

a. red
b. green
c. yellow
d. blue
e. far red

Answer: d
Difficulty: Moderate
Bloom's Taxonomy: Knowledge

57. Statoliths, particles that move in the direction gravity pulls them, are typically _____ in plants.

a. amyloplasts
b. calcium carbonate crystals
c. nuclei
d. vacuoles
e. hormones

Answer: a
Difficulty: Moderate
Bloom's Taxonomy: Comprehension

58. Which type of plant hormone appears to play a major role in gravitropism in roots and shoots?

a. salicylic acid
b. gibberellins
c. abscisic acid
d. auxins
e. oligosaccharins

Answer: d
Difficulty: Moderate
Bloom's Taxonomy: Comprehension

59. Thigmotropism is a growth response to

a. light.
b. day length.
c. contact with a solid object.
d. circadian rhythms.
e. gravity.

Answer: c
Difficulty: Easy
Bloom's Taxonomy: Knowledge

60. You observe a sunflower that moves its leaves so that it tracks the sun across the sky through the day, and a sensitive plant that reversibly folds up its leaflets when you touch it. These are both examples of

a. phototropism.
b. thigmotropism.
c. stress responses.
d. circadian rhythms.
e. nastic movements.

Answer: e
Difficulty: Moderate
Bloom's Taxonomy: Knowledge

35.4 PLANT RESPONSES TO THE ENVIRONMENT: BIOLOGICAL CLOCKS

61. Plant activities that follow a biological clock so that they are based on cycles of about 24 hours are

a. gravitropisms.
b. photoperiodisms.
c. thigmotropisms.
d. circadian rhythms.
e. phototropisms.

Answer: d
Difficulty: Easy
Bloom's Taxonomy: Comprehension

62. Plant responses to the relative lengths of light and dark periods in their environment during each 24-hour period are

a. gravitropisms.
b. photoperiodisms.
c. thigmotropisms.
d. circadian rhythms.
e. phototropisms.

Answer: b
Difficulty: Easy
Bloom's Taxonomy: Comprehension

63. The switching mechanism in the photoperiodic response is often

a. phytochrome.
b. auxin.
c. cryptochrome.
d. calmodulin.
e. phytoalexin.

Answer: a
Difficulty: Easy
Bloom's Taxonomy: Knowledge

64. During daylight hours, primarily

a. P_r absorbs red light and is converted to P_{fr}.
b. P_r absorbs far-red light and is converted to P_{fr}.
c. P_{fr} absorbs red light and is converted to P_r.
d. P_{fr} absorbs far-red light and is converted to P_r.
e. P_{fr} is converted to P_r.

Answer: a
Difficulty: Moderate
Bloom's Taxonomy: Synthesis

65. At night in the dark, primarily

a. P_r absorbs red light and is converted to P_{fr}.
b. P_r absorbs far-red light and is converted to P_{fr}.
c. P_{fr} absorbs red light and is converted to P_r.
d. P_{fr} absorbs far-red light and is converted to P_r.
e. P_{fr} is converted to P_r.

Answer: e
Difficulty: Moderate
Bloom's Taxonomy: Synthesis

66. Chrysanthemums are short-day plants, requiring a night length of longer than 12 hours to flower. Under which of the following conditions should chrysanthemums flower?

1: Night length 14 hours, day length 10 hours
2: Night length 10 hours, day length 14 hours
3: Night length 14 hours, day length 10 hours; night interrupted in the middle by an intense red flash
4: Night length 10 hours, day length 14 hours; day interrupted in the middle by half an hour of darkness

a. only 1
b. only 2
c. both 1 and 3
d. both 2 and 4
e. 2, 3, and 4

Answer: a
Difficulty: Moderate
Bloom's Taxonomy: Application

67. Bearded irises are long-day plants, requiring a night length of less than 12 hours to flower. Under which of the following conditions should bearded irises flower?

1: Night length 14 hours, day length 10 hours
2: Night length 10 hours, day length 14 hours
3: Night length 14 hours, day length 10 hours; night interrupted in the middle by an intense red flash
4: Night length 10 hours, day length 14 hours; day interrupted in the middle by half an hour of darkness

a. only 1
b. only 2
c. both 1 and 3
d. both 2 and 4
e. 2, 3, and 4

Answer: e
Difficulty: Difficult
Bloom's Taxonomy: Application

68. The exposure of plants to low temperatures in order to stimulate flowering is called

a. cryptotropism.
b. thigmotropism.
c. vernalization.
d. florigen.
e. cryotropism.

Answer: c
Difficulty: Easy
Bloom's Taxonomy: Knowledge

69. Which of the following appears to promote dormancy in at least some plants?

a. dry soil
b. nitrogen deficiency
c. long nights
d. cold nights
e. all of these

Answer: e
Difficulty: Moderate
Bloom's Taxonomy: Synthesis

FOCUS ON RESEARCH: RESEARCH METHODS: USING DNA MICROARRAY ANALYSIS TO TRACK DOWN "FLORIGEN"

70. Flowering in *Arabidopsis* appears to require

a. production in leaves of CO protein and FT protein.
b. movement of CO protein from leaves to the shoot apex.
c. expression of CO protein in the shoot apex.
d. movement of FT mRNA from leaves to the shoot apex.
e. movement of FT protein from leaves to the shoot apex.

Answer: d
Difficulty: Moderate
Bloom's Taxonomy: Knowledge

35.5 SIGNAL RESPONSES AT THE CELLULAR LEVEL

71. Where are ethylene receptors located in plant cells?

a. cell wall
b. cytoplasm
c. endoplasmic reticulum
d. plasma membrane
e. nucleus

Answer: c
Difficulty: Difficult
Bloom's Taxonomy: Knowledge

72. Where is the auxin receptor TIR1 located in plant cells?

a. cell wall
b. cytoplasm
c. endoplasmic reticulum
d. plasma membrane
e. nucleus

Answer: b
Difficulty: Difficult
Bloom's Taxonomy: Knowledge

73. The second messenger inositol triphosphate (IP$_3$) appears to be involved in stomatal closure triggered by

a. auxins.
b. abscisic acid.
c. oligosaccharins.
d. salicylic acid.
e. gibberellins.

Answer: b
Difficulty: Moderate
Bloom's Taxonomy: Comprehension

UNANSWERED QUESTIONS

74. At least some plants appear to have a backup genome that is

a. in the form of DNA.
b. a combination of RNA and protein.
c. in the form of RNA.
d. in the form of protein.
e. of unknown composition.

Answer: e
Difficulty: Easy
Bloom's Taxonomy: Comprehension

Choice

Choose the type of plant hormone most closely associated with the action(s) or researcher(s) given below.

a. gibberellins
b. salicylic acid
c. auxins
d. abscisic acid
e. ethylene

75. _____ Nam-Hai Chua

76. _____ apical dominance

77. _____ fruit ripening

78. _____ bolting

79. _____ Charles and Francis Darwin

80. _____ systemic acquired resistance

81. _____ breaking seed dormancy

82. _____ promoting leaf abscission

83. _____ Eiichi Kurosawa

84. _____ phototropism

85. _____ leaf bud dormancy

86. _____ Frits Went

Answers: *75. d*   *76. c*   *77. e*   *78. a*   *79. c*
          80. b   *81. a*   *82. e*   *83. a*   *84. c*
          85. d   *86. c*

Difficulty: Moderate
Bloom's Taxonomy: Knowledge

Source: Why It Matters; Sections 35.1 and 35.2

Short Answer

87. Describe the various effects of auxins in plants.

Answer: Auxins are involved in a variety of plant growth responses. The elongation of stems is promoted by auxins, as is the growth of stems so that they bend toward light. Auxins also promote fruit development and inhibit leaf abscission. Auxins also promote dormancy in lateral buds. In the roots auxin promotes lateral growth. Auxin also is involved in orientation of growth with respect to gravity in both shoots and roots.
Difficulty: Moderate
Bloom's Taxonomy: Synthesis
Source: Sections 35.1 and 35.3

36

INTRODUCTION TO ANIMAL ORGANIZATION AND PHYSIOLOGY

Multiple-Choice

WHY IT MATTERS

1. Maintenance of the body's internal environment at relatively constant levels is called

a. feedback.
b. homeostasis.
c. physiology.
d. natural selection.
e. none of the above

Answer: b
Difficulty: Easy
Bloom's Taxonomy: Knowledge

2. An examination of the arrangement of bones in the hand is an example of

a. an ecological study.
b. a taxonomic study.
c. an anatomical study.
d. a physiological study.
e. none of the above

Answer: c
Difficulty: Moderate
Bloom's Taxonomy: Knowledge, Comprehension

3. An examination of the processes of muscle contraction in fishes is an example of

a. an ecological study.
b. a taxonomic study.
c. an anatomical study.
d. a physiological study.
e. none of the above

Answer: d
Difficulty: Moderate
Bloom's Taxonomy: Knowledge, Comprehension

36.1 ORGANIZATION OF THE ANIMAL BODY

4. Which of the following levels of organization are correctly arranged from smallest to largest?

a. tissue, organ, organ system
b. tissue, organ system, organ
c. organ system, organ, tissue
d. organ system, tissue, organ
e. organ, organ system, tissue

Answer: a
Difficulty: Moderate
Bloom's Taxonomy: Knowledge, Analysis

5. A group of specialized cells of similar structure and function is a(n)

a. organ system.
b. cell cluster.
c. organ.
d. tissue.
e. none of the above

Answer: d
Difficulty: Easy
Bloom's Taxonomy: Knowledge

6. Why was the evolution of multicellularity important for organisms?

a. It allows for the creation of a sustainable internal fluid environment.
b. It allows for occupation of diverse habitats.
c. It allows for cellular specialization.
d. It allows for an increase in the size of an organism.
e. All of the above.

Answer: e
Difficulty: Difficult
Bloom's Taxonomy: Knowledge, Comprehension

7. Two or more different tissues arranged to carry out a specific function is called a(n)

a. organ system.
b. organ.
c. tissue.
d. organism.
e. none of the above

Answer: b
Difficulty: Easy
Bloom's Taxonomy: Knowledge

36.2 ANIMAL TISSUES

8. The structure of a tissue is determined by

a. the structure and organization of the cell cytoskeleton.
b. the type and organization of the extracellular matrix.
c. the junctions holding cells together.
d. all of the above
e. none of the above

Answer: d
Difficulty: Moderate
Bloom's Taxonomy: Knowledge

9. Which of the following cell junctions are responsible for allowing the bladder lining to stretch, thus accommodating increasing volumes of urine?

a. anchoring junctions
b. gap junctions
c. tight junctions
d. all of the above
e. none of the above

Answer: a
Difficulty: Moderate
Bloom's Taxonomy: Knowledge, Comprehension

10. You observe a tissue with one free surface with several layers of cells attached to a basal lamina. This is an example of

a. connective tissue.
b. epithelial tissue.
c. muscle tissue.
d. nervous tissue.
e. none of the above

Answer: b
Difficulty: Moderate
Bloom's Taxonomy: Analysis

11. The presence of cilia is a common feature of some

a. nervous tissues.
b. muscle tissues.
c. connective tissues.
d. epithelial tissues.
e. none of the above

Answer: d
Difficulty: Easy
Bloom's Taxonomy: Knowledge, Knowledge, Comprehension

12. You would describe an epithelial tissue with a single layer of flattened cells as

a. simple columnar epithelium.
b. stratified cuboidal epithelium.
c. stratified columnar epithelium.
d. stratified squamous epithelium.
e. none of the above

Answer: e
Difficulty: Easy
Bloom's Taxonomy: Knowledge

13. If you were presented with a tissue sample that contained long rows of cells surrounded by parallel bundles of collagen and elastin fibers, that tissue would most likely be

a. bone tissue.
b. fibrous conective tissue.
c. cartilage.
d. blood.
e. adipose tissue.

Answer: b
Difficulty: Moderate
Bloom's Taxonomy: Analysis

14. Which connective tissue contains fibroblasts?

a. loose connective tissue
b. fibrous connective tissue
c. cartilage
d. bone
e. a and b

Answer: e
Difficulty: Easy
Bloom's Taxonomy: Knowledge, Comprehension

15. Which connective tissue is best at withstanding compression due the elasticity of its matrix?

a. loose connective tissue
b. bone tissue
c. cartilage
d. fibrous connective tissue
e. none of the above

Answer: c
Difficulty: Easy
Bloom's Taxonomy: Knowledge, Comprehension

16. Which tissue has little extracellular matrix, acts as an insulator, and is used to store chemical energy?

a. loose connective tissue
b. blood
c. cartilage
d. fibrous connective tissue
e. none of the above

Answer: e
Difficulty: Moderate
Bloom's Taxonomy: Comprehension

17. Which tissue acts to transport nutrients and wastes throughout the body?

a. blood
b. adipose tissue
c. loose connective tissue
d. fibrous connective tissue
e. none of the above

Answer: a
Difficulty: Easy
Bloom's Taxonomy: Knowledge, Comprehension

18. Which tissue is best described as a sparse distribution of cells surrounded by an open network of collagen and elastin fibers?

a. fibrous connective tissue
b. adipose tissue
c. bone tissue
d. loose connective tissue
e. blood

Answer: d
Difficulty: Moderate
Bloom's Taxonomy: Knowledge, Comprehension

19. Which major tissue category often has more extracellular matrix material than cellular material?

a. epithelial tissue
b. connective tissue
c. muscle tissue
d. nervous tissue
e. none of the above

Answer: b
Difficulty: Moderate
Bloom's Taxonomy: Knowledge

20. Which type of connective tissue forms tendons and ligaments?

a. loose connective tissue
b. blood
c. cartilage
d. fibrous connective tissue
e. none of the above

Answer: d
Difficulty: Easy
Bloom's Taxonomy: Knowledge

21. You would expect the bones of a growing child to have extremely active

a. osteoclasts.
b. fibroblasts.
c. osteoblasts.
d. adipocytes.
e. chondrocytes.

Answer: c
Difficulty: Moderate
Bloom's Taxonomy: Synthesis

22. Which major tissue category has cells that can contract?

a. epithelium
b. connective tissue
c. muscle tissue
d. nervous tissue
e. none of the above

Answer: c
Difficulty: Easy
Bloom's Taxonomy: Knowledge

23. Which tissue is striated and can rapidly contract, producing voluntary movements?

a. skeletal muscle
b. smooth muscle
c. bone tissue
d. cardiac muscle
e. none of the above

Answer: a
Difficulty: Easy
Bloom's Taxonomy: Knowledge, Comprehension

24. Which tissue has contractile cells that are joined by intercalated disks, allowing it to contract as a unit?

a. smooth muscle
b. cartilage
c. skeletal muscle
d. cardiac muscle
e. none of the above

Answer: d
Difficulty: Easy
Bloom's Taxonomy: Knowledge, Comprehension

25. Which tissue is composed of spindle-shaped contractile cells that are commonly found in the walls of tubular organs?

a. squamous epithelium
b. smooth muscle
c. skeletal muscle
d. cardiac muscle
e. none of the above

Answer: b
Difficulty: Easy
Bloom's Taxonomy: Knowledge, Comprehension

26. The cells of which tissue are connected by gap junctions?

a. skeletal muscle
b. cardiac muscle
c. smooth muscle
d. all of the above
e. b and c

Answer: e
Difficulty: Difficult
Bloom's Taxonomy: Analysis

27. What cell type(s) are found in nervous tissue?

a. neurons
b. glial cells
c. squamous cells
d. a and b
e. b and c

Answer: d
Difficulty: Easy
Bloom's Taxonomy: Knowledge

28. The part of a neuron that conducts an electrical signal towards other neurons is/are the

a. dendrites.
b. cell body.
c. axon.
d. all of the above
e. none of the above

Answer: c
Difficulty: Easy
Bloom's Taxonomy: Knowledge

29. Glial cells function to

a. conduct electrical impulses.
b. support and nourish neurons.
c. secrete collagen fibers into the extracellular matrix.
d. release secretions on to a free surface.
e. none of the above

Answer: b
Difficulty: Easy
Bloom's Taxonomy: Knowledge

30. A person takes a medication that blocks their neurons' ability to be stimulated by a specific chemical signal. What part of the neuron is most likely being affected by the medication?

a. axon terminals
b. the axon
c. dendrites
d. the cell body
e. a and b only

Answer: c
Difficulty: Difficult
Bloom's Taxonomy: Evaluation

31. Which of the following human tissue cell types can be over a meter in length?

a. neurons
b. chondrocytes
c. columnar epithelial cells
d. skeletal muscle cells
e. adipocytes

Answer: a
Difficulty: Moderate
Bloom's Taxonomy: Knowledge, Comprehension

32. Which of the following major tissue types can transmit electrical impulses from one part of an organism to another?

a. epithelial tissue
b. connective tissue
c. muscle tissue
d. nervous tissue
e. none of the above

Answer: d
Difficulty: Easy
Bloom's Taxonomy: Knowledge

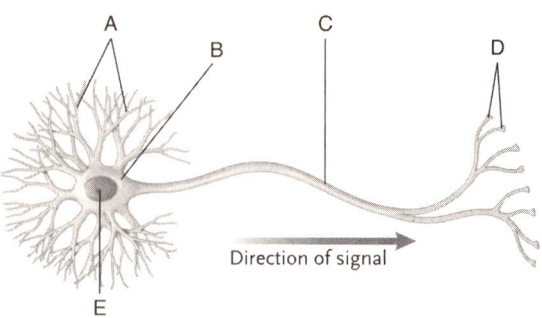

Use this figure to answer questions 33 through 34.

33. Identify neuron part B.

a. axon terminal
b. cell body
c. axon
d. glial cell
e. dendrites

Answer: b
Difficulty: Easy
Bloom's Taxonomy: Knowledge

34. Identify neuron part D.

a. axon terminal
b. cell body
c. axon
d. glial cell
e. dendrites

Answer: a
Difficulty: Easy
Bloom's Taxonomy: Knowledge

36.3 COORDINATION OF TISSUES IN ORGANS AND ORGAN SYSTEMS

35. Which of the following organ systems are involved in temperature regulation?

a. integumentary system
b. circulatory system
c. muscular system
d. all of the above
e. none of the above

Answer: d
Difficulty: Moderate
Bloom's Taxonomy: Knowledge, Application

36. Which organ system includes the spleen, lymph nodes, and thymus?

a. circulatory system
b. endocrine system
c. nervous system
d. digestive system
e. none of the above

Answer: e
Difficulty: Easy
Bloom's Taxonomy: Knowledge, Analysis

37. Which of the following organ systems is responsible for eliminating metabolic wastes from the body?

a. excretory system
b. respiratory system
c. nervous system
d. a and b
e. all of the above

Answer: d
Difficulty: Moderate
Bloom's Taxonomy: Knowledge

38. Which organ system includes the pituitary, thyroid, and adrenal glands?

a. endocrine system
b. reproductive system
c. lymphatic system
d. respiratory system
e. none of the above

Answer: a
Difficulty: Easy
Bloom's Taxonomy: Comprehension

36.4 HOMEOSTASIS

39. The two most important organ systems for the maintenance of homeostasis are

a. circulatory and nervous systems.
b. endocrine and circulatory systems.
c. lymphatic and circulatory systems.
d. nervous and endocrine systems.
e. lymphatic and circulatory systems.

Answer: d
Difficulty: Easy
Bloom's Taxonomy: Knowledge

40. Which component of a negative feedback mechanism is responsible for producing the effect that counteracts the original environmental change away from homeostasis?

a. sensor
b. response
c. effector
d. stimulus
e. integrator

Answer: c
Difficulty: Easy
Bloom's Taxonomy: Knowledge, Comprehension

41. What component of a negative feedback mechanism does the brain or spinal cord represent?

a. response
b. sensor
c. integrator
d. effector
e. stimulus

Answer: c
Difficulty: Easy
Bloom's Taxonomy: Application

42. The process in which an environmental stimulus triggers a response that compensates for changes in the internal/external environment is called

a. positive feedback.
b. physiological feedback.
c. reflex feedback.
d. negative feedback.
e. anatomical feedback.

Answer: d
Difficulty: Easy
Bloom's Taxonomy: Knowledge

43. Sweating in response to elevated temperature is an example of a(n) _____ of negative feedback.

a. response
b. integrator
c. effector
d. sensor
e. stimulus

Answer: a
Difficulty: Easy
Bloom's Taxonomy: Application

44. A response to change in the internal/external environment that adds to the change is called.

a. positive feedback.
b. physiological feedback.
c. reflex feedback.
d. negative feedback.
e. anatomical feedback.

Answer: a
Difficulty: Easy
Bloom's Taxonomy: Knowledge

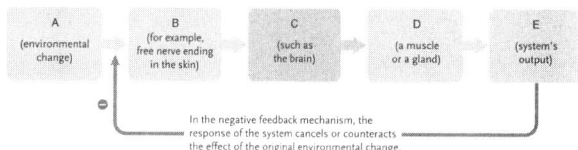

Use this figure to answer questions 45 through 46.

45. Identify negative feedback component B.

a. integrator
b. sensor
c. stimulus
d. response
e. effector

Answer: b
Difficulty: Easy
Bloom's Taxonomy: Knowledge

46. Identify negative feedback component D.

a. integrator
b. sensor
c. stimulus
d. response
e. effector

Answer: e
Difficulty: Easy
Bloom's Taxonomy: Knowledge

INSIGHTS FROM THE MOLECULAR REVOLUTION: CULTURED STEM CELLS

47. The research of James A. Thompson and his coworkers shows that sustainable human stem cell colonies can be established. As experimental evidence of this, they showed that

a. human stem cell colonies could survive outside the body if exposed to a bed of mouse fibroblasts.
b. stem cell telomerase remained active in established cell colonies.
c. cultured stem cells could differentiate when injected into mice.
d. All of the above.
e. None of the above.

Answer: d
Difficulty: Moderate
Bloom's Taxonomy: Knowledge

48. Why do we care about the formation of sustainable human stem cell cultures?

a. Organs could be grown for transplantation.
b. Defective cells could be replaced with functional versions derived by stem cells (e.g., Parkinson disease).
c. They could cure cancer.
d. All of the above.
e. a and b.

Answer: e
Difficulty: Moderate
Bloom's Taxonomy: Evaluation

49. The major disadvantage of using stem cells derived from human embryos was:

a. they did not differentiate.
b. they were difficult to maintain in culture.
c. they did not function in humans other than those that they were obtained from.
d. all of the above
e. none of the above

Answer: b
Difficulty: Moderate
Bloom's Taxonomy: Knowledge

Matching

Match each of the following organ systems with its correct description.

50. _____ nervous system
51. _____ endocrine system
52. _____ muscular system
53. _____ skeletal system
54. _____ integumentary system
55. _____ circulatory system
56. _____ lymphatic system
57. _____ respiratory system
58. _____ digestive system
59. _____ excretory system
60. _____ reproductive system

A. Exchanges gases with the environment

B. Coordinates body activities through the conduction of electrical impulses

C. Helps regulate internal water balance and pH

D. Provides leverage for body movements

E. Moves body parts

F. Converts ingested matter into molecules and ions that can be absorbed into the body

G. Covers external body surfaces and protects against injury and infection

H. Returns excess fluid to the body and protects it against pathogens

I. Passes on genes to the next generation

J. Coordinates body activities though secretion of hormones

K. Distributes water and nutrients throughout the body

Answers: *50. B*   *51. J*   *52. E*   *53. D*   *54. G*
            55. K   *56. H*   *57. A*   *58. F*   *59. C*   *60. I*

Difficulty: Moderate
Bloom's Taxonomy: Knowledge, Comprehension
Source: Section 36.3

Match each of the following tissues with its correct description.

61. _____ adipose tissue
62. _____ loose connective tissue
63. _____ cardiac muscle
64. _____ columnar epithelium
65. _____ smooth muscle

66. _____ fibrous connective tissue

67. _____ nervous tissue

68. _____ squamous epithelium

69. _____ blood

70. _____ cartilage

A. Wraps around internal organs, providing a lubricated surface

B. Cushions organs and stores surplus energy

C. Transport respiratory gases throughout the body

D. Has flattened cells that are optimal for diffusion

E. Withstand tension produced by muscles that are attached

F. Is composed of chondrocytes and collagen embedded in an elastic matrix

G. Controls body systems by conducting electrical impulses

H. Found exclusively in the heart

I. Lines the gut and respiratory tract

J. Produces involuntary movements associated with tubular organs

Answers: *61. B*   *62. A*   *63. H*   *64. I*   *65. J*
          66. E   *67. G*   *68. D*   *69. C*   *70. F*

Difficulty: Moderate
Bloom's Taxonomy: Knowledge, Comprehension
Source: Section 36.2

Select the Exception

71. Which of the following structures is *not* an example of an organ?

a. cartilage
b. blood vessels
c. the brain
d. the heart
e. the stomach

Answer: a
Difficulty: Easy
Bloom's Taxonomy: Synthesis, Evaluation
Source: Section Why it Matters

72. Which of the following is *not* an example of a homeostatic mechanism?

a. sweating when you get hot
b. the maintenance of blood glucose concentrations
c. body water regulation
d. heart rate increase during exercise
e. childbirth

Answer: e
Difficulty: Easy
Bloom's Taxonomy: Synthesis, Evaluation
Source: Section 36.1

73. Which of the following is *not* an example of an epithelial function?

a. allow diffusion
b. secrete products
c. contract when stimulated
d. protect the body from invasion by bacteria
e. form linings of organs

Answer: c
Difficulty: Easy
Bloom's Taxonomy: Synthesis, Evaluation
Source: Section 36.2

74. Which is *not* a component of blood?

a. erythrocytes
b. osteocytes
c. leukocytes
d. platelets
e. plasma

Answer: b
Difficulty: Easy
Bloom's Taxonomy: Synthesis, Evaluation
Source: Section 36.2

75. Which organ system is *not* required for the maintenance of homeostasis within an organism?

a. excretory system
b. reproductive system
c. skeletal system
d. endocrine system
e. lymphatic system

Answer: b
Difficulty: Hard
Bloom's Taxonomy: Synthesis, Evaluation
Source: Section 36.3

76. Which of the following is *not* an example of negative feedback control of homeostasis?

a. childbirth.
b. blood pH regulation.
c. regulation of heart rate.
d. temperature regulation.
e. blood oxygen regulation.

Answer: a
Difficulty: Easy
Bloom's Taxonomy: Synthesis, Evaluation
Source: Section 36.4

37

INFORMATION FLOW AND THE NEURON

Multiple-Choice

WHY IT MATTERS

1.   Nervous system cells that send and receive electrical signals are called

a.   receptors.
b.   neurons.
c.   sensors.
d.   glial cells.
e.   none of the above

Answer: b
Difficulty: Easy
Bloom's Taxonomy: Knowledge

2.   _____ assist and nourish cells that transmit electrical signals.

a.   Receptors
b.   Neurons
c.   Effector cells
d.   Glial cells
e.   None of the above

Answer: d
Difficulty: Easy
Bloom's Taxonomy: Knowledge

37.1 NEURONS AND THEIR ORGANIZATION IN NERVOUS SYSTEMS

3.   The component of a neural signaling pathway in which messages are sorted and interpreted is called

a.   a response.
b.   integration.
c.   transmission.
d.   reception.
e.   action.

Answer: b
Difficulty: Easy
Bloom's Taxonomy: Knowledge

4.   In humans, which category of neurons is composed of the greatest number of cells?

a.   efferent neurons
b.   afferent neurons
c.   interneurons
d.   motor neurons
e.   none of the above

Answer: c
Difficulty: Moderate
Bloom's Taxonomy: Knowledge

5.   Which of the following components of a neural signaling pathway are correctly arranged from beginning to end?

a.   integration, transmission, reception, response, transmission
b.   transmission, response, reception, transmission, integration
c.   response, reception, transmission, integration, transmission
d.   reception, transmission, integration, transmission, response
e.   transmission, integration, transmission, reception, response

Answer: d
Difficulty: Moderate
Bloom's Taxonomy: Knowledge, Synthesis

6.   _____ conduct(s) electrical signals away from the cell body of a neuron.

a.   An axon terminals
b.   An axon
c.   An axon hillock
d.   The nucleus
e.   Dendrites

Answer: b
Difficulty: Easy
Bloom's Taxonomy: Knowledge

7.   Which cells are part of the vertebrate peripheral nervous system?
a.   afferent neurons
b.   motor neurons
c.   efferent neurons
d.   all of the above
e.   none of the above

Answer: d
Difficulty: Moderate
Bloom's Taxonomy: Comprehension

8. Which cells are located in the brain and spinal cord of vertebrates?
a. interneurons
b. motor neurons
c. afferent neurons
d. efferent neurons
e. all of the above

Answer: a
Difficulty: Easy
Bloom's Taxonomy: Knowledge

9. What type of neuron directly stimulates a muscle or gland?
a. afferent neuron
b. efferent neuron
c. interneuron
d. all of the above
e. none of the above

Answer: b
Difficulty: Easy
Bloom's Taxonomy: Knowledge

10. Which glial cell is only found in the peripheral nervous system?
a. Schwann cells
b. oligodendrocytes
c. astrocytes
d. all of the above
e. none of the above

Answer: a
Difficulty: Moderate
Bloom's Taxonomy: Knowledge

11. What category of organic macromolecule allows myelinated sheaths to act as electrical insulators?
a. proteins
b. carbohydrates
c. lipids
d. nucleic acids
e. none of the above

Answer: c
Difficulty: Moderate
Bloom's Taxonomy: Knowledge, Comprehension

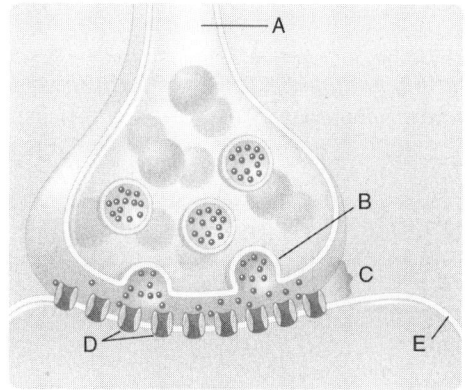

Use this figure to answer questions 12 through 14.

12. In this chemical synapse, which structure represents the synaptic cleft?
a. A
b. B
c. C
d. D
e. E

Answer: c
Difficulty: Easy
Bloom's Taxonomy: Knowledge

13. In this chemical synapse, which structure represents the axon terminal of a presynaptic cell?
a. A
b. B
c. C
d. D
e. E

Answer: a
Difficulty: Easy
Bloom's Taxonomy: Knowledge

14. When comparing electrical synapses and chemical synapses, what structure is found only in electrical synapses?
a. gap junctions
b. neurotransmitters
c. a synaptic cleft
d. postsynaptic plasma cell receptors
e. none of the above

Answer: a
Difficulty: Moderate
Bloom's Taxonomy: Knowledge, Comprehension

37.2 SIGNAL CONDUCTION BY NEURONS

15. For a neuron, the outside of the plasma membrane has a higher concentration of _____ than the inside of the plasma membrane.

a. Na^+
b. K^+
c. Ca^{2+}
d. anions
e. none of the above

Answer: a
Difficulty: Easy
Bloom's Taxonomy: Knowledge

16. The neuron resting membrane potential is established due to

a. the activity of Na^+/K^+ active transport pumps.
b. the presence of impermeable intracellular anions.
c. the buildup of Na^+ outside the plasma membrane and K^+ inside the plasma membrane.
d. all of the above
e. none of the above

Answer: d
Difficulty: Moderate
Bloom's Taxonomy: Knowledge, Comprehension

17. The typical resting potential of an isolated neuron is

a. -60 mV.
b. -70 mV.
c. -80 mV.
d. -90 mV.
e. -100 mV.

Answer: b
Difficulty: Easy
Bloom's Taxonomy: Knowledge

18. An abrupt and transient change in membrane potential is called a(n)
a. electrical potential.
b. refractory potential.
c. threshold potential.
d. resting potential.
e. action potential.

Answer: e
Difficulty: Easy
Bloom's Taxonomy: Knowledge, Comprehension

19. When a neuron membrane potential becomes less negative it becomes

a. depolarized.
b. hyperpolarized.
c. repolarized.
d. hypopolarized.
e. none of the above

Answer: a
Difficulty: Easy
Bloom's Taxonomy: Knowledge, Comprehension

20. The minimum level of depolarization required to initiate an action potential in an excitable cell is called the

a. electrical potential.
b. refractory potential.
c. membrane potential.
d. resting potential.
e. threshold potential.

Answer: e
Difficulty: Moderate
Bloom's Taxonomy: Analysis

21. When a neuron membrane potential goes below its resting value it becomes

a. depolarized.
b. hyperpolarized.
c. repolarized.
d. hypopolarized.
e. none of the above

Answer: b
Difficulty: Easy
Bloom's Taxonomy: Knowledge, Comprehension

22. The refractory period of a neuron plasma membrane is important because

a. it establishes the resting potential.
b. it initiates an action potential.
c. it ensures that the threshold potential will be reached.
d. it ensures that an impulse will travel in a one-way direction.
e. none of the above.

Answer: d
Difficulty: Moderate
Bloom's Taxonomy: Comprehension

23. During an action potential, the membrane potential can reach as high as

a. +30 mV.
b. +40 mV.
c. +50 mV.
d. +60 mV.
e. +70 mV.

Answer: a
Difficulty: Moderate
Bloom's Taxonomy: Knowledge

24. When threshold potential is reached, which of the following occurs?

a. The activation gates of Na^+ channels open.
b. The inactivation gates of Na^+ channels open.
c. The activation gates of K^+ channels open.
d. The inactivation gates of Na^+ channels open.
e. Na^+/K^+ active transport pumps are activated.

Answer: a
Difficulty: Moderate
Bloom's Taxonomy: Comprehension, Analysis

25. When threshold potential is reached, which of the following occurs?

a. The activation gates of Na^+ channels close.
b. The inactivation gates of Na^+ channels close.
c. The activation gates of K^+ channels close.
d. The inactivation gates of Na^+ channels close.
e. Na^+/K^+ active transport pumps are activated.

Answer: b
Difficulty: Moderate
Bloom's Taxonomy: Comprehension, Analysis

26. If the K^+ channels of an excitable plasma membrane were blocked by the action of a drug, which of the following would be disrupted?

a. repolarization
b. depolarization
c. hyperpolarization
d. a and b
e. a and c

Answer: e
Difficulty: Hard
Bloom's Taxonomy: Analysis, Application

27. Depolarization of the neuron plasma membrane occurs due to

a. the diffusion of K^+ into the cell.
b. the diffusion of K^+ out of the cell.
c. the diffusion of Na^+ into the cell.
d. the diffusion of Na^+ out of the cell.
e. none of the above

Answer: c
Difficulty: Easy
Bloom's Taxonomy: Knowledge, Comprehension

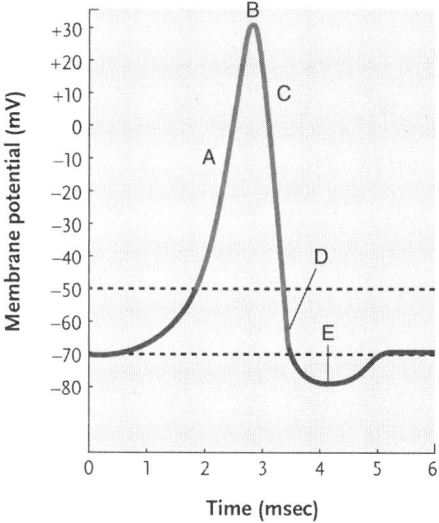

Use this figure to answer questions 28 through 29.

28. At what point in this action potential diagram are Na^+ inactivation gates opening and K^+ activation gates closing?

a. A
b. B
c. C
d. D
e. E

Answer: D
Difficulty: Hard
Bloom's Taxonomy: Comprehension, Analysis

29. At what point in this action potential diagram are many Na^+ activation gates open and K^+ activation gates closed?

a. A
b. B
c. C
d. D
e. E

Answer: a
Difficulty: Hard
Bloom's Taxonomy: Comprehension, Analysis

30. An action potential is propagated down an unmyelinated neuron plasma membrane because

a. the action potential stimulates voltage-gated Na^+ channels adjacent to it.
b. the action potential stimulates voltage-gated Ca^{2+} channels adjacent to it.
c. the action potential stimulates voltage gated K^+ channels adjacent to it.
d. a and c
e. b and c

Answer: d
Difficulty: Moderate
Bloom's Taxonomy: Knowledge, Comprehension

31. Action potentials are propagated in a one-way direction down a neuron plasma membrane because

a. the action potential is of insufficient intensity to stimulate the voltage-gated ion channels behind it.
b. the adjacent channels upstream from the action potential are ligand-gated.
c. the adjacent channels upstream from the action potential are in their refractory period.
d. all of the above
e. none of the above

Answer: c
Difficulty: Moderate
Bloom's Taxonomy: Comprehension

32. The magnitude of an action potential _____ as it is propagated down an excitable membrane. This is due to the _____.

a. increases; all-or-nothing principle
b. remains constant; all-or-nothing principle
c. increases; threshold principle
d. remains constant; threshold principle
e. none of the above

Answer: b
Difficulty: Moderate
Bloom's Taxonomy: Comprehension

33. The intensity of an electrical impulse is reflected in the

a. frequency of action potentials.
b. magnitude of action potentials.
c. duration of action potentials.
d. intensity of action potentials.
e. all of the above

Answer: a
Difficulty: Easy
Bloom's Taxonomy: Knowledge, Comprehension

34. A reasonable estimation of the duration of an action potential is

a. approximately five microseconds.
b. approximately five milliseconds.
c. approximately five centiseconds.
d. approximately five seconds.
e. none of the above

Answer: b
Difficulty: Easy
Bloom's Taxonomy: Knowledge

35. For which axon might you expect action potential propagation to be the fastest?

a. a myelinated axon with a small diameter
b. an unmyelinated axon with a small diameter
c. a myelinated axon with a large diameter
d. an unmyelinated axon with a large diameter
e. axon myelination and diameter do not affect action potential propagation rate

Answer: c
Difficulty: Difficult
Bloom's Taxonomy: Analysis

36. The spaces between adjacent Schwann cells are called

a. nodes of Ranvier.
b. axon terminals.
c. dendrites.
d. intercalated discs.
e. active gaps.

Answer: a
Difficulty: Easy
Bloom's Taxonomy: Knowledge

37. The disease multiple sclerosis causes myelin degeneration in vertebrate nervous systems. Knowing this, what symptoms/effects might you expect in a person with multiple sclerosis?

a. slowed action potential transmission
b. tissue numbness
c. muscular weakness
d. faulty coordination of movements
e. all of the above

Answer: e
Difficulty: Moderate
Bloom's Taxonomy: Comprehension, Analysis

37.3 CONDUCTION ACROSS CHEMICAL SYNAPSES

38. Place the following events in the order in which they occur for chemical synapse transmission.

(1) Ca^{2+} enters axon terminal.
(2) Ligand-gated ion channels open in postsynaptic membrane.
(3) Neurotransmitter binds to postsynaptic receptor.
(4) Action potential reaches axon terminal of presynaptic neuron.
(5) Neurotransmitter released by exocytosis.

a. 4, 1, 5, 3, 2
b. 4, 3, 1, 5, 2
c. 4, 5, 1, 3, 2
d. 4, 1, 3, 5, 2
e. 4, 5, 2, 3, 1

Answer: a
Difficulty: Moderate
Bloom's Taxonomy: Knowledge, Synthesis

39. Chemical synapses are distinctive because

a. they require the use of ligand-gated ion channels.
b. they require the use of voltage-gated ion channels.
c. they may utilize direct neurotransmitters.
d. they require the use of vesicles.
e. all of the above

Answer: e
Difficulty: Easy
Bloom's Taxonomy: Knowledge

40. Vesicles containing neurotransmitters are released into the synaptic cleft when _____ levels rise within the axon terminal.

a. Na^+
b. K^+
c. Cl^-
d. Ca^{2+}
e. H^+

Answer: d
Difficulty: Easy
Bloom's Taxonomy: Knowledge

41. What removes Ca^{2+} from the inside of an axon terminal of a chemical synapse after an electrical impulse has passed?

a. ion channels
b. active transport pumps
c. passive carrier proteins
d. exocytosis
e. simple diffusion through the plasma membrane

Answer: b
Difficulty: Easy
Bloom's Taxonomy: Knowledge, Comprehension

42. The binding of a neurotransmitter to ligand-gated K^+ channels will cause _____ in the postsynaptic membrane.

a. depolarization
b. hyperpolarization
c. repolarization
d. apolarization
e. all of the above

Answer: b
Difficulty: Moderate
Bloom's Taxonomy: Application

43. The binding of a neurotransmitter to ligand-gated Cl^- channels will cause _____ in the postsynaptic membrane.

a. depolarization
b. hyperpolarization
c. repolarization
d. apolarization
e. all of the above

Answer: b
Difficulty: Moderate
Bloom's Taxonomy: Application

44. The binding of a neurotransmitter to ligand-gated Na^+ channels will cause _____ in the postsynaptic membrane.

a. depolarization
b. hyperpolarization
c. repolarization
d. apolarization
e. all of the above

Answer: a
Difficulty: Moderate
Bloom's Taxonomy: Application

45. In vertebrates, the neurotransmitter acetylcholine

a. can act as a direct neurotransmitter between neurons and muscle cells.
b. can make it more difficult for some postsynaptic membranes to reach threshold.
c. can be released into circulation and act as a hormone.
d. can act as pain reducers and initiate euphoria.
e. a and b only

Answer: e
Difficulty: Moderate
Bloom's Taxonomy: Knowledge

46. Neuropeptide neurotransmitters

a. can act as a direct neurotransmitter between neurons and muscle cells.
b. can act as a postsynaptic membrane inhibitor.
c. directly open K+ channels to inhibit neurons.
d. can act as pain reducers and initiate euphoria.
e. none of the above

Answer: d
Difficulty: Moderate
Bloom's Taxonomy: Knowledge

INSIGHTS FROM THE MOLECULAR REVOLUTION: DISSECTING NEUROTRANSMITTER RECEPTOR FUNCTIONS

47. The neurotransmitter receptor research of Jean-Luc Eiselé and his coworkers shows that

a. neurotransmitters are dependent on the activity of Na^+.
b. neurotransmitters are dependent on the activity of K^+.
c. neurotransmitters are dependent on the activity of Cl^-.
d. a receptor binding to neurotransmitters functions independently from conducting ions.
e. none of the above

Answer: d
Difficulty: Moderate
Bloom's Taxonomy: Comprehension

37.4 INTEGRATION OF INCOMING SIGNALS BY NEURONS

48. A graded, subthreshold change in the postsynaptic membrane potential that moves it toward threshold is called a(n)

a. IPSP.
b. hyperpolarization.
c. EPSP.
d. depolarization.
e. none of the above

Answer: c
Difficulty: Moderate
Bloom's Taxonomy: Knowledge, Comprehension

49. Assume a neuron receives EPSPs and IPSPs from several adjacent neurons. How might that neuron's threshold potential be reached, causing it to produce its own action potential?

a. if EPSPs outnumber IPSPs
b. temporal summation
c. spatial summation
d. all of the above
e. none of the above

Answer: d
Difficulty: Moderate
Bloom's Taxonomy: Analysis

50. At what neuron location does summation occur?

a. dendrites
b. cell body
c. axon hillock
d. Schwann cells
e. axon terminals

Answer: c
Difficulty: Easy
Bloom's Taxonomy: Knowledge

51. What neuron location has the greatest density of voltage-gated Na$^+$ channels, resulting in the lowest threshold potential along a neuron?

a. dendrites
b. cell body
c. axon hillock
d. axon
e. axon terminals

Answer: c
Difficulty: Moderate
Bloom's Taxonomy: Knowledge

Matching

Match each of the following term with its correct description.

52. _____ graded potential

53. _____ action potential

54. _____ membrane potential

55. _____ resting potential

56. _____ IPSP

57. _____ EPSP

58. _____ threshold potential

A. An electrical potential difference across a plasma membrane

B. A subthreshold change in membrane potential toward the threshold potential of a neuron

C. The membrane potential of an excitable membrane when it is not being stimulated

D. A subthreshold change in membrane potential away from the threshold potential of a neuron

E. An abrupt, transient change in membrane potential, consisting of membrane depolarization followed by repolarization

F. Minimum membrane potential required to form an action potential

G. Any subthreshold change in membrane potential whose effects are additive

Answers: 52. G   53. E   54. A   55. C   56. D
57. B   58. F
Difficulty: Easy
Bloom's Taxonomy: Knowledge
Source: Section 37.2, 37.4

True/False

If the statement is true, write a "T" in the blank. If the statement is false, make it correct by changing the underlined word(s) and writing the correct word(s) in the answer blank.

59. _____ Depolarization occurs due to the diffusion of $\underline{K^+}$ across a neuron plasma membrane.

60. _____ EPSPs can form due to the diffusion of Na$^+$ across a neuron plasma membrane.

61. _____ Saltatory conduction is the slowest form of action potential propagation.

62. _____ The additive effect of one neuron sending many EPSPs to another neuron over a brief period of time is called spatial summation.

63. _____ Direct neurotransmitters bind to G-protein coupled receptors.

64. _____ Action potentials never vary in magnitude along a particular excitable membrane.

65. _____ Afferent neurons carry electric impulses directly to muscles and glands.

66. _____ Neurotransmitters are stored in vesicles located within axon terminals.

67. _____ Oligodendrocytes are responsible for myelinating central nervous system axons.

Answers: 59. F, Na$^+$ 60. T   61. F, fastest
62. F, temporal summation   63. F,
indirect neurotransmitters   64. T
65. F, efferent neurons   66. T
67. T

Difficulty: Moderate
Bloom's Taxonomy: Knowledge, Comprehension
Source: Section 37.1, 37.2, 37.3, 37.4

Select the Exception

68. Which of the following is *not* a component of neural signaling mechanisms?

a. reception
b. action
c. transmission
d. response
e. integration

Answer: b
Difficulty: Easy
Bloom's Taxonomy: Synthesis
Source: 37.1

69. Which of the following neurons is *not* typically located in the peripheral nervous system?

a. afferent neurons
b. sensory neurons
c. efferent neurons
d. interneurons
e. motor neurons

Answer: d
Difficulty: Easy
Bloom's Taxonomy: Synthesis, Evaluation
Source: Section 37.1

70. Which of the following structures are *not* used by chemical synapses during signal transmission?

a. gap junctions
b. ligand-gated receptors
c. Ca^{2+}
d. vesicles
e. neurotransmitters

Answer: a
Difficulty: Moderate
Bloom's Taxonomy: Synthesis, Evaluation
Source: Section 37.2

71. Which of the following does *not* involve K^+?

a. repolarization
b. hyperpolarization
c. depolarization
d. resting membrane potential
e. dampening of neural activity

Answer: c
Difficulty: Moderate
Bloom's Taxonomy: Synthesis, Evaluation
Source: Section 37.2

72. Which of the following does *not* represent a change in membrane potential?

a. EPSP
b. threshold potential
c. depolarization
d. repolarization
e. IPSP

Answer: b
Difficulty: Easy
Bloom's Taxonomy: Synthesis, Evaluation
Source: Section 37.2

73. Which of the following structures is not actively involved with saltatory conduction?

a. oligodendrocytes
b. myelin
c. Na^+/K^+ pumps
d. voltage-gated Na^+ channels
e. nodes of Ranvier

Answer: c
Difficulty: Moderate
Bloom's Taxonomy: Synthesis, Evaluation
Source: Section 37.2

74. Which of the following is *not* a possible result of EPSP & IPSP activity?

a. An unusually strong action potential forms.
b. No summation.
c. EPSPs and IPSPs cancel each other out.
d. Temporal summation.
e. Spatial summation.

Answer: a
Difficulty: Hard
Bloom's Taxonomy: Synthesis, Evaluation
Source: Section 37.4

Choice

For each of the following situations, choose the most appropriate term.

a. depolarization   b. repolarization
c. hyperpolarization

75. _____ an IPSP is an example of this

76. _____ occurs due to the diffusion of Na^+

77. _____ an EPSP is an example of this

78. _____ causes a neuron membrane potential to become more negative, allowing it to approach its resting potential

79. _____ occurs in a postsynaptic membrane when ligand-gated Na^+ channels open

80. _____ the phase of an action potential when the membrane potential becomes more positive

81. _____ begins when Na^+ channel inactivation gates close and K^+ activation gates open

82. _____ describes the change in resting membrane potential that causes it to approach threshold potential

83. _____ occurs when a membrane potential becomes more negative that its resting membrane potential

84. _____ occurs in a postsynaptic membrane when ligand-gated Cl⁻ channels open

Answers: *75. c*   *76. a*   *77. a*   *78. b*   *79. a*
            80. a   *81. b*   *82. a*   *83. c*
            84. c.

Difficulty: Easy
Bloom's Taxonomy: Knowledge, Comprehension
Source: Section 37.2, 37.4

Short Answer

85.   Explain how neuron action potential propagation rate can be increased.

Answer: Neuron action potential propagation can be increased in two ways: (1) by increasing axonal diameter, and (2) myelinating the axon. Increased axonal diameter reduces electrical resistance to current flow. Myelination causes an electrical current to "jump" from one node of Ranvier to the next by salutatory conduction, which is more rapid than conduction in an unmyelinated neuron.
Difficulty: Moderate
Bloom's Taxonomy: Knowledge, Comprehension
Source: Section 37.2

86.   Explain how the resting membrane potential and resting ion distributions are established in a neuron.

Answer: A resting membrane potential of -70 mV is established due to the unequal distribution of ions across the neuron plasma membrane. Na^+/K^+ pumps accumulate Na^+ outside the plasma membrane and K^+ inside the plasma membrane, while the presence of negatively charged anions inside the cell give the inside of the plasma membrane an overall negative charge relative to the outside .
Difficulty: Moderate
Bloom's Taxonomy: Knowledge, Comprehension
Source: Section 37.2

87.   Valium is a drug that causes voltage-gated Cl⁻ channels to open. What effects would this cause in a postsynaptic membrane?

Answer: Valium would cause hyperpolarization in the postsynaptic membrane. Thus, it would take more frequent, or a greater number, of EPSPs to generate an action potential in a postsynaptic membrane.
Difficulty: Moderate
Bloom's Taxonomy: Knowledge, Application
Source: Section 37.3

88.   A toxin interferes with the opening of Na^+ channels in the postsynaptic membrane so they open more slowly when bound to a neurotransmitter. What effect would there be on action potential formation in the postsynaptic membrane

Answer: This would cause a postsynaptic membrane action potential to form more slowly (if at all), as well as slow action potential propagation speed.
Difficulty: Moderate
Bloom's Taxonomy: Knowledge, Application
Source: Section 37.3

90.   In terms of changes in membrane potential, compare and contrast action potentials and graded potentials.

Answer: Action potentials are all-or-none phenomena that form when a membrane potential of an excitable cell reaches threshold potential. Graded potentials are additive subthreshold changes in the membrane potential of an excitable cell. The net effect of graded potentials may cause a membrane potential to move closer to, or further away from, threshold potential.
Difficulty: Moderate
Bloom's Taxonomy: Knowledge, Comprehension
Source: Section 37.4

38
Nervous Systems

Multiple-Choice

WHY IT MATTERS

1.   Detection and response to environmental cues or stimuli are primarily functions of

a.   metabolism.
b.   the nervous system.
c.   catabolism.
d.   the endocrine system.
e.   thermodynamics.

Answer: b
Difficulty: Easy
Bloom's Taxonomy: Knowledge

38.1 INVERTEBRATE AND VERTEBRATE NERVOUS SYSTEMS COMPARED

2.   In animals, cephalization is most associated with

a.   bilateral symmetry.
b.   radial symmetry.
c.   ganglia.
d.   nerve nets.
e.   nerves.

Answer: a
Difficulty: Easy
Bloom's Taxonomy: Knowledge

3.   The anatomical orientation or arrangement of the nervous system of a sea star

a.   contains a central control organ with interconnecting neurons.
b.   produces a signal that intensifies as it radiates from its point of origin.
c.   cannot coordinate movement of the arms.
d.   allows the organism to respond to stimuli approaching from any direction.
e.   sends electrical signals in only one direction.

Answer: d
Difficulty: Moderate
Bloom's Taxonomy: Comprehension

4.   Invertebrates, such as the flatworm, have groups of neurons with a common function. These nerve clusters are called

a.   nerve cords.
b.   ganglia.
c.   brains.
d.   nerve nets.
e.   radially symmetrical.

Answer: b
Difficulty: Easy
Bloom's Taxonomy: Knowledge

5.   Identify the animal with the most advanced cephalization.

a.   a flatworm
b.   an arthropod
c.   a chordate
d.   a mollusk
e.   an echinoderm

Answer: c
Difficulty: Easy
Bloom's Taxonomy: Knowledge

6.   In animals with a brain, the two major divisions of the nervous system are the

a.   autonomic and peripheral systems.
b.   central and peripheral nervous systems.
c.   brain and spinal cord.
d.   nerves from the brain and the peripheral systems.
e.   cranial and spinal nerves.

Answer: b
Difficulty: Easy
Bloom's Taxonomy: Knowledge

7.   From the following, select the animal with the least advanced nerve cord.

a.   chordate
b.   arthropod
c.   flatworm
d.   human
e.   octopus

Answer: c
Difficulty: Moderate
Bloom's Taxonomy: Comprehension

8.   Which of the following are **NOT** functions of a nervous system?

a.   communication
b.   interpretation of information
c.   detection of changes in the internal environment
d.   monitoring changes in the external environment
e.   all of these are functions of a nervous system

Answer: e
Difficulty: Easy
Bloom's Taxonomy: Knowledge

9.   A hollow neural tube will develop into all of the following **EXCEPT**

a.   fluid-filled ventricles.
b.   the forebrain, midbrain, and hindbrain.
c.   the central canal of the spinal cord.
d.   spinal nerves.
e.   the spinal cord.

Answer: d
Difficulty: Easy
Bloom's Taxonomy: Comprehension

10. Areas of the brain that are associated with higher functions, such as communication, forethought, and action, develop from the

a. hindbrain.
b. telencephalon.
c. forebrain.
d. metencephalon.
e. both b and c are correct

Answer: e
Difficulty: Moderate
Bloom's Taxonomy: Comprehension

11. Which of the following is **NOT** common to all vertebrates?

a. a cerebellum
b. a convoluted cerebrum
c. a hindbrain
d. a dorsal hollow spinal cord
e. a midbrain

Answer: b
Difficulty: Easy
Bloom's Taxonomy: Knowledge

12. Identify the region of the brain that gives rise to the shaded region in this image.

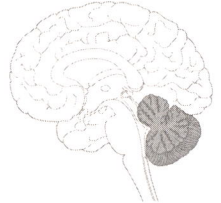

a. The diencephalon.
b. The thalamus.
c. The myelencephalon.
d. The metencephalon.
e. Both a and b are correct.

Answer: d
Difficulty: Moderate
Bloom's Taxonomy: Knowledge

38.2 THE PERIPHERAL NERVOUS SYSTEM

13. Sensory information is transmitted to the CNS by the

a. efferent neurons of the autonomic nervous system.
b. afferent neurons of the peripheral nervous system.
c. somatic neurons of the peripheral nervous system.
d. somatic neurons of the sympathetic division.
e. all of these

Answer: b
Difficulty: Moderate
Bloom's Taxonomy: Comprehension

14. All the following are characteristics of the autonomic nervous system **EXCEPT**

a. carries signals to and from visceral organs.
b. is a functional division of the peripheral nervous system.
c. includes both afferent and efferent nerve fibers.
d. controls body movements that are under conscious control.
e. is composed of parasympathetic and sympathetic divisions.

Answer: e
Difficulty: Easy
Bloom's Taxonomy: Comprehension

15. The portion of the nervous system that is responsible for increasing heart rate, as well as increasing movement of the intestines, is the

a. autonomic nervous system.
b. sympathetic division.
c. parasympathetic division.
d. somatic nervous system.
e. both a and b are correct.

Answer: a
Difficulty: Difficult
Bloom's Taxonomy: Comprehension

16. A primary difference between the somatic and autonomic nervous systems is the number of neurons between the CNS and the effector organ. If the somatic nervous system has _____, then the autonomic nervous has _____.

a. one neuron; three neurons
b. one neuron, two neurons
c. two neurons, one neuron
d. two neurons, three neurons
e. none of these

Answer: b
Difficulty: Moderate
Bloom's Taxonomy: Comprehension

17. Of the following statements concerning the sympathetic and parasympathetic branches of the nervous system, which is true?

a. The sympathetic system controls external stimuli, while the parasympathetic system controls internal stimuli.
b. The sympathetic system generally produces an increased physical activity, while the parasympathetic system produces a decrease in physical activity.
c. Both nervous systems stimulate the activities of many organs.
d. Both nervous systems release norepinephrine.
e. The sympathetic system is under voluntary control, and the parasympathetic system is under involuntary control.

Answer: b
Difficulty: Moderate
Bloom's Taxonomy: Comprehension

38.3 THE CENTRAL NERVOUS SYSTEM (CNS) AND ITS FUNCTIONS

18.   In vertebrates, the CNS consists of the

a.   brain and spinal cord.
b.   brain and gray matter.
c.   spinal cord and gray matter.
d.   brain and reflexes.
e.   spinal cord and nerves.

Answer: a
Difficulty: Easy
Bloom's Taxonomy: Knowledge

19.   If the protective coverings of the brain were damaged, the _____ would be damaged.

a.   ventricles
b.   thalamus
c.   medulla
d.   meninges
e.   reflexes

Answer: d
Difficulty: Easy
Bloom's Taxonomy: Knowledge

20.   The blood-brain barrier

a.   protects the brain and the spinal cord.
b.   is selective to which substances can enter the cerebrospinal fluid.
c.   allows oxygen, carbon dioxide, and alcohol to pass from the blood to the cerebrospinal fluid.
d.   all of these
e.   none of these

Answer: d
Difficulty: Moderate
Bloom's Taxonomy: Knowledge

21.   Which of the following brain structure–function sets are incorrectly matched?

a.   medulla oblongata – controls breathing and heart rate
b.   cerebellum – balances and coordinates muscle movement
c.   thalamus – relays sensory input to the cerebrum
d.   hypothalamus – controls voluntary movements
e.   olfactory bulbs – replays sensory information about odors to the cerebrum

Answer: d
Difficulty: Moderate
Bloom's Taxonomy: Comprehension

22.   Given the shaded region in the following illustration, if the primary motor area were activated, this region would be located in the _____ lobe.

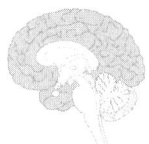

a.   frontal
b.   parietal
c.   temporal
d.   occipital
e.   cerebellar

Answer: a
Difficulty: Difficult
Bloom's Taxonomy: Application

23.   The phenomenon of brain function occurring in predominately one of the two hemispheres is

a.   consciousness.
b.   short-term memory.
c.   hemisphere association.
d.   recognition.
e.   lateralization.

Answer: e
Difficulty: Easy
Bloom's Taxonomy: Knowledge

24.   If the corpus callosum were cut, which of the following would still remain?

a.   nervous communication within each individual hemisphere
b.   nervous communication between the right and left hemispheres
c.   nervous communication between the right and left frontal lobes
d.   all of these
e.   none of these

Answer: a
Difficulty: Moderate
Bloom's Taxonomy: Comprehension

25.   The alarm rings again; you hit the snooze button, but it is broken; you try to ignore the ringing, but that sound is activating which of the following brain regions or functional groups of neurons?

a.   limbic system
b.   reticular formation
c.   blood brain activator
d.   resonance tomography system
e.   hippocampus

Answer: b
Difficulty: Moderate
Bloom's Taxonomy: Comprehension

26. From the following, identify that portion of the brain that is **NOT** considered part of the limbic system.

a. amygdala
b. olfactory bulbs
c. cerebellum
d. hippocampus
e. all of these are part of the limbic system

Answer: c
Difficulty: Moderate
Bloom's Taxonomy: Knowledge

27. One of your friends is an excellent pianist. Which of the following could be used to describe your friend?

a. The limbic system is very well developed.
b. The right hemisphere is well developed for music.
c. The left hemisphere is well developed for music.
d. The reticular formation is well developed.
e. The right hemisphere is well developed for control over the right side of the body.

Answer: b
Difficulty: Moderate
Bloom's Taxonomy: Comprehension

28. The occipital lobe of the brain is responsible for

a. speech.
b. memory.
c. smell.
d. coordination of movement.
e. vision.

Answer: e
Difficulty: Easy
Bloom's Taxonomy: Knowledge

29. With respect to the cerebrum, gray matter is

a. located in the outer region and the central region surrounding the ventricles.
b. located in the inner region.
c. composed of nerve fibers.
d. responsible for transmission of information between neurons.
e. both c and d are correct

Answer: a
Difficulty: Moderate
Bloom's Taxonomy: Knowledge

30. The limbic system plays a role in all of the following **EXCEPT**

a. the sleep wake cycle.
b. sexual behavior.
c. fighting behavior.
d. motivation.
e. evaluating rewards.

Answer: a
Difficulty: Moderate
Bloom's Taxonomy: Knowledge

31. Destruction of the motor areas in the right cerebral cortex result in the loss of

a. sensation on the right side of the body.
b. sensation on the left side of the body.
c. voluntary movement on the left side of the body.
d. voluntary movement on the right side of the body.
e. involuntary control of the right side of the body.

Answer: c
Difficulty: Difficult
Bloom's Taxonomy: Application

32. Bob is recovering from a brain injury. He is unable to speak, but he clearly understands both the written and spoken word. The area of damage that resulted in these symptoms is

a. Wernicke's area of the temporal lobe.
b. Broca's area of the frontal lobe.
c. the primary somatosensory area of the parietal lobe.
d. the cerebellular association area.
e. all of these

Answer: b
Difficulty: Difficult
Bloom's Taxonomy: Comprehension

33. The temporal lobe is associated with

a. touch and movement to that touch.
b. vision.
c. smell.
d. speech.
e. memory.

Answer: e
Difficulty: Easy
Bloom's Taxonomy: Knowledge

34. If an area of the body such as the lips or fingers is represented by a large area in the primary somatosensory area, then those areas

a. have a small number of local receptors, but a great deal of precision of touch and movement.
b. have a large number of local receptors.
c. have a great deal of precision of touch and movement.
d. have a large number of local receptors, but a small amount of precision of touch and movement.
e. both b and c

Answer: e
Difficulty: Difficult
Bloom's Taxonomy: Application

38.4 MEMORY, LEARNING, AND CONSCIOUSNESS

35. During conscious quiet rest, the brain is emitting

a. alpha waves.
b. beta waves.
c. delta waves.
d. gamma waves.
e. rapid, irregular waves.

Answer: a
Difficulty: Easy
Bloom's Taxonomy: Knowledge

36. Memories are stored in which of the following parts of the brain?

a. association areas of the cerebellum
b. limbic lobe of cerebrum
c. suprachiasmatic nucleus
d. primary sensory areas of parietal lobes
e. none of these

Answer: e
Difficulty: Easy
Bloom's Taxonomy: Knowledge

37. Bob dreamt he was biking for five hours. The actual time he was in _____ sleep was about _____ min.

a. beta; 15–20
b. EEG; 10–15
c. REM; 60–90
d. REM; 10–15
e. EEG; 90

Answer: d
Difficulty: Moderate
Bloom's Taxonomy: Comprehension

38. Memory is

a. the storage and retrieval of sensory or motor experience.
b. the response to stimuli based on experiences.
c. the awareness of ourselves and surroundings.
d. all of these
e. none of these

Answer: a
Difficulty: East
Bloom's Taxonomy: Knowledge

39. Research on memory indicates that

a. short-term memory is a product of chemical changes in neurons.
b. long-term memories are lost more frequently.
c. long-term memory is limited to a few years' duration.
d. long-term memory retention is better for skill memory than for declarative memory.
e. short-term memory is limited to several hundred bits of information.

Answer: d
Difficulty: Difficult
Bloom's Taxonomy: Comprehension

40. Your alarm clock goes off once again, and it is 7 a.m. Which of the following events leads you to hit the snooze button?

a. The number of alpha waves emitted by the brain increases.
b. The number of theta waves emitted by the brain increases.
c. The reticular formation sends stimuli to the cerebral cortex.
d. The cortex of the cerebellum is stimulated.
e. The limbic system is stimulated.

Answer: c
Difficulty: Difficult
Bloom's Taxonomy: Comprehension

41. Long-term memory

a. involves reverberating circuits.
b. is limited to multiple chunks of information over a given time.
c. requires extensive time for consolidation.
d. is dependent on the corpus callosum.
e. depends on activated receptors that are linked to G proteins.

Answer: e
Difficulty: Moderate
Bloom's Taxonomy: Knowledge

42. If an action potential persists for one minute or longer, an increase in the strength of synaptic connections develops. Which of the following describes the resulting establishment of memory between two neurons?

a. lateralization
b. consciousness potentiation
c. long-term potentiation
d. long-term association
e. cephalization association

Answer: c
Difficulty: Moderate
Bloom's Taxonomy: Comprehension

FOCUS ON RESEARCH: INVESTIGATING THE FUNCTIONS OF THE CEREBRAL HEMISPHERES

43. If the right optic nerve were cut, what would still remain?

a. the inner portion of the left visual field and the outer portion of the right visual field
b. the outer portion of the right visual field and the inner portion of the left visual field
c. the inner portion of the right visual field and the inner portion of the left visual field
d. the inner portion of the right visual field and the outer portion of the left visual field
e. the entire right visual field

Answer: d
Difficulty: Difficult
Bloom's Taxonomy: Application

44. Which of the following pertaining to a cut corpus callosum is FALSE?

a. The right hemisphere doesn't know what the left hemisphere is doing.
b. The right hemisphere directed the writing of COW but the person could not say what the word was
c. The right hemisphere directs motor activity on the left side of the body.
d. The left hemisphere processes language.
e. Information between the right and left hemispheres was shared, but reduced by 50 percent.

Answer: e
Difficulty: Moderate
Bloom's Taxonomy: Comprehension

45. If light were prevented from reaching the right side of both retinas, which of the following would occur?

a. The right half of the visual field would be lost.
b. Only the left cerebral hemisphere would be stimulated by the light.
c. No information about the light would be carried in the left optic nerve.
d. The left half of the visual field would be lost.
e. None of these.

Answer: d
Difficulty: Difficult
Bloom's Taxonomy: Application

INSIGHTS FROM THE MOLECULAR REVOLUTION: KNOCKED-OUT MICE WITH BAD MEMORY

46. If the NMDA receptor is removed

a. long-term potentiation doesn't occur.
b. short-term memory is not converted.
c. long-term memory occurs but is shortened.
d. short-term memory is lengthened.
e. both a and b are correct.

Answer: e
Difficulty: Moderate
Bloom's Taxonomy: Comprehension

47. The NMDA receptor

a. is primarily located on membranes in the amygdale.
b. is associated with a sodium ion channel.
c. inhibits the CaMK11 enzyme.
d. is involved with conversion of short-term memory to long-term memory.
e. both a and b are correct

Answer: d
Difficulty: Moderate
Bloom's Taxonomy: Knowledge

48. With an intact NMDA receptor, predict the effect if the flow of calcium were blocked.

a. Long-term potentiation would occur, but short term memory would not be converted to long-term memory.
b. Phosphate groups would not be attached to CaMKII.
c. The hippocampus would be responsible for long-term potentiation.
d. CaMKII would not be responsive to phosphate.
e. Both a and d are correct.

Answer: b
Difficulty: Difficult
Bloom's Taxonomy: Application

UNANSWERED QUESTIONS

49. With respect to other mammals, the human brain is _____ relative to body size.

a. smaller
b. larger
c. equal
d. more complex
e. none of these

Answer: b
Difficulty: Easy
Bloom's Taxonomy: Knowledge

50. The HAR1 gene region produces RNA that

a. codes for structural RNA rather than a protein.
b. codes for a protein, an enzyme that produces structural RNA.
c. produces RNA splicing.
d. both a and c
e. both b and c

Answer: a
Difficulty: Moderate
Bloom's Taxonomy: Knowledge

Integrative Multiple-Choice

51. During a fight, a boxer received a hard blow to the side of his head. The boxer was unable to recognize and interpret words. Which area of the brain was most likely injured?

a. the left occipital lobe
b. the right cerebellar hemisphere
c. the left temporal lobe
d. the right temporal lobe
e. the left prefrontal lobe

Answer: c
Difficulty: Moderate
Bloom's Taxonomy: Comprehension
Section: 38.2, 38.3

52.   During a fight, a boxer received a hard blow to the side of his head. The boxer was unable to recognize and interpret words. Which specific area of the brain must have been damaged during the fight?

a.   the cortex of the cerebellum
b.   Broca's area
c.   the corpus callosum
d.   Wernicke's area
e.   both b and d

Answer: d
Difficulty: Moderate
Bloom's Taxonomy: Comprehension
Section: 38.2, 38.3

53.   A person suffers a stroke. A blood clot reduced blood flow in the brain. The person is unable to say words, but is able to read and understand text. Which of the following best explains this person's problem?

a.   The cortex of the cerebellum was affected by the reduced blood flow.
b.   The interruption in blood flow affected the corpus callosum.
c.   The region of damage was responsible for integration of visual and auditory stimuli.
d.   The reduced blood flow damaged Wernicke's area of the temporal lobe.
e.   The reduced blood flow damaged Broca's area of the temporal lobe.

Answer: e
Difficulty: Difficult
Bloom's Taxonomy: Comprehension, Application
Source: Section 38.3, 38.4

54.   Skill memories, such as the ability to learn to play the piano, involve all of the following **EXCEPT**

a.   storage of names and places.
b.   lateralization.
c.   repetition of the skill.
d.   the cerebellum, basal ganglia, and primary motor area.
e.   conscious effort to learn the skill.

Answer: a
Difficulty: Moderate
Bloom's Taxonomy: Knowledge, Comprehension
Source: Section 38.3, 38.4

55.   If a bear walks into the room you are in, which of the following predominates and causes an increase in your heart rate and breathing to help you run away?

a.   the central nervous system.
b.   the sympathetic nervous system
c.   the parasympathetic nervous system
d.   all of these
e.   none of these

Answer: b
Difficulty: Moderate
Bloom's Taxonomy: Comprehension
Source: Section 38.1, 38.2, 38.3

56.   In an experiment, the diencephalon was prevented from developing in a rat. Predict the most likely consequences of this procedure.

a.   The rat was not able to make noises.
b.   The rat was unable to move.
c.   The rat was unable to detect touch.
d.   The rat moved in a circular fashion.
e.   The rat could not right itself.

Answer: c
Difficulty: Difficult
Bloom's Taxonomy: Comprehension
Source: Sections 38.1, 38.3

57.   Your finger touches a hot pot on the stove. Information is carried via _____ neurons to the _____, where neurons carry information to the _____, making you aware of the hot pot.

a.   afferent; basal ganglia; primary sensory area
b.   efferent; thalamus; sensory association area
c.   efferent; hippocampus; primary sensory area
d.   afferent; thalamus; primary somatosensory area
e.   interneurons; thalamus; somatomotor area

Answer: d
Difficulty: Moderate
Bloom's Taxonomy: Knowledge, Comprehension
Source: Sections 38.2, 38.3

Matching

Match each of the following terms with its correct definition.

58.   _____   peripheral nervous system

59.   _____   ventricles

60.   _____   gray matter

61.   _____   reflexes

62.   _____   brain stem

63.   _____   PET

64.   _____   learning

65.   _____   Broca's area

66.   _____   nerve net

67.   _____   somatic nervous system

68.   _____   delta waves

69.   _____   sensitization

70.   _____   theta waves

A.   Loose meshwork of neurons organized with radial symmetry

B.   Cavities in the vertebrate brain filled with cerebrospinal fluid

C.   EEG pattern that is even, slow, and common during the transition to deep sleep

D.   Increased responsiveness to mild stimuli after a strong stimulus—a form of memory

E. Efferent portion of the PNS dealing with body movements that are under conscious, voluntary control

F. Nerve cell bodies and dendrites

G. Positron emission tomography

H. EEG pattern that is pulsating and less regular

I. Response to a stimulus based on information stored in memory

J. Neuronal activity that occurs without conscious effort, often associated with protection

K. Carries nervous activity from the brain and spinal cord to effector structures

L. Composed of the midbrain, pons, and medulla

M. Import with the ability to speak

Answers: *58. K*   *59. B*   *60. F*   *61. J*   *62. L*
             63. G   *64. I*   *65. M*   *66. A*   *67. E*
             68. C   *69. D*   *70. H*

Difficulty: Moderate
Bloom's Taxonomy: Knowledge, Comprehension
Source: Sections 38.1, 38.2, 38.3, 38.4

Choice

For each of the following structures or functions, choose the most appropriate region of the neuron tube.

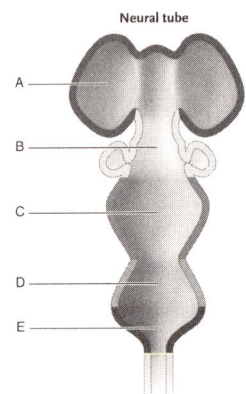

Neural tube

71. _____ cerebellum

72. _____ medulla oblongata

73. _____ midbrain

74. _____ myelencephalon

75. _____ receives information from muscles and joints, and affects balance and coordination

76. _____ bridge between the cerebellum and cerebrum

77. _____ association areas

78. _____ brain stem

79. _____ sensory relay area

80. _____ mesencephalon

81. _____ major center of homeostasis

82. _____ forebrain

83. _____ location of corpus callosum

84. _____ long-term memory

85. _____ heart rate and blood pressure, and other involuntary activities

86. _____ location of reticular formation

Answers: *71. D*   *72. E*   *73. C*   *74. E*   *75. D*
             76. D   *77. A*   *78. D,E*   *79. B*
             80. C   *81. B*   *82. A,B*   *83. A*
             84. A   *85. E*   *86. D,E*

Difficulty: Moderate
Bloom's Taxonomy: Knowledge, Comprehension
Source: Section 38.1, 38.3

Short Answer

87. Define long-term potentiation.

Answer: Series of events involved in the transfer of memory from short-term to long-term. Increased number and strength of synaptic connections occur with repeated stimulation.
Difficulty: Moderate
Bloom's Taxonomy: Knowledge, Comprehension
Source: Section 38.1. 38.2. 38.3

88. Define and explain the importance of the blood-brain barrier.

Answer: The blood-brain barrier is a structural system of tight cellular junctions. Many substances are prevented from passing from the blood to the neurons because of these cellular junctions. This provides protection for the neurons from microorganisms and toxins.
Difficulty: Moderate
Bloom's Taxonomy: Knowledge, Comprehension
Source: Section 38.3

True/False

If the statement is true, write a "T" in the blank. If the statement is false, make it correct by changing the underlined word(s) and writing the correct word(s) in the answer blank.

89. _____ The cell body of the second neuron of the parasympathetic division is located <u>outside</u> of the CNS.

90. _____ Cephalization is associated with <u>bilateral</u> symmetry.

91. _____ <u>REM</u> is the pattern of the EEG associated with being wide awake.

92. _____ If the <u>cerebellum</u> is damaged, an individual would have problems with balance.

93. _____ <u>Arthropods'</u> nervous system is composed of a nerve net.

94. _____ <u>Gray</u> matter is composed of axons with myelin sheaths.

95. _____ The autonomic nervous system is composed of the <u>afferent</u> neurons.

96. _____ The <u>limbic system</u> is involved in sleep-wake cycles.

97. _____ <u>Smell</u> is important in activating the limbic system.

Answers: **89. T   90. T   91. F, beta waves   92. T   93. F, animals with radial symmetry   94. F, white   95. F, afferent and efferent.   96. F, reticular formation   97. T**

Difficulty: Difficult
Bloom's Taxonomy: Knowledge
Source: Section 38.1. 38.2, 38.3, 38.4

Essay

98. Explain the function of the association areas of the cerebral hemispheres.

*Answer: **The primary sensory and motor areas have regions of gray matter and associated white matter that store information. This information is integrative and is involved with learning and memory of senses and motor skills. Wernicke's and Broca's areas are association areas important with language— understanding and speaking words. Other examples would be vision and smell association areas. If a person is exposed to a new sight or smell, the new information is compared to previous sights or smells stored in the association areas for faster recognition and understanding.***
Difficulty: Moderate
Bloom's Taxonomy: Knowledge, Comprehension
Source: Section 38.3

99. Distinguish between the central and peripheral nervous systems.

*Answer: **The central nervous system involves the brain and spinal cord where integration of neural activity occurs. The peripheral nervous system is involved with transmission of information to the central nervous system— the afferent division— and transmission of information away from the central nervous system— the efferent division— to target structures and organs.***
Difficulty: Moderate
Bloom's Taxonomy: Knowledge, Comprehension
Source: Section 38.1, 38.2, 38.3

39

Sensory Systems

Multiple-Choice

WHY IT MATTERS

1. Bats use high ultrasonic sounds to perceive their environment much like humans use radar. In animals, this process of detecting objects is called

a. sound reflection navigation.
b. extra sensory perception.
c. echolocation.
d. ultralocation.
e. auditory navigation.

Answer: c
Difficulty: Easy
Bloom's Taxonomy: Knowledge

2. One strategy moths have for evading bats is

a. the ability to become invisible to the bat's echolocation signals.
b. straight-line flight patterns.
c. a periodic closed-wing free-fall before resuming an erratic flight pattern.
d. an erratic flight pattern, but no instances of free-fall.
e. flying primarily in the daylight hours.

Answer: c
Difficulty: Easy
Bloom's Taxonomy: Knowledge

39.1 OVERVIEW OF SENSORY RECEPTORS AND PATHWAYS

3. Which of the following is **not** a stimulus that can be perceived by sensory signals?

a. radiation
b. heat
c. body position
d. magnetic fields
e. electrical fields

Answer: a
Difficulty: Easy
Bloom's Taxonomy: Knowledge

4. Organisms can perceive different intensities of a stimulus by

a. an increase in the frequency of action potentials of an afferent neuron.
b. an increase in intensity of the action potentials of an afferent neuron.
c. an increase in the numbers of afferent neurons generating action potentials.
d. an increase in intensity and frequency of action potentials as well as an increase in the number of cells generating action potentials.
e. an increase in the frequency of action potentials generated by an afferent neuron as well as an increase in the numbers of afferent neurons generating action potentials, but without any change in the intensity of individual action potentials generated by the neurons.

Answer: e
Difficulty: Moderate
Bloom's Taxonomy: Knowledge

5. Sensory adaptation

a. increases one's sensitivity to *changes* in environmental stimuli.
b. increases one's sensitivity to constant stimuli.
c. is primarily associated with pain receptors.
d. functions the same way in all types of sensory receptors.
e. is unique to predator animals.

Answer: a
Difficulty: Moderate
Bloom's Taxonomy: Knowledge

39.2 MECHANORECEPTORS AND THE TACTILE AND SPATIAL SENSES

6. Mechanoreceptors for touch and pressure are located in numerous locations in the vertebrate body. Which of the following is **not** a location of such mechanoreceptors in vertebrates?

a. skin
b. skeletal muscles
c. internal organs
d. walls of blood vessels
e. bones

Answer: e
Difficulty: Easy
Bloom's Taxonomy: Knowledge

7.   Many aquatic invertebrates use this structure to perceive changes in their body's position and orientation.

a.   statocysts
b.   lateral line
c.   stereocilia
d.   saccule
e.   otoliths

Answer: a
Difficulty: Moderate
Bloom's Taxonomy: Knowledge

8.   An invertebrate statocyst contains _____, which surround movable _____. Changes in body position move the latter and thus trigger action potentials.

a.   statoliths; sensory hair cells
b.   sensory hair cells; statoliths
c.   otoliths; sensory hair cells
d.   efferent neurons; otoliths
a.   efferent neurons; statoliths

Answer: b
Difficulty: Moderate
Bloom's Taxonomy: Knowledge

9.   Fishes' lateral line is comprised of multiple _____ that allow fish to detect vibrations and water currents.

a.   statoliths
b.   neuromasts
c.   statocysts
d.   stereocilia
a.   cupulas

Answer: b
Difficulty: Easy
Bloom's Taxonomy: Knowledge

10.   The loss of function of this sensory structure renders fish unable to school.

a.   vestibular apparatus
b.   statocyst
c.   lateral line
d.   eyes
a.   saccule

Answer: c
Difficulty: Easy
Bloom's Taxonomy: Knowledge

11.   The structure that detects stretch and compression forces IN the tendon are called

a.   Golgi tendon organs.
b.   Golgi apparatus.
c.   stretch receptors.
d.   acellular proprioceptor.
a.   muscle spindles.

Answer: a
Difficulty Easy
Bloom's Taxonomy: Knowledge

39.3 HEARING

12.   An earthworm detects sound

a.   though specialized tympanic membranes located near the head.
b.   through a thinned region of the exoskeleton.
c.   through general mechanoreceptors on its skin.
d.   by using a specialized organ that measures the frequency of vibration of the surrounding dirt particles.
e.   through a series of mechanoreceptors arranged in a lateral line.

Answer: c
Difficulty: Easy
Bloom's Taxonomy: Knowledge

13.   An insect such as a moth or cricket detects sound

a.   though specialized tympanic membranes located near the head.
b.   through a thinned region of the exoskeleton.
c.   through general mechanoreceptors on its skin.
d.   by using a specialized organ that measures the frequency of vibration of the surrounding air particles.
e.   through a series of mechanoreceptors arranged in a lateral line.

Answer: b
Difficulty: Easy
Bloom's Taxonomy: Knowledge

14.   In humans, where is the tympanic membrane?

a.   in the outer ear, adjacent to the pinna
b.   deep within the middle ear
c.   between the ear canal and the middle ear's cavity
d.   adjacent to the stapes
e.   between the middle ear and inner ear

Answer: c
Difficulty: Easy
Bloom's Taxonomy: Knowledge

15.   What is the combined function of the malleus, incus, and stapes?

a.   convert the mechanical vibration of the tympanic membrane into a pressure wave in the cochlear duct
b.   amplify the vibrations of the ear drum
c.   equalize the pressure on both sides of the tympanic membrane
d.   convert the oval window's vibrations into vibrations of the tympanic membrane
e.   all of the above

Answer: b
Difficulty: Easy
Bloom's Taxonomy: Knowledge

16. In a healthy human ear, where is fluid found?

a. in the middle and inner ear
b. in the vestibular canal and tympanic canals only
c. in the outer ear canal, middle ear, and cochlea
d. in the vestibular canal, tympanic canal, and cochlear duct
e. in the vestibular canal, tympanic canal, and middle ear

Answer: b
Difficulty: Easy
Bloom's Taxonomy: Knowledge

Questions 17–20 refer to the figure below.

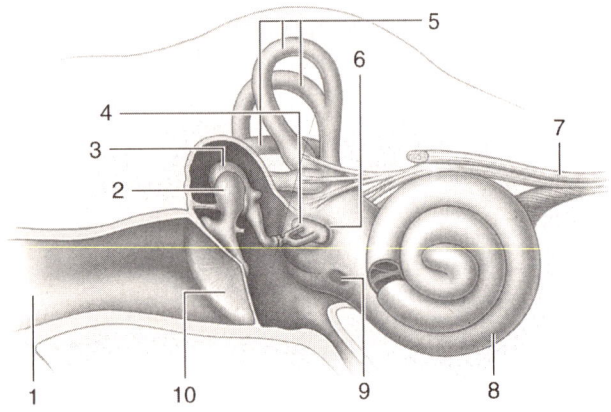

17. In the figure above, which number identifies the tympanic membrane?

a. 4
b. 5
c. 6
d. 8
e. 10

Answer: e
Difficulty: Moderate
Bloom's Taxonomy: Knowledge

18. In the figure above, which number identifies the location where the Organ of Corti might be found?

a. 2
b. 4
c. 6
d. 8
e. 10

Answer: d
Difficulty: Moderate
Bloom's Taxonomy: Knowledge

19. In the figure above, which number identifies the structure that vibrates and *directly* creates the pressure waves in the cochlea's fluid?

a. 1
b. 2
c. 3
d. 4
e. 5

Answer: d
Difficulty: Moderate
Bloom's Taxonomy: Knowledge

20. In the figure above, which number identifies the structure that transmits auditory signals to the brain?

a. 2
b. 6
c. 7
d. 8
e. 9

Answer: c
Difficulty: Easy
Bloom's Taxonomy: Knowledge

Questions 21–22 refer to the figure below.

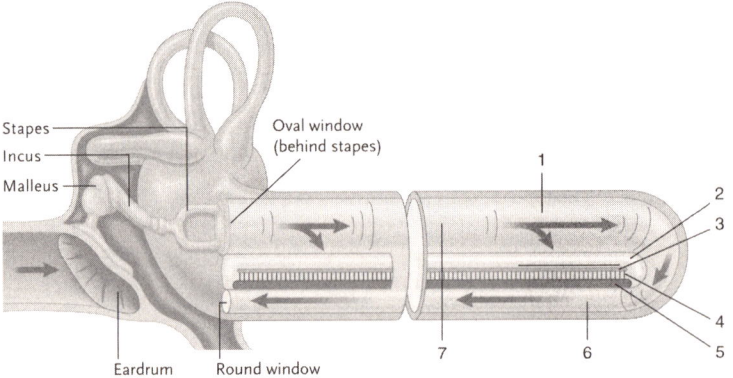

21. In the figure above, which structure vibrates and bends the sensory hairs?

a. 1
b. 2
c. 3
d. 4
e. 5

Answer: e
Difficulty: Moderate
Bloom's Taxonomy: Knowledge, Comprehension

22. In the figure above, which label identifies the sensory hair cells?

a. 1
b. 2
c. 4
d. 5
e. 6

Answer: c
Difficulty: Easy
Bloom's Taxonomy: Knowledge

23. How is it that humans can perceive multiple sounds at once?

a. Each portion of the basilar membrane vibrates at a unique frequency, and multiple regions can be stimulated at the same time.
b. The more frequencies of sounds that are present, the further the sound travels into the cochlea.
c. We have two ears, each perceiving a subset of frequencies in the environment.
d. The fluid in the middle ear can transmit multiple frequencies to the oval window.
e. None of the above explain this ability.

Answer: a
Difficulty: Moderate
Bloom's Taxonomy: Knowledge, Comprehension

39.4 PHOTORECEPTORS AND VISION

24. The simplest eye structure, which is able to perceive light but not form an image, is called a

a. compound eye.
b. single-lens eye.
c. ocellus.
d. ommatidia.
e. photoreceptor.

Answer: c
Difficulty: Easy
Bloom's Taxonomy: Knowledge

25. Which eye is the best at detecting motion?

a. compound eye
b. single-lens eye
c. ocellus
d. ommatidia
e. photoreceptor

Answer: a
Difficulty: Easy
Bloom's Taxonomy: Knowledge

26. What is the function of the iris?

a. control the amount of light entering the eye
b. focus the image on the retina
c. respond to particular colors
d. keep the lens under tension to allow for image focusing
e. prevent light scattering upon entry to the eye

Answer: a
Difficulty: Easy
Bloom's Taxonomy: Knowledge

27. In the context of vision, what does accommodation mean?

a. lack of receptor response to continued light signal
b. changing the lens shape
c. moving the lens closer to or further away from the retina in order to focus the image
d. keeping the lens under tension to allow for image focusing
e. preventing light from scattering upon entry to the eye

Answer: b
Difficulty: Easy
Bloom's Taxonomy: Knowledge

28. Sensory input from eyes and ears travels in neurons to the brain. How is it that the different stimuli are perceived as light and sound, respectively?

a. Light receptors send neural impulses more frequently than sound receptors.
b. The sensory receptors themselves are different.
c. The information from each type of receptor is sent to a different region of the brain.
d. The physical location of the sensory receptors is different, ensuring a different "sense" is perceived in the brain.
e. The neurons involved in the sensory pathways differ in shape and size.

Answer: c
Difficulty: Moderate
Bloom's Taxonomy: Knowledge, Comprehension

29. What would be the predicted outcome for a person having eyes that lack a lens but are otherwise fully functional?

a. They would be totally blind.
b. They would be able to perceive light and even respond to different light intensities, but that is all.
c. They would be able to perceive light and even colors (solid color over a large area), but not details.
d. They would be able to see some details and read print, but only if the text were written in large, black letters.
e. They would have normal vision if given glasses.

Answer: c
Difficulty: Moderate
Bloom's Taxonomy: Knowledge, Comprehension, Application

30. Where does the lens *focus* an image in the mammalian eye?

a. on the retina, using the entire surface of the retina
b. primarily on the fovea
c. on the rods only
d. slightly in front of the retina, so the neural cells don't distort the image
e. it depends on where you are looking; sometimes on the fovea, sometimes elsewhere on the retina

Answer: b
Difficulty: Moderate
Bloom's Taxonomy: Knowledge, Comprehension

31. Which of the following is a correct statement?

a. Cones respond to particular wavelengths (colors), rods perceive light at low intensity without color perception.
b. Rods respond to particular wavelengths (colors), cones perceive light at low intensities without color perception.
c. Cones are more sensitive to single photons of light than are rods.
d. Nocturnal animals have a high concentration of cones in their eyes and relatively few rods.
e. Humans have approximately equal numbers of rods and cones, but the cones are primarily localized in the fovea.

Answer: a
Difficulty: Moderate
Bloom's Taxonomy: Knowledge, Comprehension

32. This portion of the cell responds to red light photons only.

a. outer segment of rod
b. outer segment cone
c. inner segment of rod
d. inner segment of cone
e. synaptic terminal of rod or cone

Answer: b
Difficulty: Easy
Bloom's Taxonomy: Knowledge

33. Which of the following is a correct statement?

a. Rhodopsin is active in the dark and thereby allows the release of neurotransmitters from the synaptic terminal in the absence of light.
b. Rhodopsin is inactive in the dark and thereby inhibits the release of neurotransmitters from the synaptic terminal unless light is present.
c. Rhodopsin is active in the dark and thereby inhibits the release of neurotransmitters from the synaptic terminal in the absence of light.
d. Rhodopsin is inactive in the dark and thereby allows the release of neurotransmitters from the synaptic terminal unless light is present.
e. Rhodopsin's activity is different between rods and cones, but leads always to increasing amounts of neurotransmitters being release.

Answer: d
Difficulty: Moderate
Bloom's Taxonomy: Knowledge, Comprehension

34. How is active retinal returned to the inactive state?

a. It isn't; rather, the cell degrades the active form and synthesizes new, inactive proteins.
b. It is converted back to the inactive confirmation by a series of enzymes.
c. It reverts back to the inactive state on its own.
d. It is transported to a lysosome.
e. The mechanism is not known.

Answer: b
Difficulty: Moderate
Bloom's Taxonomy: Knowledge, Comprehension

Use the figure below for questions 35–37.

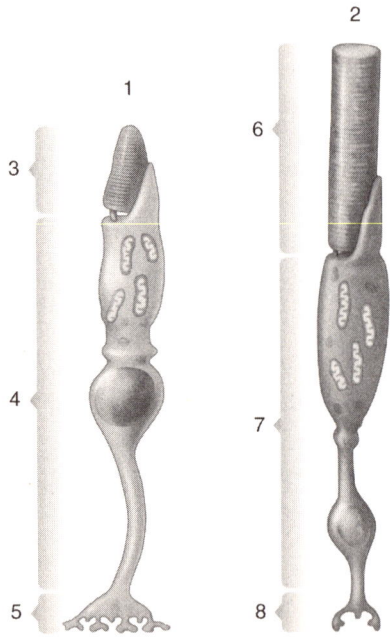

35. In the figure above, which number identifies the a portion of the cell(s) that will respond to wavelengths of *red light only*?

a. 3
b. 4
c. 5
d. 6
e. 7

Answer: a
Difficulty: Moderate
Bloom's Taxonomy: Knowledge, Comprehension

36. In the figure above, which number identifies the cell(s) or portion of the cell(s) that will respond to a single photon?

a. 1 only
b. 2 only
c. 4 and 7
d. 5 and 8
e. 3 and 6

Answer: b
Difficulty: Moderate
Bloom's Taxonomy: Knowledge, Comprehension

37. In the figure above, which number identifies the portion of the cell(s) that will release a neurotransmitter when stimulated by light?

a. 1 only
b. 2 only
c. 1 and 2
d. 5 and 8
e. Neither; photons *inhibit* neurotransmitter release by both of these types of cells.

Answer: e
Difficulty: Difficult
Bloom's Taxonomy: Knowledge, Comprehension

38. Which cell is closest to the center/interior of the human eye?

a. optic ganglion cells
b. amacrine cells
c. bipolar cells
d. rods and cones
e. horizontal cells

Answer: a
Difficulty: Easy
Bloom's Taxonomy: Knowledge

39. Which cells connect different photoreceptor cells to each other?

a. optic ganglion cells
b. amacrine cells
c. bipolar cells
d. rods and cones
e. horizontal cells

Answer: e
Difficulty: Moderate
Bloom's Taxonomy: Knowledge

40. Which cells relay messages from photoreceptor cells to non-photoreceptor cells?

a. optic ganglion cells
b. amacrine cells
c. bipolar cells
d. rods and cones
e. horizontal cells

Answer: c
Difficulty: Moderate
Bloom's Taxonomy: Knowledge

41. Each cone responds primarily to a single color. What kinds of cones do normal humans have?

a. Three cones that perceive red, blue, and yellow light.
b. Three cones that perceive red, blue, and green light.
c. Three cones that perceive orange, green, and purple light.
d. Four cones that perceive red, blue, yellow, and green light.
e. All cones respond equally to all colors of the visible light spectrum.

Answer: b
Difficulty: Moderate
Bloom's Taxonomy: Knowledge, Comprehension

42. Some animals, like humans, have two eyes on the front of the head that perceive a mostly overlapping visual field. In contrast, some herbivores (rabbits, for example) have eyes on the sides of their head and visual fields that may not overlap at all. How does this impact perception of the environment?

a. Humans have better depth perception and a wider field of vision than rabbits.
b. Humans have poorer depth perception and a narrower field of vision than rabbits.
c. Humans have better depth perception and a narrower field of vision than rabbits.
d. Humans have poorer depth perception and a wider field of vision than rabbits.
e. Due to integration by the brain, there is really no difference in depth perception or visual field size when comparing humans to rabbits.

Answer: c
Difficulty: Moderate
Bloom's Taxonomy: Knowledge, Comprehension

39.5 CHEMORECEPTORS

43. A sensillum is

a. a hollow hair-like tube characterized by a pore on the end and multiple chemoreceptor cells inside.
b. located on the antennae, foot, or mouthparts of aquatic insects.
c. a specialized smell receptor in invertebrates.
d. common to earthworms, insects, and cnidarians.
e. all of the above

Answer: a
Difficulty: Difficult
Bloom's Taxonomy: Knowledge

44. The difference between taste and smell

a. is how the receptor comes into contact with the molecule, bc it dirct touch or through diffusion in the air.
b. depends on the cellular structure of the receptor, specifically, whether the cells are derived from microvilli or cilia and have microfilaments and microtubules, respectively.
c. is essentially absent in some invertebrates that use one receptor for both functions.
d. a and b only
e. all of the above

Answer: e
Difficulty: Moderate
Bloom's Taxonomy: Knowledge

45. Taste buds are

a. really just modified sensilla.
b. found only on the tongue.
c. collections of papilla that each respond to different stimuli.
d. small capsules of cells, each with a pore at the top though which sensory hairs project and interact with environmental molecules.
e. *each* able to respond to sweet, sour, salty, bitter, and savory (umami) stimuli.

Answer: d
Difficulty: Moderate
Bloom's Taxonomy: Knowledge

46. The only receptors to make direct contact with brain interneurons rather than afferent neurons are

a. olfactory receptors.
b. taste receptors.
c. chemoreceptors.
d. mechanoreceptors.
e. photoreceptors.

Answer: a
Difficulty: Moderate
Bloom's Taxonomy: Knowledge

39.6 THERMORECEPTORS AND NOCIRECEPTORS

47. What is the function of pain?

a. It causes an organism to move away from or otherwise decrease exposure to a damaging stimulus.
b. It elicits a reflex response.
c. It protects an organism from encountering harmful conditions.
d. It works faster than conscious thought.
e. All of the above.

Answer: a
Difficulty: Easy
Bloom's Taxonomy: Knowledge

48. What are thermoreceptors used for?

a. Some animals use thermoreceptors to locate warm-blooded prey.
b. Some animals use thermoreceptors to help regulate their own body temperature.
c. Some animals use thermoreceptors to reduce exposure to extreme temperatures that are damaging to cells and tissues.
d. Some animals use thermoreceptors to trigger involuntary responses related to body temperature.
e. All of the above.

Answer: e
Difficulty: Easy
Bloom's Taxonomy: Knowledge

INSIGHTS FROM THE MOLECULAR REVOLUTION: HOT NEWS IN TASTE RESEARCH

49. Capsicum binds to a calcium channel in the nociceptors cell and triggers an influx of calcium into the cell. What happens if the same receptors are placed in an environment that is unusually warm (such as 48°C) but lacks capsicum?

a. The channels again open, allowing an influx of calcium.
b. The channels stay closed.
c. The channels open, but much more slowly than they would if capsicum is also present.
d. The channels open and are unable to close.
e. The channels are unable to open.

Answer: a
Difficulty: Easy
Bloom's Taxonomy: Knowledge

39.7 ELECTRORECEPTORS AND MAGNETORECEPTORS

50. Electroreceptors _____.

a. are only used by fishes for navigational purposes
b. can be used for communication between primates
c. are only able to receive, not produce, electrical signals
d. can be used to generate electrical fields in some invertebrates having special electric organs
e. are usually used in communication, navigation, and hunting in the organisms that possess them

Answer: e
Difficulty: Moderate
Bloom's Taxonomy: Knowledge

51. Magnetoreceptors _____.

a. are well characterized and understood
b. are known to rely on the mineral magnetite
c. have been theorized to help many types of animals navigate huge distances in their annual migrations
d. have been studied primarily in homing pigeons
e. all of the above

Answer: c
Difficulty: Easy
Bloom's Taxonomy: Knowledge

FOCUS ON RESEARCH: EXPERIMENTAL RESEARCH: DEMONSTRATION THAT MAGNETORECEPTORS PLAY A KEY ROLE IN LOGGERHEAD SEA TURTLE MIGRATION

52. How was the navigation ability of young loggerhead sea turtles studied by Kenneth Lohmann?

a. He placed radio transmitters on the shells of numerous hatchlings and tracked their migration patterns in the Atlantic ocean.
b. He genetically engineered turtles so they lacked magnetoreceptors and then studied their behavior compared to a control group.
c. He subjected turtles to an electromagnetic field opposite that of Earth's magnetic field, and then tracked the direction the hatchlings swam.
d. The loggerhead turtles were injected with magnetite to see what impact it had on their navigation.
e. Small magnets were glued to the hatchlings' shells, and their behaviors were monitored.

Answer: c
Difficulty: Easy
Bloom's Taxonomy: Knowledge

Integrative Multiple-Choice

53. Jalapeño peppers contain the chemical capsaicin. If you lack the nociceptors that bind capsaicin, what would be the predicted outcome?

a. You would be especially sensitive to spicy foods containing capsaicin.
b. You would be able to eat foods containing high concentrations of capsaicin and not feel any burning in your mouth.
c. You would have the same sensitivity to capsaicin as most other people.
d. You would experience a feeling of cold when eating foods containing capsaicin.
e. You would be more easily addicted to capsaicin-containing foods than normal individuals.

Answer: b
Difficulty: Moderate
Bloom's Taxonomy: Knowledge, Application
Source: Sections 7.1, 39.6

Matching

Match each of the following terms with its correct definition.

54. _____ sensory receptor
55. _____ sensory transduction
56. _____ sensory adaptation
57. _____ proprioceptors
58. _____ vestibular apparatus
59. _____ otoliths
60. _____ muscle spindles
61. _____ oval window
62. _____ tympanic membrane
63. _____ Eustachian tube
64. _____ ommatidia
65. _____ rhodopsin
66. _____ aqueous humor
67. _____ vitreous humor
68. _____ fovea
69. _____ amacrine cells
70. _____ lateral inhibition
71. _____ horizontal cell
72. _____ photopsins
73. _____ optic chiasm
74. _____ lateral geniculate nuclei
75. _____ sensilla
76. _____ electroreceptors
77. _____ magnetoreceptors

A. A single unit of a compound eye

B. Allow animals to detect and use Earth's magnetic field for navigation

C. Creates pressure waves in the cochlea as it vibrates

D. A reduction in the frequency of action potentials generated by afferent neurons even though the stimulus has not changed

E. Commonly known as the ear drum

F. Specialized receptors that detect electrical fields.

G. The clear, viscous liquid filling the human eye cavity between the retina and lens

H. The terrestrial vertebrates' method of perceiving motion and position of the head

I. The process of converting a stimulus into a change in the membrane potential

J. Cells in the human eye that connect bipolar cells and ganglion cells

K. Dendrites of afferent neurons, or specialized cells that synapse with afferent neurons, that can gather information about the environment

L. Calcium carbonate crystals that bend stereocilia as a result of movement of the head

M. Eliminates pressure differentials that can occur between the outer and middle ear

N. The process by which the contrast between dark and light is increased in visual processing in the mammalian eye

O. A family of photopigments consisting of opsins combined with retinal

P. A photopigment found in vertebrate and invertebrate eyes alike

Q. The liquid between the cornea and lens of a human eye

R. Cells in the human eye that connect different photoreceptor cells to each other

S. Where a portion of each optic nerve crosses to the opposite side of the brain

T. Detect stimuli used for balance and the perception of body position

U. A small region in the retina where cones are concentrated

V. Specialized muscle cells surrounded by afferent neuron dendrites; detect the degree of stretching a muscle is undergoing

W. The site in the thalamus where most optic axons make synapses with interneurons leading to the visual cortex

X. hollow sensory bristles in insects

Answers: *54. K   55. I   56. D   57. T   58. H*
59. L   60. V   61. C   62. E   63. M
64. A   65. P   66. Q   67. G   68. U
69. J   70. N   71. R   72. O   73. S
74. W   75. X   76. F   77. B

Difficulty: Moderate
Bloom's Taxonomy: Knowledge
Source: Section 39.2–39.7

Classification

Use the five receptor types listed below for questions 78–87.

    a. mechanoreceptors
    b. photoreceptors
    c. chemoreceptors
    d. thermoreceptors
    e. nociceptors

78. _____ stimulus may result in a sensation of pain

79. _____ detects tissue damage

80. _____ used to detect motion of an organism's own body

81. _____ perceives specific chemical conditions, such as alkalinity

82. _____ detects specific molecules, such as glucose

83. _____ detects the flow of heat energy

84. _____ perceives temperature changes on the body's surface

85. _____ would detect an increase in pressure against the skin

86. _____ perceives light energy

87. _____ perceives damage in some internal organs

Answers: *78. e   79. e   80. a   81. c   82. c*
83. d   84. d   85. a.   86. b   87. e.

Difficulty: Easy
Bloom's Taxonomy: Knowledge
Source: Section 39.1

Short Answer

88. Define sensory adaptation.

Answer: Sensory adaptation is when a constant stimulus results in a reduction of the frequency of the action potentials generated by the afferent neurons.
Difficulty: Moderate
Bloom's Taxonomy: Knowledge
Source: Section 39.1

89. Explain the general process of sensory transduction.

Answer: Sensory transduction is when a particular stimulus of a receptor cell results in a change in membrane potential. This membrane potential can then generate action potentials that relay the signal to the central nervous system.
Difficulty: Moderate
Bloom's Taxonomy: Knowledge
Source: Section 39.1

True/False

If the statement is true, write a "T" in the blank. If the statement is false, make it correct by changing the underlined word(s) and writing the correct word(s) in the answer blank.

90. _____ A particular type of sensory neuron will only relay information about that sensory input. (A pressure sensor will not respond to a different temperature.)

91. _____ Chronic pain is due, in part, to the underline{adapting} nature of pain receptors that diminishes the frequency of action potentials over time.

92. _____ Neuromasts are always located on the fish's underline{body surface}.

93. _____ The human vestibular apparatus is a underline{non-adapting} sensory system.

94. _____ A deaf dolphin lacks the ability to perceive its environment through underline{echolocation}.

95. _____ All animals perceive light using different forms of a pigment called underline{retinal}.

96. _____ In mammalian eyes, light must pass underline{through} neural cells to reach the photoreceptors in the retina.

97. _____ There is a network of blood vessels covering the surface of the retina through which light must pass to reach the rods and cones of the eye.

98. _____ Rods and cones both have underline{four} segments and each segment has a specialized function.

99. _____ Many invertebrates have separate and differentiated receptors for taste and smell, while others have a single receptor for both.

100. _____ Human thermoreceptors which respond at temperatures of underline{43°C} and above produce a pain response rather than participating in thermoregulation.

101. _____ Humans have different thermoreceptors for heat and cold.

Answers: **90. T   91. F, nonadapting   92. F, or in recessed canals   93. F, adapting   94. T   95. T   96. T   97. T   98. F, three   99. T   100. F, 52°C   101. T**

Difficulty: Difficult
Bloom's Taxonomy: Knowledge
Source: Sections 39.2–39.7.

Essay

102.   Explain why adaptation of pain receptors would not be desirable and yet the adaptation other sensory systems is beneficial.

*Answer: **The purpose of pain is to reduce damage of the body. In adaptation, a constant stimulus results in fewer action potentials being sent to the central nervous system. This allows the brain to focus on changes in the environment while ignoring stimuli that remain constant. With pain, however, adaptation could lead to a continuation in or an increase of damage. For this reason, pain receptors do not adapt or reduce the frequency of action potentials.***
Difficulty: Moderate
Bloom's Taxonomy: Analysis
Source: Sections 39.1, 39.6

Labeling

Identify each numbered part of a neuromast shown in the figure below.

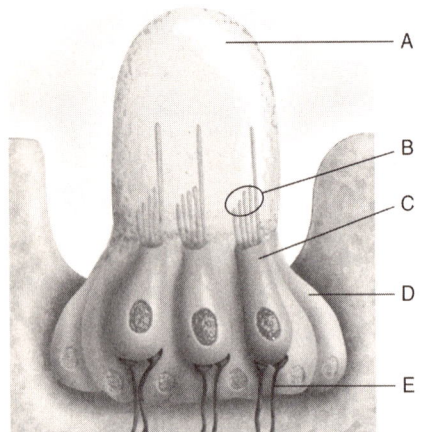

103. _____ Stereocilia
104. _____ Supporting cell
105. _____ Gelatinous cupula
106. _____ Sensory hair cell
107. _____ Afferent nerve fiber

Answers: 103. B   104. D   105.A   106. C
*   107. E*
Difficulty: Moderate
Bloom's Taxonomy: Knowledge
Source: Figure 39.4

Identify each numbered part of a vestibular apparatus shown in the figure below (Figure 39.5 a & b).

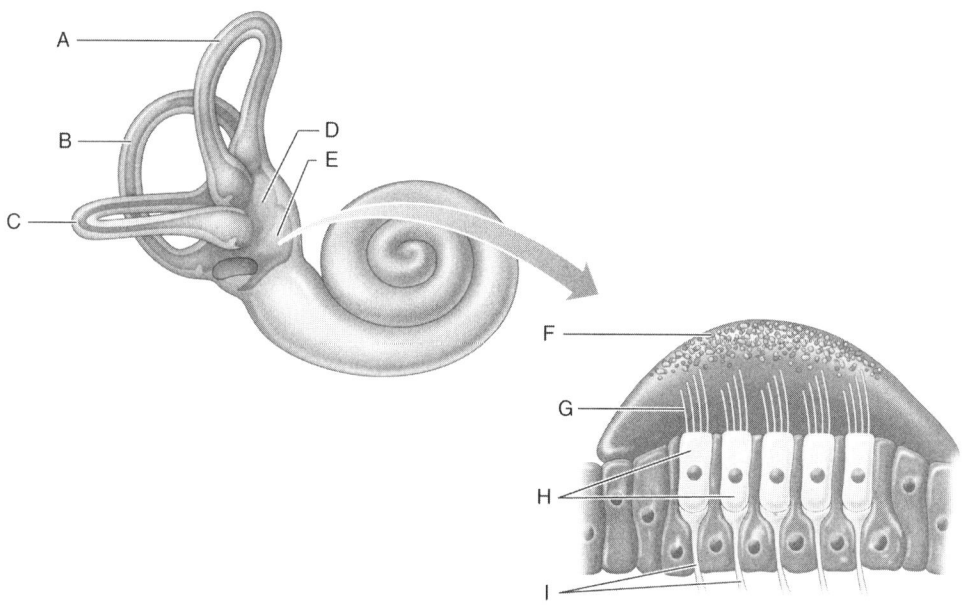

108. _____ posterior semicircular canal
109. _____ lateral semicircular canal
110. _____ otoliths
111. _____ sensory hair cells
112. _____ afferent semicircular canal
113. _____ utricle
114. _____ stereocilia
115. _____ saccule
116. _____ afferent neurons

Answers: 108. B 109. C   110. F   111. H   112. A
*   113. D   114. G 115. E   116. I*
Difficulty: Moderate
Bloom's Taxonomy: Knowledge
Source: Figure 39.5

Identify each numbered part of the human eye shown in Figure 39.6.

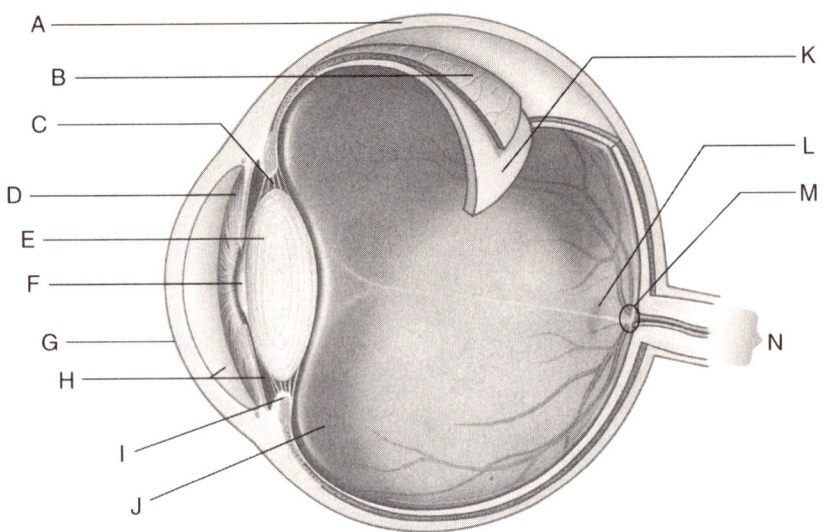

117. _____ blind spot
118. _____ retina
119. _____ lens
120. _____ choroid
121. _____ sclera
122. _____ pupil
123. _____ iris
124. _____ aqueous humor
125. _____ cornea
126 _____ ciliary body
127. _____ ciliary muscle
128. _____ optic nerve
129. _____ fovea
130. _____ vitreous humor

***Answers: 117. M 118. K   119. E   120. B   121. A
122. F   123. D   124. H   125. G   126. C   127. I
128. N   129. L   130. J***

Difficulty: Moderate
Bloom's Taxonomy: Knowledge
Source: Figure 39.12

Identify each numbered cell in the human eye shown in Figure 39.7.

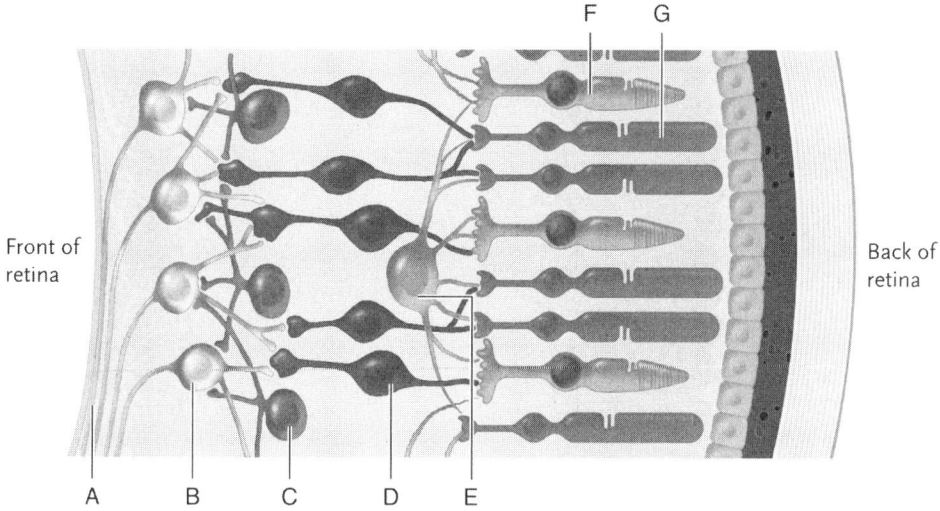

Front of retina

Back of retina

A   B   C   D   E

F   G

131. _____ ganglion
132. _____ cone
133. _____ optic nerve
134. _____ bipolar cell
135. _____ rod
136. _____ horizontal cell
137. _____ amacrine cell

Answers: *131. D*   *132. A*   *133. C*   *134. F*   *135. B*
      136. G   *137. E*

Difficulty: Moderate
Bloom's Taxonomy: Knowledge
Source: Figure 39.16

Identify each numbered structure, cell, or secretion in the human system for smell shown in the figure below (Figure 39.8).

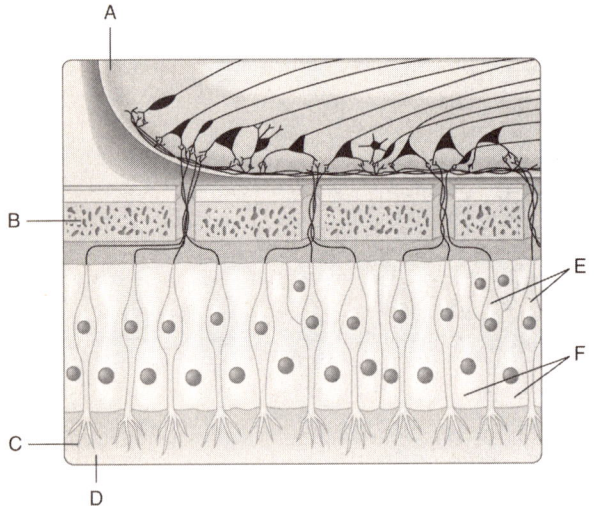

138. _____ mucus
139. _____ bone
140. _____ olfactory bulb
141. _____ olfactory receptor cells
142. _____ supporting cells
143. _____ sensory hairs

Answers: 138. D 139. B   140. A   141. E   142. F
143. C

Difficulty: Moderate
Bloom's Taxonomy: Knowledge
Source: Figure 39.21

40

The Endocrine System

Multiple-Choice

WHY IT MATTERS

1. In nature, most animals have young during seasons of abundant food. The timing of these events are dictated by environmental cues and

a. metabolism.
b. endocrine activity.
c. catabolism.
d. ecological events.
e. thermodynamics.

Answer: b
Difficulty: Easy
Bloom's Taxonomy: Knowledge, Comprehension

2. Neural activity involves transmission of information through electrical events, in contrast endocrine activity involves

a. transport of hormones in circulation to target tissues.
b. transport of hormones in neurons to target tissues.
c. release of hormones in body fluids for transport.
d. changes in neural activity for transport of hormones to target tissues.
e. both a and b are correct.

Answer: a
Difficulty: Moderate
Bloom's Taxonomy: Comprehension

40.1 HORMONES AND THEIR SECRETION

3. An important functional role of the endocrine system is to

a. provide mechanisms for rapid responses to changes in the body.
b. allow for mechanisms to control gene action.
c. inhibit nervous system action on muscles and glands.
d. maintain a constant, yet dynamic internal fluid environment.
e. all of these

Answer: d
Difficulty: Easy
Bloom's Taxonomy: Knowledge, Comprehension

4. When a cell releases chemical signals into the extracellular fluid to affect cells, this is an example of

a. autocrine regulation.
b. endocrine regulation.
c. paracrine regulation.
d. systemic regulation.
e. exocrine regulation.

Answer: c
Difficulty: Moderate
Bloom's Taxonomy: Knowledge, Comprehension

5. If a cell or tissue has a receptor for a hormone, that cell or tissue is

a. an endocrine gland.
b. a paracrine tissue.
c. a neuroendocrine tissue.
d. a target tissue.
e. a signaling tissue.

Answer: d
Difficulty: Easy
Bloom's Taxonomy: Knowledge

6. Target tissues for steroid hormones and thyroid hormones do not have membrane receptors for these hormones because they

a. stimulate the cell by changes in electrical activity.
b. are water soluble.
c. enter the cell by ion channels.
d. are small enough to diffuse directly into the cell.
e. are soluble in the lipid bilayer.

Answer: e
Difficulty: Moderate
Bloom's Taxonomy: Comprehension

7. Prostaglandins are

a. protein molecules that act either by paracrine or autocrine mechanisms.
b. fatty acid derivatives that act either by paracrine or autocrine mechanisms.
c. neurotransmitters that act by paracrine mechanisms.
d. amine molecules that act by neuroendocrine mechanisms.
e. steroid molecules that are lipid soluble.

Answer: b
Difficulty: Easy
Bloom's Taxonomy: Knowledge

8. If a hormone is hydrophobic and binds to membrane receptors, the type of molecule is most likely

a. a fatty acid.
b. a steroid.
c. a protein.
d. an amine.
e. both c and d

Answer: e
Difficulty: Difficult
Bloom's Taxonomy: Application

9. Hormone levels and action are determined by

a. other hormones or releasing agents.
b. neural activity.
c. negative feedback systems.
d. both a and c
e. a, b and c

Answer: e
Difficulty: Moderate
Bloom's Taxonomy: Comprehension

10. Select the chemical group that is **NOT** associated with any type of hormone.

a. peptide and proteins
b. fatty acid derivatives
c. steroids
d. polysaccharides
e. amino acid derivatives

Answer: d
Difficulty: Easy
Bloom's Taxonomy: Knowledge, Comprehension

11. Oxytocin is a hormone that is involved with increasing contractions during childbirth. The action is binding to membrane receptors and the result is increasing sodium and calcium. Oxytocin is most likely a

a. peptide or protein.
b. steroid.
c. carbohydrate.
d. prostaglandin.
e. fatty acid derivative.

Answer: a
Difficulty: Moderate
Bloom's Taxonomy: Comprehension

12. The products of endocrine glands

a. only affect their own tissue.
b. provide feedback by paracrine action.
c. always stimulate the target tissue.
d. reach their target tissues' circulation.
e. are released into body fluids, then into circulation.

Answer: d
Difficulty: Easy
Bloom's Taxonomy: Knowledge

40.2 MECHANISMS OF HORMONE ACTION

13. Which type of hormones enter cells and has the primary action to either increase or decrease mRNA production?

a. steroid and peptide hormones
b. steroid and phospholipid hormones
c. thyroid and peptide hormones
d. peptide and phospholipid hormones
e. steroid and thyroid hormones

Answer: e
Difficulty: Moderate
Bloom's Taxonomy: Comprehension

14. Cyclic AMP is a common second messenger, often associated with _____ hormones.

a. peptide
b. steroid
c. protein
d. fatty acids
e. both a and b

Answer: a
Difficulty: Moderate
Bloom's Taxonomy: Knowledge

15. Hormone action is associated with small quantities of hormone resulting in profound responses in target tissues. This type of response is

a. amplification.
b. a second messenger response.
c. a gene activation response.
d. a cyclic AMP response.
e. a growth factor response.

Answer: a
Difficulty: Moderate
Bloom's Taxonomy: Comprehension

16. Select the appropriate order of hormone action.

a. transduction, reception, response
b. reception, response, transduction
c. transduction, response, reception
d. reception, transduction, response
e. all of these

Answer: d
Difficulty: Easy
Bloom's Taxonomy: Knowledge

17. Your job is to determine if a steroid hormone is having an effect on its target tissue. Which of the following would be an indication of activation of the target tissue by the hormone in question?

a. increased cyclic AMP levels
b. activation of G-proteins
c. increased levels of mRNA
d. increased calcium level in the cytoplasm
e. either a or b

Answer: c
Difficulty: Difficult
Bloom's Taxonomy: Comprehension, Application

18. Glucose metabolism is regulated by several hormones. The predominant second messenger in these mechanisms is

a. glucagon.
b. insulin.
c. adenylyl cyclase.
d. cyclic AMP.
e. mRNA.

Answer: d
Difficulty: Moderate
Bloom's Taxonomy: Knowledge, Comprehension

19. Calcium ions can act as second messengers by binding to _____, which then regulates other proteins.

a. cyclic AMP
b. calmodulin
c. protein kinase
d. phosphodiesterase
e. adenylyl cyclase

Answer: b
Difficulty: Easy
Bloom's Taxonomy: Knowledge

20. Which of the following is **incorrectly** paired?

a. pancreas – insulin
b. thyroid gland – thyroxine
c. hypothalamus – melatonin
d. adrenal cortex – glucocorticoids
e. posterior pituitary – oxytocin

Answer: c
Difficulty: Easy
Bloom's Taxonomy: Knowledge

21. Which of the following hormones is **incorrectly** paired with its target tissue?

a. prolactin – mammary glands
b. growth hormone – bone
c. calcitonin – bone
d. aldosterone – uterus
e. pancreas – liver

Answer: d
Difficulty: Easy
Bloom's Taxonomy: Knowledge

40.3 THE HYPOTHALAMUS AND PITUITARY

22. The pituitary gland is regulated most directly by the

a. pineal gland.
b. hypothalamus.
c. pancreas.
d. adrenal gland.
e. thyroid gland.

Answer: b
Difficulty: Easy
Bloom's Taxonomy: Knowledge

23. Which of the following glands has a neurosecretory component as well as a component that is not under direct neural control?

a. the hypothalamus
b. the pineal gland
c. the adrenal gland
d. the pituitary gland
e. the parathyroid gland

Answer: d
Difficulty: Moderate
Bloom's Taxonomy: Comprehension

24. The posterior lobe of the pituitary gland secretes

a. ADH.
b. TSH.
c. oxytocin.
d. both a and c
e. a, b and c

Answer: d
Difficulty: Moderate
Bloom's Taxonomy: Knowledge

25. One of the main functions of the hypothalamus is to produce _____ hormones, which affect the _____ pituitary.

a. releasing; posterior
b. releasing; anterior
c. activating; posterior
d. amplification; anterior
e. none of these

Answer: b
Difficulty: Easy
Bloom's Taxonomy: Knowledge

26. Gigantism is caused by

a. hypersecretion of growth hormone during adulthood.
b. hyposecretion of thyroxine during adulthood.
c. hyposecretion of thyroxine during childhood.
d. hypersecretion of growth hormone during childhood.
e. hypersecretion of gonadotropin during childhood.

Answer: d
Difficulty: Easy
Bloom's Taxonomy: Knowledge

27. Which of the following hormones does **not** promote growth of cells or tissues?

a. ADH
b. growth hormone
c. insulin-like growth factors
d. thyroid hormones
e. prolactin

Answer: a
Difficulty: Moderate
Bloom's Taxonomy: Knowledge, Comprehension

28. Given the following, select the correct sequence of events.

   1. adrenal gland secretes more glucocorticoids
   2. anterior pituitary secretes more ACTH
   3. hypothalamus secretes ACTH
   4. plasma levels of glucocorticoids increase
   5. anterior pituitary secretes less ACTH

a. 1, 2, 4, 5
b. 2, 1, 4, 5
c. 3, 1, 4, 5
d. 3, 4, 1, 2
e. 2, 4, 5, 1

Answer: b
Difficulty: Difficult
Bloom's Taxonomy: Application

29. Question 28 is an example of a type of hormone control.

a. amplification
b. positive feedback
c. negative feedback
d. cascading feedback
e. both a and b

Answer: c
Difficulty: Moderate
Bloom's Taxonomy: Comprehension

30. If melanocyte-stimulating hormone were increased, you would expect

a. changes in kidney function for retention of water.
b. increase in reproductive or estrus cycles.
c. increase in heat lose from the body surface.
d. increase in pigmentation in the skin.
e. decrease in milk production in females.

Answer: d
Difficulty: Moderate
Bloom's Taxonomy: Knowledge, Comprehension

31. The regulation of milk release, water balance and uterine contractions during childbirth are mediated by the _____ gland.

a. parathyroid
b. pineal
c. anterior pituitary
d. posterior thyroid
e. posterior pituitary

Answer: e
Difficulty: Moderate
Bloom's Taxonomy: Knowledge, Comprehension

32. Anterior pituitary hormones have effects in

a. ovaries and testes.
b. mammary glands.
c. thyroid glands.
d. bone and muscle.
e. all of these

Answer: e
Difficulty: Easy
Bloom's Taxonomy: Knowledge

33. The anterior pituitary hormone which has a generalized effect, having receptors on almost every cell in the body, is

a. prolactin.
b. growth hormone.
c. gonadotropin.
d. ADH.
e. ACTH.

Answer: b
Difficulty: Easy
Bloom's Taxonomy: Knowledge

40.4 OTHER MAJOR ENDOCRINE GLANDS OF VERTEBRATES

34. Which of the following does **NOT** affect blood glucose levels?

a. glucagon
b. insulin
c. epinephrine
d. parathyroid hormone
e. glucocorticoids

Answer: d
Difficulty: Moderate
Bloom's Taxonomy: Comprehension

35. The actions of insulin and glucagon are

a. synergistic.
b. permissive.
c. cooperative.
d. antagonistic.
e. mutualistic.

Answer: d
Difficulty: Moderate
Bloom's Taxonomy: Knowledge, Comprehension

36. You are studying the hormonal control of metamorphosis in amphibians. If the anterior pituitary were prevented from releasing TSH, predict the effect on tadpoles.

a. Conversion to frogs would be increased five-fold.
b. Metamorphosis would be extremely slow.
c. Calcium secretion in tadpoles would be inhibited.
d. Metabolism in tadpoles would be greatly decreased.
e. Both b and d

Answer: e
Difficulty: Difficult
Bloom's Taxonomy: Comprehension, Application

37. The primary mineralocorticoid is

a. ACTH.
b. epinephrine.
c. growth hormone.
d. aldosterone.
e. ADH.

Answer: d
Difficulty: Easy
Bloom's Taxonomy: Knowledge

38. During stressful situations, epinephrine initiates all of the following actions **except**

a. constriction of blood vessels to the skin.
b. constriction of blood vessels to the heart.
c. dilation of blood vessels to the brain.
d. an increase in metabolic rate.
e. an increase in blood glucose levels.

Answer: b
Difficulty: Moderate
Bloom's Taxonomy: Comprehension

39. _____ is produced by the _____, which influences biological rhythms and the onset of sleep.

a. Cortisol; pineal gland
b. Melatonin; pituitary gland
c. Parathyroid hormone; parathyroid gland
d. Melatonin; pineal gland
e. Aldosterone; adrenal gland

Answer: d
Difficulty: Easy
Bloom's Taxonomy: Knowledge

40. An adult who is suffering from hyperthyroidism

a. is always tired.
b. shows lack of emotions.
c. is often hungry.
d. sleeps a significant amount of the time.
e. both a and d

Answer: c
Difficulty: Difficult
Bloom's Taxonomy: Application

41. Calcium levels are primarily under the control of

a. calcitonin and thyroxine.
b. parathyroid hormone and calcitonin.
c. prolactin and oxytocin.
d. parathyroid hormone and thyroxine.
e. progesterone and ADH.

Answer: b
Difficulty: Easy
Bloom's Taxonomy: Knowledge

42. Which of the following is **not** produced by either the ovaries or testes?

a. testosterone
b. estrogen
c. progesterone
d. follicle-stimulating hormone
e. androgens

Answer: d
Difficulty: Moderate
Bloom's Taxonomy: Knowledge, Comprehension

43. One of the actions of the birth control pill is to

a. prevent FSH and LH from being released from the hypothalamus.
b. prevent ovulation.
c. stop development of the uterine lining.
d. increase plasma testosterone levels.
e. both a and b

Answer: b
Difficulty: Moderate
Bloom's Taxonomy: Comprehension

44. Which of the following correctly describes the main difference between type 1 and type 2 diabetes?

a. One causes hyperglycemia, while the other causes hypoglycemia.
b. One demonstrates protein breakdown, while the other demonstrates protein synthesis.
c. One leads to blindness, while the other leads to kidney disorder.
d. One has insulin hypersecretion, while the other has insulin hyposecretion.
e. One has insulin deficiency, while the other has insulin resistance.

Answer: e
Difficulty: Moderate
Bloom's Taxonomy: Comprehension

40.5 ENDOCRINE SYSTEMS IN INVERTEBRATES

45. Select the hormone(s) that is/are involved in crustacean molting.

a. molt-inhibiting hormone
b. juvenile hormone
c. brain hormone
d. ecdysone
e. all of these

Answer: e
Difficulty: Easy
Bloom's Taxonomy: Knowledge

46. When molt-inhibiting hormone is prevented from being secreted

a. ecdysone levels increase.
b. ecdysone levels decrease.
c. molting is inhibited.
d. both a and b
e. both b and c

Answer: a
Difficulty: Moderate
Bloom's Taxonomy: Application

EXPERIMENTAL RESEARCH: DEMONSTRATION THAT BINDING OF EPINEPHRINE TO BETA RECEPTORS TRIGGERS A SIGNAL TRANSDUCTION PATHWAY WITHIN CELLS

47.   If adenylyl cyclase were inhibited in the experimental cells

a.   binding of epinephrine to beta receptors would have no affect on cyclic AMP levels.
b.   binding of epinephrine to beta receptors would increase cyclic AMP levels.
c.   an increase in mRNA for cyclic AMP would occur.
d.   an increase in mRNA for more beta receptors would occur.
e.   epinephrine would not be able to bind to beta receptors..

Answer: a
Difficulty: Moderate
Bloom's Taxonomy: Knowledge

48.   The transduction event in Question 47 was

a.   activation of epinephrine.
b.   protein kinase inhibition.
c.   activation of adenylyl cyclase.
d.   protein kinase amplification.
e.   both c and d

Answer: e
Difficulty: Moderate
Bloom's Taxonomy: Comprehension

INSIGHTS FROM THE MOLECULAR REVOLUTION: TWO RECEPTORS FOR ESTROGENS

49.   Antiestrogens can bind to estrogen receptors sites

a.   without activation.
b.   without inhibition.
c.   and amplify the response.
d.   and prevent reception of hormone action.
e.   none of these

Answer: a
Difficulty: Easy
Bloom's Taxonomy: Knowledge

50.   Tamoxifen blocked the activity of estrogen in _____, but increased the activity of estrogen in _____.

a.   breast tissue; ovarian tissue
b.   uterine tissue; ovarian tissue
c.   uterine, tissue; breast tissue
d.   breast tissue; uterine tissue
e.   none of these

Answer: d
Difficulty: Easy
Bloom's Taxonomy: Knowledge

51.   Different responses of the same hormone, or in Question 49 antiestrogen agent, is due to

a.   differences in luciferase activity.
b.   receptors sites for estrogen.
c.   differences in gene amplification.
d.   cyclic AMP levels.
e.   both b and d

Answer: b
Difficulty: Moderate
Bloom's Taxonomy: Comprehension

FOCUS ON RESEARCH: BASIC RESEARCH: NEUROENDOCRINE AND BEHAVIORAL EFFECTS OF ANABOLIC-ANDROGENIC STEROIDS IN HUMANS

52.   Anabolic-androgenic steroids

a.   are synthetic forms of testosterone.
b.   are synthetic forms of estrogens.
c.   are potent tissue building molecules.
d.   both a and c
e.   a, b and c

Answer: d
Difficulty: Easy
Bloom's Taxonomy: Knowledge

53.   Excess intact of AAS were found to

a.   increase TSH levels.
b.   decrease gonadotrophins from the hypothalamus.
c.   decrease natural testosterone.
d.   both a and c
e.   a, b and c

Answer: d
Difficulty: Moderate
Bloom's Taxonomy: Comprehension

54.   High level of hormones as a result of exogenous treatment will

a.   often produce excess blood levels of the hormones, causing feedback system to cause further increases.
b.   often cause negative feedback system to stop functioning in a normal fashion.
c.   convert negative feedback systems to positive feedback systems.
d.   have no effect on normal feedback systems.
e.   both a and c

Answer: b
Difficulty: Moderate
Bloom's Taxonomy: Comprehension

UNANSWERED QUESTIONS

55. Insulin resistance

a. is caused by inadequate levels of insulin.
b. is inadequate response of insulin receptors to physiological levels of insulin.
c. is excessive response of insulin receptors to physiological levels of insulin.
d. is associated with a beta-islet abnormality.
e. all of these

Answer: b
Difficulty: Easy
Bloom's Taxonomy: Knowledge

56. Blood glucose levels arc the current standard for diagnosis of both type 1 and type 2 diabetes. Current research suggests that a better indicator of type 2 diabetes might be

a. blood insulin levels.
b. insulin dependent glucose metabolism in muscle.
c. exercise level.
d. muscle to fat ratio.
e. none of these

Answer: b
Difficulty: Moderate
Bloom's Taxonomy: Comprehension

Integrative Multiple-Choice

57. The reason some individual hormones have multiple effects on different target tissues is that

a. gene transcription is altered.
b. second messengers trigger cascade or amplification effects.
c. many different cells in different tissues that have specific receptors for the hormone.
d. only a small amount of the hormone is required for its effect.
e. all of these

Answer: c
Difficulty: Difficult
Bloom's Taxonomy: Application
Section: 40.1, 40.2, 40.3, 40.4

58. Target tissues often have receptors for multiple hormones. The response of the target tissue depends on

a. blood levels of and interactions between the hormones that affect it.
b. blood levels of the hormones that affect it.
c. interactions between hormones.
d. blood levels of and interactions between the hormones that affect it and the effects of other signaling molecules, such as prostaglandins.
e. all of these

Answer: d
Difficulty: Difficult
Bloom's Taxonomy: Application
Section: 40.1, 40.2, 40.3, 40.4

59. Stress can have a significant effect on the body. Chronic stress

a. may lead to excessive stimulation of neurons.
b. is associated with elevated levels of epinephrine from the adrenal cortex.
c. is associated with elevated levels of cortisol from the adrenal cortex.
d. may lead to increased blood flow to the brain, especially the hippocampus.
e. all of these

Answer: c
Difficulty: Difficult
Bloom's Taxonomy: Application
Section: 38.3, 40.3, 40.4

60. Hormone A binds to intracellular receptors in its target tissue. As a result of Hormone A, the target tissue is now sensitive or responsive to Hormone B. The action of Hormone A was to

a. decrease mRNA for Hormone A receptors.
b. increase receptor sites for Hormone B.
c. increase receptor sits for Hormone A.
d. increased cyclic AMP levels which activated receptors for hormone B.
e. none of these

Answer: b
Difficulty: Difficult
Bloom's Taxonomy: Application
Source: Section 40.2, 40.3

61. Negative feedback loops to the hypothalamus and the pituitary operate to regulate the levels of all of the following hormones **except**

a. cortisol.
b. thyroid hormones.
c. insulin.
d. testosterone.
e. progesterone.

Answer: c
Difficulty: Moderate
Bloom's Taxonomy: Knowledge, Comprehension
Source: Section 40.1, 40.3, 40.4

62. Select the endocrine gland whose secretory function is under direct control of the sympathetic pre-ganglionic neuron.

a. pancreas
b. thyroid
c. adrenal medulla
d. adrenal cortex
e. posterior pituitary

Answer: c
Difficulty: Difficult
Bloom's Taxonomy: Application
Source: Section 38.3, 40.3, 40.4

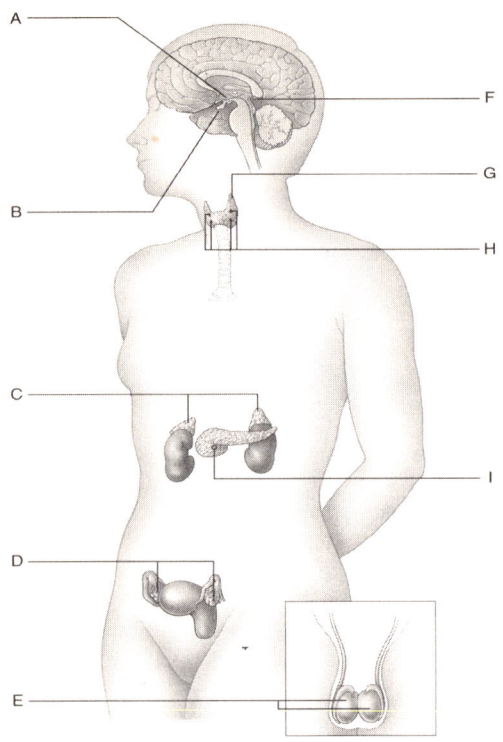

Choice

Select from the figure the appropriate endocrine gland. (Answers may be used more than once.)

63. _____ production and release of insulin

64. _____ a portion is neuroendocrine and a portion is under hormonal control of the hypothalamus

65. _____ production site of androgens

66. _____ primary calcium regulation

67. _____ ADH production

68. _____ a portion of this gland is a postganglionic sympathetic neuron

69. _____ controls biological rhythms

70. _____ hormones from this gland play an important role in metamorphosis in amphibians

71. _____ Hormones from these glands stimulate other glands to produce hormones (two answers)

72. _____ production and release of calcitonin

73. _____ gland that produces progesterone that is under the control of GnRH

74. _____ the hormone produced by this gland works in a synergistic fashion with Vitamin D

75. _____ production site of the hormone that promotes growth in bones and muscles

76. _____ production site of hormone involved in milk production

77. _____ production site of hormone involved in water balance by increasing nephron permeability to water

78. _____ site of release of oxytocin

Answers:　**63. I　64. B　65. E　66. H　67. A
68. C　69. F　70. G　71. A, B 72. G
73. D　74. H　75. A　76. B　77. B
78. B**

Difficulty: Moderate
Bloom's Taxonomy: Knowledge, Comprehension
Source: Section 40.1, 40.2, 40.3

True/False

If the statement is true, write a "T" in the blank. If the statement is false, make it correct by changing the underlined word(s) and writing the correct word(s) in the answer blank.

79. _____ Oxytocin secreted by the posterior pituitary <u>only</u> in <u>females</u>.

80. _____ Insulin is produced and released from the <u>exocrine</u> pancreas.

81. _____ <u>Ecdysone</u> inhibitors could be used as insecticides.

82. _____ Prostaglandins are under <u>endocrine</u> regulation.

83. _____ Cyclic AMP activity is associated with <u>hydrophobic</u> hormones.

84. _____ <u>Increases</u> in MSH could result in increased pigmentation.

Answers:　**79. F, both males and females 80. F, endocrine　81. T　82. F, autocrine or paracrine　83. F, hydrophilic　84. T**

Difficulty: Moderate
Bloom's Taxonomy: Knowledge
Source: Section 40.1, 40.2, 40.3, 40.4, 40.5

Matching

Match each of the following concepts with the most appropriate response.

85. _____ aldosterone

86. _____ cyclic AMP

87. _____ paracrine regulation

88. _____ tropic hormones

89. _____ thyroxine

90. _____ ecdysone

91. _____ hydrophobic hormone

92. _____ amplification

93. _____ type 1 diabetes

94. _____ suprachiasmatic nucleus

95. _____ hyposecretion of growth hormone

96. _____ prostaglandins

A. Contains four iodine atoms

B. Second messenger

C. Opposite effect of molt-inhibiting hormone

D. Hormone produced by the adrenal cortex that regulates sodium and potassium levels

E. Insulin deficiency

F. Typically steroids and affect gene activation

G. Hormones that either stimulate or inhibit another cell or tissue

H. Pituitary dwarfism

I. Small activation, cascading response

J. Target tissue of melatonin

K. Signal molecule released by a cell into the extracellular fluid having a localized response

L. Paracrine or autocrine regulation

Answers: *85. D   86. B   87. K   88. G   89. A*
              90. C   91. F   92. I   93. E   94. J
              95. H   96. L

Difficulty: Moderate
Bloom's Taxonomy: Knowledge
Source: Sections 40.1, 40.2, 40.3, 40.4, 40.5

Short Answer

97. Describe amplification of hormone action.

Answer: A hormone receptor complex is the reception component of hormone action. This complex can activate multiple copies of transduction molecules or events, such as cyclic Amp as a second messenger. Each of the second messenger molecules can activate multiple copies of the next component of the mechanism (such as multiple copies of protein kinase). One hormone receptor complex can result in multiple levels of response, since each component of the mechanism produces a cascading or amplifying response.
Difficulty: Moderate
Bloom's Taxonomy: Knowledge, Comprehension
Source: Section 40.1, 40.2

98. Explain negative feedback control using thyroid hormones as an example.

Answer: Thyroid hormones (T3 & T4) are produced by the thyroid gland. Levels are maintained within a physiological range. When the T3 and T4 levels drop due to metabolism, the hypothalamus produces thyroid releasing hormone (TRH), which acts on the anterior pituitary to produce thyroid stimulating hormone(TSH). TSH stimulates the thyroid to produce and release T3 and T4. As the levels of T3 and T4 increase, the hypothalamus decreases the production of TRH and thus TSH from the anterior pituitary also decreases.
Difficulty: Moderate
Bloom's Taxonomy: Knowledge, Comprehension
Source: Section 4.5

99. Compare and contrast receptor recognition between a hydrophobic and hydrophilic hormone.

Answer: Hydrophobic hormones are typically lipid soluble hormones—steroids or fatty acid derivatives. These will diffuse into the cell and bind to intracellular receptors. Typical action of these hormones is to affect gene activity. Hydrophilic hormones are non-lipid soluble and thus cannot easily enter the cell. Receptors are typically membrane bound. The binding of the hormone to the receptor results in activation of a type of second messenger, which then causes a variety of effects from permeability changes to changes in enzyme activity.
Difficulty: Moderate
Bloom's Taxonomy: Knowledge, Comprehension
Source: Section 40.1, 40.2

Essay

100.   A patient is having blood glucose problems. Plasma glucose levels are very low, a condition called hypoglycemia. The problem is not associated with insulin, but with the secretion of glucocorticoids. How would you evaluate the patient to determine where in the feedback system the problem is located?

Answer: The adrenal cortex secretes glucocorticoids under the control of ACTH from the anterior pituitary. The anterior pituitary is under the control of CRH from the hypothalamus. The patient could be given an injection of ACTH and then monitor blood glucose levels. Next, the patient could be given an injection of CRH and blood glucose levels evaluated. Another way to address the problem would be to measure the levels of ACTH and CRH in the patient. If the CRH were elevated, then the problem would be ACTH production from the pituitary.
Difficulty: Difficult
Bloom's Taxonomy: Comprehension, Application
Source: Section 40.3, 40.4

101.   If a hormone caused an increase in protein kinase activity, predict the type of hormone responsible and the intracellular signaling mechanism of its action.

Answer: The hormone is most likely a peptide or protein—or other hydrophobic or non-lipid soluble molecule. The receptor is a membrane bound receptor and a type of second messenger mechanism is involved. The second messenger could either be cyclic AMP or G-proteins. In either case, the hormone-receptor complex would either activate adenylyl cyclase or G-proteins. As a result, protein kinase would be activated by the second messenger products.
Difficulty: Moderate
Bloom's Taxonomy: Comprehension
Source: Section 4.1, 4.2

41

MUSCLES, BONES, AND BODY MOVEMENTS

Multiple-Choice

WHY IT MATTERS

1.   About how long is required for a Mexican leaf frog to capture a cricket, from the beginning of movement by the frog to the closing of the frog's mouth?

a.   50 milliseconds
b.   1 second
c.   80 milliseconds
d.   260 milliseconds
e.   3 seconds

Answer: d
Difficulty: Difficult
Bloom's Taxonomy: Knowledge

2.   Which muscle type is found in the walls of body tubes and cavities of vertebrates, such as blood vessels and the intestines?

a.   cardiac
b.   smooth
c.   skeletal
d.   none of these

Answer: b
Difficulty: Easy
Bloom's Taxonomy: Knowledge

41.1 VERTEBRATE SKELETAL MUSCLE: STRUCTURE AND FUNCTION

3.   Skeletal muscles cells have _____ and are controlled by _____.

a.   many nuclei; the somatic nervous system
b.   many nuclei; the autonomic nervous system
c.   one nucleus each; the autonomic nervous system
d.   one nucleus each; the somatic nervous system

Answer: a
Difficulty: Moderate
Bloom's Taxonomy: Knowledge

4.   A typical human body has _____ skeletal muscles.

a.   78
b.   168
c.   206
d.   417
e.   over 600

Answer: e
Difficulty: Moderate
Bloom's Taxonomy: Knowledge

5.   Skeletal muscles are connected to bones by cords of connective tissue called

a.   muscle fibers.
b.   sarcomeres.
c.   tendons.
d.   ligaments.
e.   myofibrils.

Answer: c
Difficulty: Easy
Bloom's Taxonomy: Knowledge

6.   Skeletal muscles are made up of bundles of elongated, cylindrical cells called

a.   muscle fibers.
b.   sarcomeres.
c.   tendons.
d.   ligaments.
e.   myofibrils.

Answer: a
Difficulty: Moderate
Bloom's Taxonomy: Comprehension

7.   Individual skeletal muscle cells are packed with cylindrical contractile elements about 1 mm in diameter called

a.   muscle fibers.
b.   sarcomeres.
c.   tendons.
d.   ligaments.
e.   myofibrils.

Answer: e
Difficulty: Moderate
Bloom's Taxonomy: Comprehension

8.   Thick filaments are parallel bundles of _____ molecules.

a.   actin
b.   tropomyosin
c.   acetylcholine
d.   myosin
e.   troponin

Answer: d
Difficulty: Moderate
Bloom's Taxonomy: Knowledge

9.   "A bands" in skeletal muscle are composed of

a.   discs to which thin filaments are anchored.
b.   stacked thick filaments along with parts of thin filaments that overlap both ends.
c.   thin filaments but no thick filaments.
d.   thick filaments but no thin filaments.
e.   discs to which thick filaments are anchored.

Answer: b
Difficulty: Moderate
Bloom's Taxonomy: Knowledge

10. "H zones" in skeletal muscle are composed of

a. discs to which thin filaments are anchored.
b. stacked thick filaments along with parts of thin filaments that overlap both ends.
c. thin filaments but no thick filaments.
d. thick filaments but no thin filaments.
e. discs to which thick filaments are anchored.

Answer: d
Difficulty: Moderate
Bloom's Taxonomy: Knowledge

11. "Z lines" in skeletal muscle are composed of

a. discs to which thin filaments are anchored.
b. stacked thick filaments along with parts of thin filaments that overlap both ends.
c. thin filaments but no thick filaments.
d. thick filaments but no thin filaments.
e. discs to which thick filaments are anchored.

Answer: a
Difficulty: Difficult
Bloom's Taxonomy: Knowledge

12. "I bands" in skeletal muscle are composed of

a. discs to which thin filaments are anchored.
b. stacked thick filaments along with parts of thin filaments that overlap both ends.
c. thin filaments but no thick filaments.
d. thick filaments but no thin filaments.
e. discs to which thick filaments are anchored.

Answer: c
Difficulty: Moderate
Bloom's Taxonomy: Knowledge

13. The region between two adjacent Z lines is a

a. muscle fiber.
b. sarcomere.
c. tendon.
d. ligament.
e. myofibril.

Answer: b
Difficulty: Moderate
Bloom's Taxonomy: Comprehension

14. The _____ is a system of vesicles that wraps around each A band and I band and stores ions that are used in muscle contractions.

a. neuromuscular junction
b. Golgi apparatus
c. endolemma
d. microvillus
e. sarcoplasmic reticulum

Answer: e
Difficulty: Moderate
Bloom's Taxonomy: Comprehension

15. In skeletal muscle contraction, which of the following acts to allow the crossbridge cycle to occur when it flows into the cytosol through open ion channels?

a. troponin
b. Ca^{2+}
c. tropomyosin
d. ATP
e. acetylcholine

Answer: b
Difficulty: Easy
Bloom's Taxonomy: Comprehension

16. In skeletal muscle contraction, which of the following is the neurotransmitter released at the axon terminal to trigger an action potential in the muscle cell?

a. troponin
b. Ca^{2+}
c. tropomyosin
d. ATP
e. acetylcholine

Answer: e
Difficulty: Moderate
Bloom's Taxonomy: Comprehension

17. In skeletal muscle contraction, which of the following must be moved to the grooves in the actin double helix to uncover the crossbridge binding site?

a. troponin
b. Ca^{2+}
c. tropomyosin
d. ATP
e. acetylcholine

Answer: c
Difficulty: Difficult
Bloom's Taxonomy: Comprehension

18. In skeletal muscle contraction, which of the following undergoes a conformational change when it binds to a specific ion and then uncovers the crossbridge binding site by causing another factor to be moved to the grooves in the actin double helix?

a. troponin
b. Ca^{2+}
c. tropomyosin
d. ATP
e. acetylcholine

Answer: a
Difficulty: Difficult
Bloom's Taxonomy: Comprehension

19. In skeletal muscle contraction, which of the following binds to myosin?

a. troponin
b. Ca^{2+}
c. tropomyosin
d. ATP
e. acetylcholine

Answer: d
Difficulty: Difficult
Bloom's Taxonomy: Comprehension

20. What action causes the shape change in the myosin crossbridge that directly triggers the power stroke in skeletal muscle contraction?

a. binding of myosin to actin
b. binding of tropomyosin to myosin
c. binding of myosin to troponin
d. binding of tropomyosin to actin
e. binding of troponin to tropomyosin

Answer: a
Difficulty: Moderate
Bloom's Taxonomy: Knowledge

21. A defect in transport of which ion would have the most direct effect on muscle contraction?

a. Cl^-
b. K^+
c. Ca^{2+}
d. Na^+
e. Zn^{2+}

Answer: c
Difficulty: Easy
Bloom's Taxonomy: Evaluation

22. *Clostridium botulinum* produces a deadly toxin that stops muscle contractions by

a. destroying cell membranes.
b. killing mitochondria.
c. disrupting the structure of actin filaments.
d. preventing ATP synthesis.
e. blocking acetylcholine release.

Answer: e
Difficulty: Moderate
Bloom's Taxonomy: Comprehension

23. The frozen contraction of muscle cells called rigor mortis occurs because after death

a. ions are not available in muscle cells.
b. tropomyosin breaks down quickly.
c. neurotransmitters are no longer released.
d. ATP production stops.
e. muscle cells are flooded with neurotransmitters.

Answer: d
Difficulty: Moderate
Bloom's Taxonomy: Knowledge

24. A single, weak contraction of a muscle fiber is called

a. an action potential.
b. a muscle twitch.
c. fatigue.
d. tetanus.
e. constriction.

Answer: b
Difficulty: Moderate
Bloom's Taxonomy: Knowledge

25. A muscle contraction where fibers cannot relax at all between stimuli is called

a. an action potential.
b. a muscle twitch.
c. fatigue.
d. tetanus.
e. constriction.

Answer: d
Difficulty: Moderate
Bloom's Taxonomy: Comprehension

26. Myoglobin content is high in

a. fast anaerobic fibers.
b. fast aerobic and fast anaerobic muscle fibers.
c. slow muscle fibers and fast aerobic muscle fibers.
d. slow muscle fibers.
e. fast aerobic fibers.

Answer: c
Difficulty: Moderate
Bloom's Taxonomy: Synthesis

27. A rapid, powerful movement of short duration that could not be sustained for long would likely involve mainly

a. fast anaerobic fibers.
b. fast aerobic and fast anaerobic muscle fibers.
c. slow muscle fibers and fast aerobic muscle fibers.
d. slow muscle fibers.
e. fast aerobic fibers.

Answer: a
Difficulty: Moderate
Bloom's Taxonomy: Synthesis

28. The main role of myoglobin in muscle fibers is to

a. magnify responses to neurotransmitters.
b. sequestering ions.
c. synthesize ATP.
d. store oxygen.
e. enhance the strength of the power stroke.

Answer: d
Difficulty: Moderate
Bloom's Taxonomy: Evaluation

29. A motor unit is

a. a single muscle fiber activated by one neuron.
b. the complete set of muscle fibers in a single muscle.
c. a single sarcomere.
d. a single muscle fiber activated by many neurons.
e. a group of muscle fibers activated by one neuron.

Answer: e
Difficulty: Easy
Bloom's Taxonomy: Knowledge

INSIGHTS FROM THE MOLECULAR REVOLUTION: A SUBSTITUTE PLAYER THAT MAY BE A BIG WINNER IN MUSCULAR DYSTROPHY

30. Which of the following is similar in structure and function to the protein that is defective in Duchenne muscular dystrophy?

a. calmodulin
b. dystrophin
c. actin
d. utrophin
e. myosin

Answer: d
Difficulty: Moderate
Bloom's Taxonomy: Knowledge

41.2 SKELETAL SYSTEMS

31. Tube feet of sea stars, erectile tissue of the penis in vertebrates, and the bodies of cnidarians are all supported by

a. an exoskeleton.
b. only nonskeletal structures.
c. a hydrostatic skeleton.
d. an endoskeleton.
e. joined exo- and endoskeletons.

Answer: c
Difficulty: Moderate
Bloom's Taxonomy: Synthesis

32. The bodies of arthropods are supported mainly by

a. an exoskeleton.
b. only nonskeletal structures.
c. a hydrostatic skeleton.
d. an endoskeleton.
e. joined exo- and endoskeletons.

Answer: a
Difficulty: Easy
Bloom's Taxonomy: Synthesis

33. For humans and most vertebrates the body is supported primarily by

a. an exoskeleton.
b. only nonskeletal structures.
c. a hydrostatic skeleton.
d. an endoskeleton.
e. joined exo- and endoskeletons.

Answer: d
Difficulty: Easy
Bloom's Taxonomy: Synthesis

34. A typical human body has _____ bones.

a. 78
b. 168
c. 206
d. 417
e. over 600

Answer: c
Difficulty: Moderate
Bloom's Taxonomy: Knowledge

35. The ribs and sternum are considered to be part of

a. both the axial skeleton and the appendicular skeleton.
b. the axial skeleton.
c. neither the axial skeleton nor the appendicular skeleton.
d. the appendicular skeleton.

Answer: b
Difficulty: Easy
Bloom's Taxonomy: Application

36. The radius and ulna are considered to be part of

a. both the axial skeleton and the appendicular skeleton.
b. the axial skeleton.
c. neither the axial skeleton nor the appendicular skeleton.
d. the appendicular skeleton.

Answer: d
Difficulty: Moderate
Bloom's Taxonomy: Application

37. Bones are organs with more than one tissue type present. Which of the following is NOT found as part of any bones?

a. blood vessels
b. bone tissue
c. adipose tissue
d. nerves
e. all of these can be part of bones

Answer: e
Difficulty: Moderate
Bloom's Taxonomy: Knowledge

38. The primary source of new red blood cells in mammals is the

a. liver.
b. blood, itself.
c. spleen.
d. heart.
e. bone marrow.

Answer: e
Difficulty: Easy
Bloom's Taxonomy: Knowledge

39. Bone plays a critical role in providing _____ for the blood.

a. sodium and calcium ions
b. magnesium and phosphate ions
c. carbon dioxide
d. phosphate and calcium ions
e. calcium ions and oxygen

Answer: d
Difficulty: Easy
Bloom's Taxonomy: Knowledge

41.3 VERTEBRATE MOVEMENT: THE INTERACTIONS BETWEEN MUSCLES AND BONES

40. These joints are somewhat moveable, but they do not have a fluid-filled capsule surrounding them. They have fibrous connective tissue covering the ends of the bones involved.

a. synovial joints
b. cartilaginous joints
c. fibrous joints
d. none of these

Answer: b
Difficulty: Moderate
Bloom's Taxonomy: Comprehension

41. These joints have bones joined by stiff fibers of connective tissue and are essentially immobile.

a. synovial joints
b. cartilaginous joints
c. fibrous joints
d. none of these

Answer: c
Difficulty: Easy
Bloom's Taxonomy: Comprehension

42. These joints are usually highly moveable. They have a fluid-filled capsule of connective tissue surrounding them.

a. synovial joints
b. cartilaginous joints
c. fibrous joints
d. none of these

Answer: a
Difficulty: Easy
Bloom's Taxonomy: Comprehension

43. Synovial joints are held together by

a. muscle fibers.
b. sarcomeres.
c. tendons.
d. ligaments.
e. myofibrils.

Answer: d
Difficulty: Easy
Bloom's Taxonomy: Comprehension

44. The human elbow is an example of which kind of joint?

a. synovial joint
b. cartilaginous joint
c. fibrous joint
d. none of these

Answer: a
Difficulty: Moderate
Bloom's Taxonomy: Application

45. Cranial bones are held together by which kind of joint?

a. synovial joint
b. cartilaginous joint
c. fibrous joint
d. none of these

Answer: c
Difficulty: Moderate
Bloom's Taxonomy: Knowledge

46. Human vertebrae are held together by which kind of joint?

a. synovial joint
b. cartilaginous joint
c. fibrous joint
d. none of these

Answer: b
Difficulty: Moderate
Bloom's Taxonomy: Knowledge

47. A muscle that causes any type of movement in a joint when it contracts is called a(n)

a. flexor.
b. agonist.
c. depressor.
d. antagonist.
e. extensor.

Answer: b
Difficulty: Difficult
Bloom's Taxonomy: Knowledge

48. A muscle that increases the angle between two bones at a joint is called a(n)

a. flexor.
b. agonist.
c. depressor.
d. antagonist.
e. extensor.

Answer: e
Difficulty: Moderate
Bloom's Taxonomy: Knowledge

49. A muscle that decreases the angle between two bones at a joint is called a(n)

a. flexor.
b. agonist.
c. depressor.
d. antagonist.
e. extensor.

Answer: a
Difficulty: Easy
Bloom's Taxonomy: Knowledge

50. A muscle that has the opposite effect to that of another muscle at the same joint is called a(n)

a. flexor.
b. agonist.
c. depressor.
d. antagonist.
e. extensor.

Answer: d
Difficulty: Moderate
Bloom's Taxonomy: Application

51. Which of the following is an example of an antagonist muscle pair in humans?

a. calf and gluteus maximus
b. deltoid and pectoral
c. hamstring and biceps
d. calf and hamstring
e. biceps and triceps

Answer: e
Difficulty: Easy
Bloom's Taxonomy: Comprehension

UNANSWERED QUESTIONS

52. Skeletal muscle growth and development is inhibited by _____ produced in muscle cells.

a. erythropoietin
b. acetylcholine
c. epidermal growth factor
d. epinephrine
e. myostatin

Answer: e
Difficulty: Moderate
Bloom's Taxonomy: Knowledge

53. Which of the following is thought to be a promising means of improving conditions for patients with type-2 diabetes?

a. injections with myostatin
b. increasing muscle mass
c. ketoacidosis
d. injections with acetylcholine
e. increasing consumption of simple carbohydrates

Answer: b
Difficulty: Moderate
Bloom's Taxonomy: Knowledge

54. Which of the following typically occurs if someone survives a heart attack?

a. Surviving muscle cells compensate by increasing the force generated by individual myofibers.
b. Atrophy of heart muscle spreads until another heart attack is imminent.
c. New muscle cells are rapidly grown to replace damaged cells.
d. Surviving muscle cells are weakened as they extend their roles to compensate for destroyed cells.
e. Function is fully restored by the growth of extra sarcomeres inside damaged cells.

Answer: a
Difficulty: Moderate
Bloom's Taxonomy: Knowledge

Matching

Match each of the following molecules with its correct description.

55. _____ myosin

56. _____ tropomyosin

57. _____ actin

58. _____ troponin

59. _____ acetylcholine

A. Molecules that form a twisted double helix that makes up most of the thin filaments

B. Neurotransmitters involved in muscle cell contraction

C. Molecules that block the crossbridge binding sites when the muscle is not contracting

D. Molecules that makes up thick filaments

E. Molecules that bind to Ca^{2+}, change their shape, and then uncover the crossbridge binding sites

Answers: 55. D   56. C   57. A   58. E   59. B

Difficulty: Moderate
Bloom's Taxonomy: Synthesis
Source: Section 41.1

Choice

Match the muscle fiber type(s) with the characteristic described. Each choice may be used once, more than once, or not at all.

a. slow
b. fast aerobic
c. fast anaerobic
d. both slow and fast aerobic
e. both fast aerobic and fast anaerobic

60. _____ intermediate glycogen content

61. _____ slow contraction speed

62. _____ high contraction intensity

63. _____ intermediate fatigue resistance

64. _____ high myosin-ATPase activity

65. _____ low oxidative phosphorylation capacity

66. _____ high fatigue resistance

67. _____ many mitochondria

68. _____ low myoglobin content

69. _____ red fiber color

Answers: *60. b*   *61. a*   *62. c*   *63. b*   *64. e*
               65. c   *66. a*   *67. d*   *68. c*   *69. d*

Difficulty: Moderate
Bloom's Taxonomy: Synthesis
Source: Section 41.1, Table 41.1

For each bone type listed, choose the skeleton type(s) to which it belongs. Each choice may be used once, more than once, or not at all.

a. axial skeleton
b. appendicular skeleton
c. both the axial and the appendicular skeleton
d. neither the axial nor the appendicular skeleton

70. _____ cranial bones

71. _____ phalanges

72. _____ vertebrae

73. _____ scapula

74. _____ sternum

75. _____ facial bones

76. _____ humerus

77. _____ femur

Answers: *70. a*   *71. b*   *72. a*   *73. b*   *74. a*
               75. a   *76. b*   *77. b*

Difficulty: Moderate
Bloom's Taxonomy: Comprehension
Source: Section 41.2, Figure 41.11

42
THE CIRCULATORY SYSTEM

Multiple Choice

WHY IT MATTERS

1. Augustus Waller invented a machine that measures

a. blood pressure when the ventricles are contracting.
b. blood pressure when the ventricles are relaxed.
c. cardiac output.
d. electrical activity of the brain.
e. electrical activity of the heart.

Answer: e
Difficulty: Easy
Bloom's Taxonomy: Knowledge

2. In Waller's experiment, he placed one of his dog's front paws and one of his back paws in pans containing

a. carbon dioxide.
b. iron.
c. mercury.
d. pure water.
e. salt solution.

Answer: e
Difficulty: Easy
Bloom's Taxonomy: Knowledge

42.1 ANIMAL CIRCULATORY SYSTEMS: AN INTRODUCTION

3. Which of the following animals has specialized air passages called tracheae throughout the body?

a. earthworm
b. fish
c. hydra
d. insect
e. sponge

Answer: d
Difficulty: Moderate
Bloom's Taxonomy: Knowledge

4. An animal's circulatory system should contain

a. blood to transport carbon dioxide and oxygen.
b. blood vessels leading away from the heart.
c. blood vessels leading to the heart.
d. a heart to pump the blood.
e. all of the above

Answer: e
Difficulty: Easy
Bloom's Taxonomy: Comprehension

5. The gastrovascular cavity of a sponge serves important functions in both the circulatory system and the

a. digestive system.
b. endocrine system.
c. muscular system.
d. reproductive system.
e. urinary system.

Answer: a
Difficulty: Moderate
Bloom's Taxonomy: Knowledge

6. Which of the following statements is correct concerning an open circulatory system?

a. Blood is contained in blood vessels.
b. It contains two circulatory systems.
c. Hemolymph and interstitial fluids mix together.
d. It is more efficient than a closed system.
e. Sinuses are never present.

Answer: c
Difficulty: Moderate
Bloom's Taxonomy: Application

7. Amphibians can accomplish gas exchange by using

a. gills.
b. lungs.
c. respiration through the skin.
d. both a and b
e. all of the above

Answer: e
Difficulty: Difficult
Bloom's Taxonomy: Knowledge

8. The purpose(s) of the pulmonary circuit in a closed circulatory system is to

a. collect and distribute blood to heart itself.
b. take blood to and from all parts of the body.
c. take blood to and from the lungs.
d. both a and b
e. both b and c

Answer: c
Difficulty: Moderate
Bloom's Taxonomy: Knowledge

42.2 BLOOD AND ITS COMPONENTS

9. Plasma approximately makes up what percent of the total blood volume?

a. 10 percent
b. 25 percent
c. 50 percent
d. 75 percent
e. 90 percent

Answer: c
Difficulty: Moderate
Bloom's Taxonomy: Knowledge

10. In humans, blood cells are produced in the

a. bone marrow.
b. heart.
c. kidneys.
d. liver.
e. lymph nodes.

Answer: a
Difficulty: Easy
Bloom's Taxonomy: Knowledge

11. Which component of plasma constitutes antibodies?

a. albumin
b. dissolved gases
c. fibrinogen
d. hormones
e. immunoglobulins

Answer: e
Difficulty: Difficult
Bloom's Taxonomy: Knowledge

12. Which two components of blood involve the formation of clots?

a. albumin and fibrinogen
b. albumin and leukocytes
c. fibrinogen and erythrocytes
d. fibrinogen and platelets
e. platelets and albumin

Answer: d
Difficulty: Difficult
Bloom's Taxonomy: Application

13. What percent of plasma is water?

a. 10 percent
b. 27 percent
c. 55 percent
d. 77 percent
e. 92 percent

Answer: e
Difficulty: Moderate
Bloom's Taxonomy: Knowledge

14. Which two ions are the most abundant in plasma?

a. HCO_3^- and Ca^{2+}
b. HCO_3^- and Cl^-
c. Na^+ and Ca^{2+}
d. Na^+ and Cl^-
e. Na^+ and K^+

Answer: d
Difficulty: Difficult
Bloom's Taxonomy: Knowledge

15. What is the function of erythrocytes?

a. carry oxygen
b. form clots
c. fight off infections
d. produce blood cells
e. secrete erythropoietin

Answer: a
Difficulty: Easy
Bloom's Taxonomy: Knowledge

16. Which components of blood are the smallest in size?

a. erythrocytes
b. leukocytes
c. platelets
d. both a and b
e. all components are the same size

Answer: c
Difficulty: Difficult
Bloom's Taxonomy: Application

17. Which type of leukocyte is the most prevalent in a normal individual?

a. basophils
b. eosinophils
c. lymphocytes
d. monocytes
e. neutrophils

Answer: e
Difficulty: Difficult
Bloom's Taxonomy: Knowledge

18. Erythropoietin is produced in the

a. bone marrow.
b. heart.
c. kidneys.
d. liver.
e. lymph nodes.

Answer: c
Difficulty: Moderate
Bloom's Taxonomy: Knowledge

42.3 THE HEART

19. The function of valves in the heart is to

a. form blood cells.
b. participate in immunity.
c. prevent backflow of blood.
d. prevent clot formation.
e. push blood ahead.

Answer: c
Difficulty: Moderate
Bloom's Taxonomy: Comprehension

20. When blood returns from the pulmonary circuit, it first enters which chamber of the heart?

a. left atrium
b. left ventricle
c. right atrium
d. right ventricle
e. both a and c

Answer: a
Difficulty: Moderate
Bloom's Taxonomy: Comprehension

21. The average heart of a human beats _____ times per minute.

a. 10
b. 53
c. 72
d. 95
e. 120

Answer: c
Difficulty: Easy
Bloom's Taxonomy: Knowledge

22. The "lub" heart sound represents

a. a heart murmur.
b. closure of the AV valves.
c. closure of the SL valves.
d. hypertension.
e. movement of actual heart muscle.

Answer: b
Difficulty: Moderate
Bloom's Taxonomy: Knowledge

23. Hypertension can be caused by

a. abnormalities of the cortex of the adrenal glands.
b. advanced age.
c. certain medications.
d. kidney disease.
e. all of the above

Answer: e
Difficulty: Easy
Bloom's Taxonomy: Knowledge

24. An atrioventricular valve would be located

a. between the left atrium and the left ventricle.
b. between the right atrium and the right ventricle.
c. where blood is entering the aorta.
d. where blood is entering the pulmonary arteries.
e. both a and b

Answer: e
Difficulty: Moderate
Bloom's Taxonomy: Knowledge

25. Where is the sinoatrial node located?

a. left atrium
b. left ventricle
c. pulmonary artery
d. right atrium
e. right ventricle

Answer: d
Difficulty: Moderate
Bloom's Taxonomy: Knowledge

26. Which is the proper order of the components of the electrical activity of the heart?

a. AV node – Purkinje fibers – SA node
b. AV node – SA node – Purkinje fibers
c. Purkinje fibers – AV node – SA node
d. Purkinje fibers – SA node – AV node
e. SA node – AV node – Purkinje fibers

Answer: e
Difficulty: Moderate
Bloom's Taxonomy: Application

27. While running,

a. only your diastolic blood pressure would elevate.
b. only your systolic blood pressure would elevate.
c. both your diastolic and systolic blood pressure would elevate.
d. both your diastolic and systolic blood pressure would decrease.
e. there would be no change in either your diastolic or your systolic blood pressure.

Answer: b
Difficulty: Difficult
Bloom's Taxonomy: Knowledge

42.4 BLOOD VESSELS OF THE CIRCULATORY SYSTEM

28. In which type of blood vessel is the pressure the greatest?

a. arterioles
b. arteries
c. capillaries
d. veins
e. venules

Answer: b
Difficulty: Easy
Bloom's Taxonomy: Application

29. Which of the following would be the most likely cause of deep venous thrombosis?

a. anemia
b. decreased skeletal muscle action
c. high blood pressure
d. increased skeletal muscle action
e. low blood pressure

Answer: b
Difficulty: Moderate
Bloom's Taxonomy: Application

30. Which structure(s) could be considered to be blood reservoirs?

a. arteries
b. arterioles
c. capillaries
d. heart
e. veins

Answer: e
Difficulty: Moderate
Bloom's Taxonomy: Knowledge

31. Atherosclerosis is

a. a clot in an artery.
b. build up of plaque in an artery.
c. build up of plaque in a vein.
d. high blood pressure.
e. low blood pressure.

Answer: b
Difficulty: Moderate
Bloom's Taxonomy: Knowledge

32. Actual gas exchange occurs through the walls of which type of vessel?

a. arteries
b. arterioles
c. capillaries
d. veins
e. venules

Answer: c
Difficulty: Easy
Bloom's Taxonomy: Knowledge

33. Atherosclerosis

a. decreases the size of the lumen of arteries.
b. effects the diameter of veins.
c. has no effect on the size of the lumen of arteries.
d. increases the size of the lumen of arteries.
e. presents many clinical symptoms immediately.

Answer: a
Difficulty: Difficult
Bloom's Taxonomy: Comprehension

42.5 MAINTAINING BLOOD FLOW AND PRESSURE

34. The main force that moves blood throughout the system of blood vessels is

a. arterial blood pressure.
b. skeletal muscle action.
c. smooth muscle action.
d. valve action.
e. venous blood pressure.

Answer: a
Difficulty: Moderate
Bloom's Taxonomy: Knowledge

35. Where are baroreceptors located?

a. adrenal glands
b. brainstem
c. heart
d. erythrocytes
e. walls of blood vessels

Answer: e
Difficulty: Moderate
Bloom's Taxonomy: Knowledge

36. Which of the choices below illustrates the control of cardiac output?

a.   baroreceptors – brainstem – heart
b.   baroreceptors – heart - brainstem
c.   brainstem – baroreceptors – heart
d.   brainstem – heart – baroreceptors
e.   heart – baroreceptors – brainstem

Answer: a
Difficulty: Difficult
Bloom's Taxonomy: Comprehension

42.6 THE LYMPHATIC SYSTEM

37. Lymph moves throughout the lymph vessels due to

a.   action of the heart.
b.   conscious actions.
c.   constriction of blood vessels.
d.   dilation of blood vessels.
e.   skeletal muscle and breathing movements.

Answer: e
Difficulty: Difficult
Bloom's Taxonomy: Knowledge

38. How much fluid does the lymphatic system return to the bloodstream each day?

a.   1 to 2 liters
b.   3 to 4 liters
c.   6 to 7 liters
d.   10 to 13 liters
e.   more than 50 liters

Answer: b
Difficulty: Difficult
Bloom's Taxonomy: Knowledge

39. What is the structure of a lymph capillary?

a.   endothelial cells surrounded by collagen fibers
b.   endothelial cells surrounded by elastic fibers
c.   collagen fibers only
d.   elastic fibers only
e.   endothelial cells only

Answer: a
Difficulty: Difficult
Bloom's Taxonomy: Knowledge

40. Which of the following structures all contain valves?

a.   arteries and veins
b.   arteries and heart
c.   arteries, capillaries, heart
d.   lymph capillaries, arteries, heart
e.   lymph capillaries, veins, heart

Answer: e
Difficulty: Difficult
Bloom's Taxonomy: Application

Unanswered Questions

41. What is thought to help explain the evolution of the blood clotting mechanism?

a.   constant mutations
b.   direct response to environmental factors
c.   gene duplication
d.   sharing of gene pools between species
e.   all of the above

Answer: c
Difficulty: Difficult
Bloom's Taxonomy: Comprehension

Matching

Match the animals to the appropriate description of their circulatory system:

42. _____ amphibians

43. _____ fish

44. _____ mammals

45. _____ snakes

A.   Heart has two chambers and one circuit

B.   Heart has three chambers and two circuits

C.   Heart has three chambers (and a partially formed septum in the ventricle) and two circuits

D.   Heart has four chambers and two circuits

Answers: 42. B   43. A   44. D   45. C

Difficulty: Difficulty
Bloom's Taxonomy: Comprehension
Source: Section 42.1

Short Answer Questions

46. Why would an iron deficiency result in anemia?

Answer: A hemoglobin molecule has an iron atom at its center. Therefore, if inadequate amounts of iron are present, hemoglobin cannot be produced at an adequate rate. Hemoglobin is the oxygen-carrying substance in an erythrocyte, so a deficiency of hemoglobin results in anemia.
Difficulty: Moderate
Bloom's Taxonomy: Application
Source: Section 42.2

47. Why do veins have less elastic components than arteries?

Answer: When blood enters the arteries, it is traveling fast and with a great deal of pressure. The arteries must have a lot of elasticity so that they can swell and recoil as a response to the pressure exerted by the blood flow. By the time blood enters veins, the pressure is greatly dissipated and not as much elasticity is required.
Difficulty: Moderate
Bloom's Taxonomy: Application
Source: Section 42.4

48. What is meant by the term *blood-brain barrier*?

Answer: The capillary walls in the brain are more restrictive than those in other areas of the body. This limits the substances that can pass from the bloodstream into brain tissue.
Difficulty: Moderate
Bloom's Taxonomy: Comprehension
Source: Section 42.4

49. What are the three factors that regulate blood pressure?

Answer: They are cardiac output, the constriction of blood vessels, and the total blood volume.
Difficulty: Moderate
Bloom's Taxonomy: Knowledge
Source: Section 42.5

50. Why is exposure to carbon monoxide so hazardous to the body?

Answer: Hemoglobin has a high affinity for carbon monoxide. Therefore, hemoglobin that combines with carbon monoxide is unavailable to carry oxygen to supply to the needy tissues of the body.
Difficulty: Difficult
Bloom's Taxonomy: Application
Source: Section 42.2

Sequence

As deoxygenated blood (blood lacking oxygen) enters the heart, list the following structures in the correct order.

51. _____ first structure
52. _____ second structure
53. _____ third structure
54. _____ fourth structure
55. _____ fifth structure
56. _____ sixth structure
57. _____ seventh structure
58. _____ eighth structure

A. aorta
B. left atrium
C. left ventricle
D. lungs
E. pulmonary arteries
F. pulmonary veins
G. right atrium
H. right ventricle

Answers: 51. G   52. H   53. E   54. D
55. F   56. B   57. C   58. A

Difficulty: Difficult
Bloom's Taxonomy: Application
Source: Section 42.3

In the human systemic circulation, list the blood vessels that are encountered as the blood leaves the heart.

59. _____ first vessel category
60. _____ second vessel category
61. _____ third vessel category
62. _____ fourth vessel category
63. _____ fifth vessel category

A. arteries
B. arterioles
C. capillaries
D. veins
E. venules

Answers: 59. A   60. B   61. C   62. E 63. D

Difficulty: Moderate
Bloom's Taxonomy: Application
Source: Section 42.4

True/False Questions

If the statement is true, write a "T" in the blank. If the statement is false, make it correct by changing the underlined word(s) and writing the correct word(s) in the answer blank.

64. _____ The <u>systolic</u> blood pressure is the time of the most pressure in the arteries.

65. _____ When experiencing stress, the sympathetic nervous system <u>decreases</u> your blood pressure.

66. _____ Lymph capillaries are <u>smaller</u> in diameter than blood capillaries.

67. _____ The primary function of <u>erythrocytes</u> is to participate in immune reactions.

68. _____ <u>Cardiac output</u> is defined as the pressure and amount of blood pumped by the ventricles.

Answers: 64. T   65. F, increases   66. F, larger 67. F, leukocytes 68. T

Difficulty: Moderate
Bloom's Taxonomy: Knowledge
Source: Sections 42.2, 42.3, 42.5, 42.6

Select the Exception

69. Which two animals listed below do <u>not</u> contain an actual circulatory system?

a. grasshopper and fish
b. grasshopper and hydra
c. hydra and fish
d. sponge and fish
e. sponge and hydra

Answer: e
Difficulty: Moderate
Bloom's Taxonomy: Knowledge
Source: Section 42.1

70. Which type of blood cell is <u>not</u> produced from myeloid stem cells?

a. basophils
b. B lymphocytes
c. eosinophils
d. erythrocytes
e. neutrophils

Answer: b
Difficulty: Difficult
Bloom's Taxonomy: Knowledge
Source: Section 42.2

71. Which of the following vessels does <u>not</u> carry deoxygenated (lacking oxygen) blood?

a. inferior vena cava
b. pulmonary artery
c. pulmonary vein
d. superior vena cava
e. all of the vessels listed above carry deoxygenated blood

Answer: c
Difficulty: Difficult
Bloom's Taxonomy: Application
Source: Section 42.4

72. Which of the following would <u>not</u> be likely to pass easily through capillary walls?

a. erythrocytes
b. glucose
c. ions
d. water
e. carbon dioxide

Answer: a
Difficulty: Moderate
Bloom's Taxonomy: Knowledge
Source: Section 42.4

Essays

73. Why must an animal such as an amphibian remain moist?

Answer: **Even though an amphibian may contain lungs and or gills, those structures do not entirely meet the respiration needs of the animal. In addition, respiration through the skin occurs, and a moist environment is necessary for gas exchange to take place effectively through the skin.**
Difficulty: Difficult
Bloom's Taxonomy: Comprehension
Source: Section 42.1

74. A hemophiliac contains adequate numbers of platelets. Why does he therefore have trouble with clot formation?

Answer: **In addition to platelets, specific clotting factors are necessary for clot formation. Most hemophiliacs have a Factor VIII deficiency, which greatly decreases their ability to form effective clots.**
Difficulty: Difficult
Bloom's Taxonomy: Comprehension
Source: Section 42.2

75. Describe several reasons why blood flows more slowly in capillaries than in arteries?

Answer: **When blood first leaves the heart, it enters large arteries and is traveling very quickly. By the time it reaches the capillary level, it is like traffic that goes from a large superhighway down to many small one-way roads. In addition, the blood must travel more slowly in capillaries to allow for adequate gas exchange to occur. If blood entered the capillary network at the same speed and force that it travels in arteries, the individual capillaries would burst.**
Difficulty: Difficult
Bloom's Taxonomy: Application
Source: Section 42.4

76. Why is the movement of blood in veins sometimes described as "milking a vein"?

Answer: **When the contractions of skeletal muscles in conjunction with valve action pushes in on the walls of veins, the blood is forced upward. This action is somewhat similar to squeezing the teat on a cow to express the milk.**
Difficulty: Difficult
Bloom's Taxonomy: Application
Source: Section 42.4

77. How do lymph nodes aid in a cancer diagnosis?

Answer: **The lymphatic system filters extracellular fluids. Therefore, if cancerous cells have metastasized (spread from the initial location), it is likely that some cancerous cells may be located in the lymph nodes. The examination of numerous lymph nodes surrounding the primary site of the cancer may enable the physician to determine if localized treatment will be effective or if a broader spectrum approach is necessary.**
Difficulty: Difficult
Bloom's Taxonomy: Application
Source: Section 42.6

Labeling

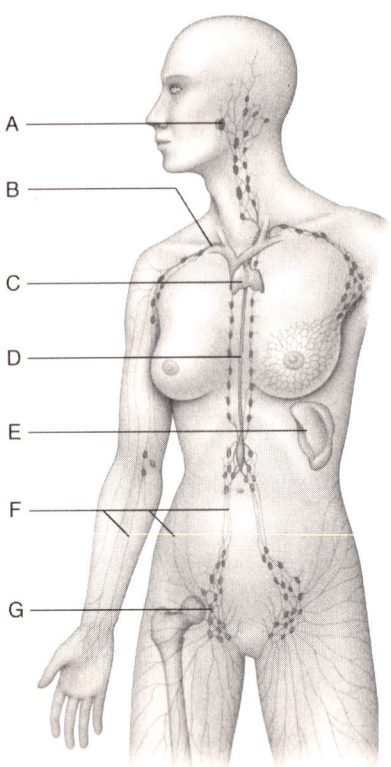

Label these important structures of the lymphatic system.

78. _____ lymph nodes
79. _____ lymph vessels
80. _____ right lymphatic duct
81. _____ spleen
82. _____ thoracic duct
83. _____ thymus
84. _____ tonsils

Answers: 78. G   79. F   80. B   81. E
82. D   83. C   84. A

Difficulty: Difficult
Bloom's Taxonomy: Knowledge
Source: Figure 42.20

Use the list of structures below to label the
diagram of the human heart.

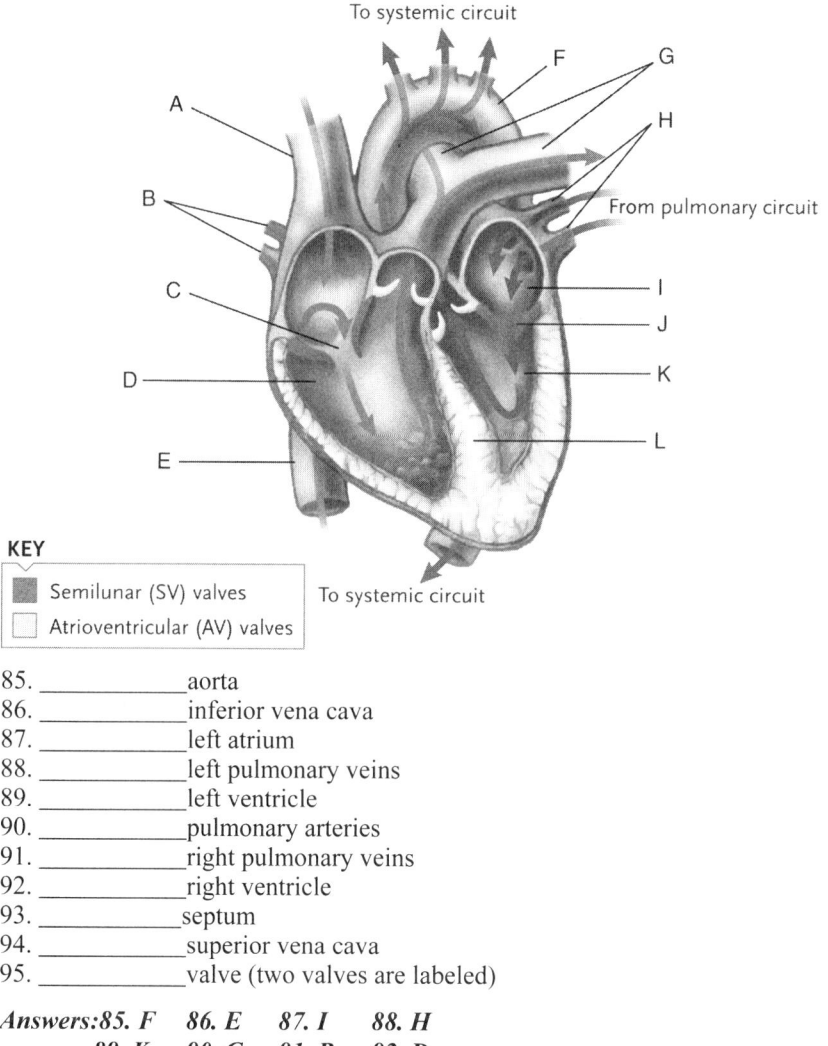

To systemic circuit

F
G
H

A

From pulmonary circuit

B

C

I
J

D

K

E

L

To systemic circuit

KEY

■ Semilunar (SV) valves
□ Atrioventricular (AV) valves

85. _____aorta
86. _____inferior vena cava
87. _____left atrium
88. _____left pulmonary veins
89. _____left ventricle
90. _____pulmonary arteries
91. _____right pulmonary veins
92. _____right ventricle
93. _____septum
94. _____superior vena cava
95. _____valve (two valves are labeled)

Answers:85. F   *86. E*   *87. I*   *88. H*
      89. K   *90. G*   *91. B*   *92. D*
      93. L   *94. A*   *95. C, J*

Difficulty: Difficult
Bloom's Taxonomy: Knowledge
Source: Figure 42.9

43

DEFENSES AGAINST DISEASE

Multiple-Choice

WHY IT MATTERS

1.   Acquired immune deficiency syndrome (AIDS) now infects about

a.   20 million people; however, the number is decreasing because of an effective vaccination campaign.
b.   20 million people; however, the number is decreasing because of effective antibiotic treatments.
c.   40 million people; however, the number is decreasing because of an effective vaccination campaign.
d.   40 million people; however, the number is decreasing because of effective antibiotic treatments.
e.   40 million people, and climbing.

Answer: e
Difficulty: Moderate
Bloom's Taxonomy: Knowledge

2.   Edward Jenner injected individuals with _____ lesions to protect them from _____.

a.   chickenpox; smallpox
b.   smallpox; chickenpox
c.   smallpox; cowpox
d.   cowpox; smallpox
e.   cowpox; chickenpox

Answer: d
Difficulty: Easy
Bloom's Taxonomy: Knowledge

3.   The term immunity comes from the Latin *immunis*, which means

a.   protect.
b.   exempt.
c.   attack.
d.   sword.
e.   cow.

Answer: b
Difficulty: Easy
Bloom's Taxonomy: Knowledge

43.1 THREE LINES OF DEFENSE AGAINST INVASION

4.   Arrange the following three lines of defense from earliest to latest.

        1 = Adaptive immune system
        2 = Physical barriers
        3 = Innate immune system
a.   1, 2, 3
b.   2, 1, 3
c.   2, 3, 1
d.   3, 1, 2
e.   3, 2, 1

Answer: c
Difficulty: Easy
Bloom's Taxonomy: Knowledge

5.   The enzyme lysozyme offers protection from

a.   viruses.
b.   bacteria.
c.   fungi.
d.   protists.
e.   viruses and bacteria.

Answer: b
Difficulty: Easy
Bloom's Taxonomy: Knowledge

6.   Adaptive immunity is also referred to as _____ immunity.

a.   acquired
b.   active
c.   passive
d.   latent
e.   hidden

Answer: a
Difficulty: Easy
Bloom's Taxonomy: Knowledge

7.   Which of the following is *not* involved in the body's first line of defense?

a.   ciliated cells
b.   acidity of stomach
c.   lysozyme
d.   digestive enzymes of small intestine
e.   complement

Answer: e
Difficulty: Easy
Bloom's Taxonomy: Knowledge

8. How is adaptive immunity different from innate immunity?

a. adaptive immunity is specific, while innate immunity is not.
b. adaptive immunity has memory, while innate immunity does not.
c. adaptive immunity is immediate, while innate immunity is not.
d. a and b only.
e. a, b, and c.

Answer: d
Difficulty: Moderate
Bloom's Taxonomy: Knowledge

43.2 NONSPECIFIC DEFENSES: INNATE IMMUNITY

9. The major phagocytic cells in the body are

a. macrophages and monocytes.
b. macrophages and neutrophils.
c. macrophages and mast cells.
d. monocytes and mast cells.
e. monocytes and basophils.

Answer: b
Difficulty: Moderate
Bloom's Taxonomy: Knowledge

10. _____ are molecules that regulate defense responses via signal transduction pathways.

a. Complement
b. Histamines
c. Prostaglandins
d. Cytokines
e. Opsonins

Answer: d
Difficulty: Moderate
Bloom's Taxonomy: Knowledge

11. Which of the following is true of histamine?

a. causes blood vessels to contract
b. causes capillaries to lose their permeability
c. causes an outward flow of fluids from the capillaries
d. reduces tissue swelling
e. none of the above

Answer: c
Difficulty: Moderate
Bloom's Taxonomy: Comprehension

12. Which of the following is *not* a symptom of the inflammatory response?

a. pain
b. swelling
c. tedness
d. heat
e. All of the above are symptoms of inflammation

Answer: e
Difficulty: Moderate
Bloom's Taxonomy: Knowledge

13. While running in a garden, a girl steps on an old board and a nail goes through her shoe and skin, causing a deep puncture wound. Which of the following non-specific defense mechanisms will come to her rescue?

a. epithclial cells with cilia
b. lysozyme
c. acidic skin secretions
d. inflammation
e. natural killer cells

Answer: d
Difficulty: Moderate
Bloom's Taxonomy: Application

14. _____ are most responsible for the symptoms of inflammation.

a. Cytotoxic T cells
b. Macrophages
c. Mast cells
d. Neutrophils
e. Eosinophils

Answer: c
Difficulty: Moderate
Bloom's Taxonomy: Knowledge

15. Interferons

a. make holes through bacterial cell membranes.
b. make holes through bacterial cell walls.
c. prevent entry of viruses into cells.
d. block the replication of viruses within cells.
c. bind iron and make it unavailable for bacteria and viruses.

Answer: d
Difficulty: Moderate
Bloom's Taxonomy: Knowledge

16. Natural killer cells destroy target cells by

a. releasing antibiotics, which breakdown the cell membrane.
b. releasing perforin, which creates holes in the cell membrane.
c. releasing histamine, which signals other blood cells to aggregate.
d. releasing lysozyme, which breaks down the cell wall.
e. releasing antibodies, which break down the cell wall.

Answer: b
Difficulty: Moderate
Bloom's Taxonomy: Knowledge

17. "Complement" refers to a set of specific

a. vaccines.
b. defense proteins.
c. antibodies.
d. antibiotics.
e. lymphocytes.

Answer: b
Difficulty: Moderate
Bloom's Taxonomy: Knowledge

18. Which of the following is a function of activated complement proteins?

a. lysis of pathogens
b. enhancing inflammation
c. coating pathogens to enhance phagocytosis
d. a and b only
e. a, b, and c

Answer: e
Difficulty: Moderate
Bloom's Taxonomy: Comprehension

19. All of the following are nonspecific defense mechanisms *except*

a. acid secretions of the stomach.
b. inflammation.
c. antibodies.
d. mucus in the respiratory tract.
e. skin.

Answer: c
Difficulty: Easy
Bloom's Taxonomy: Knowledge

20. Which of the following terms is most characteristic of innate immunity?

a. specificity
b. memory
c. gene rearrangement
d. phagocytosis
e. clonal selection

Answer: d
Difficulty: Easy
Bloom's Taxonomy: Knowledge

43.3 SPECIFIC DEFENSES: ADAPTIVE IMMUNITY

21. The leukocytes that are central to adaptive immunity are the

a. lymphocytes.
b. macrophages.
c. neutrophils.
d. mast cells.
e. eosinophils.

Answer: a
Difficulty: Easy
Bloom's Taxonomy: Knowledge

22. How is antibody-mediated immunity different from cell-mediated immunity?

a. Antibody-mediated immunity requires lymphocytes; cell-mediated immunity does not.
b. Antibody-mediated immunity is innate; cell-mediated immunity is not.
c. Antibody-mediated immunity is adaptive; cell-mediated immunity is not.
d. Antibody-mediated immunity requires both B and T cells; cell-mediated immunity requires T cells but not B cells.
e. Antibody-mediated immunity requires T cells; cell-mediated immunity requires B cells.

Answer: d
Difficulty: Moderate
Bloom's Taxonomy: Comprehension

23. Which of the following can serve as antigens?

a. proteins
b. carbohydrates
c. nucleic acids
d. a and b only
e. a, b, and c

Answer: e
Difficulty: Easy
Bloom's Taxonomy: Knowledge

24. An antibody molecule consists of a total of

a. two polypeptide chains.
b. four polypeptide chains.
c. two polysaccharide chains.
d. four polysaccharide chains.
e. four phospholipids chains.

Answer: b
Difficulty: Easy
Bloom's Taxonomy: Knowledge

25. Antibodies are classified as

a. nucleoproteins.
b. immunoglobulins.
c. steroids.
d. carbohydrates.
e. lipids.

Answer: b
Difficulty: Easy
Bloom's Taxonomy: Knowledge

26. Which of the following is *not* an antibody class in humans?

a. IgA
b. IgD
c. IgE
d. IgF
e. IgG

Answer: d
Difficulty: Easy
Bloom's Taxonomy: Knowledge

Use the figure shown below for questions 27–33.

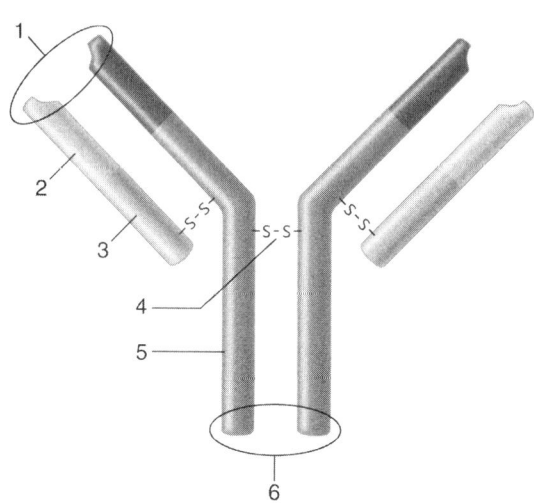

27. The antigen-binding site is indicated by number

a. 1.
b. 2.
c. 4.
d. 5.
e. 6.

Answer: a
Difficulty: Easy
Bloom's Taxonomy: Knowledge

28. The constant region of a light chain is indicated by number

a. 1.
b. 2.
c. 3.
d. 4.
e. 5.

Answer: c
Difficulty: Easy
Bloom's Taxonomy: Knowledge

29. The variable region of a light chain is indicated by number

a. 1.
b. 2.
c. 3.
d. 4.
e. 5.

Answer: b
Difficulty: Easy
Bloom's Taxonomy: Knowledge

30. The constant region of a heavy chain is indicated by number

a. 1.
b. 2.
c. 3.
d. 4.
e. 5.

Answer: e
Difficulty: Easy
Bloom's Taxonomy: Knowledge

31. The hinge region is indicated by number

a. 1.
b. 2.
c. 4.
d. 5.
e. 6.

Answer: c
Difficulty: Easy
Bloom's Taxonomy: Knowledge

32. The region that determines the class of antibody is indicated by number

a. 1.
b. 2.
c. 3.
d. 4.
e. 5.

Answer: e
Difficulty: Moderate
Bloom's Taxonomy: Comprehension

33. If this antibody serves as a B-cell receptor, it will bind to the host cell membrane via region number

a. 1.
b. 2.
c. 3.
d. 4.
e. 6.

Answer: e
Difficulty: Easy
Bloom's Taxonomy: Comprehension

34. Antibody diversity is due to genetic influences on the

a. hinge region.
b. heavy chain.
c. constant region.
d. light chain.
e. variable region.

Answer: e
Difficulty: Easy
Bloom's Taxonomy: Comprehension

35. The major source of diversity for B and T cell receptors is

a. mitosis.
b. meiosis.
c. transcript processing.
d. random recombination of gene segments.
e. mutation.

Answer: d
Difficulty: Easy
Bloom's Taxonomy: Comprehension

36. Consider two IgG molecules, each of which reacts with a different epitope. Which of the following statement(s) is(are) *true*?

a. They will have the same amino acid sequence in their constant regions.
b. They will have different amino acid sequences in their variable regions.
c. They can bind the same antigen.
d. a and b only.
e. a, b, and c.

Answer: e
Difficulty: Moderate
Bloom's Taxonomy: Application

37. The antibodies that activate the complement system are

a. IgA and IgD.
b. IgG and IgM.
c. IgG and IgD.
d. IgE and IgM.
e. IgA and IgE.

Answer: b
Difficulty: Easy
Bloom's Taxonomy: Knowledge

38. How are B-cell receptors different from antibodies?

a. B-cell receptors bind antigen, while antibodies do not.
b. B-cell receptors do not contain heavy chains, while antibodies do.
c. B-cell receptors include a variable region, while antibodies do not.
d. B-cell receptors include a transmembrane domain, while antibodies do not.
e. B-cell receptors are produced by B cells, while antibodies are not.

Answer: d
Difficulty: Moderate
Bloom's Taxonomy: Comprehension

39. How are B-cell receptors different from T-cell receptors?

a. B-cell receptors bind antigen, while T-cell receptors do not.
b. B-cell receptors have a more complex structure.
c. B-cell receptors have a more simple structure.
d. B-cell receptors are bound to the cell membrane, while T-cell receptors are bound to the nuclear membrane.
e. None of the above.

Answer: b
Difficulty: Moderate
Bloom's Taxonomy: Comprehension

40. Which of the following *best* describes an epitope?

a. the antigen binding site of an antibody molecule
b. found only in IgG and IgM
c. part of an antigen that stimulates production of, and binds to, specific antibody
d. too small to stimulate an immune response
e. a complement protein that initiates lysis of target cells

Answer: c
Difficulty: Easy
Bloom's Taxonomy: Knowledge

41. Dendritic cells enter the _____ and alert T cells to threats.

a blood
b. lymph nodes
c. bone marrow
d. liver
e. lung

Answer: b
Difficulty: Moderate
Bloom's Taxonomy: Knowledge

42. A CD4+ T cell binds to an antigen only if this antigen

a is in a complex with a class I MHC protein.
b. is in a complex with a class II MHC protein.
c. is in a complex with a class III MHC protein.
d. is *not* in a complex with MHC proteins.
e. is presented on the surface of a B cell.

Answer: b
Difficulty: Moderate
Bloom's Taxonomy: Knowledge

43. Helper T cells secrete _____, which activate B cells and stimulates their proliferation.

a antigens
b. allergens
c. antibodies
d. interferons
e. interleukins

Answer: e
Difficulty: Moderate
Bloom's Taxonomy: Knowledge

Use the figure shown below for questions 44–48.

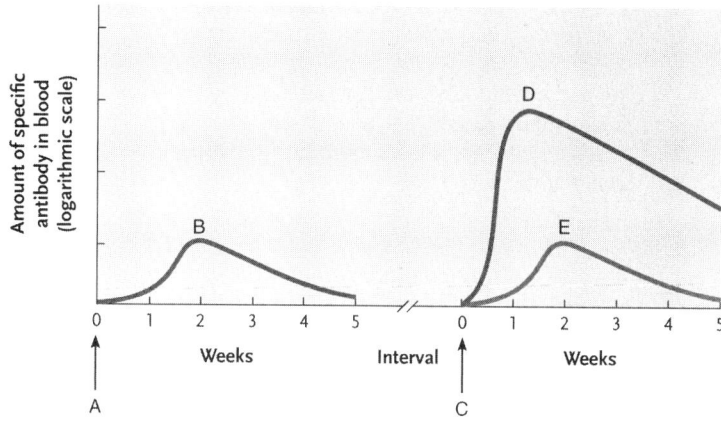

44. The interval between points A and C may be as long as several

a hours.
b. days.
c. weeks.
d. months.
e. years.

Answer: e
Difficulty: Easy
Bloom's Taxonomy: Comprehension

45. In the graph labeled B, effector B cells are produced between weeks

a 0 and 1.
b. 0 and 2.
c. 2 and 3.
d. 2 and 4.
e. 2 and 5.

Answer: b
Difficulty: Moderate
Bloom's Taxonomy: Comprehension

46. Which of the following describes graph D?

a a primary response to the initial antigen
b. a secondary response to the initial antigen
c a primary response to an antigen distinctly different from the initial antigen
d. a secondary response to an antigen distinctly different from the initial antigen
e. none of the above

Answer: b
Difficulty: Moderate
Bloom's Taxonomy: Comprehension

47. Which of the following describes graph E?

a   a primary response to the initial antigen
b.   a secondary response to the initial antigen
c   a primary response to an antigen distinctly
     different from the initial antigen
d.   a secondary response to an antigen
     distinctly different from the initial antigen
e.   none of the above

Answer: c
Difficulty: Moderate
Bloom's Taxonomy: Comprehension

48. Which immune system component has a principal role in graph D?

a   IgM
b.   IgA
c   cytotoxic T cells
d.   memory cells
e.   NK cells

Answer: d
Difficulty: Moderate
Bloom's Taxonomy: Comprehension

49. Arrange the steps of antibody-mediated immunity in the proper order.

     1) Plasma cells secrete antibodies.
     2) Helper T cells interact with B cells
        displaying the same antigen-MHC
        complex.
     3) Contact with antigen.
     4) Clones of B cells produced.
     5) Fragmentation of antigen.
a   3, 2, 5, 1, 4
b.   3, 5, 2, 4, 1
c.   1, 3, 5, 2, 4
d.   4, 3, 2, 1, 5
e.   5, 3, 4, 2, 1

Answer: b
Difficulty: Moderate
Bloom's Taxonomy: Comprehension

50. Which of the following statements best describes the clonal selection theory?

a.   B cells differentiate into plasma cells and T
     cells.
b.   An antibody is flexible and can change
     shape to bind different antigens.
c.   An individual contains a single type of B
     cell.
d.   Each B cell can produce many types of
     antibodies, each of which reacts with a
     different antigen.
e.   An individual contains many types of B
     cells, and each secretes only one kind of
     antibody.

Answer: e
Difficulty: Moderate
Bloom's Taxonomy: Comprehension

51. Which of the following can be present in a vaccine?

a.   killed pathogen
b.   weakened pathogen
c.   unmodified antigen from a pathogen
d.   a and b only
e.   a, b, and c

Answer: d
Difficulty: Moderate
Bloom's Taxonomy: Knowledge

52. Which of the following is an example of active immunity?

a.   vaccination
b.   immunity to chicken pox, acquired as a
     result of a childhood disease
c.   passage of antibody across the placenta
d.   a and b only
e.   a, b, and c

Answer: d
Difficulty: Moderate
Bloom's Taxonomy: Comprehension

53. How is active immunity different from passive immunity?

a.   Active immunity is slower in onset and
     shorter-lived.
b.   Active immunity is slower in onset and
     longer-lived.
c.   Active immunity is faster in onset and
     shorter-lived.
d.   Active immunity is faster in onset and
     longer-lived.
e.   Only active immunity involves antibodies.

Answer: b
Difficulty: Moderate
Bloom's Taxonomy: Comprehension

54. A patient's serum is analyzed and found to have an abnormally low level of CD8 molecules. This suggests that the patient has insufficient numbers of

a   macrophages.
b. B cells.
c. cytotoxic T cells.
d. helper T cells.
e. NK cells.

Answer: c
Difficulty: Moderate
Bloom's Taxonomy: Application

55. A typical primary response to an antigen peaks in about

a   one–two days.
b. four–five days.
c. two weeks.
d. two months.
e. four months.

Answer: c
Difficulty: Moderate
Bloom's Taxonomy: Knowledge

56. Which of the following statements about effector cells is *true*?

a   They are fully differentiated lymphocytes.
b. They can develop from either B cells or T cells.
c. They secrete immunoglobulins or interleukins depending on their origin.
d. a and b only.
e. a, b, and c.

Answer: e
Difficulty: Moderate
Bloom's Taxonomy: Comprehension

57. Which of the following statements about body cells is *true*?

a   All lymphocytes are leukocytes, but not all leukocytes are lymphocytes.
b. All leukocytes are lymphocytes, but not all lymphocytes are leukocytes.
c. Leukocyte and lymphocyte are different types of red blood cells.
d. Leukocyte and lymphocyte are different names for the same cell.
e. None of the above.

Answer: a
Difficulty: Moderate
Bloom's Taxonomy: Comprehension

58. A CD8+ T cell binds to an antigen only if this antigen

a   is in a complex with a class I MHC protein.
b. is in a complex with a class II MHC protein.
c. is in a complex with a class III MHC protein.
d. is *not* in a complex with MHC proteins.
e. is presented on the surface of a B cell.

Answer: a
Difficulty: Moderate
Bloom's Taxonomy: Knowledge

59. Which of the following characteristics does *not* apply to both B and T cells?

a   mature in the bone marrow
b. can differentiate into memory cells
c. secrete proteins
d. participate in acquired immunity
e. all of the above

Answer: e
Difficulty: Moderate
Bloom's Taxonomy: Knowledge

60. Hybridomas are cells formed by the fusion of

a   B cells and melanomas.
b. B cells and myelomas.
c   T cells and melanomas.
d. T cells and myelomas.
e. B cells, T cells, and myelomas.

Answer: b
Difficulty: Moderate
Bloom's Taxonomy: Knowledge

61. A monoclonal antibody reacts against

a. a single antigenic determinant.
b. different epitopes on the same antigen.
c. different antigens of the same organism.
d. a single type of pathogen.
e. a single type of macromolecule.

Answer: a
Difficulty: Easy
Bloom's Taxonomy: Knowledge

FOCUS ON RESEARCH—RESEARCH ORGANISMS: THE MIGHTY MOUSE

62. Which of the following statements about the mouse is *false*?

a. Its scientific name is *Mus musculus*.
b. Females give birth to offspring about three weeks after mating.
c. It has been used to disprove the "inheritance of acquired characters" hypothesis.
d. Knockout mice are those that carry a lethal allele.
e. All of the above are true.

Answer: d
Difficulty: Moderate
Bloom's Taxonomy: Comprehension

INSIGHTS FROM THE MOLECULAR REVOLUTION: SOME CANCER CELLS KILL CYTOTOXIC T CELLS TO DEFEAT THE IMMUNE SYSTEM

63. Cells that have the FasL molecule

a. must also possess the Fas protein.
b. must be cytotoxic T cells.
c. must be melanocytes.
d. will attack each other.
e. will attack cells that have the Fas protein.

Answer: e
Difficulty: Moderate
Bloom's Taxonomy: Comprehension

43.4 MALFUNCTIONS AND FAILURES OF THE IMMUNE SYSTEM

64. B cells and T cells that react with self-antigens are destroyed by

a. apoptosis.
b. phagocytosis.
c. cytokinesis.
d. diakinesis.
e. endocytosis.

Answer: a
Difficulty: Easy
Bloom's Taxonomy: Knowledge

65. Autoimmune diseases result from a breakdown in the ability of the body to

a. produce antibodies.
b. produce memory cells.
c. destroy major histocompatibility proteins.
d. distinguish self from non-self.
e. distinguish between harmless and pathogenic microorganisms.

Answer: d
Difficulty: Moderate
Bloom's Taxonomy: Knowledge

66. Rheumatoid arthritis is caused by

a. a deficiency of antibodies.
b. inability to respond to antigenic stimulation.
c. a virus.
d. a self-attack on blood tissue.
e. a self-attack on joints.

Answer: e
Difficulty: Easy
Bloom's Taxonomy: Knowledge

67. Which autoimmune disease is correctly matched with the tissue it affects?

a. lupus – blood
b. type I diabetes – pancreas
c. multiple sclerosis – nerves
d. a and b only
e. a, b, and c

Answer: e
Difficulty: Moderate
Bloom's Taxonomy: Knowledge

68. Allergy is a(n) _____ response to _____ antigens.

a. overactive; environmental
b. underactive; environmental
c. overactive; cellular
d. underactive; cellular
e. normal level; microbial

Answer: a
Difficulty: Easy
Bloom's Taxonomy: Knowledge

69. Arrange the steps of an allergic response in the proper order.

    1) Allergic symptoms appear.
    2) Allergen binds with IgE.
    3) IgE combines with mast cell receptors.
    4) Mast cells release histamine.
    5) Plasma cells sensitized.

a. 5, 3, 2, 4, 1
b. 1, 5, 2, 4, 3
c. 5, 2, 4, 3, 1
d. 1, 4, 5, 2, 3
e. 4, 5, 2, 3, 1

Answer: a
Difficulty: Moderate
Bloom's Taxonomy: Comprehension

FOCUS ON RESEARCH: APPLIED RESEARCH: HIV AND AIDS

70. Which of the following behaviors is not associated with a high risk of HIV transmission?

a. unprotected sex
b. exposure to contaminated blood products
c. intravenous drug use
d. sharing bathroom facilities
e. being born to an HIV-infected mother

Answer: d
Difficulty: Easy
Bloom's Taxonomy: Knowledge

71. The primary targets for HIV are

a. helper T cells and macrophages.
b. helper T cells and cytotoxic T cells.
c. B cells, helper T cells, and cytotoxic T cells.
d. B cells and helper T cells.
e. B cells and macrophages.

Answer: a
Difficulty: Easy
Bloom's Taxonomy: Knowledge

72. The development of an HIV vaccine has been very difficult because

a. the virus has no antigens to target.
b. the viral antigens are hidden by the capsid.
c. the viral coat antigens change frequently as the virus replicates.
d. the virus produces enzymes that destroy the vaccine.
e. the virus is a DNA virus, and vaccines only work against RNA viruses.

Answer: c
Difficulty: Easy
Bloom's Taxonomy: Comprehension

43.5 DEFENSES IN OTHER ANIMALS

73. How do the antibodies of sharks differ from those of mammals?

a. Shark antibodies do not have light chains.
b. Shark antibodies do not have heavy chains.
c. Shark antibodies have a different arrangement of embryonic gene segments.
d. Shark antibodies are produced without rearrangement of gene segments.
e. None of the above.

Answer: c
Difficulty: Moderate
Bloom's Taxonomy: Knowledge

74. Which of the following defense mechanisms are found in invertebrates?

a. phagocytes
b. lysozyme
c. antibodies
d. a and b only
e. a, b and c

Answer: d
Difficulty: Moderate
Bloom's Taxonomy: Knowledge

75. Moths have an Ig-family protein called _____, which binds to pathogens and enhances their phagocytosis.

a. hemolysin
b. cadherin
c. bradykinin
d. hemolin
e. defensin

Answer: d
Difficulty: Moderate
Bloom's Taxonomy: Knowledge

UNANSWERED QUESTIONS

76. Which of the following statements about pharyngitis is *false*?

a. It is also termed "strep throat."
b. It is caused by the Gram-negative bacterium, group A streptococcus (GAS).
c. It occurs in three phases: colonization, acute, asymptomatic.
d. The pattern of GAS gene expression changes over the course of the disease.
e. All of the above are true.

Answer: b
Difficulty: Moderate
Bloom's Taxonomy: Knowledge

77. The Nef protein helps HIV evade the adaptive immune system by

a. destroying cytotoxic T cells.
b. destroying class I MHC proteins as soon as they get to the surface of infected cells.
c. destroying class I MHC proteins in the cytoplasm.
d. binding to class I MHC proteins at the cell surface, thereby blocking their availability to cytotoxic T cells.
e. binding to class I MHC proteins in the cytoplasm, thereby preventing their transport to the cell surface.

Answer: e
Difficulty: Moderate
Bloom's Taxonomy: Comprehension

Integrative Multiple-Choice

78. Which of the following cells can serve as antigen-presenting cells?

a. macrophages, basophils, and B cells
b. dendritic cells, B cells, and T cells
c. macrophages, dendritic cells, and B cells
d. B cells, T cells, and basophils
e. macrophages, T cells, and basophils

Answer: c
Difficulty: Moderate
Bloom's Taxonomy: Comprehension
Source: Sections 43.2 and 43.3

79. Which of the following molecules is *mismatched* with its source?

a. interferons – virus-infected cells
b. histamine – mast cells
c. lysozyme – epithelial cells
d. immunoglobulin – B cells
e. interleukins – cytotoxic T cells

Answer: e
Difficulty: Moderate
Bloom's Taxonomy: Comprehension
Source: Sections 43.1–43.4

80. You had a cold last month and you caught another one this month from your brother. Why could this happen?

a. Memory cells were not formed as a response to the first cold.
b. The antibodies formed as a response to the first cold were degraded.
c. The two cold viruses are significantly different.
d. Your macrophages are not working properly.
e. Your antigen-presenting cells do not recognize the second cold virus.

Answer: c
Difficulty: Moderate
Bloom's Taxonomy: Comprehension
Source: Sections 43.2 and 43.3

81. Which of the following defends against *both* bacteria and viruses?

a. interferons
b. lysozyme
c. NK cells
d. plasma cells
e. cytotoxic T cells

Answer: d
Difficulty: Moderate
Bloom's Taxonomy: Comprehension
Source: Sections 43.1–43.3

82. Which of the following is *not* a characteristic of the acquired immune response?

a. memory
b. specificity
c. ability to respond to millions of different antigens
d. ability to distinguish self from nonself
e. all of the above are characteristics of the acquired immune response

Answer: c
Difficulty: Moderate
Bloom's Taxonomy: Comprehension
Source: Section 43.3

83. Which of the following molecules is *mismatched* with its function?

a. T cell – cell-mediated immunity
b. neutrophil – phagocytosis
c. B cell – antibody-mediated immunity
d. basophil – allergy
e. All of the above are correctly matched.

Answer: e
Difficulty: Moderate
Bloom's Taxonomy: Comprehension
Source: Sections 43.2–43.4

Matching

Match each cell with its correct description.

84. _____ CD4+ cell

85. _____ CD8+ cell

86. _____ dendritic cell

87. _____ eosinophil

88. _____ hybridoma

89. _____ mast cell

90. _____ monocyte

91. _____ neutrophil

92. _____ NK cell

93. _____ plasma cell

A. Part of the innate immune system, but functions primarily to stimulate an adaptive immune response

B. Functions in the defense against parasitic worms

C. Differentiates into helper T cell

D. Differentiates into macrophage

E. Differentiates into cytotoxic T cell

F. Induces apoptosis in virus-infected cells

G. Produces antibodies

H. Used to produce monoclonal antibodies

I. Kills bacteria and then usually dies

J. Releases histamine

Answers: *84. C   85. E   86. A   87. B   88. H   89. J   90. D   91. I   92. F   93. G*

Difficulty: Difficult
Bloom's Taxonomy: Comprehension
Source: Sections 43.2–43.4

Select the Exception

94. Which of the following statements about innate immunity is *false*?

a. It involves a variety of chemicals, including defensins and complement.
b. It involves a variety of cells, including phagocytes and NK cells.
c. It protects against viruses and other pathogens.
d. It must be primed by the presence of antigens.
e. It has no memory of prior exposure to an organism.

Answer: d
Difficulty: Moderate
Bloom's Taxonomy: Comprehension
Source: Section 43.2

95. Which of the following is *not* a mechanism by which antibodies can protect the body from foreign antigens?

a. neutralization
b. agglutination
c. stimulation of cytotoxic T cells
d. enhancing of phagocytosis
e. activation of complement

Answer: c
Difficulty: Moderate
Bloom's Taxonomy: Knowledge
Source: Section 43.3

96. Which of the following is *not* a part of an antibody molecule?

a. light chain
b. heavy chain
c. antigen-binding site
d. C and V regions
e. epitope

Answer: e
Difficulty: Easy
Bloom's Taxonomy: Comprehension
Source: Section 43.3

97. HIV replication requires all of the following *except*

a. RNA genome.
b. integrase.
c. reverse transcriptase.
d. cytokine.
e. gp120.

Answer: d
Difficulty: Moderate
Bloom's Taxonomy: Knowledge
Source: Focus on Research (page 990)

98. Which of the following is *not* an autoimmune disease?

a. type I diabetes
b. rheumatoid arthritis
c. hepatitis
d. lupus
e. multiple sclerosis

Answer: c
Difficulty: Moderate
Bloom's Taxonomy: Knowledge
Source: Section 43.4

Choice

Choose the class of immunoglobulin that corresponds to each of the characteristics listed below.

a. IgA   b. IgD   c. IgE
d. IgG   e.   IgM

99. _____ the first to be produced in response to an antigenic stimulus

100. _____ the most abundant in secretions

101. _____ crosses the placenta

102. _____ the most abundant in blood

103. _____ defends against parasitic worms

104. _____ the major component of the secondary immune response

105. _____ plays major role in the allergic response

106. _____ may exist as a pentamer

107. _____ has an uncertain function

***Answers: 99. e 100. a 101. d 102. d
103. c 104. d 105. c 106. e 107 .b***

Difficulty: Difficult
Bloom's Taxonomy: Comprehension
Source: Section 43.3

True/False

If the statement is true, write a "T" in the blank. If the statement is false, make it correct by changing the underlined words(s) and writing the correct word(s) in the answer blank.

108. _____ In the respiratory tract, <u>pileated</u> cells constantly sweep mucus with its trapped foreign matter.

109. _____ <u>Opsonins</u> are antimicrobial peptides that play an important role in the innate immunity of the mammalian intestinal tract.

110. _____ <u>B cells</u> differentiate into plasma cells, which produce antibodies.

111. _____ RNA interference (RNAi) defends primarily against viruses with <u>single-stranded DNA</u> genomes.

112. _____ <u>Cephalosporin</u> is a drug routinely used to reduce the rejection of transplanted organs.

113. _____ T cells are produced in the bone marrow but mature in the <u>thyroid</u>.

114. _____ <u>Antibody-mediated</u> immunity is also called humoral immunity.

115. _____ Anaphylactic shock can be controlled by an immediate injection of <u>histamine</u>.

116. _____ Upon antigenic stimulation, CD8+ T cells differentiate into <u>helper T cells</u>.

117. _____ HIV uses the protein <u>integrase</u> to insert its genome into the host cell's DNA.

Answers: 108. F, ciliated 109. F, defensins 110. T 111. F, double-stranded RNA 112. F, cyclosporine 113. F, thymus 114. T 115. F, epinephrine 116. F, cytotoxic T cells 117. T

Difficulty: Difficult
Bloom's Taxonomy: Knowledge, Comprehension
Source: Sections 43.1–43.4

Labeling

Identify each labeled part of the following illustration of the human immunodeficiency virus (HIV).

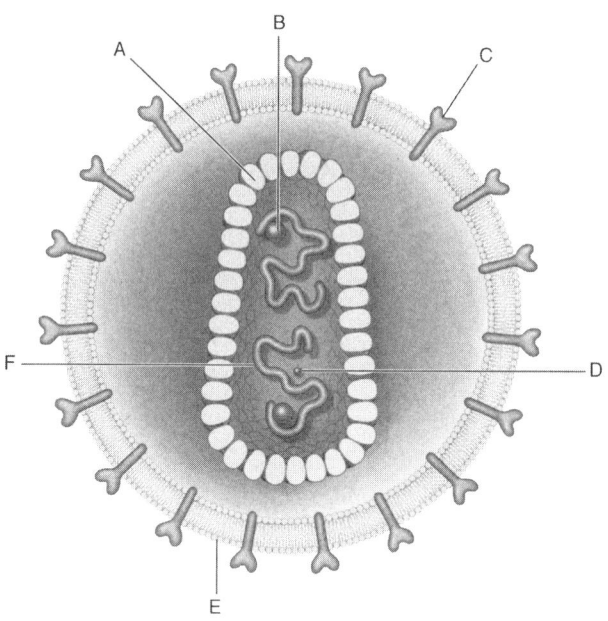

118. _____ membrane coat
119. _____ viral RNA genome
120. _____ viral coat protein
121. _____ gp
122. _____ reverse transcriptase
123. _____ integrase

Answers: 118. E 119. F 120 A 121. C 122. B 123. D

Difficulty: Moderate
Bloom's Taxonomy: Knowledge
Source: Figure a (page 990)

Essay

124. What is clonal selection? How does it occur?

Answer: Clonal selection is the process by which a particular B cell is specifically selected for cloning when it recognizes a particular foreign antigen. An individual has an enormous number of unactivated B cells, each displaying a specific B-cell receptor (BCR) on its surface. After binding and processing the antigen that matches its BCR, and under the stimulation of interleukins from helper T cells, the B cell proliferates to produce a clone of cells. Some of the B-cell clones differentiate into plasma cells, which are short-lived cells that produce antibodies. A few B-cell clones differentiate into memory B cells, which remain in circulation ready to mount a more vigorous response against the same antigen at a later time.
Difficulty: Moderate
Bloom's Taxonomy: Knowledge, Comprehension
Source: Sections 43.3

125. Discuss the phenomena of immunological tolerance and autoimmune disease.

Answer: Immunological tolerance is the phenomenon that protects the body's own tissues from attack by the immune system. Self-antigens are typically not recognized as foreign by B and T cells, and do not induce an immune response. Those B and T cells that do react with self-antigens are destroyed during their differentiation in the bone marrow and thymus, respectively. This occurs throughout the lifetime of the individual. Failure of this immunological tolerance results in autoimmune disease, which is the production of antibodies against the molecules of the body. In most cases, the effects of such anti-self antibodies are mild. However, in about 5-10 percent of the human population, they can cause serious problems. Autoimmune diseases include: type I diabetes (where the insulin-producing cells of the pancreas are attacked); systemic lupus erythematosus (antibodies against blood cells and platelets, and potentially the heart and kidneys); rheumatoid arthritis (self-attack on joints); and multiple sclerosis (antibodies against a protein in the myelin sheaths surrounding neurons).
Difficulty: Moderate
Bloom's Taxonomy: Knowledge, Comprehension
Source: Sections 43.4

44

GAS EXCHANGE: THE RESPIRATORY SYSTEM

Multiple-Choice

WHY IT MATTERS

1. Of the following cells, which are most sensitive to reduced O_2 levels?

a. heart cells
b. kidney cells
c. liver cells
d. brain cells
e. pancreatic cells

Answer: d
Difficulty: Moderate
Bloom's Taxonomy: Knowledge

2. What happens to atmospheric O_2 levels as elevation increases?

a. O_2 levels increase.
b. O_2 levels decrease.
c. O_2 levels remain constant.
d. O_2 levels increase, but there is none present above 20,000 feet.
e. O_2 levels decrease, but there is none present above 20,000 feet.

Answer: b
Difficulty: Easy
Bloom's Taxonomy: Knowledge, Comprehension

44.1 THE FUNCTION OF GAS EXCHANGE

3. Which of the following is an example of a respiratory surface?

a. amphibian skin
b. human lungs
c. fish gills
d. ctenophore body surfaces
e. all of the above

Answer: e
Difficulty: Easy
Bloom's Taxonomy: Knowledge

4. In humans, which of the following is involved in both cellular respiration and physiological respiration?

a. the circulatory system
b. the respiratory surface
c. mitochondria
d. the respiratory medium
e. all of the above

Answer: a
Difficulty: Moderate
Bloom's Taxonomy: Knowledge, Comprehension

5. Through what process is O_2 transported from air in the lungs into the blood?

a. active transport
b. simple diffusion
c. facilitated diffusion
d. osmosis
e. none of the above

Answer: b
Difficulty: Easy
Bloom's Taxonomy: Knowledge

6. Which of the following respiratory surface properties enhances the diffusion of respiratory gases?

a. large surface area
b. thick epithelial layer
c. cool temperatures
d. the presence of active transport pumps
e. none of the above

Answer: a
Difficulty: Moderate
Bloom's Taxonomy: Knowledge, Comprehension

7. Which of the following respiratory structures are moist evaginations?

a. tracheal tubes
b. lungs
c. gills
d. all of the above
e. none of the above

Answer: c
Difficulty: Easy
Bloom's Taxonomy: Knowledge

8. Which of the following air-filled respiratory system structures are common in insects?
a. gills
b. tracheal tubes
c. lungs
d. integumentary surfaces
e. none of the above

Answer: b
Difficulty: Easy
Bloom's Taxonomy: Knowledge

9. Which of the following describes the exchange of gases between a respiratory surface and the circulatory system?
a. ventilation
b. cellular respiration
c. perfusion
d. bulk flow
e. physiological respiration

Answer: e
Difficulty: Easy
Bloom's Taxonomy: Knowledge

10. Which of the following adaptations would increase the surface area of a respiratory surface?
a. skin folds on an amphibian
b. a flatworm having a broad, flat body
c. branched gill filaments
d. internal folds within lungs
e. all of the above

Answer: e
Difficulty: Easy
Bloom's Taxonomy: Comprehension, Application

11. Why is proper ventilation important for a respiring animal?
a. It directly increases gas exchange between the blood and cells.
b. It maintains proper O_2 and CO_2 levels on the external side of the respiratory surface so diffusion across it can be maintained.
c. It maintains blood circulation within the body of an animal.
d. It maintains proper O_2 and CO_2 levels on the internal side of the respiratory surface so diffusion across it can be maintained.
e. All of the above.

Answer: b
Difficulty: Moderate
Bloom's Taxonomy: Comprehension

12. In which of the following animals would perfusion be *least* efficient during gas exchange?
a. a bird
b. a frog
c. a fish
d. a flatworm
e. a human

Answer: d
Difficulty: Moderate
Bloom's Taxonomy: Knowledge, Comprehension

13. Which of the following is a disadvantage of water as a respiratory medium?
a. Water holds less O_2 than air.
b. In water O_2 concentration can be affected by solute concentration.
c. The density of water makes it more difficult to ventilate over the respiratory surface.
d. In water O_2 concentration can be affected by temperature.
e. All of the above.

Answer: e
Difficulty: Moderate
Bloom's Taxonomy: Knowledge

14. Which is a disadvantage of air as a respiratory medium?
a. Air evaporates water from the respiratory surface.
b. Air contains 30 times as much O_2 as water.
c. Air is 1000 times less dense than water.
d. Air requires less energy to ventilate over the respiratory surface.
e. O_2 diffuses more rapidly in air than in water.

Answer: a
Difficulty: Easy
Bloom's Taxonomy: Knowledge

44.2 ADAPTATIONS FOR RESPIRATION

15. Which respiratory surface lacks physical protection from the external environment?
a. tracheal system tubes
b. lungs
c. external gills
d. internal gills
e. all of the above

Answer: c
Difficulty: Easy
Bloom's Taxonomy: Knowledge

16. Why does countercurrent exchange optimize gas exchange across some gills?
a. It sustains an optimal thermal environment for gas exchange.
b. It optimizes perfusion to the respiratory surface.
c. It sustains an optimal respiratory gas concentration gradient, allowing diffusion to occur across the entire respiratory surface.
d. It optimizes delivery of the respiratory medium to the respiratory surface
e. All of the above.

Answer: c
Difficulty: Moderate
Bloom's Taxonomy: Knowledge, Comprehension

17. Countercurrent exchange allows for the removal of _____ of O_2 from water flowing over gills.
a. 50–60 percent
b. 60–70 percent
c. 70–80 percent
d. 80–90 percent
e. 90–100 percent

Answer: d
Difficulty: Easy
Bloom's Taxonomy: Knowledge

18. Which of the following lists the correct order of tracheal system structures, from most external to most internal?

a. spiracles, tracheae, tracheal branches, tracheoles
b. tracheoles, tracheae, tracheal branches, spiracles
c. tracheoles, tracheal branches, tracheae, spiracles
d. spiracles, tracheal branches, tracheae, tracheoles
e. spiracles, tracheoles, tracheae, tracheal branches

Answer: a
Difficulty: Easy
Bloom's Taxonomy: Knowledge

19. Where in an insect tracheal system does gas exchange occur between the air and a cell?
a. trachea
b. tracheal branches
c. gill filaments
d. tracheoles
e. spiracles

Answer: d
Difficulty: Easy
Bloom's Taxonomy: Knowledge

20. Which of the following animals uses positive pressure breathing to ventilate its lungs?

a. a bird
b. a frog
c. a fish
d. a reptile
e. a dog

Answer: b
Difficulty: Easy
Bloom's Taxonomy: Knowledge

21. Which of the following animals use air sacs to ventilate its lungs?

a. humans
b. birds
c. frogs
d. reptiles
e. horses

Answer: b
Difficulty: Easy
Bloom's Taxonomy: Knowledge

22. Among air breathing animals, birds have the most effective mechanism of extracting oxygen from air because

a. birds utilize countercurrent exchange.
b. bird lungs have more surface area.
c. bird lungs are the largest among vertebrates.
d. the air sacs provide additional respiratory surfaces.
e. all of the above

Answer: a
Difficulty: Easy
Bloom's Taxonomy: Knowledge, Comprehension

23. During which part of the bird respiratory cycle is O_2 extracted from the air?

a. the first inhalation
b. the first exhalation
c. the second inhalation
d. the second exhalation
e. all of the above

Answer: e
Difficulty: Moderate
Bloom's Taxonomy: Knowledge, Comprehension

44.3 THE MAMMALIAN RESPIRATORY SYSTEM

24. The respiratory surfaces of mammalian lungs are called

a. alveoli.
b. bronchi.
c. tracheae.
d. bronchioles.
e. air sacs.

Answer: a
Difficulty: Easy
Bloom's Taxonomy: Knowledge

25. The _____ is/are responsible for closing off the airway during swallowing.

a. larynx
b. epiglottis
c. pharynx
d. intercostal muscles
e. pleura

Answer: b
Difficulty: Easy
Bloom's Taxonomy: Knowledge

26. The _____ are the initial branches of the trachea that lead into each lung.

a. bronchioles
b. alveoli
c. pleura
d. bronchi
e. none of the above

Answer: d
Difficulty: Easy
Bloom's Taxonomy: Knowledge

27. Which of the following occurs during human inhalation?

a. The diaphragm contracts and the external intercostal muscles relax.
b. The diaphragm relaxes and the external intercostal muscles contract.
c. The diaphragm contracts and the external intercostal muscles contract.
d. The diaphragm relaxes and the external intercostal muscles relax.
e. None of the above.

Answer: c
Difficulty: Moderate
Bloom's Taxonomy: Knowledge, Comprehension

28. During exhalation, the chest cavity _____ in size, which causes air pressure in the lungs to _____.

a. increases; increase
b. decreases; decrease
c. increases; decrease
d. decreases; increase
e. none of the above

Answer: d
Difficulty: Easy
Bloom's Taxonomy: Knowledge, Comprehension

29. During resting breathing in humans, what causes exhalation to occur?

a. The diaphragm contracts and the internal intercostal muscles relax.
b. The diaphragm relaxes and the internal intercostal muscles contract.
c. The diaphragm contracts and the internal intercostal muscles contract.
d. The diaphragm relaxes and the external intercostal muscles contract.
e. None of the above.

Answer: e
Difficulty: Moderate
Bloom's Taxonomy: Knowledge, Comprehension

30. The primary muscle(s) involved in human breathing is/are called the

a. external intercostals.
b. internal intercostals.
c. diaphragm.
d. abdominal wall muscles.
e. none of the above

Answer: c
Difficulty: Easy
Bloom's Taxonomy: Knowledge

31. Which of the following muscles contract during forceful exhalation?

a. the diaphragm
b. external intercostals
c. internal intercostals
d. a and b
e. b and c

Answer: c
Difficulty: Easy
Bloom's Taxonomy: Knowledge

32. Airways of the mammalian respiratory system function to

a. filter external air.
b. moisten external air.
c. warm external air.
d. all of the above
e. none of the above

Answer: d
Difficulty: Moderate
Bloom's Taxonomy: Knowledge, Comprehension

33. Which of the following mammalian airway structures acts as the "windpipe"?

a. the bronchus
b. the trachea
c. the larynx
d. the bronchiole
e. the pharynx

Answer: b
Difficulty: Easy
Bloom's Taxonomy: Knowledge

34. Air movement to the mammalian respiratory surfaces can be reduced when smooth muscle in the walls of the _____ contract(s).

a. epiglottis
b. bronchioles
c. large bronchi
d. trachea
e. all of the above

Answer: b
Difficulty: Moderate
Bloom's Taxonomy: Knowledge, Comprehension

35. Which of the following lung volumes keep the lungs from completely deflating, even after a maximal exhalation?

a. residual volume
b. tidal volume
c. vital capacity
d. all of the above
e. none of the above

Answer: a
Difficulty: Easy
Bloom's Taxonomy: Knowledge

36. Which of the following lung volumes change during exercise?

a. residual volume
b. tidal volume
c. vital capacity
d. all of the above
e. none of the above

Answer: b
Difficulty: Moderate
Bloom's Taxonomy: Knowledge, Analysis

37. Which of the following chemicals has the *least* influence on the regulation of breathing rate?

a. O_2.
b. CO_2.
c. H^+.
d. These chemicals have equal influence.
e. These chemicals do not influence breathing rate.

Answer: a
Difficulty: Moderate
Bloom's Taxonomy: Evaluation

38. Which respiratory control center sets basic breathing rate by acting as the primary stimulator of inhalation?

a. the ventral interneuron group
b. the pons interneuron groups
c. the carotid bodies
d. the dorsal respiratory group
e. the aortic bodies

Answer: d
Difficulty: Easy
Bloom's Taxonomy: Knowledge

39. Which respiratory control center is responsible for forceful inhalation and exhalation?

a. the ventral interneuron group
b. the pons interneuron groups
c. the carotid bodies
d. the dorsal respiratory group
e. the aortic bodies

Answer: a
Difficulty: Easy
Bloom's Taxonomy: Knowledge

40. Which respiratory control center is most sensitive to changes in CO_2?

a. aortic bodies
b. carotid bodies
c. receptors of the medulla
d. these receptors are equally sensitive to CO_2
e. none of these receptors are sensitive to CO_2

Answer: c
Difficulty: Easy
Bloom's Taxonomy: Knowledge

41. Which of the following respiratory control centers allows you to hold your breath before diving into a swimming pool?

a. aortic bodies
b. carotid bodies
c. receptors of the medulla
d. the limbic center of the brain
e. higher brain centers of the cerebrum

Answer: e
Difficulty: Moderate
Bloom's Taxonomy: Knowledge, Application

42. Which of the following is an example of local control over breathing?

a. automated lung ventilation and lung perfusion adjustments maximizing O_2 and CO_2 exchange
b. the aortic bodies signaling an increase in breathing rate when blood pH is acidic
c. the interneuron groups of the pons refining the contractions involved with inhalation and exhalation
d. the carotid bodies signaling an increase in breathing rate when O_2 levels are low
e. the ventral group of interneurons signaling an increase in breathing rate and depth during exercise

Answer: a
Difficulty: Moderate
Bloom's Taxonomy: Application

44.4 MECHANISMS OF GAS EXCHANGE AND TRANSPORT

43. If N_2 gas comprises 79 percent of atmospheric air at sea level, what is P_{N2}?

a. about 300 mm Hg
b. about 400 mm Hg
c. about 500 mm Hg
d. about 600 mm Hg
e. about 700 mm Hg

Answer: d
Difficulty: Hard
Bloom's Taxonomy: Application

44. What would you predict would happen in an area of the body where blood was relatively cool?

a. O_2 would bind more readily to hemoglobin.
b. CO_2 would bind more readily to hemoglobin.
c. O_2 would release from hemoglobin.
d. CO_2 would release from hemoglobin.
e. None of the above apply.

Answer: a
Difficulty: Moderate
Bloom's Taxonomy: Comprehension

45. What would you predict would happen in an area of the body where blood was relatively acidic (low pH)?

a. O_2 would bind more readily to hemoglobin.
b. CO_2 would bind more readily to hemoglobin.
c. O_2 would release from hemoglobin.
d. CO_2 would release from hemoglobin.
e. None of the above apply.

Answer: c
Difficulty: Moderate
Bloom's Taxonomy: Comprehension

46. Most O_2 is transported in the blood

a. dissolved in the plasma.
b. as bicarbonate ions.
c. bound to hemoglobin.
d. all of the above
e. none of the above

Answer: c
Difficulty: Easy
Bloom's Taxonomy: Knowledge

47. CO_2 is transported in the blood

a. dissolved in the plasma.
b. as bicarbonate ions.
c. bound to hemoglobin.
d. all of the above
e. none of the above

Answer: d
Difficulty: Moderate
Bloom's Taxonomy: Knowledge

48. The function of carbonic anhydrase is

a. to speed up the conversion of CO_2 and H_2O into HCO_3^- and H^+.
b. to speed up the conversion of HCO_3^- and H^+ into CO_2 and H_2O.
c. to speed up the rate at which CO_2 binds to hemoglobin.
d. a and b
e. b and c

Answer: d
Difficulty: Moderate
Bloom's Taxonomy: Knowledge

49. Which of the following events occurs in lung blood?

a. Free H^+ binds to hemoglobin.
b. Plasma HCO_3^- enters erythrocytes.
c. In the plasma, CO_2 and H_2O are converted into HCO_3^- and H^+.
d. In erythrocytes, carbonic anhydrase converts CO_2 and H_2O into HCO_3^- and H^+.
e. CO_2 binds to hemoglobin.

Answer: b
Difficulty: Moderate
Bloom's Taxonomy: Comprehension

50. Which of the following events occurs in blood in body tissue?

a. free H^+ releases from hemoglobin.
b. erythrocyte HCO_3^- enters the plasma.
c. in the plasma, HCO_3^- and H^+ convert into CO_2 and H_2O.
d. in erythrocytes, carbonic anhydrase converts HCO_3^- and H^+ into CO_2 and H_2O.
e. CO_2 binds to hemoglobin.

Answer: e
Difficulty: Moderate
Bloom's Taxonomy: Comprehension

44.5 RESPIRATION AT HIGH ALTITUDES AND OCEAN DEPTHS

51. Which of the following are examples of adaptations that can result from living at high altitudes?

a. elevated erythrocyte count
b. increased number of alveoli
c. increased number of lung capillaries
d. hemoglobin molecules with greater affinity for O_2
e. all of the above

Answer: e
Difficulty: Moderate
Bloom's Taxonomy: Knowledge, Comprehension

52. What is the main physiological concern regarding respiration at high altitude?

a. P_{O_2} increases
b. P_{O_2} decreases
c. P_{CO_2} increases
d. P_{CO_2} decreases
e. P_{N_2} increases

Answer: b
Difficulty: Easy
Bloom's Taxonomy: Knowledge, Comprehension

53. During a dive, the partial pressure of dissolved gases increases by about 1 atm for every _____ in depth.

a. meter
b. 5 meters
c. 10 meters
d. 15 meters
e. 20 meters

Answer: c
Difficulty: Easy
Bloom's Taxonomy: Knowledge

INSIGHTS FROM THE MOLECULAR REVOLUTION: GIVING HEMOGLOBIN AND MYOGLOBIN AIR

54. The number-one risk factor for sudden infant death syndrome (SIDS) is

a. an infant lying on their back.
b. breast feeding an infant.
c. premature feeding of solid foods.
d. exposure to tobacco smoke.
e. holding the infant.

Answer: d
Difficulty: Moderate
Bloom's Taxonomy: Knowledge

55. Studies indicate that in utero nicotine exposure can cause breathing abnormalities in neonatal mammals. These abnormalities can include

a. elevated breathing rate while awake.
b. increased frequency of apneas during sleep.
c. delayed arousal from sleep when hypoxic.
d. a and b
e. b and c

Answer: e
Difficulty: Moderate
Bloom's Taxonomy: Knowledge

Matching

Match each of the following term with its correct description.

56. _____ breathing

57. _____ ventilation

58. _____ perfusion

59. _____ physiological respiration

60. _____ cellular respiration

61. _____ inhalation

62. _____ exhalation

A. The flow of blood (or body fluids) on the internal side of the respiratory surface

B. Gas exchange between the blood and body tissue cells

C. The general term used to describe an animal's exchange of gases with the respiratory medium

D. Specifically describes the release of air from the lungs to the atmosphere

E. The flow of the respiratory medium over the external side of the respiratory surface

F. Occurs due to the expansion of the lungs and thoracic cavity

G. The exchange of gases between the respiratory medium and the blood (or internal body fluids)

Answers: 56. C   57. E   58. A   59. G   60. B   61. F   62. D

Difficulty: Easy
Bloom's Taxonomy: Knowledge
Source: Section 44.1, 44.3

True/False

If the statement is true, write a "T" in the blank. If the statement is false, make it correct by changing the underlined word(s) and writing the correct word(s) in the answer blank.

63. _____ Birds utilize countercurrent mechanisms to optimize diffusion of O_2 across their respiratory surfaces.

64. _____ A larynx is found at the beginning of each insect trachea.

65. _____ CO_2 and H_2O are converted to HCO_3^- and H^+ in the blood of the lungs.

66. _____ Alveoli are the respiratory surfaces of birds.

67. _____ Humans ventilate their lungs by negative pressure breathing.

68. _____ Alveoli are surrounded by pulmonary capillaries.

69. _____ When the diaphragm contracts, it moves up into the thoracic cavity, thereby decreasing its volume.

70. _____ Blood O_2 levels have the greatest influence on mammalian breathing rate.

71. _____ Each molecule of hemoglobin can bind to six molecules of oxygen.

Answers: 63. T   64. F, spiracle   65. F, body tissues   66. F, bronchi   67. T   68. T   69. F, relaxes   70. F, CO_2 71. F, four

Difficulty: Moderate
Bloom's Taxonomy: Knowledge, Comprehension
Source: Section 44.2, 44.3, 44.4

Select the Exception

72.   Which of the following mechanisms is *not* used to encourage the movement of water over gills?

a.   cilia
b.   contraction of the diaphragm
c.   contraction of a muscular mantle
d.   a fish swimming with its mouth open

Answer: b
Difficulty: Easy
Bloom's Taxonomy: Synthesis
Source: 44.2

73.   Which of the following animals does *not* use negative pressure breathing to ventilate its lungs?

a.   a human
b.   a lizard
c.   a bird
d.   a cat
e.   a salamander

Answer: e
Difficulty: Easy
Bloom's Taxonomy: Synthesis, Evaluation
Source: Section 44.2

74.   Which of the following is *not* an adaptation used by marine mammals to extend dive times?

a.   size-relative increase in blood volume.
b.   increase in erythrocyte count
c.   size-relative increase in lung volume
d.   elevated concentration of muscle myoglobin
e.   reduced circulation to nonessential organs

Answer: c
Difficulty: Moderate
Bloom's Taxonomy: Synthesis, Evaluation
Source: Section 44.5

75.   Which of the following organs is not used by marine mammals during deep dives?

a.   the lungs
b.   the heart
c.   the brain
d.   the blood vessels

Answer: a
Difficulty: Moderate
Bloom's Taxonomy: Synthesis, Evaluation
Source: Section 44.5

Labeling

For questions 76–87, consult the diagram of the human respiratory system below. Write your answer in the spaces provided.

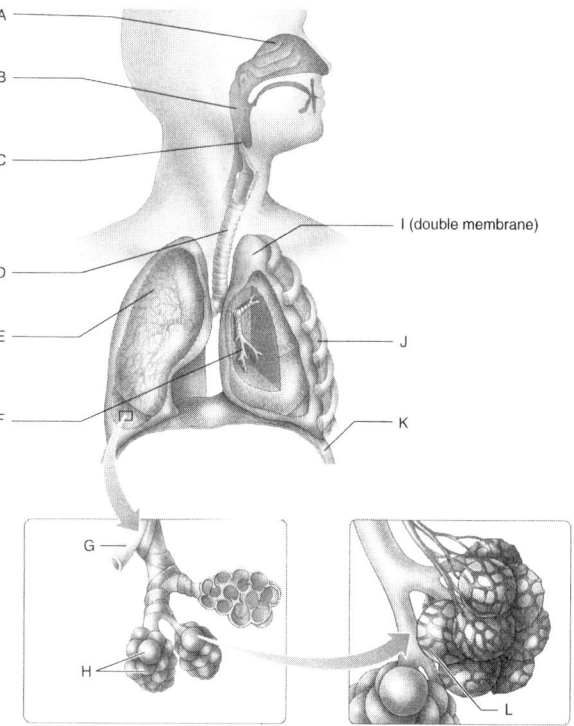

76.   _____   Identify structure A.
77.   _____   Identify structure B.
78.   _____   Identify structure C.
79.   _____   Identify structure D.
80.   _____   Identify structure E.
81.   _____   Identify structure F.
82.   _____   Identify structure G.
83.   _____   Identify structure H.
84.   _____   Identify structure I.
85.   _____   Identify structure J.
86.   _____   Identify structure K.
87.   _____   Identify structure L.

Answers: **76. nasal passages   77. pharynx 78. epiglottis   79. trachea 80. lung 81. bronchi 82. bronchiole   83.alveoli 84. pleura   85. intercostal muscles. 86. diaphragm   87. pulmonary** capillaries

Difficulty: Easy
Bloom's Taxonomy: Knowledge
Source: Section 44.3

Short Answer

88. Describe three adaptations that enhance the function of a respiratory surface.

Answer: **Gas diffusion can be enhanced across a respiratory membrane by: (1) increasing its surface area so more gases diffuse, (2) making the respiratory surface thin so gas diffusion is more rapid, and (3) making the respiratory surface moist so gases diffuse across cells more readily.**
Difficulty: Moderate
Bloom's Taxonomy: Knowledge, Comprehension
Source: Section 44.1

89. Explain how a diving marine mammal increases its supply of stored O_2, in spite of the fact that its lungs are collapsed at depths below 25 meters.

Answer: **Marine mammals increase their stores of O_2 by: (1) increasing size-relative blood volume, (2) increasing erythrocyte concentration, and (3) increasing their concentration of myoglobin in their muscle.**
Difficulty: Moderate
Bloom's Taxonomy: Knowledge, Comprehension
Source: Section 44.5

90. Llamas, which are customarily known as mountain dwellers, have hemoglobin with higher O_2 binding affinity than that of humans. That being said, how would you predict the llama hemoglobin-O_2 dissociation curve would differ compared to that of a human (see diagram below)?

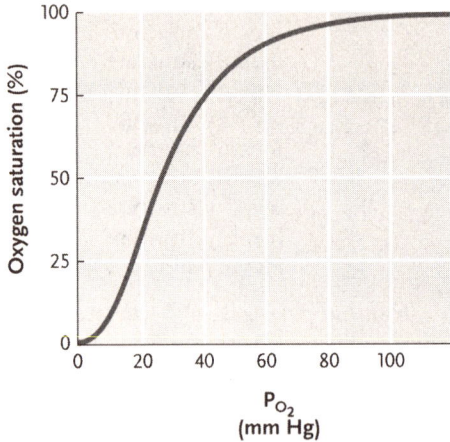

Answer: **The slope of the llama hemoglobin-O_2 dissociation curve would be steeper, effectively shifting the curve to the left. Thus, the llama hemoglobin is shown to bind to O_2 at lower partial pressures than that of a human.**
Difficulty: Moderate
Bloom's Taxonomy: Knowledge, Application
Source: Section 44.5

45

ANIMAL NUTRITION

Multiple-Choice

WHY IT MATTERS

1. The anglerfish obtains food by

a. catching the fish with its appendages.
b. emitting a scent into the water.
c. swimming to the surface.
d. inverting its stomach.
e. luring the fish near its mouth.

Answer: e
Difficulty: Easy
Bloom's Taxonomy: Knowledge

45.1 FEEDING AND NUTRITION

2. Which of the following groups consume animals?

a. carnivores and herbivores
b. carnivores and omnivores
c. omnivores and herbivores
d. carnivores only
e. herbivores only

Answer: b
Difficulty: Easy
Bloom's Taxonomy: Knowledge

3. How does a calorie (lower case c) differ from a Calorie (upper case C)?

a. One calorie = 100 Calories.
b. One calorie = 1000 Calories.
c. One Calorie = 100 calories.
d. One Calorie = 1000 calories.
e. They are the same.

Answer: d
Difficulty: Moderate
Bloom's Taxonomy: Knowledge

4. An example of a suspension feeder is a

a. bird.
b. lamprey.
c. mosquito.
d. snake.
e. whale.

Answer: e
Difficulty: Moderate
Bloom's Taxonomy: Knowledge

5. Humans are an example of

a. bulk feeders.
b. deposit feeders.
c. fluid feeders.
d. suspension feeders.
e. all of the above

Answer: a
Difficulty: Easy
Bloom's Taxonomy: Knowledge

6. Which two animals are both classified as suspension feeders?

a. clams and aphids
b. clams and fiddler crabs
c. lampreys and rabbits
d. lampreys and rabbits
e. whales and clams

Answer: e
Difficulty Moderate
Bloom's Taxonomy: Knowledge

45.2 DIGESTIVE PROCESSES

7. Choanocytes are found in _____ and participate in intracellular digestion.

a. birds
b. earthworms
c. grasshoppers
d. humans
e. sponges

Answer: e
Difficulty: Moderate
Bloom's Taxonomy: Knowledge

8. Which of the following statements is correct?

a. Intracellular digestion is seen in mammals.
b. Intracellular digestion never involves choanocytes.
c. Extracellular digestion must involve choanocytes.
d. Extracellular digestion may take place in a tube contained inside the body.
e. Most animals experience both intracellular and extracellular digestion.

Answer: d
Difficulty: Difficult
Bloom's Taxonomy: Comprehension

9.	What is a lysosome's role in intracellular digestion?

a.	absorption
b.	enzymatic digestion
c.	filtration
d.	mechanical digestion
e.	produces energy

Answer: b
Difficulty: Moderate
Bloom's Taxonomy: Application

10.	Which statement is true about an earthworm?

a.	An earthworm has a single digestive opening and typhlosoles.
b.	An earthworm has a single digestive opening and no typhlosoles.
c.	An earthworm has two digestive openings and typhlosoles.
d.	An earthworm has two digestive openings and no typhlosoles.
e.	An earthworm undergoes intracellular digestion.

Answer: c
Difficulty: Difficult
Bloom's Taxonomy: Knowledge

11.	Place the following digestive structures of an Annelid in the proper order.

a.	mouth – esophagus – crop – gizzard – intestine – anus
b.	mouth – esophagus – intestine – crop – gizzard – anus
c.	mouth – esophagus – gizzard – crop – intestine – anus
d.	mouth – esophagus – crop – intestine – gizzard – anus
e.	mouth – esophagus – intestine – gizzard – crop – anus

Answer: a
Difficulty: Difficult
Bloom's Taxonomy: Knowledge

12.	The gastrovascular cavity of a flatworm serves as both the _____ and _____ systems of the body.

a.	digestive; circulatory
b.	digestive; reproductive
c.	digestive; respiratory
d.	circulatory; respiratory
e.	respiratory; reproductive

Answer: a
Difficulty: Easy
Bloom's Taxonomy: Knowledge

13.	Place the five steps of digestion in an upper level animal in the proper order.

a.	absorption – enzymatic hydrolysis – mechanical processing – secretion of enzymes – elimination
b.	absorption – mechanical processing – enzymatic hydrolysis – secretion of enzymes – elimination
c.	mechanical processing – absorption – secretion of enzymes – enzymatic hydrolysis – elimination
d.	mechanical processing – secretion of enzymes – enzymatic hydrolysis – absorption – elimination
e.	mechanical processing – secretion of enzymes – absorption – enzymatic hydrolysis – elimination

Answer: d
Difficulty: Moderate
Bloom's Taxonomy: Knowledge

14.	Which process in the human digestive system best illustrates mechanical digestion?

a.	absorption
b.	amylase
c.	bacterial action in the large intestine
d.	chewing
e.	pepsin

Answer: d
Difficulty: Easy
Bloom's Taxonomy: Comprehension

15.	Which structure in the earthworm's digestive tract is utilized for food storage?

a.	crop
b.	esophagus
c.	gizzard
d.	intestine
e.	mouth

Answer: a
Difficulty: Moderate
Bloom's Taxonomy: Knowledge

16.	Which part of a bird's digestive system contains sand and rocks to aid in mechanical digestion?

a.	crop
b.	esophagus
c.	gizzard
d.	pharynx
e.	proventriculus

Answer: c
Difficulty: Moderate
Bloom's Taxonomy: Knowledge

45.3 DIGESTION IN HUMANS AND OTHER MAMMALS

17. Which structure of the human digestive tract produces bile?

a. liver
b. gallbladder
c. pancreas
d. small intestine
e. stomach

Answer: a
Difficulty: Moderate
Bloom's Taxonomy: Knowledge

18. A diet of primarily grains would likely be deficient in which two amino acids?

a. isoleucine and leucine
b. isoleucine and lysine
c. methionine and tryptophan
d. methionine and lysine
e. valine and lysine

Answer: b
Difficulty: Difficult
Bloom's Taxonomy: Knowledge

19. Which layer of the digestive tract in mammals directly contacts the lumen?

a. mucosa
b. muscularis
c. submucosa
d. serosa
e. all of the above

Answer: a
Difficulty: Moderate
Bloom's Taxonomy: Comprehension

20. Which of the following vitamins could be toxic in large amounts?

a. A
b. B_1
c. B_2
d. biotin
e. folic acid

Answer: a
Difficulty: Difficult
Bloom's Taxonomy: Application

21. Which pair of vitamins is produced within the human body?

a. A and C
b. A and D
c. C and D
d. C and K
e. D and K

Answer: e
Difficulty: Moderate
Bloom's Taxonomy: Application

22. Which vitamin is produced in humans when exposed to sunlight?

a. A
b. B_1
c. C
d. D
e. folic acid

Answer: d
Difficulty: Moderate
Bloom's Taxonomy: Knowledge

23. Individuals with a vitamin _____ deficiency may develop _____.

a. D; bone softening
b. D; anemia
c. K; increased blood clotting time
d. K; bacterial resistance
e. a and c only

Answer: e
Difficulty: Difficult
Bloom's Taxonomy: Comprehension

24. Place the layers of a mammalian gut in order from outside inward, toward the lumen.

a. mucosa – submucosa – muscularis – serosa
b. muscularis – mucosa – submucosa – serosa
c. serosa – muscularis – submucosa – mucosa
d. serosa – mucosa – submucosa – muscularis
e. serosa – submucosa – mucosa – muscularis

Answer: c
Difficulty: Moderate
Bloom's Taxonomy: Application

25. Which organ in the human digestive system contains an oblique layer of muscle as a part of its muscularis?

a. esophagus
b. large intestine
c. liver
d. small intestine
e. stomach

Answer: e
Difficulty: Easy
Bloom's Taxonomy: Knowledge

26. A deficiency of vitamin C can result in

a. anemia.
b. beriberi.
c. night blindness.
d. rickets.
e. scurvy.

Answer: e
Difficulty: Difficult
Bloom's Taxonomy: Knowledge

27. Which layer of the mammalian gut performs peristalsis?

a. mucosa
b. muscularis
c. serosa
d. submucosa
e. submucosa and muscularis

Answer: b
Difficulty: Easy
Bloom's Taxonomy: Knowledge

28. A diet that does not contain any dark green vegetables, whole grains, yeast, or lean meats may result in a deficiency of which vitamin?

a. B_1
b. B_2
c. B_6
d. folic acid
e. niacin

Answer: d
Difficulty: Difficult
Bloom's Taxonomy: Knowledge

29. What is the purpose of the watery fluid secreted by the serosa of the human digestive tract?

a. emulsifies fats
b. initiates peristalsis
c. kills bacteria
d. neutralizes acidic conditions
e. reduces friction between nearby organs

Answer: e
Difficulty: Moderate
Bloom's Taxonomy: Application

30. Of the 13 essential vitamins, _____ are water soluble and _____ are fat soluble.

a. 3; 10
b. 4; 9
c. 5; 8
d. 9; 4
e. 10; 3

Answer: d
Difficulty: Difficult
Bloom's Taxonomy: Knowledge

31. A deficiency of which mineral produces a goiter in humans?

a. calcium
b. iodine
c. magnesium
d. manganese
e. zinc

Answer: b
Difficulty: Difficult
Bloom's Taxonomy: Knowledge

32. Deficiencies of which two minerals may result in muscular weakness?

a. iron and copper
b. iron and manganese
c. iodine and iron
d. phosphorus and calcium
e. phosphorus and potassium

Answer: e
Difficulty: Difficult
Bloom's Taxonomy: Application

33. Anemia can result from a deficiency of

a. calcium.
b. chlorine.
c. iodine.
d. iron.
e. phosphorus.

Answer: d
Difficulty: Easy
Bloom's Taxonomy: Knowledge

34. The _____ regulates the movement of food into the stomach and the _____ regulates the passage of food into the small intestine.

a. epiglottis; gastroesophageal sphincter
b. epiglottis; pyloric sphincter
c. gastroesophageal sphincter; epiglottis
d. gastroesophageal sphincter; pyloric sphincter
e. pyloric sphincter; anal sphincter

Answer: d
Difficulty: Moderate
Bloom's Taxonomy: Comprehension

35. Human saliva contains which of the following enzymes?

a. aminopeptidase
b. amylase
c. chymotrypsin
d. pepsin
e. trypsin

Answer: b
Difficulty: Easy
Bloom's Taxonomy: Knowledge

36. Which digestive structures in humans secrete amylase?

a. liver and gallbladder
b. liver and pancreas
c. pancreas and gallbladder
d. pancreas and salivary glands
e. pancreas, liver, and salivary glands

Answer: d
Difficulty: Easy
Bloom's Taxonomy: Knowledge

37. Food is classified as a bolus in which digestive structure?

a. large intestine
b. liver
c. mouth
d. small intestine
e. stomach

Answer: c
Difficulty: Easy
Bloom's Taxonomy: Knowledge

38. Which cells produce pepsinogen in the stomach?

a. brush border cells
b. chief cells
c. choanocytes
d. parietal cells
e. villi

Answer: b
Difficulty: Moderate
Bloom's Taxonomy: Knowledge

39. The digestion of proteins in humans begins in the

a. large intestine.
b. mouth.
c. pancreas.
d. small intestines.
e. stomach.

Answer: e
Difficulty: Moderate
Bloom's Taxonomy: Knowledge

40. Heartburn is caused by _____.

a. the gastroesophageal sphincter remaining too tightly closed
b. the gastroesophageal sphincter remaining a little open
c. the pyloric sphincter remaining too tightly closed
d. the pyloric sphincter remaining a little open
e. both the gastroesophageal and the pyloric sphincters remaining too tightly closed

Answer: b
Difficulty: Difficult
Bloom's Taxonomy: Application

41. The function of the epiglottis is _____ while swallowing.

a. to move the tongue out of the way
b. to cover the esophagus
c. to cover the glottis
d. to cover both the esophagus and the glottis
e. to secrete mucus

Answer: c
Difficulty: Easy
Bloom's Taxonomy: Knowledge

42. The pH of stomach acids is usually

a. 2.
b. 4.
c. 7.
d. 10.
e. 12.

Answer: a
Difficulty: Easy
Bloom's Taxonomy: Knowledge

43. Most peptic ulcers are caused by

a. antibiotics.
b. bacteria.
c. heartburn.
d. lactose intolerance.
e. malnutrition.

Answer: b
Difficulty: Moderate
Bloom's Taxonomy: Knowledge

44. The digestion of lipids in humans begins in the

a. large intestine.
b. mouth.
c. pancreas.
d. small intestine.
e. stomach.

Answer: d
Difficulty: Moderate
Bloom's Taxonomy: Knowledge

45. What function(s) does the presence of alkaline components in pancreatic juice serve?

a. Creates an optimum environment for digestion in the small intestine
b. Neutralizes the acidity of stomach contents
c. Stimulates peristalsis in the small intestine
d. Both a and b are correct
e. Both b and c are correct

Answer: a
Difficulty: Moderate
Bloom's Taxonomy: Comprehension

46. Which of the following statements are true about bile?

a. It emulsifies fats.
b. It converts fats into droplets called micelles.
c. It makes it easier for lipase to digest fats.
d. It is yellow in color due to the presence of bilirubin.
e. All of the statements above are correct.

Answer: e
Difficulty: Easy
Bloom's Taxonomy: Knowledge

47. The pancreas is located between _____.

a. the gallbladder and the liver
b. the liver and the stomach
c. the liver and the small intestine
d. the stomach and the small intestine
e. the stomach and the liver

Answer: d
Difficulty: Moderate
Bloom's Taxonomy: Knowledge

48. In which organ do we classify the food as chyme?

a. large intestine
b. liver
c. mouth
d. small intestine
e. stomach

Answer: e
Difficulty: Moderate
Bloom's Taxonomy: Knowledge

49. Which digestive structure in humans is active in both the digestion of food and the absorption of nutrients?

a. mouth
b. liver
c. pancreas
d. small intestine
e. stomach

Answer: d
Difficulty: Easy
Bloom's Taxonomy: Application

50. The digestive structure that stores concentrated bile is the

a. duodenum.
b. gallbladder.
c. liver.
d. pancreas.
e. stomach.

Answer: b
Difficulty: Easy
Bloom's Taxonomy: Knowledge

51. The two proteases in the small intestine that complete protein digestion are

a. aminopeptidase and dipeptidase.
b. aminopeptidase and disaccharidase.
c. dipeptidase and disaccharidase.
d. dipeptidase and phosphatase.
e. nucleotidase and nucleosidase.

Answer: a
Difficulty: Difficult
Bloom's Taxonomy: Knowledge

52. The digestion of starches begins in which structure of the digestive tract?

a. esophagus
b. large intestine
c. mouth
d. small intestine
e. stomach

Answer: c
Difficulty: Moderate
Bloom's Taxonomy: Comprehension

53. The most digestion in the human body takes place in the

a. liver.
b. mouth.
c. large intestine.
d. small intestine.
e. stomach.

Answer: d
Difficulty: Difficult
Bloom's Taxonomy: Application

54. Veins from the _____ join together and form the hepatic portal vein.

a. large intestine
b. liver
c. pancreas
d. small intestine
e. stomach

Answer: d
Difficulty: Moderate
Bloom's Taxonomy: Knowledge

55. The appendix is a vestigial structure that extends off of the

a. cecum.
b. colon.
c. duodenum.
d. rectum.
e. stomach.

Answer: a
Difficulty: Easy
Bloom's Taxonomy: Knowledge

56. Which of the following statements is correct regarding the human digestive tract?

a. The large intestine contains villi and is smaller in diameter and shorter in length than the small intestine.
b. The large intestine contains villi and is smaller in diameter and longer in length than the small intestine.
c. The large intestine does not contain villi and is smaller in diameter and longer in length than the small intestine.
d. The large intestine does not contain villi and is larger in diameter and smaller in length than the small intestine.
e. The large intestine does not contain villi and is larger in diameter and longer in length than the small intestine.

Answer: d
Difficulty: Difficult
Bloom's Taxonomy: Application

57. The primary function of the large intestine is to

a. absorb water.
b. absorb nutrients.
c. complete the digestion of carbohydrates.
d. complete the digestion of fats.
e. complete the digestion of protein.

Answer: a
Difficulty: Easy
Bloom's Taxonomy: Knowledge

58. The main function of the large intestine is to

a. absorb nutrients.
b. absorb water.
c. digest fats.
d. kill bacteria.
e. neutralize acids.

Answer: b
Difficulty: Easy
Bloom's Taxonomy: Knowledge

45.4 REGULATION OF THE DIGESTIVE PROCESS

59. Which structure in the human body regulates hunger and satiety (a sense of fullness)?

a. hypothalamus
b. liver
c. pancreas
d. small intestine
e. stomach

Answer: a
Difficulty: Moderate
Bloom's Taxonomy: Knowledge

60. When fat enters the duodenum, it stimulates release of which hormone?

a. amylase
b. cholecystokinin
c. gastrin
d. secretin
e. trypsin

Answer: b
Difficulty: Difficult
Bloom's Taxonomy: Knowledge

61. Which of these structures produces a hormone that activates secretion of enzymes within it?

a. mouth
b. liver
c. large intestine
d. small intestine
e. stomach

Answer: e
Difficulty: Moderate
Bloom's Taxonomy: Comprehension

62. A drug that inhibits the Y5 receptor has been shown to

a. have an effect on body weight in normal rats.
b. have an effect on body weight in obese rats.
c. have an effect on food intake in normal rats.
d. have an effect on food intake in obese rats.
e. have no effect on food intake or body weight in any rats.

Answer: e
Difficulty: Difficult
Bloom's Taxonomy: Comprehension

45.5 DIGESTIVE SPECIALIZATIONS IN VERTEBRATES

63. Which tooth type may not be present in herbivores yet prominent in carnivores?

a. canines
b. incisors
c. molars
d. premolars
e. The teeth types of herbivores and carnivores are the same.

Answer: a
Difficulty: Easy
Bloom's Taxonomy: Knowledge

64. An example of an animal with a long digestive tract and molars with a large, ridged surface is

a. cat.
b. deer.
c. dog.
d. shark.
e. tiger.

Answer: b
Difficulty: Easy
Bloom's Taxonomy: Application

65. Which teeth type would be used to crush and grind food?

a. canines
b. incisors
c. molars
d. canines and incisors
e. incisors and molars

Answer: c
Difficulty: Easy
Bloom's Taxonomy: Knowledge

66. Ruminants have how many chambers in their stomachs?

a. one
b. two
c. three
d. four
e. five

Answer: d
Difficulty: Difficult
Bloom's Taxonomy: Knowledge

67. When a ruminant "chews his cud," it involves regurgitation from which two digestive chambers?

a. abomasum and omasum
b. abomasum and rumen
c. omasum and rumen
d. rumen and reticulum
e. reticulum and omasum

Answer: d
Difficulty: Difficult
Bloom's Taxonomy: Knowledge

Select the Exception

68. Which of these functions does not relate to the liver?

a. detoxifies toxic substances
b. produces bile
c. produces amylase
d. synthesizes lipoproteins
e. turns glucose into glycogen

Answer: c
Difficulty: Moderate
Bloom's Taxonomy: Knowledge
Source: Section 45.3

69. Which of the following enzymes is matched incorrectly with the substance it digests?

a. lipase – fats
b. pancreatic amylase – starch
c. pepsin – proteins
d. salivary amylase – starch
e. trypsin – fats

Answer: e
Difficulty: Moderate
Bloom's Taxonomy: Application
Source: Section 45.3

70.	Which of the following animals does <u>not</u> have a single digestive opening?

a.	coral
b.	earthworm
c.	flatworm
d.	hydra
e.	sea anemone

Answer: b
Difficulty: Moderate
Bloom's Taxonomy: Knowledge
Source: Section 45.2

71.	Which of the following is <u>not</u> a fat soluble vitamin?

a.	A
b.	C
c.	D
d.	E
e.	K

Answer: b
Difficulty: Moderate
Bloom's Taxonomy: Knowledge
Source: 45.3

72.	Which of the following organisms is <u>not</u> a ruminant?

a.	cows
b.	deer
c.	dogs
d.	goats
e.	sheep

Answer: c
Difficulty: Easy
Bloom's Taxonomy: Knowledge
Source: Section 45.5

73.	Which of the following statements is <u>not</u> true about fluid feeders?

a.	They may have long tongues to effectively extract nectar from plants.
b.	They may gather organic material from soil located nearby.
c.	They may have piercing mouth parts to obtain blood from a host.
d.	They may secrete enzymes on a food source and consume it once it is liquefied.
e.	They may inject anticoagulants into a host to inhibit blood from clotting.

Answer: b
Difficulty: Easy
Bloom's Taxonomy: Comprehension
Source: Section 45.1

74.	The presence of hydrochloric acid in the stomach serves all of the following purposes <u>except</u>

a.	activates pepsinogen.
b.	emulsifies fats.
c.	inactivates amylase.
d.	kills bacteria.
e.	unfolds proteins.

Answer: b
Difficulty: Moderate
Bloom's Taxonomy: Knowledge
Source: Section 45.3

75.	Which of the following is <u>not</u> an enzyme?

a.	amylase
b.	bile
c.	lipase
d.	pepsin
e.	trypsin

Answer: b
Difficulty: Easy
Bloom's Taxonomy: Comprehension
Source: Section 45.3

76.	Which of the following enzymes is <u>not</u> produced by the pancreas?

a.	amylase
b.	carboxypeptidase
c.	lipase
d.	pepsin
e.	trypsin

Answer: d
Difficulty: Moderate
Bloom's Taxonomy: Knowledge
Source: Section 45.3

77.	Which of the following conditions is <u>not</u> caused by a deficiency of a water-soluble vitamin?

a.	beriberi
b.	pellagra
c.	rickets
d.	scurvy
e.	All of the diseases listed above are caused by water-soluble vitamin deficiencies.

Answer: c
Difficulty: Difficult
Bloom's Taxonomy: Application
Source: Section 45.3

Integrative Multiple-Choice

78. Which two structures (found in different organisms) both serve to increase the absorptive surfaces of the intestines?

a. choanocytes and gastric ceca
b. choanocytes and proventriculus
c. proventriculus and villi
d. typhlosoles and gastric ceca
e. typhlosoles and villi

Answer: e
Difficulty: Difficult
Bloom's Taxonomy: Application
Source: Sections 45.2, 45.3

79. Which two structures of the human digestive tract can be classified as endocrine glands since they produce hormones?

a. liver and gallbladder
b. liver and pancreas
c. liver and small intestine
d. pancreas and small intestine
e. pancreas and stomach

Answer: e
Difficulty: Difficult
Bloom's Taxonomy: Application
Source: Sections 45.3, 45.4

80. A byproduct of digestion in the stomach of some mammals is _____, and it may be produced in large quantities in _____.

a. methane; carnivores
b. methane; ruminants
c. mucus; carnivores
d. mucus; ruminants
e. mucus; both carnivores and ruminants

Answer: b
Difficulty: Moderate
Bloom's Taxonomy: Application
Source: Sections 45.3, 45.5

True/False

81. _____ A bird cracks open the seed coat <u>with his teeth</u> before swallowing.

82. _____ The mucosa of the digestive tract may have <u>villi</u> as an adaptation.

83. _____ In the human diet, <u>linoleic acid and linolenic acid</u> are the only essential fatty acid acids.

84. _____ The appendix is a <u>vestigial</u> structure in humans.

85. _____ The "small" in the small intestine refers to its <u>length</u>.

86. _____ The digestive system of an earthworm is <u>more advanced</u> than that of a flatworm.

Answers: 81. F, with his beak 82. T   83. T   84. T   85. F, diameter   86. T

Difficulty: Moderate
Bloom's Taxonomy: Knowledge, Comprehension
Source: Sections 45.2, 45.3

Labeling

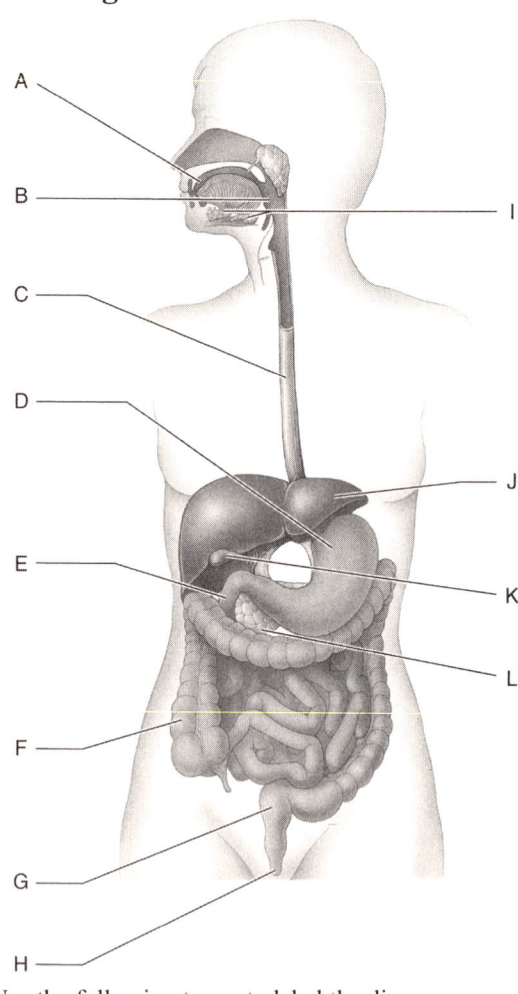

Use the following terms to label the diagram.

87. _____ stomach
88. _____ mouth
89. _____ large intestines
90. _____ esophagus

91. _____ anus
92. _____ liver
93. _____ pharynx
94. _____ pancreas
95. _____ gallbladder
96. _____ small intestines
97. _____ salivary glands
98. _____ rectum

Answers: 87. D   88. A   89. F   90. C
91. H   92. J   93. B   94. L
95. K   96. E   97. I   98. G

Difficulty: Moderate
Bloom's Taxonomy: Knowledge
Source: Figure 45.5

Generalized mammalian dentition

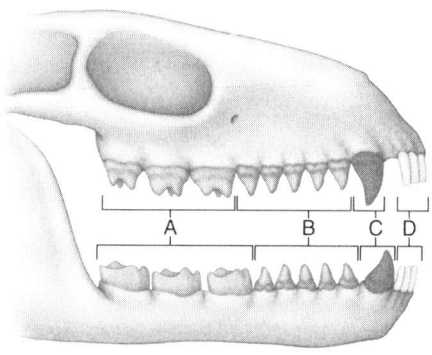

Label each of the following teeth types in the diagram.

99. _____ incisors
100. _____ molars
101. _____ premolars
102. _____ canines

Answers: 99. D   100. A   101. B   102. C

Difficulty: Easy
Bloom's Taxonomy: Knowledge
Source: Figure 45.17

Matching

Match the digestive substances to the structures that produce it. (Note: An answer may be used more than once.)

103. _____ amylase (two answers)

104. _____ bile

105. _____ lipase

106. _____ pepsin

107. _____ trypsin

A.   Liver

B.   Pancreas

C.   Salivary glands

D.   Stomach

Answers: 132. B & C   104. A   105. B
106. D   107. B

Difficulty: Moderate
Bloom's Taxonomy: Knowledge
Source: Section 45.3

Short Answer

108. What purpose do villi and microvilli serve in the small intestine?

Answer: The mucosa of the small intestine has villi and microvilli as adaptations to greatly increase surface area for absorption.
Difficulty: Moderate
Bloom's Taxonomy: Application
Source: Section 45.3

109. How does undernutrition impact the human body?

Answer: Once nutrient stores in the body are fully utilized, components of the body are broken down as energy sources. This process leads to the destruction of muscles and organs, and can eventually lead to death.
Difficulty: Moderate
Bloom's Taxonomy: Knowledge
Source: Section 45.1

110. How are the gastric glands in the stomach protected from digestion by the enzyme they produce?

Answer: *The chief cells in the gastric glands produce the inactive form of the enzyme called pepsinogen. The pepsinogen does not become active, in the form of pepsin, until it encounters hydrochloric acid in the stomach. At this stage, the mucus lining of the stomach protects it from enzymatic action.*
Difficulty: Difficult
Bloom's Taxonomy: Comprehension
Source: Section 45.3

111. Why is a digestive system with a single digestive opening inefficient?

Answer: *Since nutrients enter and wastes exit through the same opening, wastes may be taken into the animal's body only to be expelled again.*
Difficulty: Easy
Bloom's Taxonomy: Comprehension
Source: Section 45.2

112. How does the fact that the mosquito is often a vector of disease relate directly to its method of obtaining food?

Answer: *The mosquito is classified as a fluid feeder because the females obtain the blood of other animals as one of its food sources. It is this practice of obtaining blood from various animals that has the capability of spreading bloodborne pathogens from one animal to another.*
Difficulty: Difficult
Bloom's Taxonomy: Application
Source: Section 45.1

113. Accessory structures in humans contribute to digestion, but food does not actually pass through them. Name all of the accessory structures and the major digestive substances that they produce.

Answer: *salivary glands – salivary amylase; pancreas – pancreatic amylase, trypsin, lipase; liver – bile; gallbladder – bile (does not produce bile, but stores and releases bile that was produced in the liver)*
Difficulty: Moderate
Bloom's Taxonomy: Knowledge
Source: Section 45.3

Essay

114. The pyloric sphincter, the circular muscle that controls the entrance of food from the stomach to the small intestine, usually opens only slightly. What is the purpose of this?

Answer: *Since the small intestine is small in diameter (approximately 4 cm), it cannot accommodate all the food from the stomach at once. Since the pyloric sphincter is only slightly open, small amounts of the watery chyme can enter the small intestine as the stomach churns. These small aliquots of chyme will not overly distend the small intestine.*
Difficulty: Difficult
Bloom's Taxonomy: Application
Source: Section 45.3

115. Describe the enzymatic digestion that takes place in each structure of the digestive tract.

Answer: *Enzymatic digestion in humans starts in the mouth. Amylase from the salivary glands begins the digestion of starches. This digestion only takes place while the food is in the mouth and during the quick trip down the esophagus. When the amylase contacts strong stomach acids, it is denatured. In the stomach, however, pepsin begins the digestion of proteins. Once the food reaches the small intestine, digestion is completed, primarily due to the action of the enzymes produced by accessory structures. The liver secretes bile that serves to emulsify fats, which makes it easier for the lipase (secreted by the pancreas) to digest the fats. Pancreatic juice also contains trypsin to complete the digestion of proteins and amylase to complete the digestion of starches. At this point, the nutrients are in a simplified, usable form and ready to be absorbed by the small intestine.*
Difficulty: Moderate
Bloom's Taxonomy: Comprehension
Source: Section 45.2

46

REGULATING THE INTERNAL ENVIRONMENT

Multiple-Choice

WHY IT MATTERS

1. What happened to the crew of the World War II bomber *Lady Be Good*?
a. They died from heat exposure and dehydration in the desert.
b. They died after crashing into the ocean.
c. They survived for 10 days after crashing in Siberia and then they were rescued.
d. They died from trying to survive on salt water while on an island.
e. They survived for two weeks after crashing into the ocean and then they were rescued.

Answer: a
Difficulty: Easy
Bloom's Taxonomy: Comprehension

46.1 INTRODUCTION TO OSMOREGULATION AND EXCRETION

2. Osmolarity is the outcome of the
a. total solute plus solvent concentration in a solution.
b. total solvent concentration in a solution.
c. ratio of solvent to solute in a solution.
d. total solute concentration in a solution.
e. ratio of solute to solvent in a solution.

Answer: d
Difficulty: Moderate
Bloom's Taxonomy: Comprehension

3. Suppose that you have a cell with a membrane that is permeable to water but not to Na^+ or Cl^- or any other solute. There is more NaCl outside the cell than inside the cell, and the osmolarity of the cell is higher than that of the solution surrounding the cell. Which of the following should occur?
a. Overall, the cell should lose water.
b. Overall, the cell should take up water, Na^+, and Cl^-.
c. Overall, the cell should take up water.
d. Overall, the cell should take up water and lose Na^+ and Cl^-.
e. Overall, the cell should lose water and take up Na^+ and Cl^-.

Answer: c
Difficulty: Moderate
Bloom's Taxonomy: Application

4. The typical osmolarity of most body fluids in humans and other mammals is
a. ~5 mOsm/L.
b. ~225 mOsm/L.
c. ~300 mOsm/L.
d. ~1000 mOsm/L.
e. ~1500 mOsm/L.

Answer: c
Difficulty: Moderate
Bloom's Taxonomy: Knowledge

5. Fresh water has an osmolarity of
a. ~5 mOsm/L.
b. ~225 mOsm/L.
c. ~300 mOsm/L.
d. ~1000 mOsm/L.
e. ~1500 mOsm/L.

Answer: a
Difficulty: Easy
Bloom's Taxonomy: Knowledge

6. The typical osmolarity of body fluids in most freshwater invertebrates is
a. ~5 mOsm/L.
b. ~225 mOsm/L.
c. ~300 mOsm/L.
d. ~1000 mOsm/L.
e. ~1500 mOsm/L.

Answer: b
Difficulty: Difficult
Bloom's Taxonomy: Knowledge

7. The typical osmolarity of body fluids in most marine invertebrates is
a. ~5 mOsm/L.
b. ~225 mOsm/L.
c. ~300 mOsm/L.
d. ~1000 mOsm/L.
e. ~1500 mOsm/L.

Answer: d
Difficulty: Difficult
Bloom's Taxonomy: Knowledge

8. Seawater typically has an osmolarity of
a. ~5 mOsm/L.
b. ~225 mOsm/L.
c. ~300 mOsm/L.
d. ~1000 mOsm/L.
e. ~1500 mOsm/L.

Answer: d
Difficulty: Moderate
Bloom's Taxonomy: Knowledge

9.   Which of the following are typically osmoconformers?

a.   freshwater teleost fishes
b.   sharks
c.   birds
d.   marine teleost fishes
e.   freshwater invertebrates

Answer: b
Difficulty: Moderate
Bloom's Taxonomy: Comprehension

10.   In all but the simplest animals, osmoregulation and excretion is carried out using tubules that are

a.   specialized blood vessels.
b.   part of the lining of the digestive tract.
c.   modifications of the respiratory and digestive systems.
d.   associated with the surface of the skin.
e.   formed from transport epithelium.

Answer: e
Difficulty: Moderate
Bloom's Taxonomy: Comprehension

11.   The nonselective movement of water and a number of solutes into excretory system tubules is

a.   reabsorption.
b.   release.
c.   filtration.
d.   secretion.
e.   none of these

Answer: c
Difficulty: Moderate
Bloom's Taxonomy: Comprehension

12.   The exiting of wastes from the excretory system is

a.   reabsorption.
b.   release.
c.   filtration.
d.   secretion.
e.   none of these

Answer: b
Difficulty: Easy
Bloom's Taxonomy: Comprehension

13.   The selective movement of specific small molecules and ions into excretory system tubules is

a.   reabsorption.
b.   release.
c.   filtration.
d.   secretion.
e.   none of these

Answer: d
Difficulty: Difficult
Bloom's Taxonomy: Comprehension

14.   The movement of some molecules and ions out of excretory system tubules and back into body fluids is

a.   reabsorption.
b.   release.
c.   filtration.
d.   secretion.
e.   none of these

Answer: a
Difficulty: Moderate
Bloom's Taxonomy: Comprehension

15.   Suppose sugar is taken out of an excretory system tubule and back into the blood. This would be an example of

a.   reabsorption.
b.   release.
c.   filtration.
d.   secretion.
e.   none of these

Answer: a
Difficulty: Moderate
Bloom's Taxonomy: Application

16.   Which of the following is the main form of nitrogenous wastes released by aquatic invertebrates to their environment?

a.   nitrate
b.   ammonia
c.   uric acid
d.   amino acids
e.   urea

Answer: b
Difficulty: Moderate
Bloom's Taxonomy: Comprehension

17.   Which of the following is the main form of nitrogenous wastes released by mammals to their environment?

a.   nitrate
b.   ammonia
c.   uric acid
d.   amino acids
e.   urea

Answer: e
Difficulty: Easy
Bloom's Taxonomy: Comprehension

18.   Which of the following is the main form of nitrogenous wastes released by birds to their environment?

a.   nitrate
b.   ammonia
c.   uric acid
d.   amino acids
e.   urea

Answer: c
Difficulty: Moderate
Bloom's Taxonomy: Comprehension

19. Compared to uric acid, using urea as the main form of nitrogenous waste requires

a. more energy and water.
b. less energy and water.
c. more energy and less water.
d. the same amount of energy and water.
e. less energy and more water.

Answer: e
Difficulty: Moderate
Bloom's Taxonomy: Synthesis

46.2 OSMOREGULATION AND EXCRETION IN INVERTEBRATES

20. The relatively simple excretory system used by flatworms consists of tubules called

a. nephrons.
b. hepatic tubules.
c. metanephridia.
d. Malpighian tubules.
e. protonephridia.

Answer: e
Difficulty: Easy
Bloom's Taxonomy: Knowledge

21. The smallest branches of _____ end with a large flame cell.

a. nephrons
b. hepatic tubules
c. metanephridia
d. Malpighian tubules
e. protonephridia

Answer: e
Difficulty: Moderate
Bloom's Taxonomy: Synthesis

22. Protonephridia are found in

a. annelids.
b. mammals.
c. flatworms.
d. insects.
e. none of these

Answer: c
Difficulty: Moderate
Bloom's Taxonomy: Comprehension

23. The excretory system used by most annelids and adult mollusks consists of tubules called

a. nephrons.
b. hepatic tubules.
c. metanephridia.
d. Malpighian tubules.
e. protonephridia.

Answer: c
Difficulty: Moderate
Bloom's Taxonomy: Knowledge

24. Metanephridia are found in

a. annelids.
b. mammals.
c. flatworms.
d. insects.
e. none of these

Answer: a
Difficulty: Moderate
Bloom's Taxonomy: Comprehension

25. You examine an animal and find that it has excretory tubules that have a closed proximal end immersed in hemolymph and a distal end that empties into the gut. Which term should you use to describe these tubules?

a. nephrons
b. hepatic tubules
c. metanephridia
d. Malpighian tubules
e. protonephridia

Answer: d
Difficulty: Moderate
Bloom's Taxonomy: Application

26. Malpighian tubules are found in

a. annelids.
b. mammals.
c. flatworms.
d. insects.
e. none of these

Answer: d
Difficulty: Moderate
Bloom's Taxonomy: Comprehension

27. Nitrogenous waste is excreted from Malpighian tubules mainly in the form of

a. nitrate.
b. ammonia.
c. uric acid.
d. amino acids.
e. urea.

Answer: c
Difficulty: Difficult
Bloom's Taxonomy: Synthesis

46.3 OSMOREGULATION AND EXCRETION IN MAMMALS

28. True kidneys have tubules called

a. nephrons.
b. hepatic tubules.
c. metanephridia.
d. Malpighian tubules.
e. protonephridia.

Answer: a
Difficulty: Moderate
Bloom's Taxonomy: Knowledge

29. Nephrons are found in

a. annelids.
b. mammals.
c. flatworms.
d. insects.
e. none of these

Answer: b
Difficulty: Moderate
Bloom's Taxonomy: Comprehension

30. The descending segment of the longest loops of Henle descend into the

a. renal pelvis.
b. renal cortex.
c. Bowman's capsule.
d. collecting ducts.
e. renal medulla.

Answer: e
Difficulty: Difficult
Bloom's Taxonomy: Synthesis

31. The tube through which urine leaves the urinary bladder is the

a. collecting duct.
b. urethra.
c. renal pelvis.
d. ureter.
e. renal vein.

Answer: b
Difficulty: Moderate
Bloom's Taxonomy: Knowledge

32. In mammals, the filtrate leaves the blood by exiting from the

a. efferent arteriole.
b. peritubular capillaries.
c. renal artery.
d. afferent arteriole.
e. glomerulus.

Answer: e
Difficulty: Moderate
Bloom's Taxonomy: Synthesis

33. In mammals, the glomerulus is located within the

a. distal convoluted tubule.
b. Bowman's capsule.
c. descending segment of the loop of Henle.
d. proximal convoluted tubule.
e. ascending segment of the loop of Henle.

Answer: b
Difficulty: Moderate
Bloom's Taxonomy: Knowledge

34. In mammals, urine drains out of the _____ into a collecting duct.

a. distal convoluted tubule
b. Bowman's capsule
c. descending segment of the loop of Henle
d. proximal convoluted tubule
e. ascending segment of the loop of Henle

Answer: a
Difficulty: Moderate
Bloom's Taxonomy: Knowledge

35. Because it has no aquaporins, water is generally trapped in the _____ in mammals.

a. distal convoluted tubule
b. Bowman's capsule
c. descending segment of the loop of Henle
d. proximal convoluted tubule
e. ascending segment of the loop of Henle

Answer: e
Difficulty: Difficult
Bloom's Taxonomy: Comprehension

36. As filtrate moves through the _____ in mammals it generally undergoes a dramatic decrease in osmolarity.

a. distal convoluted tubule
b. Bowman's capsule
c. descending segment of the loop of Henle
d. proximal convoluted tubule
e. ascending segment of the loop of Henle

Answer: e
Difficulty: Difficult
Bloom's Taxonomy: Synthesis

37. Which of the following excretory tubule segments in mammals allows water to exit but does not allow ions or urea to exit?

a. distal convoluted tubule
b. Bowman's capsule
c. descending segment of the loop of Henle
d. proximal convoluted tubule
e. ascending segment of the loop of Henle

Answer: c
Difficulty: Difficult
Bloom's Taxonomy: Synthesis

38. In which of the following excretory tubule segments in mammals does about 65 percent of the water from the filtrate get reabsorbed?

a. distal convoluted tubule
b. Bowman's capsule
c. descending segment of the loop of Henle
d. proximal convoluted tubule
e. ascending segment of the loop of Henle

Answer: d
Difficulty: Moderate
Bloom's Taxonomy: Comprehension

39. In which of the following excretory tubule segments in mammals does essentially all of the glucose and amino acids from the filtrate get reabsorbed?

a. distal convoluted tubule
b. Bowman's capsule
c. descending segment of the loop of Henle
d. proximal convoluted tubule
e. ascending segment of the loop of Henle

Answer: d
Difficulty: Moderate
Bloom's Taxonomy: Comprehension

40. Which of the following segments of the mammalian nephron receives the filtrate first?

a. distal convoluted tubule
b. Bowman's capsule
c. descending segment of the loop of Henle
d. proximal convoluted tubule
e. ascending segment of the loop of Henle

Answer: b
Difficulty: Moderate
Bloom's Taxonomy: Comprehension

41. Molecules and ions reabsorbed from the nephron reenter the blood at the

a. efferent arteriole.
b. peritubular capillaries.
c. renal artery.
d. afferent arteriole.
e. glomerulus.

Answer: b
Difficulty: Moderate
Bloom's Taxonomy: Synthesis

42. In a typical adult human about 180 L of fluid leaves the blood as filtrate each day, and _____ is reabsorbed.

a. ~50 percent
b. ~65 percent
c. ~80 percent
d. ~90 percent
e. over 99 percent

Answer: e
Difficulty: Easy
Bloom's Taxonomy: Knowledge

43. Aquaporins are

a. cells that are specialized for water transport.
b. transport channels for water.
c. the entry point of filtrate into a nephron.
d. transport channels for ions.
e. the exit channel for urine out of a nephron.

Answer: b
Difficulty: Easy
Bloom's Taxonomy: Knowledge

44. Which of the following excretory tubule segments in mammals consumes the most ATP?

a. distal convoluted tubule
b. Bowman's capsule
c. descending segment of the loop of Henle
d. proximal convoluted tubule
e. ascending segment of the loop of Henle

Answer: e
Difficulty: Difficult
Bloom's Taxonomy: Synthesis

45. Where in the mammalian kidney should you expect to find cells with the highest concentration of osmolytes such as sorbitol?

a. renal pelvis
b. renal cortex
c. Bowman's capsule
d. collecting ducts
e. renal medulla

Answer: e
Difficulty: Moderate
Bloom's Taxonomy: Application

46. The kangaroo rat gets the bulk of its water from

a. drinking liquids.
b. absorption through its skin.
c. ingesting food.
d. oxidative reactions in its cells.
e. reuptake of water from urine.

Answer: d
Difficulty: Moderate
Bloom's Taxonomy: Knowledge

47. The urine of marine mammals is _____ when compared to _____.

a. hypoosmotic; seawater
b. hypoosmotic; their body fluids
c. hyperosmotic; seawater
d. isoosmotic; their body fluids
e. isoosmotic; seawater

Answer: c
Difficulty: Moderate
Bloom's Taxonomy: Synthesis

INSIGHTS FROM THE MOLECULAR REVOLUTION: AN ORE SPELLS RELIEF FOR OSMOTIC STRESS

48. The osmotic response element discovered by Ferraris and her colleagues came from the promoter for

a. hexokinase.
b. sorbitol dehydrogenase.
c. aldose reductase.
d. luciferase.
e. sorbitase.

Answer: c
Difficulty: Moderate
Bloom's Taxonomy: Knowledge

49.   Suppose that the osmotic response element is in a promoter joined to the coding region of luciferase, and this composite gene is in cultured renal medulla cells. The cells should glow when they are placed in a medium that is _____ to the cells.

a.   hypertonic
b.   hypotonic
c.   hypotonic or isotonic
d.   isotonic
e.   hypertonic or isotonic

Answer: a
Difficulty: Moderate
Bloom's Taxonomy: Application

46.4 REGULATION OF MAMMALIAN KIDNEY FUNCTION

50.   The filtration rate at the Bowman's capsule is kept constant in response to small variations in blood pressure by the actions of

a.   the renin-angiotensin-aldosterone system.
b.   osmoreceptors.
c.   atrial naturiuretic factor.
d.   the juxtaglomerular apparatus.
e.   antidiuretic hormone.

Answer: d
Difficulty: Moderate
Bloom's Taxonomy: Comprehension

51.   Which of the following is secreted by cells in the juxtaglomerular apparatus in response to a significant drop in blood pressure or blood volume?

a.   atrial naturiuretic factor
b.   renin
c.   aldosterone
d.   antidiuretic hormone
e.   angiotensin

Answer: b
Difficulty: Difficult
Bloom's Taxonomy: Comprehension

52.   Which of the following is a hormone produced in the adrenal cortex that increases Na^+ reabsorption in the kidneys?

a.   atrial naturiuretic factor
b.   renin
c.   aldosterone
d.   antidiuretic hormone
e.   angiotensin

Answer: c
Difficulty: Moderate
Bloom's Taxonomy: Comprehension

53.   Which of the following, produced from a blood protein as part of the chain of events in response to a significant drop in blood pressure, quickly raises blood pressure by constricting many arterioles?

a.   atrial naturiuretic factor
b.   renin
c.   aldosterone
d.   antidiuretic hormone
e.   angiotensin

Answer: e
Difficulty: Difficult
Bloom's Taxonomy: Comprehension

54.   Which of the following is a hormone produced in the posterior pituitary that increases water absorption in the kidneys?

a.   atrial naturiuretic factor
b.   renin
c.   aldosterone
d.   antidiuretic hormone
e.   angiotensin

Answer: d
Difficulty: Easy
Bloom's Taxonomy: Comprehension

55.   Which of the following is a hormone produced by specialized cells in the heart in response to high blood pressure?

a.   atrial naturiuretic factor
b.   renin
c.   aldosterone
d.   antidiuretic hormone
e.   angiotensin

Answer: a
Difficulty: Easy
Bloom's Taxonomy: Comprehension

56.   Osmoreceptors help the body deal with dehydration in part by stimulating release of

a.   atrial naturiuretic factor.
b.   renin.
c.   aldosterone.
d.   antidiuretic hormone.
e.   angiotensin.

Answer: d
Difficulty: Moderate
Bloom's Taxonomy: Comprehension

57.   The osmoreceptors that the body uses to detect and react to situations such as dehydration are located in the

a.   hypothalamus.
b.   adrenal glands.
c.   efferent arteriole.
d.   kidneys.
e.   posterior pituitary.

Answer: a
Difficulty: Difficult
Bloom's Taxonomy: Knowledge

46.5 KIDNEY FUNCTION IN NONMAMMALIAN VERTEBRATES

58. Marine teleost fishes deal with the osmotic stress of living in seawater mainly by

a. active transport of ions into the body through the gills.
b. excreting nitrogenous wastes through their kidneys as urea.
c. active transport of ions out of the body by chloride cells.
d. excreting excess salt through the actions of specialized salt glands in their heads.
e. maintaining high levels of urea and trimethylamine oxide in body fluids.

Answer: c
Difficulty: Moderate
Bloom's Taxonomy: Comprehension

59. Sharks and rays deal with the osmotic stress of living in seawater mainly by

a. active transport of ions into the body through the gills.
b. excreting nitrogenous wastes through their kidneys as urea.
c. active transport of ions out of the body by chloride cells.
d. excreting excess salt through the actions of specialized salt glands in their heads.
e. maintaining high levels of urea and trimethylamine oxide in body fluids.

Answer: e
Difficulty: Moderate
Bloom's Taxonomy: Comprehension

60. Teleost fishes deal with the osmotic stress of living in seawater in part by

a. active transport of ions into the body through the gills.
b. excreting nitrogenous wastes through their kidneys as urea.
c. active transport of ions out of the body by chloride cells.
d. excreting excess salt through the actions of specialized salt glands in their heads.
e. maintaining high levels of urea and trimethylamine oxide in body fluids.

Answer: a
Difficulty: Easy
Bloom's Taxonomy: Comprehension

61. Many birds, such as seagulls which rarely drink fresh water, deal with the osmotic stress of taking in large quantities of salt in their food mainly by

a. active transport of ions into the body through their feet.
b. excreting nitrogenous wastes through their kidneys as urea.
c. active transport of ions out of the body by chloride cells.
d. excreting excess salt through the actions of specialized salt glands in their heads.
e. maintaining high levels of urea and trimethylamine oxide in body fluids.

Answer: d
Difficulty: Moderate
Bloom's Taxonomy: Comprehension

46.6 INTRODUCTION TO THERMOREGULATION

62. Animals exchange heat with their environment by which of the following processes?

a. conduction
b. radiation
c. evaporation
d. convection
e. all of these

Answer: e
Difficulty: Easy
Bloom's Taxonomy: Knowledge

63. Which of the following processes is the flow of heat between atoms or molecules in direct contact?

a. conduction
b. radiation
c. evaporation
d. convection
e. all of these

Answer: a
Difficulty: Moderate
Bloom's Taxonomy: Knowledge

64. Which of the following processes is the transfer of heat from a body to a fluid that passes over its surface?

a. conduction
b. radiation
c. evaporation
d. convection
e. all of these

Answer: d
Difficulty: Moderate
Bloom's Taxonomy: Knowledge

65. Organisms that obtain heat primarily from the external environment are called

a. endotherms.
b. isotherms.
c. exotherms.
d. ectotherms.
e. allotherms.

Answer: d
Difficulty: Easy
Bloom's Taxonomy: Comprehension

66. Organisms that obtain heat primarily from internal physiological sources are called

a. endotherms.
b. isotherms.
c. exotherms.
d. ectotherms.
e. allotherms.

Answer: a
Difficulty: Easy
Bloom's Taxonomy: Comprehension

67. As it gets colder, the metabolic rate of an endotherm typically goes _____ and the metabolic rate of an ectotherm typically goes _____.

a. down; down
b. up; down
c. down; up
d. up; up

Answer: b
Difficulty: Moderate
Bloom's Taxonomy: Application

46.7 ECTOTHERMY

68. Which of the following is a way that many terrestrial ectotherm regulates their temperature?

a. moving into or out of the shade
b. panting
c. changing the angle of the body relative to the sun
d. changing blood flow to the skin
e. all of these

Answer: e
Difficulty: Easy
Bloom's Taxonomy: Synthesis

69. Physiological changes in ectotherms that change their temperature tolerance as the seasons change from winter to summer are called

a. estivation.
b. hyperthermia.
c. thermal acclimatization.
d. hypothermia.
e. hibernation.

Answer: c
Difficulty: Moderate
Bloom's Taxonomy: Comprehension

70. Which of the following organisms is an ectotherm?

a. elephant
b. lizard
c. owl
d. bat
e. all of these

Answer: b
Difficulty: Easy
Bloom's Taxonomy: Application

46.8 ENDOTHERMY

71. Which of the following organisms is an endotherm?

a. mouse
b. salamander
c. turtle
d. frog
e. all of these

Answer: a
Difficulty: Easy
Bloom's Taxonomy: Application

72. Which of the following is a reaction by mammals to cold that limits the amount of heat lost to the surroundings?

a. shivering
b. going off to lay alone so that others do not take their heat
c. pressing hair shafts close to the body
d. reducing blood flow to the skin
e. losing brown adipose tissue

Answer: d
Difficulty: Moderate
Bloom's Taxonomy: Synthesis

73. In humans brown adipose tissue is used for

a. nitrogenous waste processing.
b. controlling blood pressure.
c. nonshivering thermogenesis.
d. digestion of fats.
e. cooling the body when exposed.

Answer: c
Difficulty: Moderate
Bloom's Taxonomy: Knowledge

74. In humans, an increase of body temperature a few degrees above normal for a prolonged period produces a state called

a. estivation.
b. hyperthermia.
c. thermal acclimatization.
d. hypothermia.
e. hibernation.

Answer: b
Difficulty: Easy
Bloom's Taxonomy: Comprehension

75. The human temperature set point typically

a. ranges from about 95.9°F in the morning to about 99.9°F in the evening.
b. stays right at about 98.6°F.
c. ranges from about 94.5°F in the morning to about 99.2°F in the evening.
d. ranges from about 99.9°F in the morning to about 95.9°F in the evening.
e. ranges from about 99.2°F in the morning to about 94.5°F in the evening.

Answer: a
Difficulty: Moderate
Bloom's Taxonomy: Knowledge

76. An extended period of torpor entered by an animal during the winter when the environment is too cold and food is too scarce is called

a. estivation.
b. hyperthermia.
c. thermal acclimatization.
d. hypothermia.
e. hibernation.

Answer: e
Difficulty: Easy
Bloom's Taxonomy: Comprehension

77. An extended period of torpor entered by an animal during the summer when the environment is too hot and water is too scarce is called

a. estivation.
b. hyperthermia.
c. thermal acclimatization.
d. hypothermia.
e. hibernation.

Answer: a
Difficulty: Moderate
Bloom's Taxonomy: Comprehension

UNANSWERED QUESTIONS

78. The temperature preferred by *Paralvinella* worms is

a. ~12°C.
b. ~50°C.
c. ~37°C.
d. ~99°C.
e. ~55°C.

Answer: b
Difficulty: Moderate
Bloom's Taxonomy: Knowledge

79. Sharks are able to survive having high levels of urea in their cells because

a. urea stabilizes proteins.
b. urea does not affect proteins.
c. trimethylamine oxide counteracts the negative effects of urea.
d. urea counteracts the unfolding of proteins by trimethylamine oxide.
e. urea unfolds proteins that were improperly folded by trimethylamine oxide.

Answer: c
Difficulty: Difficult
Bloom's Taxonomy: Comprehension

Choice

Choose the nephron tubule portion that best matches the description.

a. ascending segment of the loop of Henle
b. proximal convoluted tubule
c. Bowman's capsule
d. descending segment of the loop of Henle
e. distal convoluted tubule

80. _____ site where water and small substances are first passed into the nephron

81. _____ tubule segment that usually has the largest decrease in osmolarity

82. _____ may start in the cortex and end in the medulla

83. _____ empties into a collecting duct

84. _____ site of reabsorption of nearly all glucose

85. _____ may start in the medulla and end in the cortex

86. _____ surrounds the glomerulus

87. _____ Na^+ and Cl^- are actively transported out here, but water and urea are not moved across the membrane

88. _____ tubule segment that usually has the largest increase in osmolarity

89. _____ site of over half of the water reabsorption from the filtrate

Answers: *80. c*   *81. a*   *82. d*   *83. e*   *84. b*
           85. a   *86. c*   *87. a*   *88. d*   *89. b*

Difficulty: Moderate
Bloom's Taxonomy: Comprehension
Source: Section 46.3

47
ANIMAL REPRODUCTION

Multiple-Choice

WHY IT MATTERS

1. Which of the following statements about the palolo worm is *false*?

a. Its scientific name is *Eunice viridis*.
b. It is a polychaete annelid.
c. It lives in coral reefs.
d. It produces gamete-filled tails in its anterior end.
e. Its tails break off once a year, in October.

Answer: d
Difficulty: Moderate
Bloom's Taxonomy: Comprehension

47.1 ANIMAL REPRODUCTIVE MODES: ASEXUAL AND SEXUAL REPRODUCTION

2. Which of the following statements about asexual reproduction is *true*?

a. The offspring are genetically identical to each other.
b. The offspring are genetically identical to their parent.
c. It is rare in vertebrates.
d. a and b only.
e. a, b, and c.

Answer: e
Difficulty: Easy
Bloom's Taxonomy: Knowledge

3. Asexual reproduction can involve all of the following processes *except*

a. fission.
b. fertilization.
c. fragmentation.
d. budding.
e. parthenogenesis.

Answer: b
Difficulty: Easy
Bloom's Taxonomy: Knowledge

4. If humans reproduced by parthenogenesis,

a. males would need to produce more sperm.
b. the number of males in the population would most likely increase.
c. the number of females in the population would most likely increase.
d. the children would be identical to their parents.
e. there would be no need for mitosis.

Answer: c
Difficulty: Moderate
Bloom's Taxonomy: Application

5. In _____, the parent separates into two or more offspring of approximately the same size.

a. fission
b. fertilization
c. fragmentation
d. budding
e. parthenogenesis

Answer: a
Difficulty: Easy
Bloom's Taxonomy: Knowledge

6. Which of the following mechanisms increases genetic diversity?

a. crossing-over between chromosomes during meiosis
b. independent assortment of chromosomes during meiosis
c. random DNA mutations
d. a and b only
e. a, b, and c

Answer: e
Difficulty: Easy
Bloom's Taxonomy: Comprehension

7. Which of the following conditions favors sexual reproduction?

a. stable environment
b. uniform environments
c. sparsely settled populations
d. sessile animals
e. none of the above

Answer: e
Difficulty: Moderate
Bloom's Taxonomy: Comprehension

47.2 CELLULAR MECHANISMS OF SEXUAL REPRODUCTION

8.   Which of the following statements best describes the gametes of most animals?

a.   A sperm is large and motile; an egg is small and nonmotile.
b.   A sperm is small and nonmotile ; an egg is large and motile.
c.   A sperm is large and nonmotile ; an egg is small and motile.
d.   A sperm is small and motile ; an egg is large and nonmotile.
e.   The sperm and egg are about the same size, but only the sperm is motile.

Answer: d
Difficulty: Easy
Bloom's Taxonomy: Knowledge

9.   In humans, spermatogenesis differs from oogenesis in that

a.   spermatogenesis involves meiosis, while oogenesis involves mitosis.
b.   spermatogenesis begins at birth, while oogenesis begins at puberty.
c.   spermatogenesis has unequal cytoplasmic divisions, while oogenesis does not.
d.   spermatogenesis produces fewer numbers of gametes.
e.   spermatogenesis continues throughout a male's life, while oogenesis ends at menopause.

Answer: e
Difficulty: Easy
Bloom's Taxonomy: Knowledge

10.   Sperm pass through several stages during their development. The earliest stage in which the cell becomes haploid is the

a.   sperm.
b.   spermatid.
c.   primary spermatocyte.
d.   secondary spermatocyte.
e.   spermatogonium.

Answer: d
Difficulty: Moderate
Bloom's Taxonomy: Knowledge

11.   One secondary spermatocyte ultimately produces _____ mature sperm.

a.   1
b.   2
c.   3
d.   4
e.   8

Answer: b
Difficulty: Easy
Bloom's Taxonomy: Knowledge

12.   The mitochondria present in the sperm provide

a.   energy for the movement of the flagellum.
b.   energy for the acrosome to penetrate the egg.
c.   energy for the maturation of the sperm.
d.   nutrients for the sperm.
e.   nutrients for the developing embryo.

Answer: a
Difficulty: Easy
Bloom's Taxonomy: Comprehension

13.   Which of the following is a correct sequence of cells in oogenesis?

a.   oogonium → primary oocyte → secondary oocyte
b.   oogonium → secondary oocyte → primary oocyte
c.   primary oocyte → secondary oocyte → oogonium
d.   primary oocyte → oogonium →secondary oocyte
e.   none of the above

Answer: a
Difficulty: Easy
Bloom's Taxonomy: Knowledge

14.   In females, meiosis II is completed

a.   before birth.
b.   at the time of birth.
c.   when eggs are ovulated each month.
d.   at puberty.
e.   only if the egg is fertilized.

Answer: e
Difficulty: Easy
Bloom's Taxonomy: Knowledge

15. If the number of chromosomes in an animal's gametes is 20, then a somatic cell from that animal would contain _____ chromosomes.

a. 5
b. 10
c. 20
d. 40
e. 80

Answer: d
Difficulty: Moderate
Bloom's Taxonomy: Application

16. Which of the following is *not* a mammalian group?

a. monotremes
b. eutherians
c. polytherians
d. metatherians
e. All of the above are mammalian groups

Answer: c
Difficulty: Easy
Bloom's Taxonomy: Knowledge

17. In some species of sharks, babies hatch from shells within their mother's body. This is an example of

a. oviparity.
b. viviparity.
c. ovoviviparity.
d. budding.
e. fission.

Answer: c
Difficulty: Easy
Bloom's Taxonomy: Application

18. During mating in frogs, the male clasps the female tightly around the body with his forelimbs. This is termed

a. amplexus.
b. fertilization.
c. copulation.
d. fusion.
e. marriage.

Answer: a
Difficulty: Easy
Bloom's Taxonomy: Knowledge

19. External fertilization is most common in animals living in

a. mountains.
b. prairies.
c. forests.
d. oceans.
e. deserts.

Answer: d
Difficulty: Moderate
Bloom's Taxonomy: Application

20. Which of the following is an advantage of internal over external fertilization?

a. protection of gametes from predators
b. protection of gametes from drying
c. production of more offspring
d. a and b only
e. a, b, and c

Answer: d
Difficulty: Moderate
Bloom's Taxonomy: Application

21. Which of the following statements about the fast block to polyspermy is *false*?

a. It occurs within seconds after fertilization.
b. It involves a wave of electrical depolarization that spreads over the egg surface.
c. It is not mediated by Ca^{2+} ions.
d. It is more pronounced in vertebrates than in invertebrates.
e. All of the above are true.

Answer: d
Difficulty: Moderate
Bloom's Taxonomy: Comprehension

22. The presence of both male and female reproductive tissue in the same individual is termed

a. parthenogenesis.
b. hermaphroditism.
c. ovipary.
d. vivipary.
e. ovovivipary.

Answer: b
Difficulty: Easy
Bloom's Taxonomy: Knowledge

47.3 SEXUAL REPRODUCTION IN HUMANS

23. In humans, the gonads

a. produce gametes.
b. secrete hormones.
c. nourish the embryo.
d. a and b only
e. a, b and c

Answer: d
Difficulty: Moderate
Bloom's Taxonomy: Knowledge

24. A human female could ovulate about _____ times in her lifetime.

a. 40
b. 400
c. 4,000
d. 40,000
e. 400,000

Answer: b
Difficulty: Moderate
Bloom's Taxonomy: Comprehension

25. The birth canal is the

a. oviduct.
b. uterus.
c. cervix.
d. vagina.
e. vulva.

Answer: d
Difficulty: Easy
Bloom's Taxonomy: Knowledge

Use the figure shown below for questions 26–32,

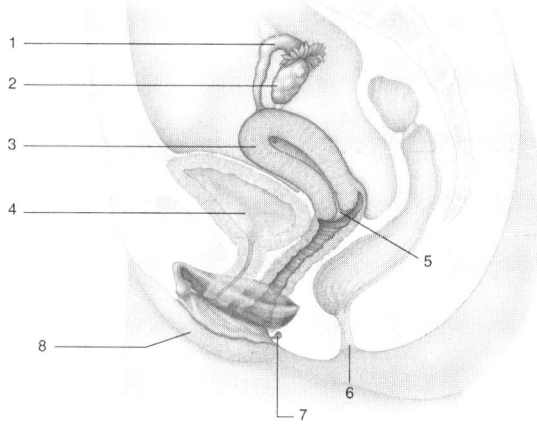

26. The cervix is labeled

a. 3.
b. 4.
c. 5.
d. 6.
e. 7.

Answer: c
Difficulty: Moderate
Bloom's Taxonomy: Comprehension

27. The urinary bladder is labeled

a. 1.
b. 2.
c. 3.
d. 4.
e. 5.

Answer: d
Difficulty: Moderate
Bloom's Taxonomy: Comprehension

28. The site of fertilization is labeled

a. 1.
b. 2.
c. 3.
d. 4.
e. 5.

Answer: a
Difficulty: Moderate
Bloom's Taxonomy: Comprehension

29. The structure that is responsible for producing hormones is labeled

a. 1.
b. 2.
c. 3.
d. 4.
e. 5.

Answer: b
Difficulty: Easy
Bloom's Taxonomy: Comprehension

30. The site of implantation is labeled

a. 1.
b. 2.
c. 3.
d. 4.
e. 5.

Answer: c
Difficulty: Easy
Bloom's Taxonomy: Comprehension

31. The structure that is covered with cilia is labeled

a. 1.
b. 2.
c. 3.
d. 4.
e. 5.

Answer: a
Difficulty: Moderate
Bloom's Taxonomy: Comprehension

32. The vulva includes the structure(s) labeled

a. 6.
b. 7.
c. 8.
d. 7 and 8 only.
e. 6, 7, and 8.

Answer: d
Difficulty: Difficult
Bloom's Taxonomy: Comprehension

33. The starting point for the ovarian cycle is a

a. primary oocyte arrested in meiosis I.
b. primary oocyte arrested in meiosis II.
c. primary oocyte having just completed meiosis I.
d. secondary oocyte arrested in meiosis I.
e. secondary oocyte arrested in meiosis II.

Answer: a
Difficulty: Easy
Bloom's Taxonomy: Knowledge

34. The ovarian cycle includes the

a. luteal and secretory phases.
b. menstrual and proliferative phases.
c. follicular and luteal phases.
d. follicular and proliferative phases.
e. proliferative and secretory phases.

Answer: c
Difficulty: Moderate
Bloom's Taxonomy: Knowledge

35. One function of the developing follicle is to

a. stimulate ovulation.
b. secrete progesterone.
c. secrete estrogens.
d. secrete gonadotropin-releasing hormone.
e. secrete glycoproteins, which promote sperm motility.

Answer: c
Difficulty: Easy
Bloom's Taxonomy: Knowledge

36. In the female, the _____ serves as a *temporary* endocrine gland and secretes the _____.

a. pituitary; luteinizing hormone and follicle-stimulating hormone
b. corpus luteum; estrogen and progesterone
c. ovaries; estrogen and luteinizing hormone
d. endometrium; estrogen and follicle-stimulating hormone
e. uterus; luteinizing hormone and follicle-stimulating hormone

Answer: b
Difficulty: Moderate
Bloom's Taxonomy: Knowledge

37. Ovulation typically occurs on day _____ of a 28-day ovarian cycle.

a. 1
b. 7
c. 10
d. 14
e. 28

Answer: d
Difficulty: Easy
Bloom's Taxonomy: Knowledge

38. In the female reproductive cycle, the endometrium is thickest

a. during the menstrual phase.
b. during the proliferative phase.
c. shortly before ovulation.
d. during ovulation.
e. in the secretory phase.

Answer: e
Difficulty: Easy
Bloom's Taxonomy: Knowledge

39. The presence of _____ in urine indicates a pregnancy.

a. estrogen
b. progesterone
c. human chorionic gonadotropin
d. follicle-stimulating hormone
e. luteinizing hormone

Answer: c
Difficulty: Easy
Bloom's Taxonomy: Knowledge

Use the figure shown below for questions 40–46.

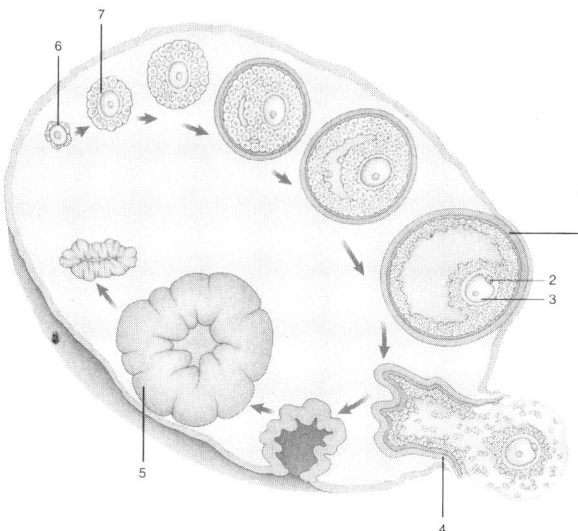

40. Label 1 indicates which of the following?

a. the zona pellucida
b. the cortical granules
c. the vitelline coat
d. the acrosome
e. none of the above

Answer: e
Difficulty: Moderate
Bloom's Taxonomy: Comprehension

41. Label 4 indicates

a. implantation.
b. ovulation.
c. menstruation.
d. fertilization.
e. foliculization.

Answer: b
Difficulty: Easy
Bloom's Taxonomy: Comprehension

42. The structure labeled 6 indicates

a. a primordial follicle.
b. a mature follicle.
c. a primordial germ cell.
d. a mature germ cell.
e. an ootid.

Answer: a
Difficulty: Easy
Bloom's Taxonomy: Comprehension

43. The structure labeled 2 indicates

a. the Barr body.
b. the first polar body.
c. the second polar body.
d. the third polar body.
e. the no body.

Answer: b
Difficulty: Easy
Bloom's Taxonomy: Comprehension

44. The structure labeled 3 indicates

a. the primary oocyte.
b. the secondary oocyte.
c. the ootid.
d. the oogonium.
e. the mature follicle.

Answer: b
Difficulty: Easy
Bloom's Taxonomy: Comprehension

45. A structure whose secretions cause the uterine endometrial tissue to thicken in preparation for pregnancy is indicated by the label

a. 1.
b. 2.
c. 3.
d. 5.
e. 7.

Answer: d
Difficulty: Moderate
Bloom's Taxonomy: Application

46. Which structures can still undergo meiosis at the time of fertilization?

a. 2
b. 3
c. 2 and 3
d. 2, 3, and 5
e. none of the above

Answer: c
Difficulty: Moderate
Bloom's Taxonomy: Comprehension

47. Menstruation begins in response to

a. rising levels of FSH and LH.
b. rising levels of GnRH.
c. rising levels of progesterone.
d. falling levels of progesterone.
e. rising levels of estrogen.

Answer: d
Difficulty: Moderate
Bloom's Taxonomy: Comprehension

48. In a human female, an inhibition of LH would also inhibit which of the following?

a. maturation of follicles
b. secretion of FSH
c. secretion of GnRH
d. secretion of estrogen
e. none of the above

Answer: d
Difficulty: Difficult
Bloom's Taxonomy: Application

49. Menstrual flow results in the discharge of

a. the follicle.
b. the corpus luteum.
c. the lining of the endometrium.
d. surface cells from the vagina.
e. blood from vessels on the outer side of the uterus.

Answer: c
Difficulty: Easy
Bloom's Taxonomy: Knowledge

50. Female dogs undergo a(n) _____ cycle, in which _____ is absent.

a. menstrual cycle; ovulation
b. menstrual cycle; implantation
c. estrous cycle; ovulation
d. estrous cycle; menstruation
e. none of the above

Answer: d
Difficulty: Difficult
Bloom's Taxonomy: Application

51. Sperm are produced in the

a. epididymis.
b. seminiferous tubules.
c. penis.
d. prostate gland.
e. vas deferen.

Answer: b
Difficulty: Easy
Bloom's Taxonomy: Knowledge

52. Sperm are stored in the _____ after they are produced.

a. epididymis
b. seminiferous tubules
c. penis
d. prostate gland
e. vas deferens

Answer: a
Difficulty: Easy
Bloom's Taxonomy: Knowledge

53. Which of the following is *not* an accessory gland in males?

a. prostate gland
b. vestibular gland
c. bulbourethal gland
d. seminal vesicle.
e. all of the above *are* accessory glands in males

Answer: b
Difficulty: Moderate
Bloom's Taxonomy: Knowledge

54. Which of the following structures in males is shared by both the urinary and reproductive systems?

a. urinary bladder
b. prostate gland
c. ureter
d. vas deferens
e. urethra

Answer: e
Difficulty: Moderate
Bloom's Taxonomy: Knowledge

55. In many mammalian species, including bats and most non-human primates, a bone in the penis, called the _____, helps to maintain it in an erect state.

a. pendulum
b. baculum
c. colostrum
d. gametogonium
e. marsupium

Answer: b
Difficulty: Moderate
Bloom's Taxonomy: Knowledge

56. The mammalian acrosome contains enzymes that

a. digest the zona pellucida.
b. digest the endometrium.
c. stimulate ovulation.
d. stimulate sperm motility.
e. inhibit premature ejaculation.

Answer: a
Difficulty: Easy
Bloom's Taxonomy: Knowledge

Use the figure shown below for questions 57–65.

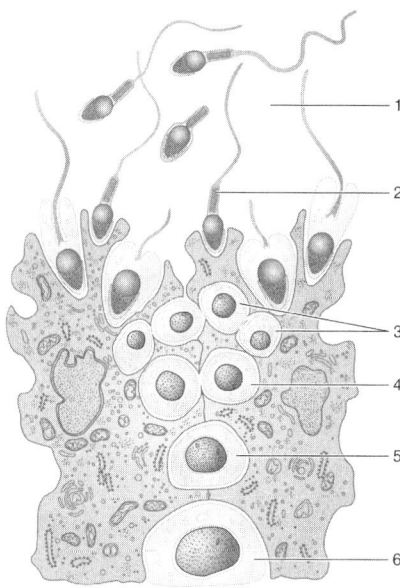

57. The gray matter that surrounds cells 3, 4, 5 and 6 indicates the

a. primordial germ cell.
b. spermatogonium.
c. Leydig cell.
d. Sertoli cell.
e. Ballack cell.

Answer: d
Difficulty: Moderate
Bloom's Taxonomy: Comprehension

58. Label 1 indicates the

a. matrix of the seminiferous tubule.
b. lumen of the seminiferous tubule.
c. epididymis.
d. vas deferens.
e. penis.

Answer: b
Difficulty: Moderate
Bloom's Taxonomy: Comprehension

59. The primary spermatocyte is cell number

a. 2.
b. 3.
c. 4.
d. 5.
e. 6.

Answer: d
Difficulty: Moderate
Bloom's Taxonomy: Comprehension

60. The spermatid is cell number

a. 2.
b. 3.
c. 4.
d. 5.
e. 6.

Answer: b
Difficulty: Moderate
Bloom's Taxonomy: Comprehension

61. The secondary spermatocyte is cell number

a. 2.
b. 3.
c. 4.
d. 5.
e. 6.

Answer: c
Difficulty: Moderate
Bloom's Taxonomy: Comprehension

62. The spermatogonium is cell number

a. 2.
b. 3.
c. 4.
d. 5.
e. 6.

Answer: e
Difficulty: Moderate
Bloom's Taxonomy: Comprehension

63. Which cell(s) is(are) diploid?

a. 4
b. 5
c. 6
d. 5 and 6 only
e. 4, 5, and 6

Answer: d
Difficulty: Moderate
Bloom's Taxonomy: Comprehension

64. Meiosis II results in the formation of cell number

a. 2.
b. 3.
c. 4.
d. 5.
e. 4.

Answer: b
Difficulty: Moderate
Bloom's Taxonomy: Comprehension

65. In a human male, how many total chromatids are present in cell number 4?

a. 2
b. 4
c. 23
d. 46
e. 92

Answer: d
Difficulty: Difficult
Bloom's Taxonomy: Application

66. Which of the following is required for the secretion of testosterone?

a. GnRH
b. LH
c. Sertoli cells
d. a and b only
e. a, b, and c

Answer: d
Difficulty: Difficult
Bloom's Taxonomy: Comprehension

INSIGHTS FROM THE MOLECULAR REVOLUTION: EGGING ON THE SPERM

67. Marc Spehr and his colleagues demonstrated which of the following?

a. Sperm can detect and swim toward attractant molecules.
b. Sperm attraction can be inhibited by chemicals.
c. Eggs release chemicals that attractant sperm.
d. a and b only.
e. a, b, and c.

Answer: d
Difficulty: Moderate
Bloom's Taxonomy: Comprehension

47.4 METHODS FOR PREVENTING PREGNANCY: CONTRACEPTION

68. Which of the following contraception techniques consists of avoiding intercourse during the time of the month when the egg can be fertilized?

a. withdrawal
b. tubal ligation
c. dilation and evacuation
d. rhythm method
e. none of the above

Answer: d
Difficulty: Moderate
Bloom's Taxonomy: Knowledge

69. The oral contraceptive pill contains

a. estrogen only.
b. estrogen and progesterone.
c. estrogen and follicle-stimulating hormone.
d. follicle-stimulating hormone and progesterone.
e. estrogen, progesterone, and follicle-stimulating hormone.

Answer: b
Difficulty: Moderate
Bloom's Taxonomy: Knowledge

70. The "pill" works primarily by

a. blocking sperm from entering the uterus.
b. causing spontaneous abortions.
c. inhibiting the release of FSH and LH.
d. preventing implantation of the embryo.
e. stimulating the release of GnRH.

Answer: c
Difficulty: Moderate
Bloom's Taxonomy: Comprehension

71. The type of contraception technique that prevents implantation (even after successful fertilization) is

a. tubal ligation.
b. spermicidal jelly.
c. birth control pills.
d. intrauterine device.
e. a diaphragm.

Answer: d
Difficulty: Moderate
Bloom's Taxonomy: Knowledge

72. The drug mifepristone (RU-486) is a contraceptive that

a. blocks passage of sperm through the male reproductive tract.
b. blocks passage of sperm through the female reproductive tract.
c. prevents ovulation.
d. causes shedding of the endometrial lining and the implanted embryo by initiating a menstrual period.
e. prevents implantation by blocking estrogen receptors in the uterus.

Answer: d
Difficulty: Moderate
Bloom's Taxonomy: Knowledge

73. Which of the following methods is the most effective as birth control?

a. condoms
b. diaphragms
c. abstinence
d. oral contraceptives
e. vasectomy

Answer: c
Difficulty: Easy
Bloom's Taxonomy: Knowledge

UNANSWERED QUESTIONS

74. Which of the following statements about prolactin is *true*?

a. It is produced by the pituitary gland.
b. It stimulates milk production in females.
c. Its levels decrease during orgasms in both sexes.
d. a and b only.
e. a, b, and c.

Answer: d
Difficulty: Moderate
Bloom's Taxonomy: Knowledge

75. The female erectile structure analogous to the male glans penis is the

a. labia majora.
b. labia minora.
c. mons pubis.
d. cervix.
e. clitoris.

Answer: e
Difficulty: Easy
Bloom's Taxonomy: Knowledge

Integrative Multiple-Choice

76. The major difference between the human male and female reproductive systems is the

a. provision of a site for fertilization and embryo development in the female.
b. production of haploid gametes only by the male.
c. diploid polar bodies of the female.
d. generation of millions of eggs but only thousands of sperm each month.
e. presence of a duct system opening to the body's exterior only in females.

Answer: a
Difficulty: Moderate
Bloom's Taxonomy: Knowledge
Source: Sections 47.2 and 47.3

77. Vasectomy causes sterility because

a. sperm is no longer be produced.
b. sperm no longer leave the testis.
c. sperm no longer reach the vas deferens.
d. sperm no longer reach the urethra.
e. semen is no longer produced.

Answer: d
Difficulty: Moderate
Bloom's Taxonomy: Knowledge
Source: Sections 47.3 and 47.4

Matching

Match each hormone with its correct description.

78. _____ estrogen

79. _____ follicle stimulating hormone (FSH)

80. _____ gonadotropin releasing hormone (GnRH)

81. _____ human chorionic gonadotropin (hCG)

82. _____ inhibin

83. _____ luteinizing hormone (LH)

84. _____ oxytocin

85. _____ progesterone

86. _____ relaxin

87. _____ testosterone

A. Triggers ovulation

B. Stimulates Sertoli cells

C. Secreted by the developing embryo; maintains function of the corpus luteum

D. Induces contraction of the uterus

E. Stimulates growth of the endometrium

F. Secreted by Sertoli cells and the corpus luteum

G. Secreted by the hypothalamus

H. Secreted by follicular cells

I. Secreted by the corpus luteum even after it regresses later in pregnancy; blocks uterine contractions until time of birth is near

J. Secreted by Leydig cells

Answers: ***78. H***   ***79. B***   ***80. G***   ***81. C***
              82. F   ***83. A***   ***84. D***   ***85. E***
              86. I   ***87. J***

Difficulty: Difficult
Bloom's Taxonomy: Comprehension
Source: Sections 47.3

Select the Exception

88.   Which of the following is *not* an advantage of asexual reproduction?

a.   In stable environments, the offspring are as well adapted as the parents.
b.   Individual organisms do not have to seek mates.
c.   It is less energetically costly than sexual reproduction.
d.   It promotes genetic diversity.
e.   It is faster than sexual reproduction.

Answer: d
Difficulty: Easy
Bloom's Taxonomy: Knowledge
Source: Section 47.1

89.   Semen contains all of the following *except*

a.   about 5 percent sperm by volume.
b.   a fast-acting clotting enzyme.
c.   a slow-acting enzyme that breaks down the semen clot.
d.   a substance that decreases the pH in the female reproductive tract.
e.   a substance that triggers contraction of muscles in the female reproductive tract.

Answer: d
Difficulty: Easy
Bloom's Taxonomy: Knowledge
Source: Section 47.3

90.   In mammalian females, which of the following is *not* essential for reproduction?

a.   ovary
b.   oviduct
c.   uterus
d.   vagina
e.   urethra

Answer: e
Difficulty: Easy
Bloom's Taxonomy: Knowledge
Source: Section 47.3

91.   All of the following animals have placentas *except*

a.   kangaroos.
b.   opposums.
c.   echidnas.
d.   koalas.
e.   wombats.

Answer: c
Difficulty: Moderate
Bloom's Taxonomy: Knowledge
Source: Section 47.2

92.   All of the following contraception techniques are aimed at preventing fertilization *except*

a.   withdrawal.
b.   intrauterine device.
c.   vasectomy.
d.   rhythm method.
e.   tubal ligation.

Answer: b
Difficulty: Moderate
Bloom's Taxonomy: Knowledge
Source: Section 47.4

True/False

If the statement is true, write a "T" in the blank. If the statement is false, make it correct by changing the underlined words(s) and writing the correct word(s) in the answer blank.

93.   _____   The cnidarian *Hydra* reproduces asexually by <u>fragmentation</u>.

94.   _____   All mammals except the <u>metatherians</u> are viviparous.

95.   _____   The thin flap of tissue that covers the opening of the vagina is called the <u>hymen</u>.

96.   _____   The slow block to polyspermy occurs when <u>polar bodies</u> discharge their contents, thereby altering the egg coats.

97.   _____   The uterine cycle is also called the <u>ovarian</u> cycle.

98. _____ After ovulation, the remaining follicle cells develop into the <u>corpus luteum</u>.

99. _____ After fertilization, the sperm that enters the egg releases <u>nitric oxide</u>, which stimulates the release of stored calcium ions in the egg.

100. _____ The <u>intrauterine device</u> is a cuplike rubber device that blocks the cervix in females.

101. _____ The egg is fertilized at the <u>secondary oocyte</u> stage.

102. _____ Menstruation begins immediately after the <u>follicular phase</u>.

103. _____ The <u>female</u> sexual response has a refractory period.

104. _____ In human males, the conversion of a spermatogonium into a sperm takes about <u>9 to 10 weeks</u>.

105. _____ Progesterone is produced in the <u>pituitary</u> and acts directly on the uterus.

Answers: *93. F, budding   94. F, monotremes 95. T   96. F, cortical granules   97. F, menstrual   98. T   99. T   100. F, diaphragm   101. T   102 F, secretory 103. F, male 104. T   105. F, ovary*

Difficulty: Difficult
Bloom's Taxonomy: Knowledge, Comprehension
Source: Sections 47.1–47.4

Labeling

Identify each labeled part of the following illustration.

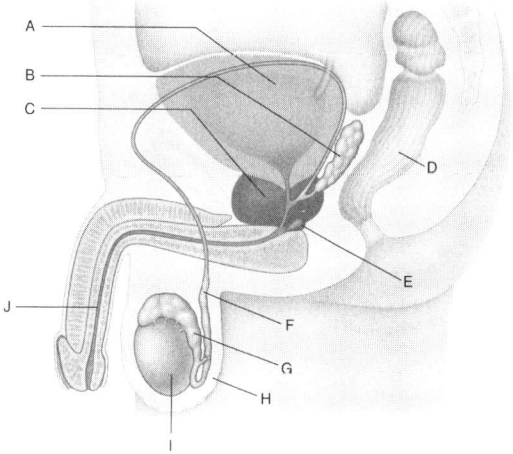

106. _____ seminal vesicle
107. _____ rectum
108. _____ bulbourethral gland
109. _____ epidiymus
110. _____ testis
111. _____ urinary bladder
112. _____ urethra
113. _____ scrotum
114. _____ prostate gland
115. _____ vas deferens

Answers: *106. B   107. D   108. E   109. G   110. I   111. A   112. J   113. H   114. C   115. F*

Difficulty: Moderate
Bloom's Taxonomy: Knowledge
Source: Figure 47.14

Essay

116.　Compare and contrast the processes of spermatogenesis and oogenesis.

Answer: Both spermatogenesis and oogenesis are types of gametogenesis, or gamete production. Spermatogenesis (sperm production) occurs in males in the seminiferous tubules of the testes, while oogenesis (egg formation) occurs in the ovaries of females. The two processes involve early mitotic and late meiotic divisions. Spermatogenesis and oogenesis differ in many respects. These include the following: 1) Spermatogenesis begins at puberty and continuous throughout the male's lifetime. Oogenesis, on the other hand, begins during embryonic development and stops at menopause. 2) While spermatogenesis is a continuous uninterrupted process, oogenesis undergoes two long "pauses" (one in meiosis I and the other in meiosis II). Indeed, oogenesis is not completed until fertilization occurs. 3) Spermatogenesis involves equal cytoplasmic divisions that result in four equivalent and functional gametes. Oogenesis results in one large functional gamete and up to three very small nonfunctional polar bodies. 4) Spermatogenesis results in the production of about 130 million sperm per day. Oogenesis, on the other hand, produces an average of one egg per month!
Difficulty: Moderate
Bloom's Taxonomy: Knowledge, Comprehension
Source: Sections 47.2, 47.3

117.　Explain the roles of the hormones GnRH, FSH, and LH, in the male and female reproductive systems.

Answer:　In both males and females, gonadotropin-releasing hormone (GnRH) is secreted by the hypothalamus gland and acts on the pituitary gland to release follicle-stimulating hormone (FSH) and luteinizing hormone (LH) into the bloodstream. While these two hormones were named for their function in females, they both exert effects on the male reproductive system as well. In females, the target of FSH and LH is the ovary: FSH stimulates follicular growth and maturation, and LH triggers ovulation and subsequent formation of the corpus luteum from the remaining follicle cells. In males, the target of the two gonadotropins is the testis. FSH stimulates Sertoli cells to secrete a protein and other molecules that are required for spermatogenesis, while LH stimulates the Leydig cells to secrete testosterone.
Difficulty: Moderate
Bloom's Taxonomy: Knowledge, Comprehension
Source: Section 47.3

48
ANIMAL DEVELOPMENT

Multiple-Choice

WHY IT MATTERS

1.   During the development of a human, which of these steps follows fertilization?

a.   cell differentiation
b.   development of organs
c.   formation of primary tissues
d.   morphogenesis
e.   rapid cell division via mitosis

Answer: e
Difficulty: Moderate
Bloom's Taxonomy: Comprehension

2.   Morphology refers to the _____ of an organism.

a.   reproductive ability
b.   appearance
c.   differentiation
d.   lifespan
e.   movement

Answer: b
Difficulty: Moderate
Bloom's Taxonomy: Knowledge

48.1 MECHANISMS OF EMBRYONIC DEVELOPMENT

3.   How does the amount of yolk relate to the nourishment of the organism?

a.   An egg with a small amount of yolk contains all of the nutrients for the embryo's development.
b.   An egg with a large amount of yolk contains all of the nutrients for the embryo's development.
c.   An egg lacking yolk contains all of the nutrients for the embryo's development.
d.   There is no relationship between the amount of yolk and the nourishment of the organism.
e.   All animals have the same amount of yolk and provide the same amount of nourishment.

Answer: b
Difficulty: Moderate
Bloom's Taxonomy: Comprehension

4.   Eggs from which of the following organisms would contain the lowest percentage of yolk?

a.   birds
b.   humans
c.   insects
d.   reptiles
e.   all of the animals above would contain the same percentage of yolk in their eggs

Answer: b
Difficulty: Easy
Bloom's Taxonomy: Knowledge

5.   Before the genes of a zygote become active, the stages of animal development are directed by

a.   the environment.
b.   the cytoplasmic determinants of the egg.
c.   the cytoplasmic determinants of the sperm.
d.   the cytoplasmic determinants of the egg and the sperm.
e.   the nucleus.

Answer: b
Difficulty: Moderate
Bloom's Taxonomy: Knowledge

6.   During mitotic cleavage

a.   no cell division occurs and the mass does not increase.
b.   no cell division occurs and the mass does increase.
c.   cell division occurs and the mass does not increase.
d.   cell division occurs and the mass does increase.
e.   the cell is in a dormant stage.

Answer: c
Difficulty: Moderate
Bloom's Taxonomy: Knowledge

7.   Put the following developmental processes in the correct order.

a.   cleavage – gastrulation – organogenesis
b.   cleavage – organogenesis – gastrulation
c.   gastrulation – cleavage – organogenesis
d.   gastrulation – organogenesis – cleavage
e.   organogenesis – cleavage – gastrulation

Answer: a
Difficulty: Moderate
Bloom's Taxonomy: Knowledge

8.   Place the three primary cell layers of an embryo in order from superficial to deep.

a.   ectoderm – endoderm – mesoderm
b.   ectoderm – mesoderm – endoderm
c.   endoderm – ectoderm – mesoderm
d.   endoderm – mesoderm – ectoderm
e.   mesoderm – ectoderm – endoderm

Answer: b
Difficulty: Moderate
Bloom's Taxonomy: Knowledge

9.   Put these developmental stages in the proper order.

a.   blastula – gastrula – morula
b.   blastula – morula – gastrula
c.   gastrula – blastula – morula
d.   morula – blastula – gastrula
e.   morula – gastrula – blastula

Answer: d
Difficulty: Moderate
Bloom's Taxonomy: Knowledge

10.   Which term represents the unequal distribution of yolk in an egg?

a.   cleavage
b.   gastrulation
c.   involution
d.   neurulation
e.   polarity

Answer: e
Difficulty: Moderate
Bloom's Taxonomy: Knowledge

48.2 MAJOR PATTERNS OF CLEAVAGE AND GASTRULATION

11.   Muscles originate from

a.   ectoderm.
b.   endoderm.
c.   mesoderm.
d.   ectoderm and endoderm.
e.   ectoderm and mesoderm.

Answer: c
Difficulty: Moderate
Bloom's Taxonomy: Knowledge

12.   Which group of adult tissues is derived from mesoderm?

a.   lining of digestive tract, liver, pancreas
b.   muscles, bones, cartilage
c.   skin, brain, retina
d.   skin, liver, pancreas
e.   lining of respiratory tract, thyroid gland, urinary bladder

Answer: b
Difficulty: Moderate
Bloom's Taxonomy: Knowledge

13.   In which animal's development does the rotation of the gray crescent play a major role?

a.   bird
b.   frog
c.   grasshopper
d.   human
e.   sea urchin

Answer: b
Difficulty: Moderate
Bloom's Taxonomy: Knowledge

14.   Which of the following organisms has a pattern of gastrulation that is the most similar to humans?

a.   amphibians
b.   birds
c.   drosophila
d.   sea urchin
e.   zebra fish

Answer: b
Difficulty: Moderate
Bloom's Taxonomy: Knowledge

15.   Which structure in the development of the amphibian is equivalent to the primitive knot in bird development?

a.   the archenteron
b.   the dorsal lip of the blastopore
c.   the ventral lip of the blastopore
d.   ectoderm
e.   endoderm

Answer: b
Difficulty: Difficult
Bloom's Taxonomy: Application

16. The extraembryonic membrane that secretes fluid around the developing embryo is the

a. allantois.
b. amnion.
c. chorion.
d. yolk sac.
e. both the allantois and the yolk sac.

Answer: b
Difficulty: Easy
Bloom's Taxonomy: Knowledge

17. The chorion is produced from which primary tissue layer(s)?

a. ectoderm and endoderm
b. ectoderm and mesoderm
c. endoderm and mesoderm
d. ectoderm only
e. endoderm only

Answer: b
Difficulty: Difficult
Bloom's Taxonomy: Knowledge

48.3 FROM GASTRULATION TO ADULT BODY STRUCTURES: ORGANOGENESIS

18. Which of the following structures of the eye is matched appropriately with its developmental material?

a. cornea – ectoderm
b. cornea – crystallin
c. lens – optic cup
d. retina – crystallin
e. retina – ectoderm

Answer: a
Difficulty: Difficult
Bloom's Taxonomy: Knowledge

19. In response to induction by the optic vesicle,

a. genes coding for crystallin and keratin are activated.
b. genes coding for only crystallin are activated.
c. genes coding for only keratin are activated.
d. genes coding for crystalline and keratin are not activated.
e. apoptosis occurs.

Answer: b
Difficulty: Difficult
Bloom's Taxonomy: Comprehension

48.4 EMBRYONIC DEVELOPMENT OF HUMANS AND OTHER MAMMALS

20. At what point in development is a human embryo considered to be a fetus?

a. 2 weeks
b. 4 weeks
c. 8 weeks
d. 10 weeks
e. 12 weeks

Answer: c
Difficulty: Moderate
Bloom's Taxonomy: Knowledge

21. Gestation takes approximately _____ days in humans.

a. 279
b. 266
c. 231
d. 206
e. 178

Answer: b
Difficulty: Moderate
Bloom's Taxonomy: Knowledge

22. In the postpartum female,

a. oxytocin stimulates the production and secretion of milk.
b. oxytocin stimulates the production of milk and prolactin stimulates the secretion of milk.
c. prolactin stimulates the production and secretion of milk.
d. prolactin stimulates the production of milk and oxytocin stimulates the secretion of milk.
e. estrogen and progesterone are directly responsible for the production and secretion of milk.

Answer: d
Difficulty: Moderate
Bloom's Taxonomy: Comprehension

23. What determines the development of male or female sex organs in the human embryo?

a. environmental factors
b. genes on the X chromosome
c. genes on the Y chromosome
d. genes on both the X and Y chromosomes
e. neither the X or the Y chromosomes

Answer: c
Difficulty: Moderate
Bloom's Taxonomy: Knowledge

24. When the SRY gene of the Y chromosome becomes active (around 7 weeks),

a. the Müllerian and Wolffian ducts both develop into male reproductive structures.
b. the Müllerian ducts develop into male reproductive structures and the Wolffian ducts disappear.
c. the Müllerian ducts disappear and the Wolffian ducts develop into male reproductive structures.
d. the Müllerian and Wolffian ducts both disappear.
e. the Müllerian and Wolffian ducts both develop into female reproductive structures.

Answer: c
Difficulty: Difficult
Bloom's Taxonomy: Comprehension

48.5 THE CELLULAR BASIS OF DEVELOPMENT

25. Which of the following cells stop dividing once fully formed?

a. bone cells
b. cheek cells
c. liver cells
d. muscle cells
e. nerve cells in the brain

Answer: e
Difficulty: Easy
Bloom's Taxonomy: Application

26. In the experiments where cytochalasin was added to a developing neural plate,

a. microfilament growth was enhanced and invagination of the ectoderm resulted.
b. microfilament growth was impaired and invagination of the ectoderm resulted.
c. microfilament growth was enhanced and invagination of the ectoderm did not result.
d. microfilament growth was impaired and invagination of the ectoderm did not result.
e. cytochalasin had no effect on microfilament growth or invagination of the ectoderm.

Answer: d
Difficulty: Difficult
Bloom's Taxonomy: Application

27. During differentiation, organs develop with different numbers of cells because

a. the cell adhesion molecules are present in some tissues.
b. the cells migrate.
c. the formation of myosin begins.
d. the length of interphase varies.
e. the microtubules rearrange.

Answer: d
Difficulty: Difficult
Bloom's Taxonomy: Knowledge

EXPERIMENTAL RESEARCH: SPEMANN AND MANGOLD'S EXPERIMENT DEMONSTRATING INDUCTION IN EMBRYOS

28. When a dorsal lip from a newt is attached to the ventral side of another embryo, it results in

a. two separate newts.
b. two attached newts.
c. survival of only the original newt.
d. death of the newt.
e. no observable result.

Answer: b
Difficulty: Difficult
Bloom's Taxonomy: Comprehension

48.6 THE GENETIC AND MOLECULAR CONTROL OF DEVELOPMENT

29. Somites in mammals become

a. digestive system components.
b. extraembryonic membranes.
c. neurons.
d. reproductive structures.
e. skeletal muscle cells.

Answer: e
Difficulty: Moderate
Bloom's Taxonomy: Knowledge

30. The key gene for head and thorax development is

a. the bicoid gene, which is a maternal-effect gene.
b. the bicoid gene ,which is a paternal-effect gene.
c. the pair-rule gene, which is a maternal-effect gene.
d. the pair-rule gene, which is a paternal-effect gene.
e. both the bicoid and pair-rule genes.

Answer: a
Difficulty: Difficult
Bloom's Taxonomy: Knowledge

FOCUS ON RESEARCH: MODEL RESEARCH ORGANISMS: THE ZEBRAFISH MAKES A BIG SPLASH AS THE VERTEBRATE FRUIT FLY

31. Zebra fish are known as the "vertebrate fruit fly" because

a. they are easy to maintain.
b. they produce multiple offspring.
c. they develop rapidly.
d. they are transparent and therefore easy to study.
e. all of the above

Answer: e
Difficulty: Easy
Bloom's Taxonomy: Knowledge

Labeling

Label each of the lettered areas with one of the developmental terms from the word bank below.

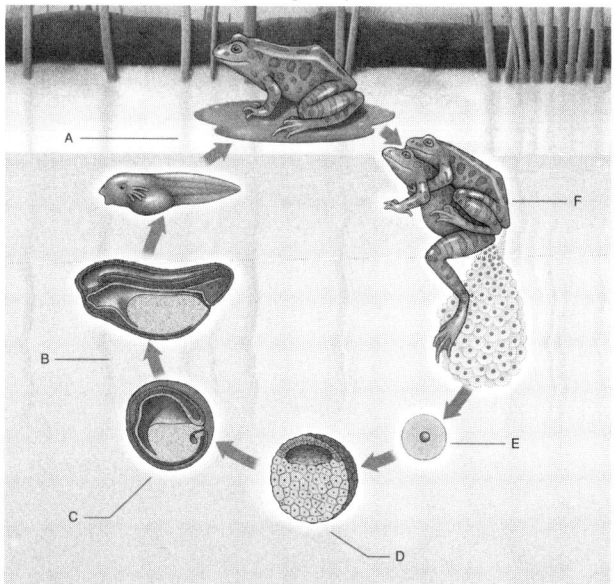
Stages in Frog Development

32. _____ cleavage
33. _____ development into adult
34. _____ fertilization
35. _____ gastrulation
36. _____ organogenesis
37. _____ sexual reproduction

Answers: *32. D   33. A   34. E   35. C*
36. B   37. F

Difficulty: Moderate
Bloom's Taxonomy: Knowledge
Source: Figure 48.3

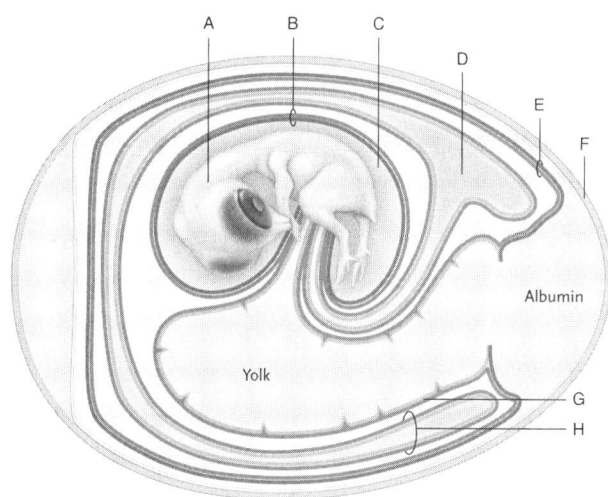
Extraembryonic Membranes in a Bird Embryo

Label the diagram of extraembryonic membranes with the terms listed below.

38. _____ allantois
39. _____ amnion
40. _____ amniotic cavity
41. _____ embryo
42. _____ chorion
43. _____ yolk sac

Answers: *38. D   39. B   40. C   41. A   42. E*
43. G

Difficulty: Moderate
Bloom's Taxonomy: Knowledge
Source: Figure 48.9

True/False Questions

44. _____ An egg's <u>polarity</u> determines the orientation of body axes.

45. _____ The material in the <u>gray crescent</u> of a frog embryo is necessary for normal development.

46. _____ Cells that undergo apoptosis <u>are essential</u> to the organism throughout their life.

47. _____ The symmetrical pattern of yolk distribution in a sea urchin represents the beginnings of <u>radial symmetry</u>.

48. _____ In terrestrial animals, such as the bird, the embryo is exposed to a <u>dry</u> environment while in the egg.

49. _____ Cleavage in a frog embryo takes place more quickly in the <u>vegetal</u> half.

50. _____ The nervous system structures are derived from <u>ectoderm</u>.

51. _____ <u>Chorionic villi</u> eventually become the placenta.

Answers: 44. T   45. T   46. F, not essential
47. T   48. F, wet   49. F, animal
50. T   51. T

Difficulty: Moderate
Bloom's Taxonomy: Knowledge, Application
Source: Sections 48.1, 48.2, 48.3, 48.4

Matching

Match the terms to the appropriate definitions.

52. _____ blastocoel

53. _____ blastopore

54. _____ blastula

55. _____ cleavage

56. _____ gastrula

57. _____ morula

58. _____ neural plate

A. A fluid filled cavity within the blastomere

B. The stage in which an embryo has three tissue layers

C. Solid ball of blastomeres

D. An opening that eventually becomes the anus or mouth

E. Process that involves increase in number of cells, but not mass

F. Early stage in formation of the nervous system

G. Second stage of cleavage; blastomeres are surrounding a hollow cavity

Answers: 52. A   53. D   54. G   55. E
56.B   57.C   58.F

Difficulty: Moderate
Bloom's Taxonomy: Knowledge
Source: Sections 48.1, 48.2, 48.3

Put the following mechanisms of development in the order they occur.

59. _____ first mechanism

60. _____ second mechanism

61. _____ third mechanism

62. _____ fourth mechanism

63. _____ fifth mechanism

64. _____ sixth mechanism

A. Cell movements

B. Determination

C. Differentiation

D. Induction

E. Mitotic cell divisions

F. Selective cell adhesions

Answers: 59. E   60. A   61.F   62.D
63. B   64. C

Difficulty: Difficult
Source: Section 48.1

Match the scientific terms with the appropriate definitions below.

65. _____ gray crescent

66. _____ blastodisc

67. _____ notochord

68. _____ crystalline

69. _____ apoptosis

70. _____ chorionic villi

71. _____ induction

72. _____ myoblast

A. Undifferentiated muscle cells

B. Pigmented layer of cytoplasm opposite from point of entry of sperm

C. Rounded layer of cells at the surface of yolk in birds

D. Solid rod of tissue that is involved in neurulation

E. Fingerlike extensions into the endometrium

F. A clear fibrous protein

G. Programmed cell death

H. Event when a group of cells influences a nearby group of cells

Answers:65. B   66. C   67. D   68. F
69. G   70. E   71. H   72. A

Difficulty: Moderate
Source: Sections 48.2, 48.3, 48.4, 48.5, 48.6

Select the Exception

73. Which of the following statements regarding polarity in an egg is *incorrect*?

a. The animal pole forms the anterior end of the animal.

b. The vegetal pole forms the posterior end of the animal.

c. The animal pole forms the animal's gut.

d. The vegetal pole forms internal structures.

e. Polarity relates to development of body axes.

Answer: c
Difficulty: Moderate
Bloom's Taxonomy: Application
Source: Section 48.1

74. Which of these structures is not derived from the neural crest?

a. bones of inner ear
b. cartilage of the face
c. cranial nerves
d. teeth
e. muscles

Answer: e
Difficulty: Moderate
Bloom's Taxonomy: Knowledge
Source: Section 48.2

75. Which of the following structures is not derived from the pharyngeal arches?

a. face
b. mouth
c. neck
d. nasal cavities
e. shoulder blades

Answer: e
Difficulty: Easy
Bloom's Taxonomy: Knowledge
Source: Section 48.2

76. The information that directs the development of the fertilized egg is present in all of the following except

a. information in egg nucleus
b. information in sperm nucleus
c. mRNA in egg cytoplasm
d. proteins in egg cytoplasm
e. sperm cytoplasm

Answer: e
Difficulty: Moderate
Bloom's Taxonomy: Comprehension
Source: Section 48.1

77. Apoptosis is responsible for all of the following <u>except</u>

a. the ability of newly born kittens to open their eyelids.
b. the generation of body axes.
c. destruction of the tadpole's tail.
d. destruction of some larval tissue in a developing butterfly.
e. removal of webbing between human fingers.

Answer: b
Difficulty: Moderate
Bloom's Taxonomy: Knowledge
Source: 48.3, 48.6

Short Answer

78. How does the location of the yolk in an egg influence its development?

Answer: Development of the egg in the embryo takes place more rapidly in areas of the egg that do not contain the yolk. In organisms whose eggs contain large amounts of yolk (e.g., chickens), the egg develops only in the small area of the egg not containing yolk.
Difficulty: Moderate
Bloom's Taxonomy: Knowledge
Source: Section 48.1

79. Why could an ultrasound performed at four weeks be unable to determine the sex of a developing fetus?

Answer: We can see from figure 48.17 of the text that the male and female structures look nearly identical until around seven weeks. At that point, hormones direct the development of distinctive male or female reproductive structures, which would later be identifiable on a sonogram.
Difficulty: Easy
Bloom's Taxonomy: Comprehension
Source: Section 48.4

80. How do stem cell studies relate to the process of cell differentiation in an embryo?

Answer: It is thought that if you harvest cells before cell differentiation takes place, perhaps these cells could be "convinced" to form replacement cells for any necessary location in the body.
Difficulty: Difficult
Bloom's Taxonomy: Application
Source: Why It Matters

81. How can your knowledge of neural tube formation provide you with an insight into congenital diseases that involve incomplete neural tube closure?

Answer: In the text, we learned that organogenesis begins with formation of the neural tube. Therefore, we can assume that diseases that result from an inappropriate neural tube closure occur at a very early gestational age.
Difficulty: Difficult
Bloom's Taxonomy: Application
Source: Section 48.3

Essay Questions

82. Describe how the selective adhesion property of cells was proven.

Answer: Easily recognizable cells from the primary cell layers were mixed with an alkaline solution that would separate the mixture into individual cells. Then the ectoderm and mesoderm from different species were added together. Initially they stayed as a homogeneous mixture, but then resulted in distinctive cell colonies of separate ectoderm and mesoderm cells. This indicates that "like" cells have an affinity for each other and adhere selectively to cells of their own type.
Difficulty: Moderate
Bloom's Taxonomy: Knowledge, Comprehension
Source: Experimental Research Figure 48.20

83. Fate mapping is currently used to determine the embryonic origin of tissues in an adult. What possible applications do you see for this procedure in the future?

Answer: At the present time, completed fate mapping has only been entirely accomplished in transparent organisms such as the nematode mentioned in the text. If fate mapping were completed for all organisms, we might find powerful embryonic evidence of a shared ancestry. In addition, one can envision in the future that the treatment for hereditary diseases could possibly begin in an embryonic stage.
Difficulty: Difficult
Bloom's Taxonomy: Application
Source: Section 48.5

84. What is the immune function of the placenta in human embryo development?

Answer: Since the placenta prevents fetal blood from entering the maternal bloodstream, the mother's immune system does not sense the presence of a foreign substance in the body. In marsupials, such as the kangaroo, there is no placenta so the fetus is expelled from the mother's body before development is complete. The young kangaroo then must complete his development in an external pouch.
Difficulty: Difficult
Bloom's Taxonomy: Comprehension
Source: Section 48.4

49

POPULATION ECOLOGY

Multiple-Choice

WHY IT MATTERS

1.   Rabbits introduced to Australia overpopulated because in their new habitat _____.

a.   people rarely hunted them
b.   they had no natural diseases
c.   they had no natural predators
d.   their food supply was larger
e.   there were few herbivorous competitors

Answer: c
Difficulty: Easy
Bloom's Taxonomy: Knowledge

2.   What was the long-term effect of the introduction of myxoma virus to control rabbits in Australia?

a.   The rabbits were all killed.
b.   Most rabbits became immune to the virus.
c.   The virus killed off some of the rabbit population but also infected kangaroos and wallabies.
d.   The virus became less virulent and the rabbits gained some immunity.
e.   The virus had no effect.

Answer: d
Difficulty: Moderate
Bloom's Taxonomy: Knowledge

49.1 THE SCIENCE OF ECOLOGY

3.   Which branch of ecology would encompass studies of the flow of energy between a community and the abiotic environment?

a.   population ecology
b.   community ecology
c.   ecosystem ecology
d.   organismal ecology
e.   biogeography

Answer: c
Difficulty: Easy
Bloom's Taxonomy: Knowledge

4.   Basic ecology is distinguished from applied ecology because basic ecology investigates _____ while applied ecology looks at _____.

a.   distribution and interaction of species; conservation and mitigation of damage
b.   conservation and mitigation of damage; distribution and interaction of species
c.   populations; ecosystems
d.   individual species; food webs
e.   mathematical models; field studies

Answer: a
Difficulty: Moderate
Bloom's Taxonomy: Knowledge, Comprehension

FOCUS ON RESEARCH: RESEARCH METHODS: MARK-RELEASE-RECAPTURE

5.   Tagging fish is often used in the _____ method of determining population size.

a.   capture-recapture
b.   direct count
c.   extrapolation of counts from quadrants to entire population
d.   none of the choices

Answer: a
Difficulty: Easy
Bloom's Taxonomy: Knowledge

6.   Marine biologists tagged and released 50 marlin. Later, fishermen caught 300 marlin, 15 of which had tags. What is the estimate for the number of marlin in the population?

a.   5,000
b.   315
c.   365
d.   2,500
e.   1,000

Answer: e
Difficulty: Difficult
Bloom's Taxonomy: Application

INSIGHTS FROM THE MOLECULAR REVOLUTION: TRACING ARMADILLO PATERNITY

7.   Investigators using molecular techniques determined that the rapid spread of armadillos into the southern U.S. was due to _____.

a.   territorial sires driving out male offspring
b.   a 92 percent reproduction rate
c.   extensive migration of individuals
d.   unexplained reasons

Answer: d
Difficulty: Moderate
Bloom's Taxonomy: Knowledge

49.2 POPULATION CHARACTERISTICS

8.   Populations in which individuals repel each other because resources are in short supply tend to have _____ distributions.

a.   random
b.   clumped
c.   uniform
d.   dynamic
e.   unpredictable

Answer: c
Difficulty: Easy
Bloom's Taxonomy: Knowledge

9. A population of mostly _____ individuals with a _____ generation time and a _____ proportion of females is expected to show the most growth in the future.

a. prereproductive; short; high
b. reproductive; short; high
c. reproductive; long; low
d. postreproductive; long; low
e. prereproductive; short; low

Answer: a
Difficulty: Moderate
Bloom's Taxonomy: Application

10. In which of the following species would the number of males **least** affect the rate of population growth?

a. northern elephant seals
b. trumpeter swans
c. Canadian geese
d. European rabbits
e. Norwegian rats

Answer: a
Difficulty: Moderate
Bloom's Taxonomy: Knowledge

49.3 DEMOGRAPHY

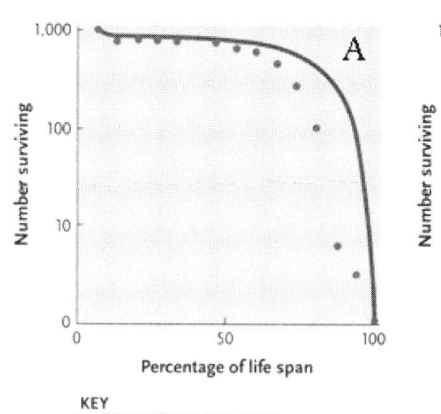
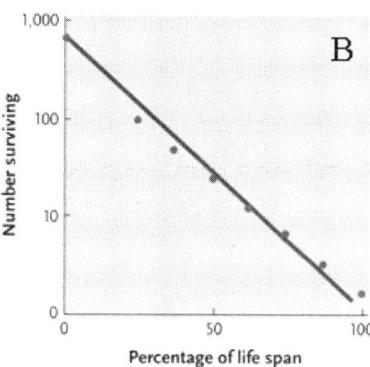
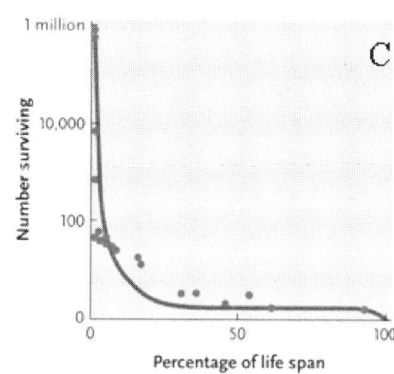

KEY
— Theoretical   • Data

Use the diagram above for questions 11–12.

11. Sea turtles are likely to die when they are young, but have a good chance of surviving once they reach large size. Which survivorship curve best fits them?

a. Type I, which is graph A
b. Type II, which is graph B
c. Type I, which is graph C
d. Type III, which is graph C
e. Type III, which is graph A

Answer: d
Difficulty: Moderate
Bloom's Taxonomy: Comprehension, Application
Source: Fig. 49.6

12. Humans living in environments with good medical care are most likely to exhibit which of the following survivorship curves?

a. Type I, which is graph A
b. Type II, which is graph B
c. Type I, which is graph C
d. Type III, which is graph C
e. Type III, which is graph A

Answer: a
Difficulty: Moderate
Bloom's Taxonomy: Comprehension, Application
Source: Fig. 49.6

49.4 THE EVOLUTION OF LIFE HISTORIES

13. The allocation of resources for survival, growth, and maximal reproduction determines the _____ of a species.

a. survivorship curve
b. life history
c. specific mortality
d. age-specific fecundity
e. dispersion

Answer: b
Difficulty: Easy
Bloom's Taxonomy: Knowledge

14. Which of the following characteristics would **not** have evolved with the others in the life history strategy of a single species?

a. low fecundity
b. active parental care
c. multiple reproductive episodes
d. low chance of survival in adulthood
e. late reproduction

Answer: d
Difficulty: Difficult
Bloom's Taxonomy: Comprehension

49.5 MODELS OF POPULATION GROWTH

15. All of the following choices are descriptive of a population experiencing exponential population growth **except** _____.

a. resources are limited
b. graph of growth produces a "J" shaped curve
c. per capita growth rate remains constant
d. $\Delta N/\Delta t$ is increasing
e. population increases at an increasing rate

Answer: a
Difficulty: Moderate
Bloom's Taxonomy: Knowledge, Comprehension

16. A population of 100 animals in logarithmic growth has $r_{max} = 0.4$ and $K = 102$. Using the fact that $r = r_{max}(K-N)/K$, calculate the value of r.

a. 0.0078
b. 0.792
c. -0.580
d. 0.078
e. 0.99

Answer: a
Difficulty: Difficult
Bloom's Taxonomy: Application

17. In the logistic model of population growth, what is expected to happen to r when N > K?

a. r will approach r_{max}
b. r will approach 1
c. r becomes negative
d. r = K
e. r > 0

Answer: d
Difficulty: Moderate
Bloom's Taxonomy: Application

49.6 POPULATION REGULATION

18. Which of the following factors is **not** a density-dependent regulator of population size?

a. competition for food
b. limitation of nesting sites
c. disease
d. parasitism
e. drought

Answer: e
Difficulty: Easy
Bloom's Taxonomy: Knowledge

19. Which of the following traits is characteristic of a K-selected species?

a. adapted to rapidly changing environments
b. small body size
c. short generation time
d. provide substantial parental care to offspring
e. single reproductive event common

Answer: d
Difficulty: Easy
Bloom's Taxonomy: Knowledge

20. Studies have determined that _____ is responsible for the cyclical rise and fall of arctic hare and lynx populations.

a. predation
b. a limited food supply for hares
c. predation by other mammals
d. intraspecies competition
e. all of the choices

Answer: e
Difficulty: Easy
Bloom's Taxonomy: Comprehension

49.7 HUMAN POPULATION GROWTH

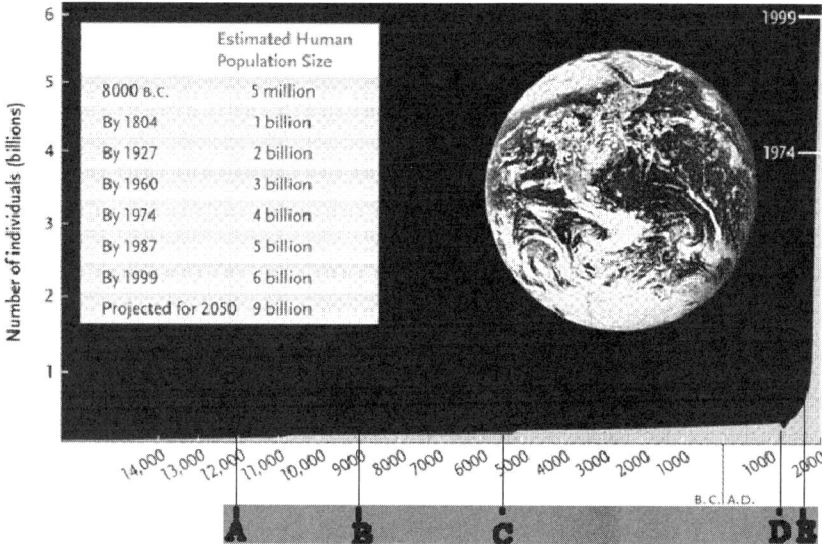

Use the diagram above for questions 21–22.

21. Which label on the graph represents the transformation of humans from hunter-gatherer societies to agriculturally based ones?

a. A
b. B
c. C
d. D
e. E

Answer: b
Difficulty: Moderate
Bloom's Taxonomy: Knowledge
Source: Fig. 49.19

22. Which label on the graph represents the beginning of the industrial and scientific revolution?

a. A
b. B
c. C
d. D
e. E

Answer: e
Difficulty: Moderate
Bloom's Taxonomy: Knowledge
Source: Fig. 49.19

23. Birth rates generally reach their highest point in which stage of the demographic transition of a country?

a. preindustrial
b. transitional
c. industrial
d. postindustrial
e. All stages have equal birth rates.

Answer: b
Difficulty: Easy
Bloom's Taxonomy: Knowledge
Source: Fig. 49.19

UNANSWERED QUESTIONS

24. Recent studies have demonstrated evolution in populations over a span of time as short as _____.

a. single years
b. decades
c. centuries
d. thousands of years
e. millions of years

Answer: a
Difficulty: Moderate
Bloom's Taxonomy: Knowledge

25. Research has shown that commercial fisheries place selective pressure on fish populations by harvesting mostly the _____ fish, resulting in _____.

a. young; aging populations
b. larger; maturity at smaller sizes
c. smaller; maturity at larger sizes
d. prereproductive; declining birth rates
e. postreproductive; no change in population size

Answer: b
Difficulty: Easy
Bloom's Taxonomy: Knowledge

Integrative Multiple-Choice

26. A human population with an age structure diagram that is narrower at the base than at the top would have _____.

a. zero population growth
b. r < 0
c. r > 0
d. been affected by density independent factors
e. (K-N)/K < 1

Answer: b
Difficulty: Moderate
Bloom's Taxonomy: Knowledge
Source: Sections 49.5, 49.6, 49.7

27. A human population living without medical care in Third-World conditions would probably have

_____.

a. a rectangle-shaped age structure diagram
b. an r value < 0
c. a pyramid-shaped age structure diagram
d. per capita birth and death rates consistent with those of a stage 4 society
e. no density dependent factors operating

Answer: c
Difficulty: Difficult
Bloom's Taxonomy: Application
Source: Sections 49.5, 49.6, 49.7

28. A stage 3 society would be characterized by

_____.

a. an r value > 0
b. industrialization
c. a clumped population distribution
d. (K-N)/K approaching 1
e. all of the choices

Answer: e
Difficulty: Moderate
Bloom's Taxonomy: Knowledge
Source: Sections 49.2, 49.5, 49.6, 49.7

Matching

Match each of the following terms with its correct definition.

29. _____ population density

30. _____ emigration

31. _____ age-specific fecundity

32. _____ exponential population growth

33. _____ the logistic model

A. Population increasing steadily by a constant ratio

B. The average number of offspring produced by females during each age interval

C. Population growth slows as population approaches K

D. The number of individuals per unit area or unit volume

E. Movement out of a population

Answers: 29. D   30. E   31. B   32. A   33. C

Difficulty: Moderate
Bloom's Taxonomy: Knowledge
Source: Sections 49.1–49.7

Classification

Use the five equations listed below for questions 34–38.

a. $\Delta N/\Delta t = B - D$
b. $dN/dt = (b - d)N$
c. $dN/dt = r_{max}N$
d. $d = (D/N)$
e. $dN/dt = r_{max}N(K-N)/K$

34. _____ population growth under ideal conditions

35. _____ logistic model of population growth

36. _____ exponential model of population growth

37. _____ per capita death rate

38. _____ change in population size

Answers: 34. c   35. e   36. b   37. d   38. a

Difficulty: Difficult
Bloom's Taxonomy: Knowledge
Source: Section 49.5

Choice

For each of the following statements, choose the most appropriate survivorship curve from the list below.

a. type I   b. type II   c. type III

39. _____ typical of humans

40. _____ typical of *Cleome droserifolia*

41. _____ constant rate of mortality in all age classes

42. _____ high juvenile mortality followed by increased chance of survival as adult

Answers: 39. a   40. c   41. b   42. c

Difficulty: Easy
Bloom's Taxonomy: Knowledge, Comprehension
Source: Section 49.3

Short Answer

43. What characteristics are consistent with a K-selected species? Where do humans fall on the r through K continuum?

Answer: K-selected species tend to have larger body size, longer generations, and a lower reproductive rate. More energy is devoted to each offspring, so individual offspring have a higher chance of survival than do r-selected offspring. Humans are K-selected, especially when compared to r-selected organisms like houseflies or plants, which produce thousands of offspring.
Difficulty: Moderate
Bloom's Taxonomy: Comprehension
Source: Sections 49.6

44. What phenotypic effects does crowding have on the migratory locust? How are these changes adaptive?

Answer: Migratory locusts can develop into either solitary or migratory forms. Crowding induces a transformation to the migratory form with more body fat and longer wings. These changes are adaptive because dispersal reduces competition in the original environment and may establish new populations in unoccupied environments.
Difficulty: Moderate
Bloom's Taxonomy: Knowledge, Comprehension
Source: Section 49.6

Classification

Categorize the following traits as belonging predominantly to an r-selected species or a K-selected species.

45. _____ short maturation time

46. _____ long life span

47. _____ low mortality rate

48. _____ one reproductive episode

49. _____ early timing of first reproduction

50. _____ small clutch or brood size

51. _____ large offspring

52. _____ little or no parental care

53. _____ extensive parental care

54. _____ fluctuating population size

55. _____ good tolerance of environmental change

A. r-selected
B. K-selected

Answers: *45. A*   *46. B*   *47. B*   *48. A*   *49. A*
        50. B   *51. B*   *52. A*   *53. B*   *54. A*
        55. B

Difficulty: Easy
Bloom's Taxonomy: Knowledge
Source: Section 49.6

True/False

If the statement is true, write a "T" in the blank. If the statement is false, make it correct by changing the underlined word(s) and writing the correct word(s) in the answer blank.

56. _____ Biologists captured and marked 100 butterflies, and later captured 115, 5 of which were marked. The population estimate is <u>2000</u> butterflies.

57. _____ Uniform dispersion usually results from individuals of a species <u>repelling</u> each other.

58. _____ A cohort is a group of individuals of the same <u>sex</u>.

59. _____ Egg yolk and seed endosperm are examples of <u>passive</u> parental care.

60. _____ <u>Logistic</u> population growth describes growth without limitation.

61. _____ Parasitism is a density-<u>independent</u> factor of population control.

62. _____ An <u>"S"</u> shaped curve is characteristic of logistic population growth.

Answers: *56. F, 2,300*   *57.T*   *58.F, age*   *59. T*
        60. F, exponential   *61. F, dependent*
        62. T

Difficulty: Moderate
Bloom's Taxonomy: Knowledge
Source: Section 49.1–49.6

Select the Exception

63.   Most of the following factors are intrinsic controls to population growth. Select the exception.

a.   increased aggression
b.   hormonal changes
c.   reduced reproduction
d.   increased dispersal
e.   increased predation

Answer: e
Difficulty: Moderate
Bloom's Taxonomy: Knowledge
Source: Section 49.6

64.   Four of the five regions listed below are experiencing less than 1.9 percent annual population growth. Select the exception.

a.   Western Europe
b.   North America
c.   Africa
d.   Australia
e.   Russia

Answer: c
Difficulty: Easy
Bloom's Taxonomy: Knowledge
Source: Section 49.7

Sequence

Write the letter of the region expected to have the highest population by 2025 next to 65, the letter of the region expected to have the lowest population by 2025 next to 70.

65.   _____   A. Europe
66.   _____   B. Asia
67.   _____   C. Latin America
68.   _____   D. North America
69.   _____   E. Oceania
70.   _____   F. Africa

Answers: 65. B   66. F   67.A   68.C   69.D
*   70.E*

Essay

71.   What are some density-dependent factors that affect humans at higher population densities, and how have these been addressed in industrial societies?

Answer: Competition for all resources increases at high population densities. This leads to an increased mortality rate. Disease, violence, and famine are some of the factors that worsen at high population densities. In industrial societies, sanitation has decreased death by disease. Poverty has been addressed less successfully in industrialized nations and can be viewed as an outcome of competition for resources. War, too, is almost invariably a result of competition for resources, although fighters are often recruited using religious or cultural propaganda.
Difficulty: Moderate
Bloom's Taxonomy: Analysis
Source: Section 49.7

72.   "Humans have two options for limiting population growth: We can make a global effort to limit our own population, or we can wait until the environment does it for us." Briefly address each scenario. What methods would result in a successful global decline in birthrate? What environmental events are the alternatives?

Answer: Government-sponsored family planning programs have had great success both in reducing birth rates, and in improving the health and status of women. Providing medical care in conjunction with such programs encourages families to have fewer children. In places with high child mortality rates, family sizes are high. Environmental events that will bring about an uncontrolled decline in population include war, famine, drought, and exposure to accumulating toxins.
Difficulty: Moderate
Bloom's Taxonomy: Synthesis
Source: Section 49.7.

Labeling

Identify each labeled part of the following illustration.

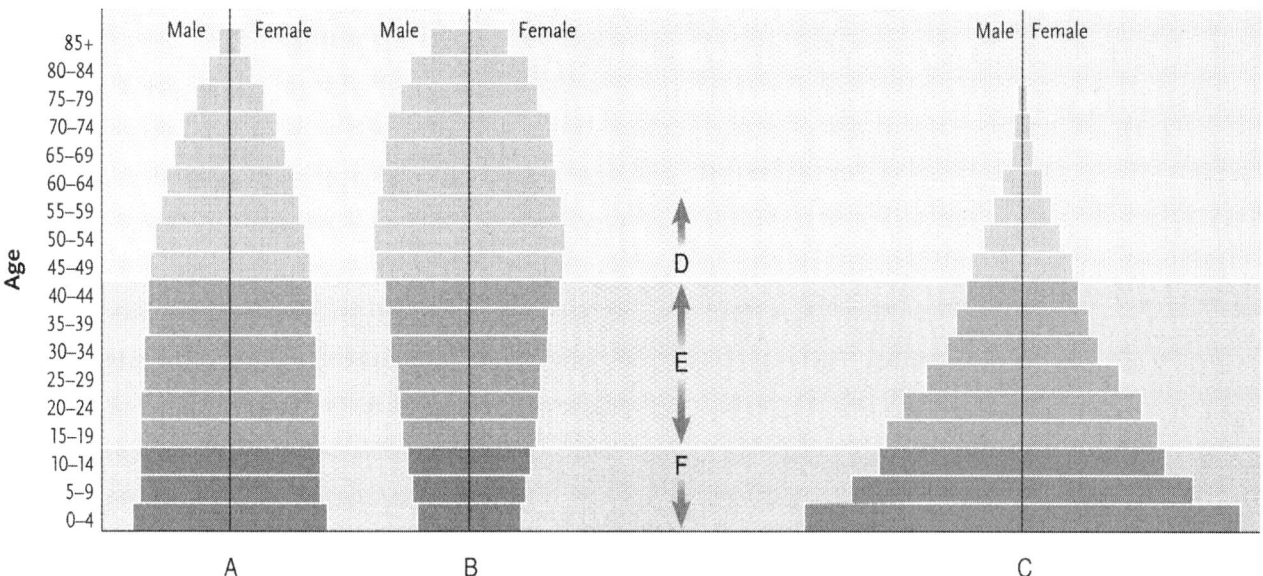

73. _____ negative growth
74. _____ rapid growth
75. _____ reproductive
76. _____ prereproductive
77. _____ postreproductive
78. _____ zero growth

Answers: 73. B   74. C   75. E   76. F   77. D
78. A
Difficulty: Moderate
Bloom's Taxonomy: Knowledge, Application
Source: Figure 49.21

50

POPULATION INTERACTIONS AND COMMUNITY ECOLOGY

Multiple Choice

WHY IT MATTERS

1.   If cowbird eggs are included in an oropendola nest in the open woodlands of Central America, what would be the expected outcome?

a.   The cowbird eggs would be ejected from the nest to eliminate the competition.
b.   The cowbird eggs would be ejected from the nest due to a reaction of the mother bird's immune system.
c.   The cowbird eggs would not be ejected from the nest because they provide an additional food source.
d.   The cowbird eggs would not be ejected from the nest because they provide protection from large predators.
e.   The cowbird eggs would not be ejected from the nest because they provide protection from parasites.

Answer: e
Difficulty: Moderate
Bloom's Taxonomy: Comprehension

2.   If cowbird eggs are included in an oropendola nest that is located near bee hives, what would be the expected outcome?

a.   The cowbird eggs would be ejected from the nest to eliminate the competition.
b.   The cowbird eggs would be ejected from the nest due to a reaction of the mother bird's immune system.
c.   The cowbird eggs would not be ejected from the nest because they provide an additional food source.
d.   The cowbird eggs would not be ejected from the nest because they provide protection from large predators.
e.   The cowbird eggs would not be ejected from the nest because they provide protection from parasites.

Answer: a
Difficulty: Moderate
Bloom's Taxonomy: Comprehension

3.   What role do bees play in the interaction between cowbirds and oropendola birds?

a.   keep away large predators
b.   keep away insect pests
c.   pollinate species that only cowbirds prefer
d.   pollinate species that only oropendola birds prefer
e.   have no effect on the relationship between cowbirds and oropendolas

Answer: b
Difficulty: Moderate
Bloom's Taxonomy: Comprehension

50.1 POPULATION INTERACTIONS

4.   Which of the following types of interactions is advantageous for one species, but has no effect on the other?

a.   commensalism
b.   herbivory
c.   mutualism
d.   parasitism
e.   predation

Answer: a
Difficulty: Easy
Bloom's Taxonomy: Knowledge

5.   What adaptation(s) allow(s) a rattlesnake to more effectively locate prey?

a.   acute sense of hearing
b.   heat sensors
c.   acute sense of smell
d.   acute sense of vision
e.   both c and d

Answer: b
Difficulty: Easy
Bloom's Taxonomy: Knowledge

6.   A panda bear who consumes a diet primarily of eucalyptus leaves would be considered a

a.   generalist and a carnivore.
b.   generalist and a herbivore.
c.   specialist and a carnivore.
d.   specialist and a generalist.
e.   specialist and a herbivore.

Answer: e
Difficulty: Moderate
Bloom's Taxonomy: Application

7. What is the factor(s) involved in the optimal foraging theory?

a. availability of parasites
b. availability of predators
c. energy needed to capture food
d. energy that the food provides
e. both c and d

Answer: e
Difficulty: Moderate
Bloom's Taxonomy: Knowledge

8. Which of the following population interactions is advantageous to both populations?

a. commensalism
b. competition
c. mutualism
d. parasitism
e. predation

Answer: c
Difficulty: Easy
Bloom's Taxonomy: Knowledge

9. In the experiment involving bluegill sunfish and their choice of diet, when equal numbers of small, medium, and large Daphnia are present,

a. the sunfish prefer small Daphnia.
b. the sunfish prefer medium Daphnia.
c. the sunfish prefer large Daphnia.
d. the sunfish show no preference.
e. sunfish do not consume Daphnia.

Answer: c
Difficulty: Moderate
Bloom's Taxonomy: Comprehension

10. An African violet plant discourages consumption by herbivores by

a. developing sap that stunts insect development.
b. developing an appearance that mimics an undesirable plant.
c. having hairy leaves.
d. producing an undesirable scent.
e. having thorns.

Answer: c
Difficulty: Difficult
Bloom's Taxonomy: Application

11. How is the grasshopper mouse able to consume the *Eleodes longicollis* beetle?

a. The grasshopper mouse avoids the undesirable secretions by burying the beetle's abdomen.
b. The grasshopper mouse has cryptic coloration so that the beetle does not notice him.
c. The grasshopper mouse is immune to the beetle's toxins.
d. The grasshopper mouse poisons the beetle first.
e. The grasshopper mouse uses heat detectors to easily locate the beetle.

Answer: a
Difficulty: Moderate
Bloom's Taxonomy: Knowledge

12. How do pancake tortoises protect themselves from predators?

a. Their coloration serves as a disguise.
b. They emit an undesirable scent.
c. They have large jaws with teeth.
d. They puff themselves up and become wedged in between cracks in rocks.
e. They retreat into their shell.

Answer: d
Difficulty: Moderate
Bloom's Taxonomy: Knowledge

13. The Monarch butterfly has an undesirable taste and the Viceroy butterfly has a selective advantage because it looks like the Monarch. This is an example of which process?

a. aposematic coloration
b. Batesian mimicry
c. competitive exclusion principle
d. cryptic coloration
e. Müllerian mimicry

Answer: b
Difficulty: Moderate
Bloom's Taxonomy: Application

14. What conclusion can be drawn from Gause's experiments on interspecific competition in Paramecium?

a. Both populations of Paramecium expire due to a build-up of waste products.
b. One population of Paramecium used the other as a food source.
c. Paramecium populations survive well together because they have no limiting resources.
d. When two populations of Paramecium use the same limiting resource, they can coexist long term.
e. When two populations of Paramecium use the same limiting resource, they cannot coexist long term.

Answer: e
Difficulty: Difficult
Bloom's Taxonomy: Comprehension

15. Which of the following statements is correct?

a. Allopatric populations look different and use different resources.
b. Allopatric populations look different and use the same resources.
c. Allopatric populations look the same and use different resources.
d. Allopatric populations look the same and use the same resources.
e. Sympatric and allopatric populations both look the same and use the same resources.

Answer: d
Difficulty: Difficult
Bloom's Taxonomy: Knowledge

16. An example of commensalism involves which two organisms?

a. acacia trees and ant colonies
b. bacteria and legumes
c. cattle egrets and cattle
d. plants and pollinators
e. tapeworms and pigs

Answer: c
Difficulty: Moderate
Bloom's Taxonomy: Application

17. What benefit do legumes derive from Rhizobium bacteria?

a. help make carbon dioxide more available
b. increase root surface area
c. nitrogen fixation
d. protection
e. retain water so it is more available to the plant

Answer: c
Difficulty: Easy
Bloom's Taxonomy: Knowledge

50.2 THE NATURE OF ECOLOGICAL COMMUNITIES

18. What is an ecotone?

a. a parasite that is external to the body
b. a parasite that is internal to the body
c. an example of an individualistic hypothesis
d. a zone between two communities
e. the combination of species in an environment that is constant

Answer: d
Difficulty: Moderate
Bloom's Taxonomy: Knowledge

19. Which of the following fact(s) is/are true concerning research on Rhizobium bacteria?

a. It looks as though all nitrogen fixing bacteria have a common ancestor.
b. It is possible that non-legume plants could eventually be made into nitrogen fixators.
c. Rhizobium bacteria reduces the need for commercial fertilizer.
d. The nod genes make an infection thread, so that Rhizobium can enter the plant root.
e. All of the above are correct.

Answer: e
Difficulty: Easy
Bloom's Taxonomy: Knowledge

50.3 COMMUNITY CHARACTERISTICS

20.   An example from the text contrasts two forests, each with 10 species and a total of 50 trees. The first forest has 39 of the 50 trees representing the dominant species. The second forest has 2 of each of the 10 different species. What conclusion can be drawn regarding these two forests?

a.   The amount of animal species in each forest would be the same.
b.   The first forest is more diverse than the second.
c.   The first forest must have had some human interference.
d.   The second forest is more diverse than the second.
e.   The second forest must have had some human interference.

Answer: d
Difficulty: Moderate
Bloom's Taxonomy: Application

21.   Which trophic level would be best represented by a rabbit?

a.   decomposer
b.   detritivore
c.   primary consumer
d.   primary producer
e.   secondary consumer

Answer: c
Difficulty: Easy
Bloom's Taxonomy: Application

22.   What conclusion can be drawn regarding a community in which many species are involved in a food web?

a.   It is more fragile since organisms have more than a single food source.
b.   It is more fragile since organisms have only a single food source.
c.   It is more stable since organisms have more than a single food source.
d.   It is more stable since organisms have only a single food source.
e.   It is very prone to damage from human interference.

Answer: c
Difficulty: Moderate
Bloom's Taxonomy: Comprehension

50.4 EFFECTS OF POPULATION INTERACTIONS ON COMMUNITY CHARACTERISTICS

23.   Which of the following are biases that could be encountered when investigating the role of competition in a community?

a.   More studies are published when competition is a factor.
b.   More studies in K-selected species than in r-related species.
c.   The importance of competition may be overestimated.
d.   The importance of competition may be underestimated.
e.   All of the above are correct.

Answer: e
Difficulty: Difficult
Bloom's Taxonomy: Application

24.   In the study involving periwinkle snails and their algal food sources, which conclusion can be drawn?

a.   Periwinkles eliminated all algae and died out.
b.   Periwinkles eliminated all of the less dominant algae species.
c.   Periwinkles eliminated the dominant algae entirely.
d.   Periwinkles had no effect on the algae populations.
e.   Periwinkles made one type of algae less dominant, which allowed other algae populations to continue.

Answer: e
Difficulty: Difficult
Bloom's Taxonomy: Application

50.5 EFFECTS OF DISTURBANCE ON COMMUNITY CHARACTERISTICS

25.   In the study relating to the growth of coral in a portion of the Great Barrier Reef, which conclusion concerning alterations in the coral colonies is the most accurate?

a.   Changes in the colony are due to external factors only.
b.   Changes in the colony are due to internal factors only.
c.   Changes in the colony are due to both external and internal factors.
d.   Changes in the colony are solely due to man's interference.
e.   The study was not able to observe any changes in the coral colony over time.

Answer: c
Difficulty: Difficult
Bloom's Taxonomy: Application

26.   When sea stars were released into a controlled plot containing mussels, other invertebrates, and algae, what happened to the richness of the other species?

a.   All mussels disappeared.
b.   Both the starfish and the mussel population did not survive.
c.   Species richness decreased.
d.   Species richness increased.
e.   The species richness was maintained.

Answer: e
Difficulty: Difficult
Bloom's Taxonomy: Application

27.   The relationship between species richness and recovery from natural disturbances can best be summarized as follows:

a.   communities with more diversity do not recover from natural disturbances.
b.   communities with more diversity experience more natural disturbances.
c.   communities with more diversity recover less rapidly from natural disturbances.
d.   communities with more diversity recover more rapidly from natural disturbances.
e.   there is no correlation between community diversity and recovery from natural disturbances.

Answer: d
Difficulty: Difficult
Bloom's Taxonomy: Application

50.6 ECOLOGICAL SUCCESSION: RESPONSES TO DISTURBANCE

28.   Which hypothesis states that new species are prevented from entering a community by the existing species?

a.   equilibrium theory of island biogeography
b.   facilitation hypothesis
c.   inhibition hypothesis
d.   intermediate hypothesis
e.   tolerance hypothesis

Answer: c
Difficulty: Moderate
Bloom's Taxonomy: Knowledge

29.   Succession results from _____.

a.   facilitation
b.   growth and maturation rates
c.   inhibition
d.   interspecific differences in dispersal
e.   all of the above

Answer: e
Difficulty: Moderate
Bloom's Taxonomy: Knowledge

50.7 VARIATIONS IN SPECIES RICHNESS AMONG COMMUNITIES

30.   Which statement best summarizes the connection between islands and species richness?

a.   The factor influencing species richness on an island is distance from the mainland.
b.   The factor influencing species richness on an island is island size.
c.   The factor influencing species richness on an island is the type of species on the island.
d.   Both a and b are correct.
e.   Both b and c are correct.

Answer: d
Difficulty: Difficult
Bloom's Taxonomy: Knowledge

UNANSWERED QUESTIONS

31. Which statement is true about the effect of stress on a population?

a. Plants compete more in an ideal environment.
b. Plants compete more in a stressful environment.
c. Only abiotic factors influence populations.
d. Stressful environments brought about by man do not effect populations.
e. Stressful environments brought about by nature do not effect populations.

Answer: a
Difficulty: Difficult
Bloom's Taxonomy: Comprehension

Matching

Match the description to the appropriate term from this chapter.

32. _____ sporting a brightly contrasting, dangerous-looking pattern

33. _____ a species closely resembling another species

34. _____ one species harms organisms of another species

35. _____ two or more species use the same limiting resourc

36. _____ defined by the resources used by a species and the conditions it requires

37. _____ use of resources in different ways

38. _____ one species derives benefits and the other is unaffected

39. _____ situation where both species benefit

40. _____ one organism feeds off of another organism, causing harm

A. Aposematic coloration

B. Commensalism

C. Ecological niche

D. Exploitative competition

E. Interference competition

F. Mimicry

G. Mutualism

H. Parasitism

I. Resource partitioning

Answers: 32. A   33. F   34. E   35. D
36. C   37. I   38. B   39. G
40. H

Difficulty: Moderate
Bloom's Taxonomy: Knowledge
Source: Section 50.1

Short Answer

41. Why does a rattlesnake open its mouth before it even strikes?

Answer: If a rattlesnake has the intention to bite a large animal, it first unhinges its jaw so that it can engulf an animal larger than its head.
Difficulty: Difficult
Bloom's Taxonomy: Application
Source: Section 50.1

42. Since all land-dwelling animals utilize oxygen for respiration, does that make oxygen a limiting factor?

Answer: Even though all land-dwelling animals require oxygen, it is not considered a limiting factor since it is in abundant supply.
Difficulty: Moderate
Bloom's Taxonomy: Application
Source: Section 50.1

43. What is the difference between a food chain and a food web?

Answer: A food web is in a straight line to indicate which species eat which other species. Rarely in nature is any relationship between organisms in a community that simple. A food web shows the interwoven relationships between species, because most organisms have more than one food source.
Difficulty: Easy
Bloom's Taxonomy: Knowledge
Source: Section 50.3

44. What is the difference between a decomposer and a detritivore?

Answer: A decomposer is a small organism such as bacteria or fungi that digests dead or dying matter. A detritivore is usually larger and digests dead matter, and plant and animal wastes. Examples of detritivores or scavengers are turkey vultures and hyenas.
Difficulty: Moderate
Bloom's Taxonomy: Application
Source: section 50.3

45. What is the definition of the intermediate disturbance hypothesis?

Answer: When disturbances of moderate intensity occur regularly, species richness is greatest.
Difficulty: Moderate
Bloom's Taxonomy: Knowledge
Source: Section 50.5

Sequence

Put the following stages of primary succession in the proper order.

46. _____ first event
47. _____ second event
48. _____ third event
49. _____ fourth event
50. _____ fifth event
51. _____ sixth event
52. _____ seventh event

A. bushes appear
B. ferns and grasses become established
C. glaciers retreat
D. lichens erode rocks
E. mosses appear
F. soil is present
G. trees are established

**Answers: 46. C   47. D   48. F   49. E
50. B   51. A   52. G**

Difficulty: Moderate
Bloom's Taxonomy: Knowledge
Source: Section 50.6

Put the following stages of secondary succession in the proper order.

53. _____ first event
54. _____ second event
55. _____ third event
56. _____ fourth event
57. _____ fifth event
58. _____ sixth event
59. _____ seventh event
A. asters and broomsedges
B. crabgrass
C. hardwood trees
D. horseweed
E. pine seedlings
F. ragweed
G. shrubs

**Answers: 53. B   54. D   55. F   56. A
57. G   58. E   59. C**

Difficulty: Moderate
Bloom's Taxonomy: Knowledge
Source: Section 50.6

True/False

If the statement is true, write a "T" in the blank. If the statement is false, make it correct by changing the underlined word(s) and writing the correct word(s) in the answer blanks.

60. _____ Two snakes that have similar bright color patterns are an example of aposematic coloration.

61. _____ In interference competition, two or more populations use the same limiting resource.

62. _____ Realized niches are larger than fundamental niches.

63. _____ Most gradient analyses point toward the individualistic view of communities.

64. _____ A tropical rain forest would most likely include vegetation that includes a canopy.

65. _____ In the 1960s, it was felt that the major factor(s) influencing community structure was competition.

66. _____ The intermediate disturbance hypothesis says that species richness is least in a community that has moderate disturbances.

Answers: 60. T   61. F, exploitative competition
62. F, smaller   63. T   64. T
65. T   66. F, greatest

Difficulty: Moderate
Bloom's Taxonomy: Knowledge, Comprehension
Source: Sections 50.1, 50.2, 50.3, 50.4, 50.5

Select the Exception

67. Which of the following is not a legume?

a. beans
b. clover
c. peanuts
d. peas
e. wheat

Answer: e
Difficulty: Easy
Bloom's Taxonomy: Knowledge
Source: Section 50.1

68. Which of the following is not classified as an ectoparasite?

a. aphids
b. bedbugs
c. leeches
d. mosquitos
e. tapeworm

Answer: e
Difficulty: Moderate
Bloom's Taxonomy: Application
Source: Section 50.1

69. Which of the following pairs of terms is not matched appropriately?

a. primary consumer – carnivore
b. primary consumer – herbivore
c. primary producer – autotroph
d. primary producer – plant
e. secondary consumer – carnivore

Answer: a
Difficulty: Easy
Bloom's Taxonomy: Knowledge
Source: Section 50.3

Essays

70. When populations of antelope and cheetah coexist, what changes in the populations are likely to occur due to natural selection?

Answer: First, the antelope population may decline due to heavy predation. Natural selection would encourage the survival of faster antelopes that may be able to escape predation by the cheetahs. Natural selection would again come into play and encourage the survival of the fastest cheetahs since they would be the best adapted to capture their food source.
Difficulty: Moderate
Bloom's Taxonomy: Application
Source: Section 50.1

71. What is the best way to investigate the effects of factors that influence populations?

Answer: The best way to determine the effects of specific changes on a population is to perform experiments in a controlled environment. The researcher could then initiate a change in only one factor. By testing different factors in this controlled environment, it is possible to postulate how things would occur in nature.
Difficulty: Moderate
Bloom's Taxonomy: Comprehension
Source: Section 50.2

72. Discuss one hypothesis that attempts to explain the high species richness in the tropics.

Answer: The climate in the tropics, which includes warm temperatures and plenty of rainfall throughout the year, promotes a long growing season. This longer growing season increases the availability of fruits and plants throughout the year, which enables many animals to easily survive in that environment.
Difficulty: Moderate
Bloom's Taxonomy: Comprehension
Source: Section 50.7

Labeling

Population Interactions and Their Effects	
Interaction	**Effects on Interacting Populations**
Predation	A
Herbivory	B
Competition	C
Commensalism	D
Mutualism	E
Parasitism	F

Match these descriptions of the effects on interacting populations with the interaction terms in the table above by labeling with the appropriate letter.

73. _____ one population obtains energy from plants; plants are destroyed
74. _____ one population obtains energy from animals; animals are killed
75. _____ one population obtains energy, while the other is damaged (usually not killed)
76. _____ both populations benefit
77. _____ one population benefits while the other is unaffected
78. _____ can be disadvantageous to both populations

Answers: 73. B   74. A   75. F   76. E
77. D   78. C

Difficulty: Moderate
Bloom's Taxonomy: Knowledge
Source: Section 50.1

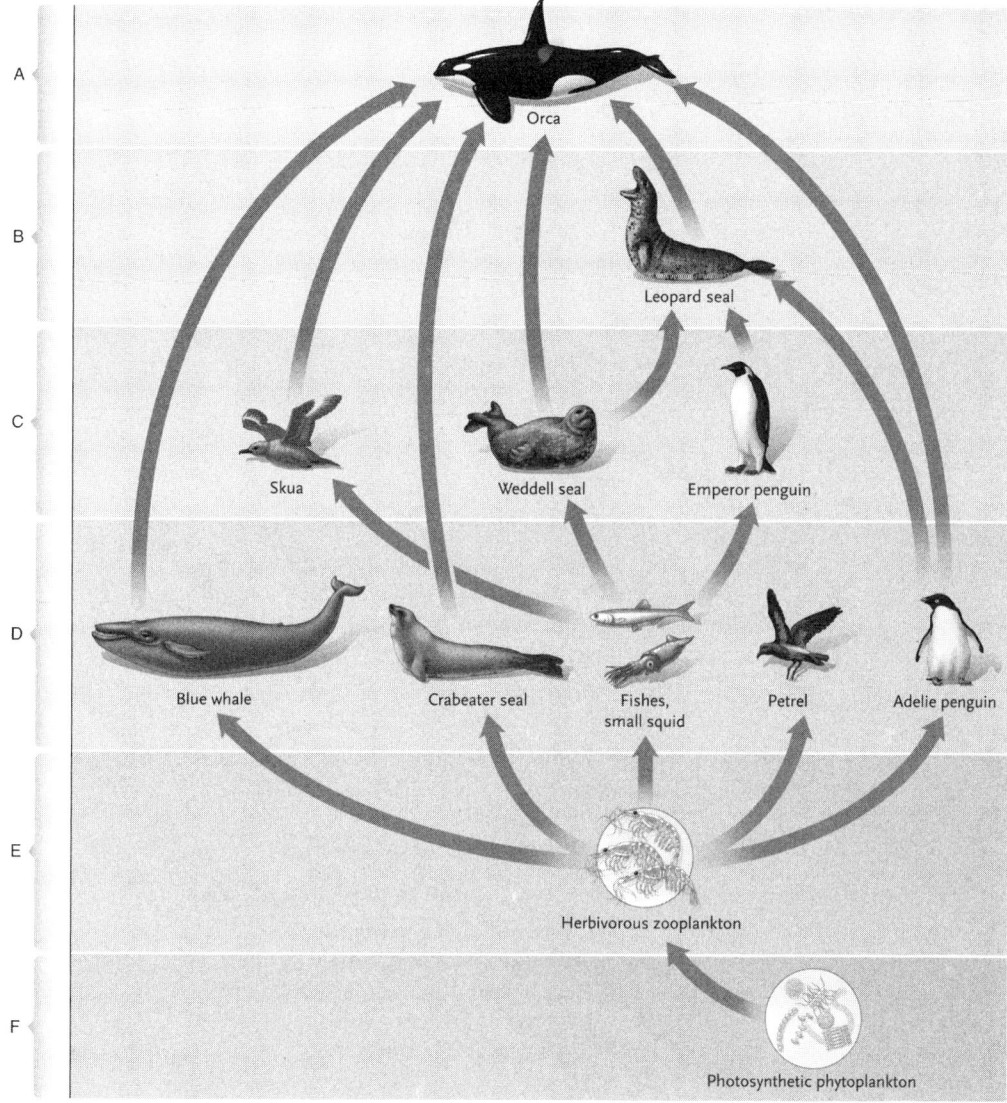

A

B

C

D

E

F

Orca

Leopard seal

Skua   Weddell seal   Emperor penguin

Blue whale   Crabeater seal   Fishes,   Petrel   Adelie penguin
                                           small squid

Herbivorous zooplankton

Photosynthetic phytoplankton

Fill in the appropriate titles for each portion of this food web:

79. _____ primary consumers
80. _____ primary producers
81. _____ quaternary consumers
82. _____ secondary consumers
83. _____ tertiary consumers
84. _____ top carnivore

Answers: 79. E   80. F   81. B   82. D
           83. C   84. A

Difficulty: Easy
Bloom's Taxonomy: Knowledge
Source: Section 50.3

51
ECOSYSTEMS

Multiple Choice

WHY IT MATTERS

1. Between the 1930s and the 1970s, the amount of _____ in Lake Erie increased.

a. coliform bacteria
b. phosphorus
c. phytoplankton
d. sewage
e. all of the above

Answer: e
Difficulty: Easy
Bloom's Taxonomy: Knowledge

2. A nonnative species that was accidentally introduced into Lake Erie by attaching itself to boats is

a. mayflies.
b. oligochaete worms.
c. phytoplankton.
d. salmon.
e. zebra mussels.

Answer: e
Difficulty: Moderate
Bloom's Taxonomy: Knowledge

51.1 ENERGY FLOW AND ECOSYSTEM ENERGETICS

3. Herbivores are _____ energy transfer(s) away from the sun.

a. one
b. two
c. three
d. four
e. five

Answer: b
Difficulty: Moderate
Bloom's Taxonomy: Knowledge

4. The overall ecological efficiency of most organisms is between

a. 1 and 2 percent.
b. 2 and 5 percent.
c. 5 and 20 percent.
d. 10 and 50 percent.
e. 50 and 75 percent.

Answer: c
Difficulty: Difficult
Bloom's Taxonomy: Knowledge

5. Nearly all ecosystems receive a constant input of energy from which outside source?

a. bacteria
b. biomass
c. nutrients
d. plants
e. the sun

Answer: e
Difficulty: Easy
Bloom's Taxonomy: Knowledge

6. _____ make up the detrital food web.

a. Carnivores and decomposers
b. Decomposers and detritivores
c. Decomposers and producers
d. Herbivores and carnivores
e. Producers and herbivores

Answer: b
Difficulty: Easy
Bloom's Taxonomy: Knowledge

7. The length of day is _____ apt to change and the sunlight is _____ near the equator.

a. less; strongest
b. less; weakest
c. more; strongest
d. more; weakest
e. latitude has no effect on amount of sunlight or length of day

Answer: a
Difficulty: Moderate
Bloom's Taxonomy: Knowledge

8. Trophic pyramids typically have a
    _____ base and a _____ top portion.

a. narrow; narrow
b. narrow; wide
c. wide; narrow
d. wide; wide
e. one cannot predict the shape of a trophic
    pyramid

Answer: c
Difficulty: Easy
Bloom's Taxonomy: Comprehension

9. Warm blooded animals use the majority of
    their energy to

a. digest.
b. grow.
c. move.
d. reproduce.
e. stay warm.

Answer: e
Difficulty: Moderate
Bloom's Taxonomy: Application

10. The conclusion from the energy flow study
    in Silver Springs was that

a. energy flows are difficult to measure.
b. most of the sun's energy is harvested.
c. only a little energy is transferred from one
    trophic level to the next.
d. the energy flow in carnivores is equal to
    that of producers.
e. trophic energy losses are not significant.

Answer: c
Difficulty: Moderate
Bloom's Taxonomy: Comprehension

11. What three factors determine the ecological
    efficiencies of consumers?

a. assimilation and harvesting efficiencies,
    and nutrient volume
b. assimilation and harvesting efficiencies,
    and water volume
c. production, harvesting, and assimilation
    efficiencies
d. production and harvesting efficiencies, and
    amount of sunlight
e. production and harvesting efficiencies, and
    percentage of decomposers

Answer: c
Difficulty: Difficult
Bloom's Taxonomy: Knowledge

12. When comparing the pyramids of biomass
    between Silver Springs, Florida, and the English
    Channel,

a. the pyramids for Silver Springs and the
    English Channel are both bottom heavy.
b. the pyramids for Silver Springs and the
    English Channel are both top heavy.
c. the pyramid for Silver Springs is bottom
    heavy and the English Channel pyramid is
    top heavy.
d. the pyramid for Silver Springs is top heavy
    and the English Channel pyramid is bottom
    heavy.
e. researchers are unable to create pyramids
    for Silver Springs and the English Channel.

Answer: c
Difficulty: Difficult
Bloom's Taxonomy: Comprehension

13. What is the definition of the standing crop
    biomass?

a. amount of sunlight absorbed by the plants
b. assessing the health of the plants
c. dry weight of the plants
d. number of plant species present
e. productivity of the plants

Answer: c
Difficulty: Difficult
Bloom's Taxonomy: Knowledge

14. Which of these following marine
    ecosystems has the highest mean net primary
    productivity?

a. continental shelf
b. estuaries
c. kelp beds and reefs
d. open ocean
e. upwelling zones

Answer: c
Difficulty: Difficult
Bloom's Taxonomy: Comprehension

51.2 NUTRIENT CYCLING IN ECOSYSTEM

15. The largest reservoir of carbon is

a. consumers.
b. nutrients in soil.
c. producers.
d. rock.
e. water.

Answer: d
Difficulty: Moderate
Bloom's Taxonomy: Knowledge

16. Nitrogen is made available to plants for their use by the process(es) of

a. ammonification.
b. nitrification.
c. nitrogen fixation.
d. b and c
e. a, b, and c

Answer: e
Difficulty: Moderate
Bloom's Taxonomy: Knowledge

17. Name the process that takes place in nodules located on legumes.

a. ammonification
b. biological magnification
c. denitrification
d. nitrification
e. nitrogen fixation

Answer: e
Difficulty: Moderate
Bloom's Taxonomy: Knowledge

18. What is the main cause of carbon dioxide build up in the atmosphere?

a. burning fossil fuels
b. deforestation
c. droughts
d. flooding
e. respiration due to overpopulation

Answer: a
Difficulty: Moderate
Bloom's Taxonomy: Knowledge

19. The main natural source of phosphorus is

a. Earth's crust.
b. fertilizer.
c. guano.
d. legumes.
e. water.

Answer: a
Difficulty: Moderate
Bloom's Taxonomy: Knowledge

20. Which of the following types of ecosystems would have the lowest percentage of total net primary productivity?

a. open ocean
b. savanna
c. swamp and march
d. temperate deciduous forests
e. tundra

Answer: e
Difficulty: Moderate
Bloom's Taxonomy: Knowledge

21. In the hydrologic cycle, water moves from the land to the air by _____ and returns to the land via _____.

a. precipitation; evaporation and transpiration
b. precipitation and evaporation; transpiration
c. precipitation and transpiration; evaporation
d. transpiration; precipitation and evaporation
e. transpiration and evaporation; precipitation

Answer: e
Difficulty: Moderate
Bloom's Taxonomy: Comprehension

22. Which of the following carbon reservoirs contains the most carbon?

a. atmosphere
b. biomass on land
c. lakes
d. oceans
e. soil

Answer: d
Difficulty: Difficult
Bloom's Taxonomy: Knowledge

23. What were the primary results from the Hubbard Brook Watershed Project?

a. Deforestation causes flooding.
b. New plant species that absorb less water were discovered.
c. Pollution is gravely affecting that water supply.
d. Primary consumers are the most important part of that energy pyramid.
e. Trees absorb too much water and are not useful in a watershed area.

Answer: a
Difficulty: Difficult
Bloom's Taxonomy: Application

24. Which forms of nitrogen are readily usable by plants?

a. NH_4^+ and NO_3^-
b. NH_4^+ and NO_2^-
c. N_2 and NH_3
d. NH_4 and NH_3
e. NO_3^- and NO_4^-

Answer: a
Difficulty: Difficult
Bloom's Taxonomy: Knowledge

51.3 ECOSYSTEM MODELING

25.   To develop an effective model for an ecosystem, a researcher must

a.   describe food webs.
b.   develop relevant equations.
c.   estimate productivity of all populations.
d.   identify major species.
e.   all of the above

Answer: e
Difficulty: Easy
Bloom's Taxonomy: Comprehension

26.   One disadvantage of a simulation model of an ecosystem is that it

a.   does not take into effect interactions between organisms.
b.   does not take into effect nutrient availability.
c.   does not take into effect water availability.
d.   may not be accurate.
e.   may predict disastrous consequences.

Answer: d
Difficulty: Easy
Bloom's Taxonomy: Comprehension

UNANSWERED QUESTIONS

27.   Which factors make predictions difficult regarding the changes of the carbon cycle?

a.   the architecture of the canopy of trees overhead
b.   leaf positioning
c.   leaf losses
d.   timing of natural events
e.   all of the above

Answer: e
Difficulty: Easy
Bloom's Taxonomy: Knowledge

Matching

Match each description with the appropriate scientific term.

28.   _____ predicting what would happen if an ecosystem was changed

29.   _____ nutrients cycle between abiotic and living organisms

30.   _____ describing how nutrients accumulate in four compartments

31.   _____ rate at which a plant converts the sun's energy into chemical energy

32.   _____ the element in short supply

33.   _____ the energy transfer from producers to consumers

34.   _____ a diagram showing the inefficient transfer of energy from one trophic level to the next

35.   _____ the total dry weight of plants at a certain time

36.   _____ gross primary productivity minus the energy used for cellular respiration

37.   _____ the ratio of net productivity at one trophic level to the productivity at the trophic level below it

38.   _____ the reduction of productivity decreases at higher trophic levels

39.   _____ the relationship between a predator and prey that effects populations in two or more trophic levels

A.   Biogeochemical cycles

B.   Ecological efficiency

C.   Ecological pyramid

D.   Generalized compartment model

E.   Gross primary productivity

F.   Limiting factor

G.   Net primary productivity

H.   Pyramid of biomass

I.   Secondary productivity

J.   Simulation modeling

K.   Standing crop mass

L.   Trophic cascade

Answers: 28. *J*   29. *A*   30. *D*   31. *E*   32. *F*
                    33. *I*   34. *C*   35. *K*   36. *G*   37. *B*
                    38. *H*   39. *L*

Difficulty: Moderate
Bloom's Taxonomy: Knowledge
Sources: Sections 51.2, 51.2

Short Answer

40.   Why is the productivity poor in the deeper parts of the ocean?

Answer: In the deeper parts of the ocean, sunlight only influences the activities happening near the surface since the sunlight does not penetrate deep into the water. In addition, many nutrients sink to the bottom of the ocean. This impacts productivity greatly since both sunlight and nutrients are necessary to improve the productivity.
Difficulty: Moderate
Bloom's Taxonomy: Comprehension
Source: Section 51.2

41.   How would you calculate the net primary productivity of an ecosystem?

Answer: You would subtract the energy used for cellular respiration from the gross primary productivity. Gross primary productivity is the rate that plants convert solar energy into chemical energy.
Difficulty: Moderate
Bloom's Taxonomy: Knowledge
Source: Section 51.1

42.   How would it change the ecological efficiency if all humans became vegetarians?

Answer: Since energy is lost at each trophic level, we would conserve more energy by consuming plants directly. By including another trophic level (an herbivore) or even two more trophic levels (an herbivore and a carnivore), more energy is lost and thus, efficiency is compromised.
Difficulty: Difficult
Bloom's Taxonomy: Application
Source: Section 51.1

43.   Why is nitrogen necessary in the human body?

Answer: Nitrogen is required in order for the body to produce proteins, nucleic acids, and other biological molecules. It is one of the nutrients necessary for life.
Difficulty: Moderate
Bloom's Taxonomy: Knowledge
Source: Section 51.2

44.   Where do plants obtain the necessary carbon for photosynthesis?

Answer: Plants obtain the majority of their carbon from carbon dioxide in the air and water.
Difficulty: Moderate
Bloom's Taxonomy: Knowledge
Source: Section 51.1

Sequence

Put the following stages of the phosphorus cycle in the proper order:

45. _____   First stage

46. _____   Second stage

47. _____   Third stage

48. _____   Fourth stage

49. _____   Fifth stage

50. _____   Sixth stage

51. _____   Seventh stage

A.   Feces and urine are released from animals into soil.

B.   Phosphorus is plentiful in rocks and soil.

C.   Phosphorus goes from rivers into the ocean.

D.   Phosphorus goes from soil into rivers.

E.   Phosphorus goes to higher trophic levels.

F.   Following uplifting, plants take in phosphorus.

G.   Ocean floor experiences an uplifting and releases phosphorus.

Answers: 45. B   46. D   47. C   48. G   49.   F   50. E   51. A

Difficulty: Easy
Bloom's Taxonomy: Application
Source: Section 51.2

True/False Questions

If the statement is true, write a "T" in the blank. If the statement is false, make it correct by changing the underlined word(s) and writing the correct word(s) in the answer blank.

52. _____ Oceans and the tropical rain forests contribute <u>equally</u> to global productivity.

53. _____ Large zooplankton <u>decrease</u> the ecosystem's main productivity when they are abundant.

54. _____ Energy is <u>conserved</u> between each trophic level.

55. _____ <u>Water</u> can be a limiting factor when measuring productivity.

56. _____ Energy losses are <u>multiplied</u> in successive energy transfers in a food web.

57. _____ Lake ecosystems have <u>many</u> trophic levels.

58. _____ The most biomass in trees is <u>leaves, which do</u> undergo photosynthesis.

59. _____ A <u>large</u> percentage of energy passes on from one trophic level to the next.

60. _____ Fossil fuels are the result of buried organisms where <u>low</u> oxygen levels existed.

Answers: 52. T   53. T   54. F, lost   55. T 56. T   57. F, few   58. F, wood which does not   59. F, small   60. T

Difficulty: Moderate
Bloom's Taxonomy: Knowledge, Application
Source: Sections 51.1, 51.2

Select the Exception

61.   In a marine environment, which of the following is <u>not</u> a limiting factor for primary productivity?

a.   depth of the water
b.   nutrients
c.   sunlight
d.   water
e.   nitrogen

Answer: d
Difficulty: Easy
Bloom's Taxonomy: Knowledge
Source: Section 51.1

62.   An urban environment may impact all of the following changes to a forest <u>except</u>

a.   amount of rainfall.
b.   rate of photosynthesis.
c.   rate of cellular respiration.
d.   seedling size.
e.   stomatal densities.

Answer: a
Difficulty: Moderate
Bloom's Taxonomy: Comprehension
Source: Unanswered Questions

63.   Each of the following statements regarding loggerhead sea turtles is correct <u>except</u>

a.   sea turtles migrate from Japan to Baja California.
b.   some turtles are killed because they become entangled in fishermen's nets.
c.   the number of sea turtles is declining.
d.   the sea turtles in Japan and those in Baja California do not appear to be related.
e.   sea turtles return to the specific location of their birth.

Answer: d
Difficulty: Moderate
Bloom's Taxonomy: Comprehension
Source: Section 51.1

64.   The Hubbard Brook Watershed project measured all of the following <u>except</u>

a.   amount of energy used by consumers.
b.   amount of nutrients washed away by streams.
c.   nutrient input.
d.   precipitation.
e.   uptake of nutrients by plants.

Answer: a
Difficulty: Difficult
Bloom's Taxonomy: Knowledge
Source: Section 51.2

65. All of the following would help to reduce the amount of greenhouse gases in the atmosphere <u>except</u>

a. planting trees.
b. reducing emissions from factories.
c. reducing water usage.
d. using solar heat in homes.
e. walking to work.

Answer: c
Difficulty: Moderate
Bloom's Taxonomy: Application
Source: Section 51.2

66. Which of the following plants <u>cannot</u> participate in nitrogen fixation?

a. beans
b. clover
c. peanuts
d. roses
e. wheat

Answer: e
Difficulty: Moderate
Bloom's Taxonomy: Comprehension
Source: Section 51.2

Essays

67. How does crop rotation aid the farmer?

Answer: If a farmer plants a different type of crop each year in a particular field, he may help prevent nutrient depletion in that field. For example, if one crop requires a great deal of phosphorus, that field will eventually be severely depleted of that nutrient. But since different crops have varying nutrient demands, crop rotations may prevent nutrient depletions.
Difficulty: Moderate
Bloom's Taxonomy: Comprehension
Source: Section 51.1

68. How do food webs relate to the idea of biological magnification of toxins?

Answer: When there are toxins present in the soil, water, etc., they are absorbed into organisms in the bottom of the food chain. As you progress up the food chain, the amount of toxins present will increase. This is why researchers often study the effects of toxins on small, seemingly insignificant species. Determining the deleterious effects of these toxins on lower level organisms may alert humans to the ultimate danger of these substances.
Difficulty: Difficult
Bloom's Taxonomy: Application
Source: Section 51.1

69. What steps have we learned in the chapter to help the environment?

Answer: Firstly, we have learned that we should work hard not to disrupt the balance of nature. Eliminating the existence of a lower level species may not seem very significant, but it may have a huge impact on that particular ecosystem. Also we must take steps to conserve water, and prevent fertilizers and other disruptive substances from becoming runoff into our water supplies. We should develop more efficient machines to reduce the amount of greenhouse gases produced. Globally, we should have zero population growth. In short, there is much we can do individually and as a species to safeguard our environment.
Difficulty: Moderate
Bloom's Taxonomy: Application
Sources: Sections 51.1, 51.2

Labeling

Fill in the blanks with the appropriate terms to correspond with the boxes in the following image.

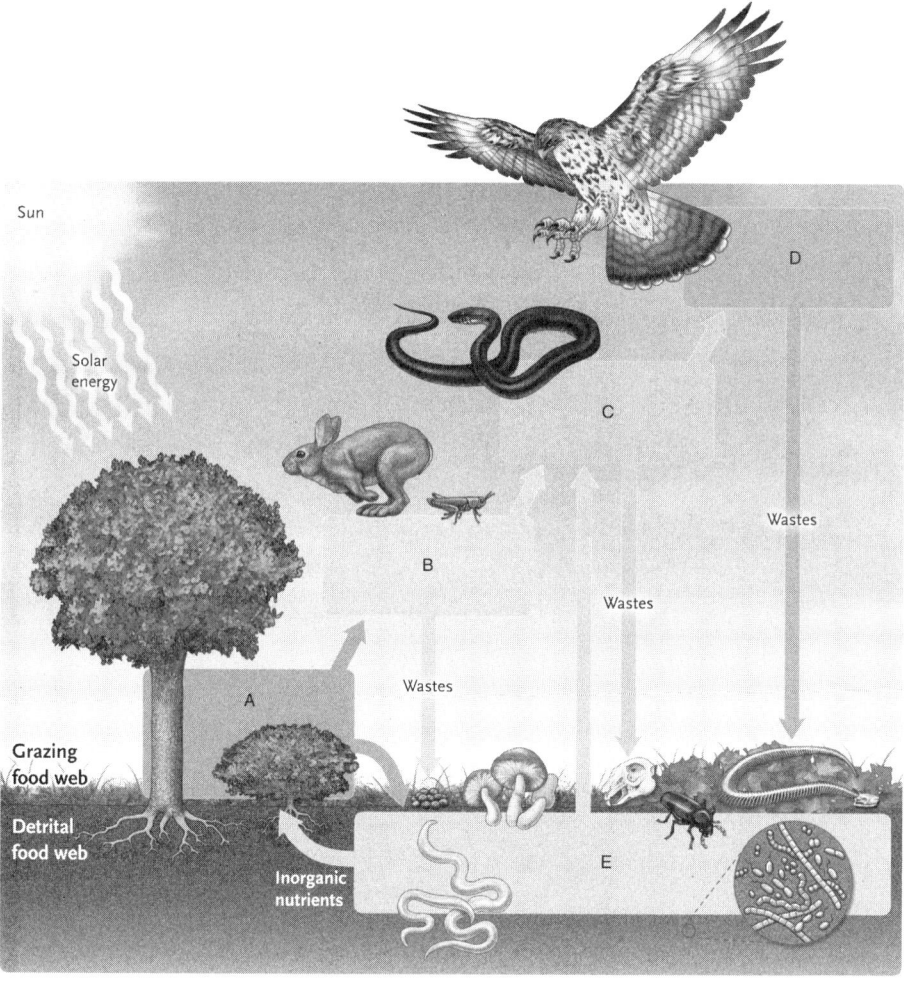

71. _____primary consumers
72. _____primary producers
73. _____secondary consumers
74. _____tertiary consumers

Answers: 70. E 71. B   72. A   73. C
74. D

Difficulty: Easy
Bloom's Taxonomy: Application
Source: Section 51.1

Fill in the blanks with the appropriate terms to correspond with the boxes in the following image.

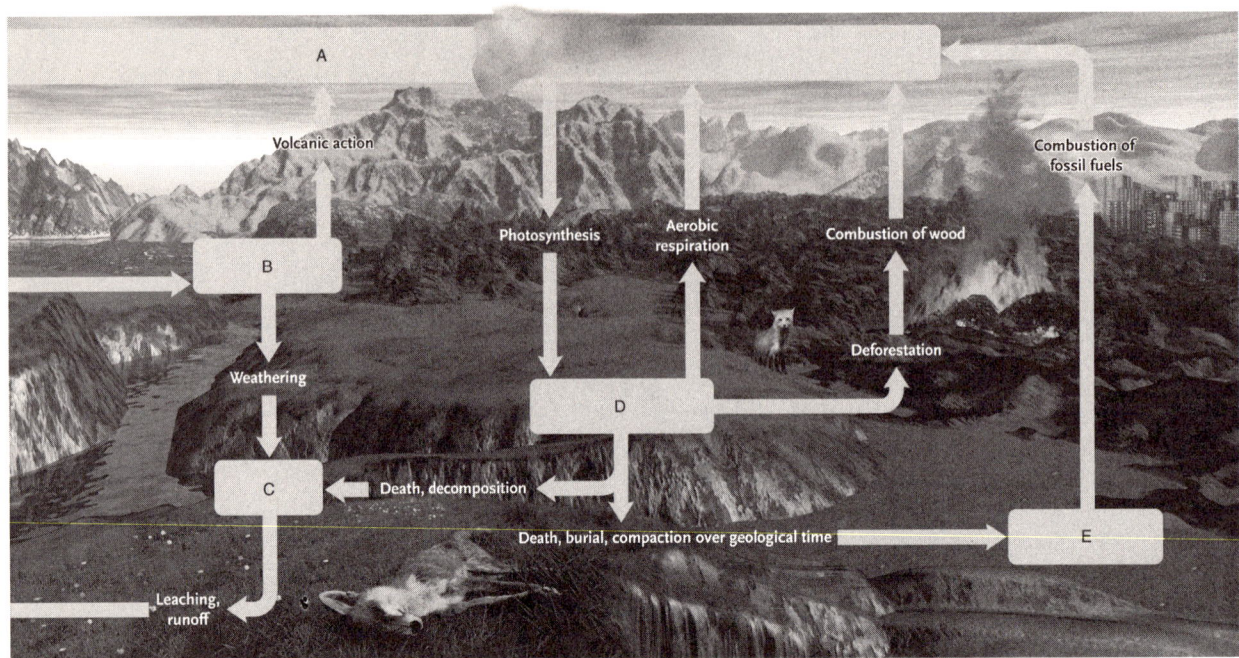

75. _____coal, oil, peat
76. _____soil water
77. _____terrestrial food webs
78. _____terrestrial rocks

Answers: 75. E   76. C   77. D   78. B

Difficulty: Moderate
Bloom's Taxonomy: Comprehension
Source: Section: 51.2

52
THE BIOSPHERE

Multiple-Choice

WHY IT MATTERS

1. Which of the following events would **not** be expected to occur during a strong El Niño year?

a. heavy rains in the central and eastern Pacific ocean
b. upwelling of cold water along the west coast of the Americas
c. death of phytoplankton in western coastal areas
d. warm surface currents flowing from west to east
e. weakened equatorial winds

Answer: b
Difficulty: Moderate
Bloom's Taxonomy: Knowledge

2. Which of the following weather patterns would be characteristic of a La Niña event?

a. high pressure over the western Pacific
b. high ocean surface temperatures
c. warm and dry weather in the southern U.S.
d. ocean surface waters moving from west to east
e. air movement from west to east

Answer: c
Difficulty: Moderate
Bloom's Taxonomy: Knowledge

52.1 ENVIRONMENTAL DIVERSITY OF THE BIOSPHERE

3. In the Northern Hemisphere, seasonal variation in temperatures is caused mainly by _____.

a. the Earth's elliptical orbit taking it closer to and farther away from the sun
b. the Earth's tilt on its axis of 23.5°
c. seasonal variation of output of energy from the sun
d. warm equatorial air masses rising and spreading north and south
e. uneven warming of the ocean's surface water

Answer: b
Difficulty: Easy
Bloom's Taxonomy: Knowledge

Use the figure below to answer questions 4 and 5.

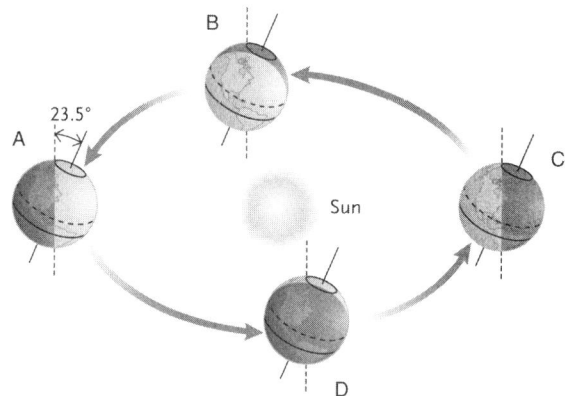

4. Which of these images represents the position of the Earth during the December solstice?

a. A
b. B
c. C
d. D
e. the answer cannot be determined from the figure provided

Answer: c
Difficulty: Moderate
Bloom's Taxonomy: Knowledge, Comprehension

5. Which of these images represents the position of the Earth during an equinox?

a. A and B
b. B and C
c. C and D
d. B and D
e. none of the choices

Answer: d
Difficulty: Moderate
Bloom's Taxonomy: Knowledge, Comprehension

OBSERVATIONAL RESEARCH: HOW LIZARDS COMPENSATE FOR ALTITUDINAL VARIATIONS IN ENVIRONMENTAL TEMPERATURE

6.   Hertz and Huey discovered which of the following compensatory mechanisms that allows lizards living at high altitudes to survive lower mean air temperatures?

a.   hibernation
b.   torpor
c.   increased time spent basking
d.   increased consumption of food
e.   shelter-seeking behavior

Answer: c
Difficulty: Easy
Bloom's Taxonomy: Knowledge

INSIGHTS FROM THE MOLECULAR REVOLUTION: FISH ANTIFREEZE PROTEINS

7.   Identify the trait that is **not** characteristic of the antifreeze protein of polar-dwelling fish.

a.   beta pleated sheet structure
b.   allows freezing to an ultrafine slush
c.   contains 37 amino acids
d.   ice binding probably by polar amino acids
e.   all of the choices are correct

Answer: a
Difficulty: Moderate
Bloom's Taxonomy: Knowledge

52.2 ORGANISMAL RESPONSES TO ENVIRONMENTAL VARIATION

8.   Which of the following are documented biological results of global warming?

a.   Warm-adapted marine species are becoming more common and cold-adapted species less common.
b.   Plants display earlier flowering and growth.
c.   Alpine butterflies are shifting their distribution 6.1 meters higher in altitude per decade.
d.   Animals display earlier migration and breeding.
e.   All of the choices.

Answer: e
Difficulty: Easy
Bloom's Taxonomy: Knowledge

52.3 TERRESTRIAL BIOMES

9.   A biome is most specifically defined as _____.

a.   a vegetation type plus its associated microorganisms, fungi, and animals
b.   a biological community and the physical environment with which it interacts
c.   the total dry weight of plants present in a specific area at any given time
d.   an assemblage of species living in the same place
e.   a group of organisms of the same species living together at one time

Answer: a
Difficulty: Moderate
Bloom's Taxonomy: Knowledge

10.   Temperate deciduous forest generally has _____ precipitation and _____ temperature than boreal forest.

a.   higher; higher
b.   the same; lower
c.   the same; higher
d.   lower; lower
e.   lower; the same

Answer: c
Difficulty: Moderate
Bloom's Taxonomy: Comprehension, Application

11.   Most deserts occur near _____ latitude because of dry descending air masses.

a.   0°
b.   10°
c.   20°
d.   30°
e.   40°

Answer: d
Difficulty: Easy
Bloom's Taxonomy: Knowledge

FOCUS ON RESEARCH: EXPLORING THE RAIN FOREST CANOPY

12.   The biologist Donald Perry discovered that rainforest birds interact in what way with the Norantea sessilis vine?

a.   Birds nest within the shelter of the thorny vine, which repels predators.
b.   Birds feed on the fruits of the vine, distributing the seeds around the forest in their droppings.
c.   Birds appear to be overfeeding on the vine's succulent fruits, probably limiting the plant's chances of long-term survival.
d.   Birds that feed on the nectar of the flowers inadvertently transfer pollen from one vine to another on their feet.
e.   Birds that feed on the nectar of the flowers inadvertently transfer pollen from one vine to another on their beaks.

Answer: d
Difficulty: Moderate
Bloom's Taxonomy: Knowledge, Comprehension

52.4 FRESHWATER BIOMES

13.   Dissolved oxygen would be expected to be highest in which of these bodies of water?

a.   warm eutrophic lake
b.   slow-moving tropical river
c.   pool within a cool forest stream
d.   white water stretch of a cold river
e.   dissolved oxygen should be the same in all waters

Answer: d
Difficulty: Easy
Bloom's Taxonomy: Knowledge

Use the figure below for questions 14–16.

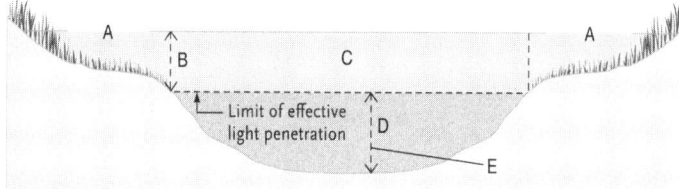

14.   Which of the zones of a lake supports detritivores and decomposers but no photosynthesizers?

a.   zones A & B
b.   zones B & C
c.   zones C & A
d.   zones D & E
e.   zones A & D

Answer: d
Difficulty: Easy
Bloom's Taxonomy: Knowledge, Comprehension

15.   Which letter in the figure above represents the limnetic zone?

a.   A
b.   B
c.   C
d.   D
e.   E

Answer: c
Difficulty: Easy
Bloom's Taxonomy: Knowledge, Comprehension

16.   Which letter in the figure above represents the littoral zone?

a.   A
b.   B
c.   C
d.   D
e.   E

Answer: a
Difficulty: Easy
Bloom's Taxonomy: Knowledge, Comprehension

17.   The addition of excess phosphorous causes lakes to become eutrophic because the nutrients trigger blooms of photosynthetic cyanobacteria. Why do these blooms cause low oxygen levels in lakes?

a.   Cyanobacteria use up more oxygen than they produce.
b.   Excess phosphorous binds to dissolved oxygen in the water.
c.   Cyanobacteria are net producers of oxygen while alive, but when they die aerobic bacteria use oxygen while decomposing them.
d.   Cyanobacterial blooms only occur in waters naturally low in oxygen.
e.   Cyanobacteria are aerobic heterotrophs, consuming oxygen and producing carbon dioxide.

Answer: c
Difficulty: Difficult
Bloom's Taxonomy: Knowledge, Comprehension

52.5 MARINE BIOMES

18.   Coastal regions where seawater mixes with fresh water from rivers, streams, and runoff are called

_____ .

a.   ncritic zones
b.   estuaries
c.   benthos
d.   wetlands
e.   intertidal zones

Answer: b
Difficulty: Moderate
Bloom's Taxonomy: Knowledge

Use the figure below for questions 19–21.

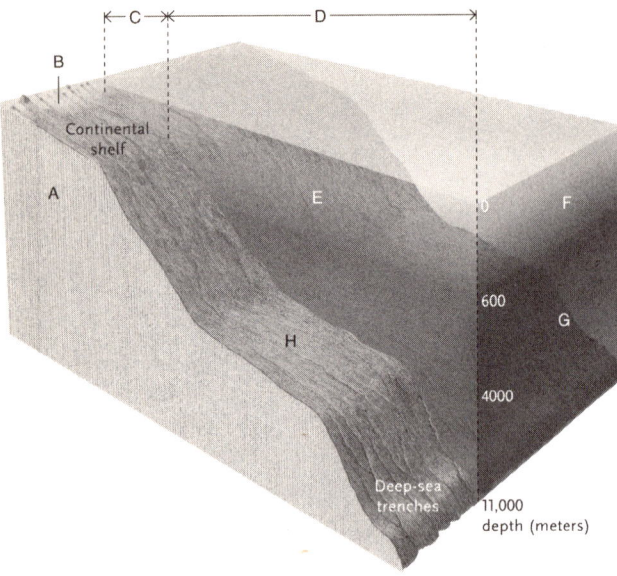

19. Letter "C" in the diagram above represents the _____ zone.

a. neritic
b. abyssal
c. oceanic
d. benthic
e. intertidal

Answer: a
Difficulty: Moderate
Bloom's Taxonomy: Knowledge

20. The benthic province includes the intertidal zone, represented by letter _____, and the abyssal zone, represented by letter _____.

a. G; H
b. B; H
c. E; H
d. B; C
e. E; F

Answer: b
Difficulty: Moderate
Bloom's Taxonomy: Knowledge

21. The oceanic zone is represented by letter _____.

a. A
b. B
c. C
d. D
e. E

Answer: d
Difficulty: Easy
Bloom's Taxonomy: Knowledge

UNANSWERED QUESTIONS

22. According to researchers Parmesan and Yohe, how are living systems responding to anthropogenic global warming?

a. All biomes are remaining intact and all are shifting their distributions northward.
b. Some species are moving northward and others are not, so communities are being disrupted.
c. All biomes are remaining intact, but some biomes are shifting their distributions and other biomes are not.
d. Most plants are shifting their distributions northward, but most animals are not.
e. Most animals are shifting their distributions northward, but most plants are not.

Answer: b
Difficulty: Difficult
Bloom's Taxonomy: Comprehension

23. Some studies suggest that, as global warming progresses, tropical lowland trees may be limited in their ability to move northward because of _____.

a. predicted desertification of temperate areas
b. temperatures in the north still being too low
c. flooding of coastal areas
d. extreme weather events like hurricanes and blizzards
e. development and habitat loss

Answer: a
Difficulty: Moderate
Bloom's Taxonomy: Knowledge

24. Global warming has already brought all of the following changes to the tundra **except** _____.

a. encroachment of shrubs
b. encroachment of trees
c. melting permafrost
d. drying soil
e. mosquitoes carrying malaria

Answer: e
Difficulty: Easy
Bloom's Taxonomy: Knowledge

Matching

Match each of these terrestrial biomes with its correct description.

25. _____ tropical rain forest

26. _____ savanna

27. _____ chaparral

28. _____ temperate deciduous forest

29. _____ boreal forest

30. _____ arctic tundra

31. _____ alpine tundra

32. _____ desert

33. _____ temperate rain forest

A. Occurs on mountaintops throughout the world; strong winds; plants form low mats; cold winter temperatures

B. North of boreal forests; permafrost; very short summers; short low-growing plants

C. Grasslands with few trees, adjacent to tropical forests; have wet and dry seasons

D. Long cold winters, wet summers; dominated by coniferous trees; northern latitudes

E. Scrubby mix of trees and low shrubs in coastal land between 30° and 40° latitude

F. Mean annual rainfall > 250 cm; mean annual temperature ≥ 25°C, humidity > 80%; tall trees

G. Warm summers, cold winters; annual precipitation 75–250 cm; most plants shed leaves in winter

H. Rainfall < 25 cm per year; near 30° latitude; little organic matter present in topsoil, which is eroded by rare deluges of rain

I. West coast of North America; heavy rain and fog; Douglas fir, Sitka spruce, and redwoods

***Answers: 25. F   26. C   27. E   28. G   29. D
30. B   31. A   32. H   33. I***

Difficulty: Moderate
Bloom's Taxonomy: Knowledge
Source: Section 52.3

Classification

Use these five terms for questions 34–38.

   a. epilimnion
   b. hypolimnion
   c. thermocline
   d. fall overturn
   e. spring overturn

34. The top layer of the limnetic zone is called the _____, and is heated by the sun in summer.

35. The epilimnion cools, becomes denser, and sinks, eliminating the thermocline during the _____.

36. The deep water of a lake's profundal zone is called the _____.

37. Water temperature is briefly uniform at all depths, and winds create vertical currents during the _____.

38. At the _____, water temperature changes abruptly over a narrow depth range.

Answers: 34. a   35. d   36. b   37. e   38. c

Difficulty: Moderate
Bloom's Taxonomy: Knowledge
Source: Section 52.4

Choice

For each of the following ecological problems, choose the most probable cause from the list below. Some options may be used more than once.

a. global warming
b. introduction of exotic species
c. phosphorous runoff from fertilizers and feedlots
d. El Niño

39. _____ eutrophication of lakes

40. _____ failure of normal upwellings to occur along western continental coasts

41. _____ distributions of polar species contracting to higher latitudes

42. _____ elimination of the American chestnut tree

43. _____ shifting geographic distributions of species

Answers: 39. c   40. d   41. a   42. b   43. a

Difficulty: Easy
Bloom's Taxonomy: Knowledge, Comprehension
Source: Sections 52.1–52.5

Short Answer

44. Communities near benthic hydrothermal vents survive in perpetual darkness, where no photosynthesis is possible. What do the producers of this ecosystem use as an energy source?

Answer: Hydrogen sulfide from the hot water vents serves as an energy source for autotrophic bacteria. These bacteria act as primary producers for the ecosystem, sometimes residing as endosymbionts in larger organisms.
Difficulty: Moderate
Bloom's Taxonomy: Knowledge, Comprehension
Source: Section 52.5

45. Many polar-dwelling fishes have "antifreeze proteins" that prevent their bodies from freezing solid even at very low temperatures. Briefly describe how these proteins function.

Answer: "Antifreeze proteins" are small molecules containing 30–50 amino acids. They bind to ice crystals and cover them with a protein coat that prevents further crystallization.
Difficulty: Difficult
Bloom's Taxonomy: Knowledge, Comprehension
Source: Section 52.2

46. In which of the three vertical intertidal zones is biodiversity the highest, and why?

Answer: Biodiversity is highest in the lower intertidal, because that zone is submerged for the greatest proportion of time. Exposure to the air is challenging to tidal organisms, not all of which can sustain long periods out of water.
Difficulty: Moderate
Bloom's Taxonomy: Knowledge, Comprehension
Source: Section 52.5.

Classification

Classify the following marine organisms into the province in which they reside.

a. pelagic
b. benthic

47. _____ humpback whale

48. _____ giant tube-dwelling worms

49. _____ phytoplankton

50. _____ *Himantolophus* sp. anglerfish

51. _____ sea anemones

52. _____ clams

53. _____ crabs

54. _____ great white sharks

Answers: 47. a   48. b   49. a   50. a   51. b
           52. b   53. b   54. a

Difficulty: Easy
Bloom's Taxonomy: Knowledge
Source: Section 52.5

True/False

If the statement is true, write a "T" in the blank. If the statement is false, make it correct by changing the underlined word(s) and writing the correct word(s) in the answer blank.

55. _____ In coastal areas, wind blows in from the sea during the night.

56. _____ Most ocean currents flow clockwise in the Northern Hemisphere.

57. _____ The Gulf Stream takes relatively cold water toward northwestern Europe.

58. _____ Behavioral responses that animals may or may not use are called facultative.

59. _____ Rain shadows form on the leeward sides of mountain ranges.

60. _____ Soils in tropical rain forests are nutrient rich.

61. _____ Crystal-clear mountain lakes are oligotrophic.

62. _____ Mangroves encourage the erosion of soils on coastlines.

Answers: 55. F, day   56. T   57. F, warm 58. T
           59. T   60. F, poor   61.T   62. F,
deposition

Difficulty: Moderate
Bloom's Taxonomy: Knowledge
Source: Sections 52.1–52.5

Select the Exception

63. Most of the following homeostatic responses of animals are obligate. Select the exception.

a. carbonic acid buffering system regulating pH of blood
b. basking in the sun
c. shivering when cold
d. kidneys regulating the osmotic content of blood
e. closing of capillaries near the skin in the cold

Answer: b
Difficulty: Difficult
Bloom's Taxonomy: Knowledge, Application
Source: Section 52.2.

64. Four of the five biomes listed below are terrestrial. Select the exception.

a. boreal forest
b. temperate deciduous forest
c. tropical forest
d. kelp forest
e. thorn forest

Answer: d
Difficulty: Easy
Bloom's Taxonomy: Knowledge
Source: Sections 52.3 and 52.5.

65. Most of the following events are characteristic of a La Niña weather pattern. Select the exception.

a. low pressure over the western Pacific
b. air and ocean surface waters moving west to east
c. low ocean surface temperatures off the coast of South America
d. cold, wet weather in the northeast U.S.
e. wet weather on the west coast of the U.S.

Answer: e
Difficulty: Moderate
Bloom's Taxonomy: Knowledge
Source: Why it Matters

66. Four of these five environments are examples of microclimates. Select the exception.

a. a fallen log on the forest floor
b. the space under a rock
c. a moist, shady forest
d. the sunny side of a rock wall
e. a crack in a sidewalk

Answer: c
Difficulty: Easy
Bloom's Taxonomy: Application
Source: Section 52.1

Sequence

Write the letter of the terrestrial biome with the lowest mean temperature next to 67. Sequence the remaining biomes by increasing temperature, finishing by writing the letter of the biome with the highest mean temperature next to 70.

A. temperate deciduous forest
B. savanna
C. tundra
D. boreal forest

67. _____
68. _____
69. _____
70. _____

Answers: 67. C   68. D   69. A   70. B

Write the letter of the terrestrial biome with the lowest mean precipitation next to 71. Sequence the remaining biomes by increasing precipitation, finishing by writing the letter of the biome with the highest mean precipitation next to 74.

A. chaparral
B. desert
C. tropical forest
D. savanna

71. _____
72. _____
73. _____
74. _____

Answers: 71. B   72. A   73. D   74. C

Essay

75. Biologists sometimes damage ecosystems in order to study them. For example, researchers have fogged rainforest trees with insecticides to search for undiscovered species that fall to the ground when poisoned. Others have deliberately added phosphates to natural lakes to study the eutrophication process. What is your ethical position on such studies? Should they be banned, or is the knowledge gained worth the harm they cause?

Answer: There are two possible perspectives in answer to this question, both equally correct.

Possible "pro" answer: It is often difficult, if not impossible, to glean information from living systems without disassembling them or altering their normal function. These studies are ethical, because the knowledge gained is well worth the limited harm caused. Damage to one ecosystem may yield information that will prevent similar damage from occurring to many other similar systems. For example, phosphates are now banned from detergents in Florida because they caused eutrophication in the Everglades. Without research, the connection between eutrophication and phosphates could not have been demonstrated.

Possible "con" answer: Biologists, more than any other scientists, should respect the natural world. Especially now, when so many systems are stressed from human activity, it is not ethical to deliberately cause widespread destruction. It will certainly take longer to gain knowledge this way, but the alternative is unacceptable. Pollution, overpopulation, and global warming are bringing the planet near the breaking point, and we must do everything in our power to reverse these trends. Ethical limits that already exist for human and animal experimentation should be extended to ecosystems as well.
Difficulty: Moderate
Bloom's Taxonomy: Analysis
Source: Sections 52.1–52.5

53

BIODIVERSITY AND CONSERVATION BIOLOGY

Multiple-Choice

WHY IT MATTERS

1. Biologists have recently been interested in Miss Waldron's red colobus monkey because it is _____.

a. increasing in numbers
b. a pest to farmers
c. thought to be extinct
d. capable of carrying HIV
e. none of the choices

Answer: c
Difficulty: Easy
Bloom's Taxonomy: Knowledge

2. Biodiversity encompasses which of the following concepts?

a. genetic variation
b. species richness
c. interactions among ecosystems
d. diversity of different ecosystems
e. all of the choices

Answer: e
Difficulty: Easy
Bloom's Taxonomy: Knowledge

53.1 THE BENEFITS OF BIODIVERSITY

3. Which of the following benefits from biodiversity would be categorized as an ecosystem service?

a. anticancer drug Taxol, derived from the yew tree
b. plants and fruits that are edible by humans
c. cotton and other useful plant-derived fibers
d. decomposition of wastes
e. wild organisms as a source of genes for genetic engineering

Answer: d
Difficulty: Moderate
Bloom's Taxonomy: Comprehension

4. The genes of teosinte, a wild relative of corn, were used in crossbreeding with domestic corn in an attempt to produce a _____ variety.

a. herbicide resistant
b. pest resistant
c. high lysine
d. perennial
e. higher yielding

Answer: d
Difficulty: Moderate
Bloom's Taxonomy: Knowledge

5. Photosynthetic organisms are essential for limiting the damage done by global warming because they _____.

a. move carbon from the atmosphere to living organisms
b. move carbon from the soil to the atmosphere
c. move carbon from the atmosphere to the soil
d. produce oxygen
e. sequester carbon in geologic sediments

Answer: a
Difficulty: Easy
Bloom's Taxonomy: Knowledge

EXPERIMENTAL RESEARCH: PREDATION ON SONGBIRD NESTS IN FORESTS AND FOREST FRAGMENTS

Use the graph below to answer questions 6 and 7.

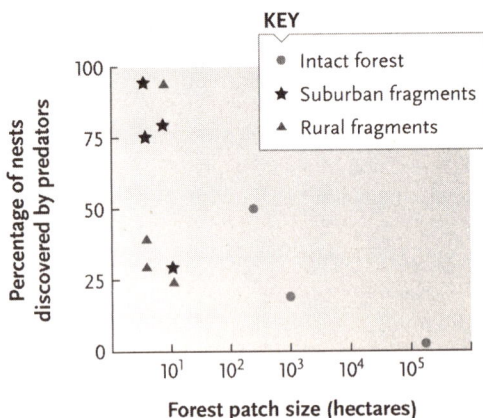

6. The logical conclusion that can be drawn from Wilcove's experimental data in the graph above is that _____.

a. songbird nests in a variety of habitats are equally likely to be found by predators
b. songbird nests in forest fragments are more likely to be found by predators than those in intact tracts of forest
c. songbird nests in forest fragments are less likely to be found by predators than those in intact tracts of forest
d. songbird nests in rural areas are more likely to be found by predators than those in suburban areas
e. songbirds that nest on the ground are more vulnerable to predators than those that nest in trees or shrubs

Answer: b
Difficulty: Moderate
Bloom's Taxonomy: Application

FOCUS ON RESEARCH: APPLIED RESEARCH: BIOLOGICAL MAGNIFICATION

7. Biological magnification is best defined as _____.

a. the increase in toxicity of a chemical compound resulting from its metabolism by a living organism
b. the higher relative proportion of biomass among producers as compared to consumers in an ecosystem
c. the concentration of nondegradable poisons within organisms at higher trophic levels
d. the concentration of degradable poisons within organisms at lower trophic levels
e. the declining relative proportion of nondegradable poisons in organisms at higher trophic levels as compared to those at lower trophic levels

Answer: c
Difficulty: Moderate
Bloom's Taxonomy: Knowledge

8. If all of the organisms below lived in an ecosystem contaminated by DDT, which one would contain the highest concentration of the chemical?

a. phytoplankton
b. zooplankton
c. small fish
d. large fish
e. osprey

Answer: e
Difficulty: Easy
Bloom's Taxonomy: Knowledge

9. Which of the following symptoms would be characteristic of a child whose mother ate fish contaminated by PCBs during her pregnancy?

a. low birth weight and neonatal behavior problems
b. low IQ and teen delinquency
c. misshapen head and facial abnormalities
d. spina bifida
e. club foot

Answer: a
Difficulty: Moderate
Bloom's Taxonomy: Application

55.2 THE BIODIVERSITY CRISIS

10. How does habitat fragmentation affect biodiversity?

a. As long as the fragments are of good quality habitat, biodiversity will not be affected.
b. Habitat fragmentation increases biodiversity because it reduces entry of exotic species.
c. Habitat fragmentation is a threat to biodiversity because small habitat patches sustain only small populations.
d. Habitat fragmentation increases biodiversity because adaptation to local conditions stimulates evolution.
e. None of the choices.

Answer: c
Difficulty: Moderate
Bloom's Taxonomy: Knowledge, Comprehension

11. Disturbance and exposure at borders of habitat fragments are collectively termed _____.

a. edge effects
b. fragmentation effects
c. anthropogenic disturbances
d. reduced perimeter diversity
e. boundary decline

Answer: a
Difficulty: Easy
Bloom's Taxonomy: Knowledge

12. Which of the following factors is believed to reduce songbird breeding success in fragmented habitats?

a. brood parasitism by cowbirds
b. nest predation by blue jays
c. lack of necessary habitat types
d. nest predation by domestic cats
e. all of the choices

Answer: e
Difficulty: Easy
Bloom's Taxonomy: Comprehension, Application

13. When subtropical forest is cleared, what biome is most likely to replace it?

a. grassland suitable for grazing
b. cropland ideal for agriculture
c. young forest with the same species
d. desert with poor eroded soil
e. chaparral with grass and low bushes

Answer: d
Difficulty: Moderate
Bloom's Taxonomy: Knowledge

14. What molecule from coal-burning power plants dissolves in water vapor in the air and falls as acid precipitation?

a. sulfuric acid
b. ozone
c. mercury
d. sulfur dioxide
e. carbonic acid

Answer: d
Difficulty: Moderate
Bloom's Taxonomy: Knowledge

15. The administration of the drug Diclofenac to livestock inadvertently poisoned _____ and led to an increase in the disease _____.

a. cattle; hoof and mouth
b. vultures; rabies
c. children; epilepsy
d. coyotes; hemolytic *E. coli*
e. cattle egrets; avian flu

Answer: b
Difficulty: Moderate
Bloom's Taxonomy: Knowledge, Comprehension

16. Atlantic cod have evolved to mature at a younger age and smaller size due to _____.

a. DNA damage from radiation released accidentally from nuclear power plants
b. unknown causes
c. food competition from other predatory fish
d. intraspecies competition
e. overexploitation by fishermen

Answer: e
Difficulty: Moderate
Bloom's Taxonomy: Knowledge, Comprehension

17. The greatest rate of extinction of all time is/was during _____ and is/was caused by _____.

a. Cambrian; an ice age
b. Devonian; unknown reasons
c. Permian; formation of Pangea
d. Quaternary; human activity
e. Cretaceous; asteroid impact

Answer: d
Difficulty: Moderate
Bloom's Taxonomy: Knowledge

53.3 BIODIVERSITY HOTSPOTS

18. Biodiversity hotspots are defined as areas where biodiversity is both _____ and _____.

a. concentrated; endangered
b. researched; protected
c. low; increasing
d. overexploited; studied
e. low; declining

Answer: a
Difficulty: Easy
Bloom's Taxonomy: Knowledge

19. Endemic species tend to have all of the following characteristics **except** _____.

a. high specific habitat requirements
b. low dispersal ability
c. restricted geographic distributions
d. high specific dietary requirements
e. high charisma to motivate conservationists

Answer: e
Difficulty: Easy
Bloom's Taxonomy: Knowledge

20. Which of the following environments would be most likely to harbor a biodiversity hotspot?

a. tropical island
b. temperate grassland
c. temperate deciduous forest
d. desert
e. savanna

Answer: a
Difficulty: Moderate
Bloom's Taxonomy: Comprehension

53.4 CONSERVATION BIOLOGY: PRINCIPLES AND THEORY

21. Scientists at the Natural History Museum in London estimate that more than _____ percent of living species have not yet been discovered.

a. 11
b. 27
c. 32
d. 56
e. 98

Answer: e
Difficulty: Moderate
Bloom's Taxonomy: Knowledge

22. Biologists believe that whooping cranes suffer from a high rate of developmental deformities of the spine and trachea because of _____.

a. biological magnification of DDT
b. exposure to hormone-mimicking pollutants from plastics
c. population bottleneck and loss of genetic variability
d. conservation biologists saving chicks that would have died at hatching
e. exposure to PCBs used as insulators in electronics

Answer: c
Difficulty: Easy
Bloom's Taxonomy: Knowledge

23. Sea otter populations were reduced by overhunting, allowing an overgrowth of one of their prey animals, the _____, which consumed populations of _____ and decimated them.

a. crabs; shrimp
b. abalone; algae
c. sea urchin; kelp
d. harbor seals; fish
e. crab; kelp

Answer: c
Difficulty: Moderate
Bloom's Taxonomy: Knowledge

24. Conservation biologists conduct a population viability analysis to determine _____.

a. whether an environment is too badly degraded to support a species
b. how large a population must be to ensure its long-term survival
c. likely source populations
d. likely sink populations
e. the annual reproduction rate of a population

Answer: b
Difficulty: Moderate
Bloom's Taxonomy: Knowledge

25. A group of neighboring populations that exchange individuals is defined as a _____.

a. metapopulation
b. megapopulation
c. community
d. source population
e. sink population

Answer: a
Difficulty: Easy
Bloom's Taxonomy: Knowledge

EXPERIMENTAL RESEARCH: EFFECT OF LANDSCAPE CORRIDORS ON PLANT SPECIES RICHNESS IN HABITAT FRAGMENTS

Use the graph below for questions 26–27.

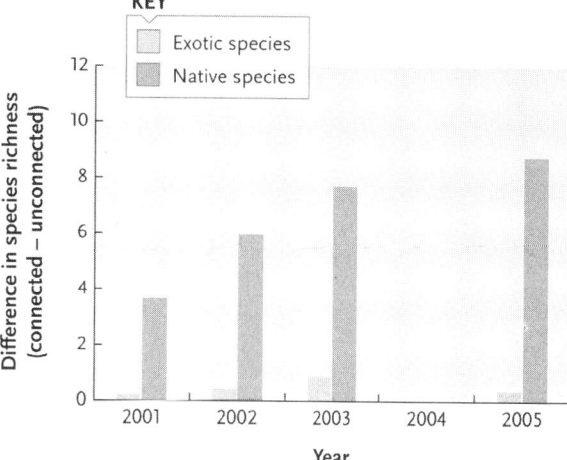

26. Which of the following choices best summarizes the results of Damschen's experiment in the graph above?

a. Habitat patches with landscape corridors are at higher risk of encroachment by exotic species than isolated habitat patches are.
b. Over time, the number of native species harbored by habitat patches with landscape corridors was increasingly greater than in isolated habitat patches.
c. The number of both native and exotic species increased over time in habitat patches that were connected by landscape corridors.
d. The number of native species increased over time in isolated habitat patches but not in those connected by corridors.
e. None of the choices.

Answer: b
Difficulty: Difficult
Bloom's Taxonomy: Application

27. In 2002, habitat patches that were connected by corridors had _____.

a. six more species than isolated habitat patches had
b. a total of six species
c. six less species than isolated habitat patches had
d. six more species than that patch had sustained the year before
e. six less species than that patch had sustained the year before

Answer: a
Difficulty: Difficult
Bloom's Taxonomy: Application

53.5 CONSERVATION BIOLOGY: PRACTICAL STRATEGIES AND ECONOMIC TOOLS

28. The Ngorongoro Conservation Area is an example of _____.

a. conservation through preservation, with people completely excluded
b. conservation through preservation, with people allowed only as temporary visitors
c. mixed use conservation, with some human residents and livestock grazing
d. mixed use conservation, with no human residents and limited livestock grazing
e. conservation through restoration, with efforts to remove contaminants underway

Answer: c
Difficulty: Moderate
Bloom's Taxonomy: Knowledge

29. The success of Chitwan National Park in Nepal is attributed primarily to _____.

a. ecotourism
b. exclusion of local residents
c. ecosystem valuation
d. benefits provided to local residents
e. income derived through hunting

Answer: d
Difficulty: Easy
Bloom's Taxonomy: Knowledge

UNANSWERED QUESTIONS

30. In mutualistic networks, species that have few links to other species tend to interact most with _____.

a. other specialists
b. generalists
c. plant species but not animals
d. animal species but not plants
e. decomposers

Answer: b
Difficulty: Moderate
Bloom's Taxonomy: Knowledge

31. Of the following food webs, _____ would probably be the most stable and persistent over time.

a. a simple one with few interactions and few trophic levels
b. a detrital food web
c. a grazing food web
d. a complex food web with many interactions and many trophic levels
e. a simple predatory food chain

Answer: d
Difficulty: Moderate
Bloom's Taxonomy: Knowledge

Matching

Match each of the following terms with its correct definition.

32. _____ non-native organisms in an ecosystem
33. _____ species found only within a limited area
34. _____ concentration of poisons in higher trophic levels of a food web
35. _____ inadvertent conversion of habitats to desert
36. _____ indirect benefits from normal activity of ecological processes
37. _____ assignment of a monetary value to the normal activity of ecological processes

A. ecosystem services
B. ecosystem valuation
C. desertification
D. endemic species
E. biological magnification
F. exotic species

Answers: 32. F   33 .D   34. E   35. C   36. A
37.B

Difficulty: Moderate
Bloom's Taxonomy: Knowledge
Source: Sections 53.1–53.5

Choice

For each of the following species, choose the most appropriate description from the list below.

a. endemic   b. exotic   c. extinct

38. _____ starling
39. _____ kudzu
40. _____ Miss Waldron's red colobus
41. _____ yellow-bellied glider
42. _____ bay checkerspot butterfly

Answers: 38. b   39. b   40. c   41. a   42. a

Difficulty: Easy
Bloom's Taxonomy: Knowledge, Comprehension
Source: Sections 53.1–53.5

Short Answer

43.   A developer putting tract housing in a previously pristine area plans numerous small open space areas where the original habitat is to remain. As a conservationist, how would you suggest changing the plan, if at all? (Canceling the development is not an option.)

Answer: One large preserve would be better than many small preserves, since populations in isolated habitat patches have a high risk of extinction. Isolated populations often become inbred as they may not be able to travel to individuals in other habitat patches.
Difficulty: Moderate
Bloom's Taxonomy: Comprehension
Source: Section 53.4

44.   Briefly describe some of the pros and cons of using ecotourism to generate income from a natural area.

Answer: Ecotourism employs local people and increases support for the preservation of ecosystems and endangered species. However, heavy tourism traffic can damage ecosystems and put pressure on ecosystem services like sewage disposal and clean water supplies.

45.   What is a population viability analysis (PVA), and what predictions does it allow population biologists to make?

Answer: A PVA is a collection of information about a threatened species, including its diet, predators, mating habits, space requirements, distribution, and response to disturbances. This allows population biologists to predict the chances of its survival given various different habitat protection scenarios.

Classification

Categorize each of the following conservation projects into one of the categories below. Some categories may be used more than once.

46.   _____   The Nature Conservancy buys up ecologically important tracts of land.

47.   _____   An electric company building a hydroelectric dam pays people upstream not to deforest the watershed.

48.   _____   Replanting of trees in habitat corridors between forest fragments.

49.   _____   U.S. National Forest land may be leased by ranchers for grazing.

50.   _____   Indigenous peoples of Belize develop an educational program to host international students as they study rainforest ecology.

51.   _____   Endangered black-footed ferrets are raised in captivity and released back into the wild.

A.   ecosystem valuation

B.   conservation through restoration

C.   conservation through preservation

D.   mixed-use conservation

Answers: 46. C   47. A   48. B   49. D   50. D
           51. B

Difficulty: Easy
Bloom's Taxonomy: Knowledge
Source: Section 53.5

True/False

If the statement is true, write a "T" in the blank. If the statement is false, make it correct by changing the underlined word(s) and writing the correct word(s) in the answer blank.

52.   _____   The yellow-bellied glider is a mammal whose continued survival depends on undisturbed eucalyptus forests.

53.   _____   A DNA barcode system uses a nuclear gene that varies greatly between species.

54.   _____   Background extinction rates once eliminated 7–8 species per year.

55.   _____   Loss of genetic variability in an endangered species can lead to deformities.

56.   _____   Most deaths among endangered Florida panthers are caused by disease.

57.   _____   A 20 m wide edge disrupts a smaller fraction of a small habitat patch than of a large one.

58.   _____   The study of how large-scale ecological factors influence local populations and communities is specifically defined as conservation biology.

Answers: 52 .T   53. F, mitochondrial 54. T   55. T
           56 .F, collisions with vehicles 57.F,
larger   58. F, landscape ecology

Difficulty: Moderate
Bloom's Taxonomy: Knowledge
Source: Sections 55.1–55.5

Select the Exception

59. Most of the following species are invasive exotics. Select the exception.

a. Atlantic cod
b. kudzu
c. starlings
d. hemlock wooly adelgid
e. Africanized honeybee

Answer: a
Difficulty: Moderate
Bloom's Taxonomy: Knowledge
Source: Section 53.2

60. Four of the five animals listed below are threatened or endangered. Select the exception.

a. whooping crane
b. sea otter
c. sea urchin
d. bay checkerspot butterfly
e. Karner blue butterfly

Answer: c
Difficulty: Easy
Bloom's Taxonomy: Knowledge
Source: Section 53.4

61. Most of these problems are accepted by the vast majority of reputable scientists as being anthropogenic, or human-caused. Select the exception.

a. global warming
b. decline in Atlantic cod size at maturity
c. desertification of the Everglades
d. background extinction rate
e. deformities among whooping cranes

Answer: d
Difficulty: Easy
Bloom's Taxonomy: Knowledge
Source: Section 53.4

62. Most of the following geographic areas are biodiversity hotspots. Select the exception.

a. New Zealand
b. Mesoamerica
c. Caribbean
d. South-Central China
e. Norway

Answer: e
Difficulty: Easy
Bloom's Taxonomy: Knowledge
Source: Section 53.3

Matching

Match each of the ecological problems labeled A–H with the immediate (most direct) cause from the choices below.

63. _____ emissions from cars and coal-fired power plants

64. _____ excessive tapping of groundwater for irrigation, landscaping, and household use

65. _____ irrigation of agricultural lands, especially using ground water

66. _____ artificial light at night

67. _____ use of Diclofenac in livestock

68. _____ transport by humans

69. _____ overexploitation

70. _____ airborne pollutants from combustion of fossil fuels

71. _____ desertification

72. _____ habitat fragmentation

73. _____ excess levels of atmospheric carbon dioxide

A. Disorientation among sea turtle hatchlings

B. Increasing rate of asthma

C. Decline in Atlantic cod population

D. Acid precipitation

E. Global spread of exotic species

F. Death of vultures and increase in populations of feral dogs, rats, and flies

G. Salinization of soil

H. Salt water intrusion to the water table

I. Development

J. Global warming

K. Deforestation

Answers: 63. D   64. H   65. G   66. A   67. F
68. E   69. C   70. B   71. K   72. I
73. J

Difficulty: Easy
Bloom's Taxonomy: Knowledge
Source: Section 53.1–53.5

Essay

74.   Under the leadership of Arata Kochi, the anti-malaria division of the World Health Organization (WHO) recently recommended the use of DDT to kill mosquitoes in malaria-prone areas. Evaluate the consequences of this decision based on your knowledge and understanding of the properties of DDT.

Answer: DDT is transmitted around the planet far from its point of application. It biomagnifies, causing disruption of physiology and reproduction of organisms at higher trophic levels. Use of DDT has pushed top carnivores in many ecosystems near extinction. It accumulates to the highest levels in the longest-lived animals, most notably humans and whales. Pest organisms develop resistance to the poison but humans and other organisms apparently do not. The use of DDT as a malarial-control agent is opposed by many scientists, who advocate the use of insecticide-impregnated bed nets to avoid mosquito bites instead.
Difficulty: Moderate
Bloom's Taxonomy: Analysis
Source: Section 53.2

75.   Summarize the human social elements necessary to address when designing a conservation plan.

Answer: To be successful, a conservation plan must have support from the local people. They must benefit in some way, often economically, from the preserve. Legislating preserves that exclude or disenfranchise local people is not a good idea. Poaching and illegal harvesting of resources are rampant in places where the locals are opposed to the conservation plan. Ecotourism or allowing limited extraction of resources from the preserve are options that have been successful. Education of the local people about ecosystem services and consequences of ecological degradation is valuable as well.
Difficulty: Moderate
Bloom's Taxonomy: Knowledge, Comprehension
Source: Section 53.5

54

THE PHYSIOLOGY AND GENETICS OF ANIMAL BEHAVIOR

Multiple-Choice

WHY IT MATTERS

1. Ethology is the study of _____.

a. how animals process information
b. genetic mechanisms that underlie behavior
c. the adaptive value of behaviors
d. how animals behave in their natural environments
e. physiological mechanisms that underlie behavior

Answer: d
Difficulty: Easy
Bloom's Taxonomy: Knowledge

2. Which of the following causes of behavior is classified as an ultimate cause?

a. genetic
b. cellular
c. physiological
d. anatomical
e. adaptive

Answer: e
Difficulty: Moderate
Bloom's Taxonomy: Knowledge

54.1 GENETIC AND ENVIRONMENTAL CONTRIBUTIONS TO BEHAVIOR

3. Young male white-crowned sparrows that were experimentally raised without ever hearing the song of their species _____.

a. produced only random vocalizations
b. never sang at all
c. sang a poorly developed version of their species' song
d. instinctively performed the song perfectly
e. sang a recognizable version of the song of an ancestral sparrow species

Answer: a
Difficulty: Easy
Bloom's Taxonomy: Knowledge

4. Instinctive behaviors are distinguished from learned behaviors because instinctive behaviors are _____ while learned behaviors are _____.

a. genetically programmed responses / dependent upon experience
b. dependent upon experience / genetically programmed responses
c. acquired from observation / acquired from practice
d. incomplete the first few times they are displayed / apparently a product of reason
e. taught by parents / acquired from trial and error

Answer: a
Difficulty: Moderate
Bloom's Taxonomy: Knowledge, Comprehension

FOCUS ON RESEARCH: THE ROLE OF SIGN STIMULI IN PARENT-OFFSPRING INTERACTIONS

5. Pecking behavior in young herring gulls is triggered primarily by _____.

a. hunger
b. the shape of the parent gull's head
c. a red spot on the bill of the parent
d. the overall body silhouette of the parent bird

Answer: c
Difficulty: Easy
Bloom's Taxonomy: Knowledge

6. Biologists categorize the cue that stimulates young herring gulls to peck as a sign stimulus because the pecking behavior _____.

a. requires learning to perfect
b. is completely instinctive
c. is triggered by presentation of the cue
d. is only triggered by the cue if it is accompanied by species-specific behaviors
e. appears both in the presence and absence of the cue

Answer: c
Difficulty: Moderate
Bloom's Taxonomy: Comprehension

INSIGHTS FROM THE MOLECULAR REVOLUTION: A KNOCKOUT BY A WHISKER

7. Knockout mice with no active copies of the *Dvl-1* gene _____.

a. were apparently normal physically
b. displayed abnormal social behavior
c. were easily startled
d. all of the choices

Answer: d
Difficulty: Moderate
Bloom's Taxonomy: Knowledge

54.2 INSTINCTIVE BEHAVIORS

8. Instinctive behaviors that are performed in exactly the same way every time they are triggered are defined as _____.

a. sign stimuli
b. fixed action patterns
c. releasers
d. habitual behaviors
e. conditioned behaviors

Answer: b
Difficulty: Easy
Bloom's Taxonomy: Knowledge

9. Which cues function as sign stimuli to trigger smiling in very young babies?

a. the outline of the smiling mouth
b. the presence of two eyes
c. a simple but complete image of a human face
d. a detailed likeness of one of the baby's caregivers
e. a detailed likeness of any person's face

Answer: b
Difficulty: Easy
Bloom's Taxonomy: Knowledge

10. When newborn coastal garter snakes and newborn inland garter snakes are exposed to banana slug chemicals, the coastal snakes responded by _____ while the inland snakes _____.

a. attacking / fled
b. tongue-flicking / did not tongue-flick
c. feeding / did not feed
d. avoiding the stimuli / tongue-flicked
e. ignoring the stimuli / performed threat displays

Answer: b
Difficulty: Easy
Bloom's Taxonomy: Knowledge

54.3 LEARNED BEHAVIORS

11. When geese or ducks are imprinted, this means that they _____.

a. will return to the place they were born to breed and raise their own young
b. have become habituated to harmless stimuli
c. have been experimentally trained by operant conditioning
d. learned the features of a suitable mate during a critical period
e. have two nonfunctional copies of a particular gene

Answer: d
Difficulty: Easy
Bloom's Taxonomy: Knowledge

12. Classical conditioning differs from operant conditioning in that classical conditioning associates _____ while operant conditioning links _____.

a. two phenomena that are usually unrelated / a voluntary activity with a reward
b. a response with punishment / a response with a reward
c. instinctive behaviors with punishments / learned behaviors with rewards
d. a voluntary activity with a reward / two phenomena that are usually unrelated
e. learned behaviors with rewards / instinctive behaviors with rewards

Answer: a
Difficulty: Moderate
Bloom's Taxonomy: Comprehension

54.4 THE NEUROPHYSIOLOGICAL CONTROL OF BEHAVIOR

13. When young male white-crowned sparrows first begin to sing, they match their vocal output to _____.

a. any birdsong they hear during their initial days of singing
b. an instinctive blueprint of their species' song
c. the memory of their species' song heard months earlier
d. the memory of whatever song they heard during an early critical period
e. the songs of male relatives heard during their initial days of singing

Answer: c
Difficulty: Moderate
Bloom's Taxonomy: Knowledge

14. Which of the following events occurs to a territory-holding male zebra finch when he hears the song of the male zebra finch that holds the neighboring territory?

a. Cells in the nucleus of his forebrain fire frequently.
b. The territory-holding male flies out to drive off the neighboring male.
c. Cells in the nucleus of his forebrain do not respond to the song of the neighboring male.
d. Testosterone levels in his blood rise measurably.
e. The territory-holding male initiates a ritualized threat display that may culminate in an expansion of his territory into the neighbor's land.

Answer: c
Difficulty: Difficult
Bloom's Taxonomy: Knowledge

54.5 HORMONES AND BEHAVIOR

15. Which of the following hormones results in the production of more neurons in the higher vocal center of the brains of male, but not female, zebra finches?

a. testosterone
b. estrogen
c. follicle-stimulating hormone
d. epinephrine
e. cortisol

Answer: b
Difficulty: Moderate
Bloom's Taxonomy: Knowledge

16. The brains of female zebra finches lack the hormonal influences that occur in the brains of males. What changes are seen in the female brains as a result?

a. Auditory processing centers are especially well-developed in females.
b. More neurons are produced in the higher vocal center.
c. Transcription is boosted in genes whose products are involved in memory retention.
d. The number of neurons in the higher vocal center declines.
e. Octopamine levels increase, stimulating neural transmissions.

Answer: d
Difficulty: Moderate
Bloom's Taxonomy: Knowledge

17. When extra juvenile hormone is experimentally administered to worker bees, they _____.

a. revert to performing the duties of younger bees
b. live longer
c. begin laying eggs
d. produce more octopamine
e. develop impaired memories

Answer: d
Difficulty: Moderate
Bloom's Taxonomy: Knowledge

18. A worker bee about a day old is most likely to perform which of the following functions in the hive?

a. cleaning cells
b. feeding nestmates
c. packing pollen
d. feeding brood
e. foraging

Answer: a
Difficulty: Moderate
Bloom's Taxonomy: Knowledge

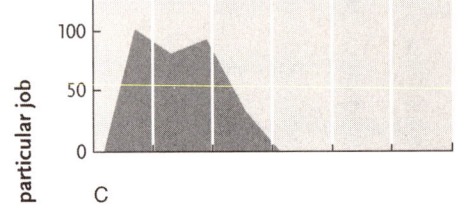
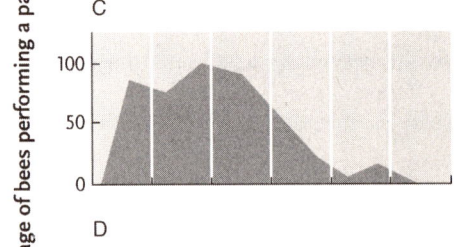
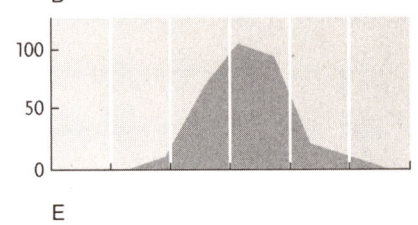
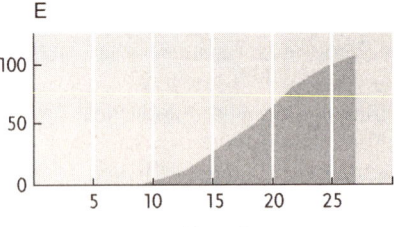

Use the diagram above for questions 19—20.

19. Which of the labeled graphs above represents the age span of worker bees primarily occupied by foraging?

a. A
b. B
c. C
d. D
e. E

Answer: e
Difficulty: Moderate
Bloom's Taxonomy: Knowledge, Application

20. Which of the labeled graphs above represents the age span of worker bees primarily occupied by feeding the brood?

a. A
b. B
c. C
d. D
e. E

Answer: b
Difficulty: Moderate
Bloom's Taxonomy: Knowledge, Application

54.6 NERVOUS SYSTEM ANATOMY AND BEHAVIOR

21. Black field crickets detect the ultrasound of predatory bats through _____.

a. ears on the sides of their heads
b. acoustic organs on the antennae
c. oil-filled organs in their foreheads
d. ears in their front legs
e. an acoustic organ on the right hind leg

Answer: d
Difficulty: Moderate
Bloom's Taxonomy: Knowledge

22. When a cricket hears the ultrasonic call of a bat coming from its right side, it responds by _____.

a. releasing a noxious chemical from specialized epidermal glands
b. interrupting the movement of the left wing
c. swerving down and to the right
d. interrupting the movement of the right wing
e. holding its hind legs close to the body

Answer: b
Difficulty: Moderate
Bloom's Taxonomy: Knowledge, Application

23. A moving object below the midline of a fiddler crab's eyes would stimulate the crab to _____.

a. ignore the object
b. perform evasive maneuvers
c. become aggressive
d. prepare for mating
e. attempt to feed on the object

Answer: a
Difficulty: Easy
Bloom's Taxonomy: Knowledge

24. Most of a mole's cerebral cortex is devoted to processing input from _____.

a. tentacles closest to the mole's mouth
b. tentacles farthest from the mole's mouth
c. front feet
d. both front and hind feet
e. both front feet and tentacles closest to the mole's mouth

Answer: e
Difficulty: Easy
Bloom's Taxonomy: Knowledge

UNANSWERED QUESTIONS

25. Which of the following events is involved in the translation of experience into new or improved behavior?

a. stimulus perception
b. activation of specific neural circuits
c. activation of certain molecular pathways in the brain
d. changes in brain structure and chemistry
e. all of the choices

Answer: e
Difficulty: Moderate
Bloom's Taxonomy: Knowledge

26. Select the option that best describes the current level of understanding about the way genes influence behavior.

a. It is now possible to screen people for genes that may influence their behavior.
b. The protein products of certain genes are being investigated to determine their effect on nervous system structure and function.
c. No genes that influence behavior have yet been identified.
d. Certain features of brain structure can now be associated with particular personality types.
e. None of the choices.

Answer: b
Difficulty: Difficult
Bloom's Taxonomy: Knowledge, Comprehension

Integrative Multiple-Choice

27. Which of the following choices is the ultimate cause of the ability of male zebra finches to discriminate between the songs of established neighbors and those of strangers?

a. Cells in a nucleus in the forebrain fire more often when the song of a new male is heard.

b. Neurons cease to respond to the sound of an established neighbor.

c. Zebra finches are genetically programmed to tolerate territory-holding neighbors but repel invaders.

d. Zebra finches are capable of selective learning.

e. This adaptation allows males to produce more offspring by saving their energy for real battles.

Answer: e
Difficulty: Difficult
Bloom's Taxonomy: Comprehension
Source: Why It Matters and Section54.4.

28. Which of the following choices is a proximate cause of the non-combative behavior of male African cichlids that do not hold territories?

a. Cichlids can detect and store information about aggressive encounters.

b. The GnRH-producing neurons in the hypothalamus are small.

c. The testes produce no testosterone.

d. GnRH production is high.

e. Noncombative behavior allows the male to build his strength for a takeover attempt on a territory.

Answer: b
Difficulty: Moderate
Bloom's Taxonomy: Application
Source: Why It Matters and Section 54.5

Matching

Match each of the following terms with its correct definition.

29. _____ ethology

30. _____ fixed action patterns

31. _____ sign stimuli

32. _____ classical conditioning

33. _____ operant conditioning

34. _____ habituation

35. _____ learned behavior

36. _____ instinctive behavior

37. _____ learning

A. An association between two phenomena that are usually unrelated

B. Linking a voluntary activity with favorable consequences

C. Stereotyped behaviors triggered by a specific cue

D. A behavior dependent on having a particular kind of experience

E. The study of how animals behave in natural environments

F. A process in which experiences change behavioral responses

G. Simple cues that trigger specific behaviors

H. A genetically programmed response

I. A learned loss of responsiveness

Answers: *29. E   30.C   31.G   32.A*
33.B34.I   35.D   36.H   37.F

Difficulty: Moderate
Bloom's Taxonomy: Knowledge
Source: Sections

Classification

Passages 38–42 contain novel information describing real animal and human behaviors. Use choices a–e to classify the type of learning each passage exemplifies.

a. imprinting
b. classical conditioning
c. operant conditioning
d. insight learning
e. habituation

38. _____ Without training, an octopus lifts the lid off his aquarium tank, climbs out, and enters a nearby tank. He eats a fish there and climbs out, replacing the tank lid. He then returns to his own tank, closing his own aquarium lid after him.

39. _____ Wild ravens note the passage of a garbage collection truck and fly off to raid trash cans before the truck arrives to empty them.

40. _____ A group of ducklings were experimentally isolated in a room containing only a large cardboard box that researchers rigged to be movable by remote control. The ducklings soon became attached to the box, following it around the enclosure.

41. _____ Jane Goodall accustomed wild chimps to her presence so that her company ceased to alter their behavior.

42. _____ A teacher only gives candy to the children in her class who finish their assignment on time. The other children receive no reward. The next day, a larger proportion of the class finishes their work on time.

Answers: 38. a 39. c   40. a   41. e   42. c
Difficulty: Difficult
Bloom's Taxonomy: Application
Source: Section 54.3

Choice

For each of the following physiological responses, choose the hormone that causes the response from the list below.

a. GnRH hormone   b. estrogen   c. juvenile

43. _____ stimulates the production of additional neurons in the higher vocal center

44. _____ stimulates genes in brain cells to produce certain proteins that affect nervous system function

45. _____ induces the testes to produce testosterone

46. _____ production of this hormone is altered by outcomes of conflicts

Answers: 43. b   44. c   45. a   46. a

Difficulty: Difficult
Bloom's Taxonomy: Knowledge, Comprehension
Source: Section 54.5

Short Answer

47. What symptoms are shared by knockout mice with nonfunctional Dvl genes and humans with certain psychiatric disorders?

Answer: Knockout mice with nonfunctional Dvl genes do not build nests or socialize with other mice. These mice also have an intensified startle reflex. Humans with certain psychiatric disorders may also startle easily, and often do not display appropriate social behavior.
Difficulty: Moderate
Bloom's Taxonomy: Comprehension
Source: Insights from the Molecular Revolution

48. Under what circumstances does the gene called zenk become active in zebra finches, and what is its effect?

Answer: Transcription of the zenk gene is stimulated when the zebra finch hears its species' song. Zenk produces an enzyme that changes the structure and function of neurons. The outcome is an auditory sensitization to slight variations in the song.
Difficulty: Difficult
Bloom's Taxonomy: Knowledge, Comprehension
Source: Section 54.4

Categorization

Categorize the following learning examples as either classical or operant conditioning.

49. _____ A trained dog sits on command and gets a treat.
50. _____ Horses hurry to their mangers when the barn door opens.
51. _____ Dairy calves recognize the vehicle of their caregiver and vocalize excitedly.
52. _____ Captive dolphins jump from the water on command and are rewarded with fish.
53. _____ A person who was shocked by an electric fence becomes reluctant to touch the wire even when informed that the power to the fence is off.
54. _____ A student receives a merit scholarship for good grades and is motivated to work even harder.
55. _____ A truck has a loud alarm that goes off if the keys are in the ignition when the door is opened. Its owner avoids the alarm by removing keys.
56. _____ A man becomes hungry when he hears the beep of the microwave oven.

A. classical conditioning
B. operant conditioning

Answers: 49. B   50. A   51. A   52. B   53. B   54. B   55. B   56. A

Difficulty: Easy
Bloom's Taxonomy: Application
Source: Section 54.3

True/False

If the statement is true, write a "T" in the blank. If the statement is false, make it correct by changing the underlined word(s) and writing the correct word(s) in the answer blank.

57. _____ Food preference in garter snakes is <u>learned</u> behavior

58. _____ The red spot on the beak of a herring gull is a <u>sign stimulus</u>.

59. _____ The zenk gene in the finch brain is activated by <u>tactile</u> stimuli.

60. _____ Neurons in the higher vocal center of a male finch's brain are stimulated to proliferate by <u>testosterone</u>.

61. _____ Younger honeybees have <u>higher</u> levels of juvenile hormone.

62. _____ GnRH-producing cells in the brains of male cichlids are <u>larger</u> in territorial individuals.

63. _____ Crickets have ears in their <u>front legs</u>.

Answers: *57. F, instinctive 58. T   59. F, acoustical   60. F, estrogen   61. F, lower 62. T   63. T*

Difficulty: Moderate
Bloom's Taxonomy: Knowledge
Source: Section 54.2–54.6

Select the Exception

64. Most of the following behaviors are learned behaviors with an instinctive component. Select the exception by choosing the behavior that is most purely instinctive.

a. The song of the male white-crowned sparrow.
b. A flying cricket jerks up its right hind leg when it hears a bat coming from the left.
c. Baby geese imprint on their mother and follow her.
d. Coyotes in urban areas forage in trash cans near humans.
e. Nonterritorial male cichlids make no attempt to court females.

Answer: b
Difficulty: Moderate
Bloom's Taxonomy: Application, Knowledge
Source: Section 54.1

65. Four of the five stimuli listed below are sign stimuli. Select the exception.

a. a smiling human face
b. the red spot on the beak of a herring gull
c. the open mouth of a peeping baby bird
d. a red-colored stimulus that triggers a territorial red-bellied stickleback fish to attack it
e. a mother rat picking up young rats and returning them to the nest

Answer: e
Difficulty: Easy
Bloom's Taxonomy: Knowledge, Application
Source: Section 54.2

Sequence

Write the letter of the event that happens first in the development of song in male zebra finches next to 66. Sequence the following events, with the final event in the pathway written next to 71.

66. _____   A. structure and function of neurons are changed
67. _____   B. neurons anticipate key acoustical events

68. _____   C. a territory owner habituates to the sound of his singing neighbor
69. _____   D. exposure of hatchling to its species' song
70. _____   E. an enzyme is produced
71. _____   F. *zenk* is activated

Answers: *66. D   67. F   68. E   69. A 70. B   71. C*

Essay

72. It will someday become possible to screen people for genes that can influence their behavior. As scientists we must consider the ethical implications of our work. What ethical issues may arise as a result of such screening?

Answer: The presence of a gene can correlate with the tendency for a behavior, but it does not guarantee the behavior will exist. For example, if a gene was discovered that increased aggression in humans, it would not be fair to assume all the carriers of this gene are potential criminals. Aggression can be channeled in both positive and negative ways. Conversely, a person could be born with normal genetics but become emotionally unstable due to trauma or abuse. Genetic screening of such a person would not reveal their instability. Widespread use of such screening could lead to ethical misconduct.
Difficulty: Moderate
Bloom's Taxonomy: Analysis
Source: Unanswered Questions.

73. Koko the gorilla was trained by scientists at Stanford University to communicate using American Sign Language. She was housed in a trailer and interacted with frequently. One day she tore the sink in her trailer out of the wall. When confronted by her trainers about this, Koko blamed her petite female trainer for the damage. What insights might this behavior provide to animal behaviorists?

Answer: Not all animals are capable of distinguishing information that they know from information that they think another individual knows. It takes a higher level of intelligence to grasp that concept. Here Koko demonstrates her awareness that her trainers did not witness the destruction of the sink. It is also interesting to note that Koko's excuse allows us to analyze her ability to reason. She was capable of knowing that she has done something wrong, and of lying about it. However, her logic was insufficient to enable her to construct a plausible lie. She was apparently incapable of reasoning that her trainer would not have been physically capable of wrenching a sink from the wall.
Difficulty: Moderate
Bloom's Taxonomy: Synthesis
Source: Sections 54.1–54.7

Labeling

Use the following illustration to answer questions 74–76.

An Eimer's organ in longitudinal section

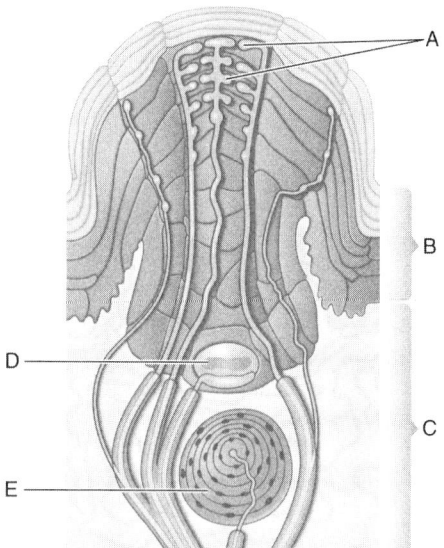

74. Which label on the graph represents the sensory nerve terminals on the Eimer's organ of the star-nosed mole?

a. A
b. B
c. C
d. D
e. E

Answer: a
Difficulty: Moderate
Bloom's Taxonomy: Knowledge
Source: Fig. 54.13

75. Which label on the graph represents the epidermis on the Eimer's organ of the star-nosed mole?

a. A
b. B
c. C
d. D
e. E

Answer: b
Difficulty: Moderate
Bloom's Taxonomy: Knowledge
Source: Fig. 54.13

76. Which label on the graph represents the dermis on the Eimer's organ of the star-nosed mole?

a. A
b. B
c. C
d. D
e. E

Answer: c
Difficulty: Moderate
Bloom's Taxonomy: Knowledge
Source: Fig. 54.13

55

THE ECOLOGY AND EVOLUTION OF ANIMAL BEHAVIOR

Multiple-Choice

WHY IT MATTERS

1.   The ultimate evolutionary benefit to a male white-crowned sparrow's singing is to _____.

a.   attract a mate
b.   defend his territory
c.   advertise his availability as a sire
d.   intimidate rival males
e.   produce surviving offspring

Answer: e
Difficulty: Moderate
Bloom's Taxonomy: Comprehension

2.   When behavioral biologists look at the ultimate evolutionary benefit to animal communication, they are primarily focusing on the effects of the communication on the individual's ability to _____.

a.   maintain a favorable position in the dominance hierarchy
b.   produce surviving offspring
c.   intimidate other animals
d.   form alliances with other animals
e.   find food

Answer: b
Difficulty: Moderate
Bloom's Taxonomy: Knowledge

55.1 MIGRATION AND WAYFINDING

3.   If an animal finds his way by the type of wayfinding known as piloting, then he is using _____.

a.   familiar landmarks to guide his journey
b.   the sun's position in the sky to orient himself
c.   both the sun's position and a mental map
d.   the Earth's magnetic field to orient himself
e.   another animal to lead the way as he learns the route

Answer: a
Difficulty: Easy
Bloom's Taxonomy: Knowledge

4.   In a 1938 experiment, Niko Tinbergen showed that female digger wasps find their nests by _____.

a.   olfactory cues
b.   visual landmarks
c.   a physiological magnetic compass
d.   the position of stars in the night sky
e.   the position of the sun

Answer: b
Difficulty: Easy
Bloom's Taxonomy: Knowledge

5.   The most complex wayfinding mechanism is _____, in which animals use both a compass and a mental map of the area.

a.   piloting
b.   orienteering
c.   compass orientation
d.   migration
e.   navigation

Answer: e
Difficulty: Moderate
Bloom's Taxonomy: Knowledge

EXPERIMENTAL RESEARCH: USING LANDMARKS TO FIND THE WAY HOME

6.   When Tinbergen arranged pinecones around the nest of a female digger wasp but moved them when the wasp was gone, the wasp _____.

a.   never found the nest until the pinecones were replaced
b.   took longer to find her nest
c.   found the nest immediately
d.   abandoned the nest
e.   dug a new nest in the center of the repositioned circle of pinecones

Answer: a
Difficulty: Easy
Bloom's Taxonomy: Knowledge

7. Tinbergen's wasp experiments demonstrated that wasps find their nests by _____ cues.

a.   olfactory
b.   auditory
c.   visual
d.   both olfactory and visual
e.   both magnetic and auditory

Answer: c
Difficulty: Easy
Bloom's Taxonomy: Knowledge

EXPERIMENTAL RESEARCH: EXPERIMENTAL ANALYSIS OF THE INDIGO BUNTING'S STAR COMPASS

8.   In an experiment on wayfinding, indigo buntings were housed in outdoor cages with blotting paper walls and inkpads on the floors. Which of the following facts demonstrated to researchers that the indigo bunting uses the positions of stars to orient its migration?

a.   The birds' inky feet made prints indicating the direction in which they attempted to fly.
b.   In spring the inky footprints were mostly on the northern side of the cage.
c.   In fall the inky footprints were mostly on the south side of the cage.
d.   On cloudy nights the footprint patterns were random.
e.   In spring and fall, the anterior pituitary of the bird's brain generated a series of hormonal changes.

Answer: d
Difficulty: Moderate
Bloom's Taxonomy: Comprehension

INSIGHTS FROM THE MOLECULAR REVOLUTION: UNADORNED TRUTHS ABOUT NAKED MOLE-RAT WORKERS

9.   Two major factors underlying the evolution and maintenance of the nonbreeding worker caste in naked mole rat colonies are _____ and _____.

a.   structural sterility of the workers; pronounced aggression of breeding males
b.   low testosterone levels among male workers; high testosterone levels among breeding males
c.   haploid genomes of workers; diploid genomes of breeders
d.   close genetic relatedness among colony members; greater chance of survival in the colony than outside it
e.   all of the choices

Answer: d
Difficulty: Moderate
Bloom's Taxonomy: Knowledge

10.   DNA fingerprinting analysis revealed that the relatedness of naked mole rats in the same colony approached the band similarity of _____.

a.   siblings
b.   parent and child
c.   half-siblings
d.   identical twins
e.   unrelated individuals

Answer: d
Difficulty: Moderate
Bloom's Taxonomy: Knowledge

55.2 HABITAT SELECTION AND TERRITORIALITY

11.   A taxis is distinguished from a kinesis because a taxis is _____ movement while a kinesis is _____ movement.

a.   ordered; a change in the rate of
b.   a migratory; a local
c.   random; directional
d.   a group; an individual
e.   slow; fast

Answer: a
Difficulty: Moderate
Bloom's Taxonomy: Knowledge

Use the figure above for question 12.

12.   The graph above illustrates the finding that _____.

a.   habitat preference is innate in blue tits but not in coal tits
b.   blue tits instinctively prefer pine trees while coal tits instinctively prefer oak trees
c.   habitat preference in both blue tits and coal tits is innate
d.   habitat preference in both blue tits and coal tits is learned
e.   habitat preference in blue tits is learned but in coal tits it is innate

Answer: c
Difficulty: Moderate
Bloom's Taxonomy: Application

13. Animals establish and defend territories only when _____.

a. the effort does not endanger the life of the individual
b. males compete for females by fighting
c. females choose among displaying males
d. some critical resource is in short supply
e. territorial defense is not energetically costly

Answer: d
Difficulty: Moderate
Bloom's Taxonomy: Knowledge

55.3 THE EVOLUTION OF COMMUNICATION

14. The courtship display of the wandering albatross is primarily a(n) _____ display.

a. acoustical
b. visual
c. chemical
d. electrical
e. tactile

Answer: b
Difficulty: Easy
Bloom's Taxonomy: Comprehension, Knowledge

A   B

Use the figure above for questions 15 and 16.

15. If the food source being communicated about is close to the hive, the bee performs the _____.

a. round dance, which is image A
b. round dance, which is image B
c. waggle dance, which is image A
d. waggle dance, which is image B
e. round dance and the waggle dance in sequence

Answer: a
Difficulty: Moderate
Bloom's Taxonomy: Knowledge
Source: Fig. 55.11

16. If a honeybee performing the dance in image B moves straight down the comb, then the other bees will seek the nectar source by flying _____.

a. directly toward the sun
b. at a 45° angle to the right of the sun
c. at a 45° angle to the left of the sun
d. directly away from the sun
e. due south

Answer: d
Difficulty: Difficult
Bloom's Taxonomy: Knowledge, Application

17. Upon finding a carcass, a raven will only call and attract other ravens to the food source if the _____.

a. nearby ravens are relatives
b. food source is large and plentiful
c. food is found in a resident pair's territory
d. food is found in his own territory
e. food is being consumed by an animal of another species

Answer: c
Difficulty: Moderate
Bloom's Taxonomy: Knowledge

55.4 THE EVOLUTION OF REPRODUCTIVE BEVAVIOR AND MATING SYSTEMS

18. Males and females often differ in their reproductive strategies due to the fact that males can increase the number of offspring they produce by _____ while females reproduce most successfully by _____.

a. mating with a high-quality female; mating with a high-quality male
b. driving off other males; defending a productive territory
c. defending a productive territory; enlisting multiple males to aid in raising her offspring
d. mating with multiple females; mating with a high-quality male
e. mating with multiple females; mating with multiple males

Answer: d
Difficulty: Moderate
Bloom's Taxonomy: Knowledge

19. All of the following traits are consistent with evolution by sexual selection *except* _____.

a. males are larger than females
b. females actively choose superior males
c. males often bear showy or defensive structures
d. males gather and defend harems of females
e. gametes are dispersed into the environment by wind or water

Answer: e
Difficulty: Moderate
Bloom's Taxonomy: Comprehension

20. Polyandry is the mating system in which _____.

a. one male and one female form a long-term association
b. both males and females mate with multiple partners
c. one male mates with many females
d. one female mates with many males
e. both males and females provide care to offspring

Answer: d
Difficulty: Easy
Bloom's Taxonomy: Knowledge

55.5 THE EVOLUTION OF SOCIAL BEHAVIOR

21.   Social groups face increased competition, parasitism, and disease. Why has natural selection failed to eliminate the tendency of animals to form social groups?

a.   Prey gain safety in numbers.
b.   Predators have better hunting success in groups.
c.   Prey in large groups satiates predators so all are not eaten.
d.   Offspring are protected and taught by both parents and other group members.
e.   All of the choices.

Answer: e
Difficulty: Easy
Bloom's Taxonomy: Knowledge, Comprehension

22.   A subordinate animal in a dominance hierarchy has limited access to food and mates. Why would he remain in a social group that dominated him?

a.   He is exhibiting altruistic behavior.
b.   His chances of survival and reproduction are better in the group than alone.
c.   He expects to become dominant in the future.
d.   He does not know that it would be an advantage for him to leave.
e.   Dominant group members will not allow him to leave.

Answer: b
Difficulty: Difficult
Bloom's Taxonomy: Comprehension

23.   Kin selection is a form of natural selection in which _____.

a.   dominant individuals share resources disproportionately with their close relatives
b.   individuals help relatives only if they are likely to return the favor in the future
c.   group members sacrifice their own reproductive success to help individuals who are not their direct descendents
d.   shared alleles for altruism are perpetuated if the assisted animals produce more offspring than the helper could have produced if he had not helped
e.   shared alleles for altruism are perpetuated if the helper produces more offspring than the relatives that he aids

Answer: d
Difficulty: Difficult
Bloom's Taxonomy: Knowledge

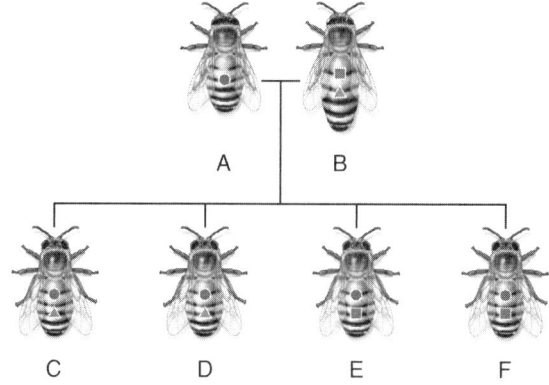

Use the figure above for questions 24 and 25.

24.   What percent of its total genome does bee A donate to its offspring?

a.   50 percent
b.   75 percent
c.   0 percent
d.   100 percent
e.   25 percent

Answer: d
Difficulty: Difficult
Bloom's Taxonomy: Knowledge, Application

25. Bees C and D differ from bees E and F because _____.

a.   bees C and D had a different father than bees E and F
b.   bees C and D had a different mother than bees E and F
c.   bees C and D inherited different alleles than bees E and F from their mother
d.   bees C and D inherited different alleles than bees E and F from their father
e.   bees C and D are haploid drones while bees E and F are diploid female workers

Answer: c
Difficulty: Difficult
Bloom's Taxonomy: Knowledge

26.   Altruism can be distinguished from reciprocal altruism because altruism is defined as _____ but reciprocal altruism involves _____.

a.   the sacrifice of an individual's reproductive success to help others; the expectation of a future favor in return for present aid
b.   aid given to nonrelatives; benefits given to close relatives only
c.   the expectation of a future favor in return for present aid; close relatives helping each other
d.   low-risk behavior that provides minor benefits to others; high-risk behavior providing major benefits to others
e.   none of the choices

Answer: a
Difficulty: Moderate
Bloom's Taxonomy: Knowledge

55.6 AN EVOLUTIONARY VIEW OF HUMAN SOCIAL BEHAVIOR

27. Wilson and Daly used an evolutionary perspective on human behavior to hypothesize that stepparents who cared for their own children as well as children not genetically related to them _____.

a. almost always will abuse their stepchildren
b. would never abuse any of the children
c. are as likely to abuse their own children as their stepchildren
d. are more likely to abuse their stepchildren
e. are more likely to abuse their own children

Answer: d
Difficulty: Easy
Bloom's Taxonomy: Knowledge

28. The chance that a young child would be criminally abused was _____ for children living with one parent and one stepparent than for children living with both biological parents.

a. 10 times lower
b. 10 times higher
c. 20 times lower
d. 40 times lower
e. 40 times higher

Answer: e
Difficulty: Moderate
Bloom's Taxonomy: Knowledge

UNANSWERED QUESTIONS

29. Ultraviolet signals in swordtail fishes and electrical signals in weakly electrical fishes are examples of _____ communication.

a. network
b. private channel
c. sensory drive
d. overheard
e. none of the choices

Answer: b
Difficulty: Moderate
Bloom's Taxonomy: Knowledge

30. Evolutionary biologists believe that females prefer attractive mates mainly because _____.

a. attractive mates confer status in hierarchical societies
b. attractiveness is a predictor of good performance and survivorship
c. ornaments such as bright plumage stimulate her visual system
d. superior ornaments deter rival males
e. a sense of aesthetics is shared by both higher animals and humans

Answer: b
Difficulty: Easy
Bloom's Taxonomy: Knowledge

Matching

Match each of the following terms with its correct definition.

31. _____ sexual selection
32. _____ courtship displays
33. _____ haplodiploidy
34. _____ migration
35. _____ piloting
36. _____ territory
37. _____ habitat
38. _____ taxis
39. _____ pheromones
40. _____ leks
41. _____ dominance hierarchies

A. A site that provides food, shelter, nesting materials, and interacting organisms

B. An unusual pattern of sex determination

C. A resource-rich site defended by its occupants

D. Natural selection for mating success

E. The use of familiar landmarks to guide a journey

F. Travel from the birth site to another destination and back

G. Grounds where males display for females

H. Volatile chemicals that influence the behavior of embers of the same species

I. Ordered movement

J. Ritualized behaviors engaged in to attract the attention of the opposite sex

K. Social systems in which individuals are ranked

Answers: 31. D   32. J   33. B   34 .F
35. E   36. C   37. A   38. I   39.H
40.G   41. K

Difficulty: Moderate
Bloom's Taxonomy: Knowledge
Source: Sections 55.1–55.5

Classification

Classify examples 42–47 into one of the following three behaviors.

    a. kin selection
    b. reciprocal altruism
    c. altruism

42. _____ A young woman delays marriage to help her mother raise her many younger brothers and sisters.

43. _____ A man stops his car on a deserted road to aid a stranded motorist he does not know.

44. _____ A young immigrant couple decide to limit their family size so they can send money to nieces and nephews in their native land.

45. _____ You skip class but borrow the notes from a friend, who skips class himself the next week.

46. _____ During a blood drive, people wear stickers announcing, "I donated!"

47. _____ You donate money anonymously to a charity and tell no one about it.

Answers: *42.a*   *43.c*   *44.a*   *45.b*   *46.b*
             47.c

Difficulty: Moderate
Bloom's Taxonomy: Application
Source: Section 55.5

Choice

For each of the following mating systems, choose the most appropriate term from the list below.

a. promiscuity   b. monogamy
c. polygyny   d. polyandry

48. _____ Prairie voles choose one mate and stay with that partner for life.

49. _____ Corals release gametes into the open sea.

50. _____ Male northern fur seals aggressively guard harems of up to 40 females.

51. _____ Female chimpanzees will mate with more than one male, eliciting protective behavior for her offspring from them all.

Answers: *48.b*   *49.a*   *50.c*   *51.d*

Difficulty: Easy
Bloom's Taxonomy: Knowledge, Comprehension
Source: Section 55.4

Short Answer

52. Why do peahens choose peacocks with the longest, showiest tails, when a long tail might be easily grabbed by a predator?

Answer: Studies have shown the ability to survive despite the long tail is predictive of the survival and health of offspring. The long tail is a handicap, so males that can thrive in spite of that handicap apparently have excellent genetics.
Difficulty: Moderate
Bloom's Taxonomy: Comprehension
Source: Section 55.4

53. What is haplodiploidy, and how is it related to altruism in honeybees?

Answer: Drones (male bees) are haploid, while queen bees are diploid. Worker bees are either 50 percent or 100 percent related to each other, depending on whether they share the same alleles from their mother. Nonreproducing workers help to pass alleles they share into the next generation by caring for their siblings.
Difficulty: Difficult
Bloom's Taxonomy: Knowledge, Comprehension
Source: Section 55.5

54. Briefly explain some of the advantages and disadvantages of living in social groups.

Answer: Social animals face more competition for food, nesting sites, and other resources. They also tend to contract diseases and parasites more easily, due to frequent contact between individuals. The advantages of social behavior include safety in numbers for prey animals. Predators can take on larger prey when working together. In higher animals, learning may take place more effectively in social groups.
Difficulty: Difficult
Bloom's Taxonomy: Knowledge, Application
Source: Section 55.5

True/False

If the statement is true, write a "T" in the blank. If the statement is false, make it correct by changing the underlined word(s) and writing the correct word(s) in the answer blank.

55. _____ <u>Individual</u> monarch butterflies migrate from Canada to Mexico each year.

56. _____ Many insects have a <u>genetically determined</u> preference for plants they eat during their larval stage.

57. _____ A change in the rate or frequency of movements in response to environmental stimuli is called <u>taxis</u>.

58. _____ Male lizards experimentally dosed with supplemental testosterone were more active, displayed more often, and had a higher rate of <u>death</u>.

59. _____ Bees surrounding a dancing bee produce a brief <u>acoustical</u> signal stimulating the dancer to regurgitate a sample of the new food.

60. _____ Females almost always have a <u>lower</u> parental investment than males.

61. _____ Female blackbirds choose a mate primarily by the quality of his <u>plumage</u>.

62. _____ Dominant male wild dogs have <u>higher</u> levels of cortisol and other stress-related hormones than submissive wild dogs.

63. _____ Full siblings share <u>25 percent</u> of their alleles.

64. _____ <u>Reciprocal altruism</u> requires that animals remember which individuals have and have not shared.

65. _____ Humans are <u>more</u> likely to abuse children that they know are not their own.

Answers: *55. F, offspring of 56. T 57. F, kinesis 58. T 59. T 60. F, higher 61. F, territory 62. T 63. F, 50% 64. T 65. T*

Difficulty: Moderate
Bloom's Taxonomy: Knowledge
Source: Section 55.1-55.6

Select the Exception

66. Most of the following choices are examples of piloting. Select the exception.

a. Gray whales use visual cues to migrate from Alaska to Baja California.
b. Pacific salmon use olfactory cues to find the stream where they hatched.
c. Female digger wasps use landmarks to find their nests.
d. Indigo buntings use the position of stars to orient during migration.
e. A hiker looks for familiar rock formations as she finds her way back to the trail head.

Answer: d
Difficulty: Moderate
Bloom's Taxonomy: Knowledge
Source: Section 55.1

67. Four of the five animals listed below are migratory. Select the exception.

a. monarch butterflies
b. indigo buntings
c. white-crowned sparrows
d. gray whales
e. wood lice

Answer: e
Difficulty: Easy
Bloom's Taxonomy: Knowledge
Source: Sections 55.1 and 55.2

68. Most of the following forms of communication are likely to be intercepted by an attentive third party. Select the exception.

a. visual
b. auditory
c. chemical
d. electrical
e. tactile

Answer: d
Difficulty: Easy
Bloom's Taxonomy: Knowledge
Source: Unanswered Questions

69. Most of the following species are known to engage in altruistic behavior. Select the exception.

a. humans
b. naked mole-rats
c. wolves
d. honeybees
e. lobsters

Answer: e
Difficulty: Easy
Bloom's Taxonomy: Knowledge
Source: Section 55.5

Sequence

Write the letter of the closest genetic relationship next to 60. The letter of the most distant genetic relationship is written next to 63.

70. _____   A.   first cousins
71. _____   B.   full siblings
72. _____   C.   half siblings
73. _____   D.   identical twins

Answers: *60.D*   *61.B*   *62.C*   *63.A*
Difficulty: Easy
Bloom's Taxonomy: Knowledge
Source: Section 55.5

Essay

74.   From an evolutionary perspective, why would humans sometimes act charitably toward strangers?

Answer: Most acts of charity are publicized in some way, if only by the giver mentioning the act to a friend. This would be classified as reciprocal altruism. From an evolutionary perspective, the giver is aided by the favorable impression others in society have of him. Presumably such a generous individual would be likely to receive help if ever it were needed. People who aid others in secret may also be reciprocal altruists if they hold religious or spiritual beliefs that they will benefit either presently or in an afterlife for their generosity.
Difficulty: Moderate
Bloom's Taxonomy: Analysis
Source: Section 55.5

75.   Sociologists have noted that women in different environments seem to find different types of men attractive. For example, a physically weak but professionally successful man may be admired by women in New York City, while in rural Wyoming the rugged cowboys are most popular. Explain this phenomenon from the perspective of an evolutionary view of human social behavior.

Answer: From an evolutionary perspective, females try to choose a male with good genes and the ability to provide for her offspring. An urban area is a different habitat from a rural one. Perhaps the genes that are best adapted for one environment are not best adapted for another. Women may instinctively evaluate their environments and choose the males that seem best adapted to that habitat.
Difficulty: Moderate
Bloom's Taxonomy: Application
Source: Section 55.6.

to 27

R-20-27